CALCIUM SIGNALING

Second Edition

METHODS IN SIGNAL TRANSDUCTION SERIES

Joseph Eichberg, Jr., Series Editor

Published Titles

CALCIUM SIGNALING

Second Edition

Edited by

James W. Putney, Jr., Ph.D.

Chief
Calcium Regulation Section
National Institute of Environmental Health Sciences
National Institutes of Health
Research Triangle Park
North Carolina

Taylor & Francis

Taylor & Francis Group
Boca Raton   London   New York

A CRC title, part of the Taylor & Francis imprint, a member of the
Taylor & Francis Group, the academic division of T&F Informa plc.

Published in 2006 by
CRC Press
Taylor & Francis Group
6000 Broken Sound Parkway NW, Suite 300
Boca Raton, FL 33487-2742

Library of Congress Cataloging-in-Publication Data

Calcium signaling / edited by James W. Putney, Jr.-- 2nd ed.
     p. cm. -- (Methods in signal transduction)
  Includes bibliographical references and index.
  ISBN 0-8493-2783-0 (alk. paper)
  1. Calcium--Physiological effect. 2. Calcium channels. 3. Cellular signal transduction. I. Putney, James W. II. Series.

QP535.C2C2663 2005
572'.516--dc22   2005051121

Taylor & Francis Group
is the Academic Division of T&F Informa plc.

Visit the Taylor & Francis Web site at
http://www.taylorandfrancis.com

and the CRC Press Web site at
http://www.crcpress.com

Series Preface

The concept of signal transduction at the cellular level is now established as a cornerstone of biological sciences. Cells sense and react to environmental cues by means of a vast panoply of signaling pathways and cascades. While the steady accretion of knowledge regarding signal transduction mechanisms is continuing to add layers of complexity, this greater depth of understanding has also provided remarkable insights into how healthy cells respond to extracellular and intracellular stimuli and how these responses can malfunction in many disease states.

Central to advances in unraveling signal transduction is the development of new methods and the refinement of existing ones. Progress in this field relies upon an integrated approach that utilizes techniques drawn from cell and molecular biology, biochemistry, genetics, immunology, and computational biology. The overall aim of this series is to bring together, and continually update the wealth of methodologies now available for research in many aspects of signal transduction. Each volume is assembled by one or more editors who are leaders in their specialty. Their guiding principle is to recruit knowledgeable authors who will present procedures and protocols in a critical, yet, reader-friendly format. Our goal is to ensure that each volume will be of maximum practical value to a broad audience, including students, seasoned investigators, and researchers who are new to the field.

The six years, since the first volume on *Calcium Signaling* edited by Dr. James Putney appeared in this series, have been marked by a rapid expansion of both knowledge and techniques applicable to this area. Like its predecessor, this second edition presents a combination of cutting edge methodology and concise reviews of both familiar and novel aspects of the field. The book presents updates of previously covered methods, such as utilization of fluorescent indicators to measure Ca^{2+}, and introduces new ones, such as, the application of high-throughput technology to follow rapid calcium changes in the cytoplasm. Particularly notable is the inclusion of approaches to studying calcium signaling in lower eukaryotes that are especially amenable to utilizing genetic approaches. In addition, attention is devoted to topics in which the importance of calcium has more recently gained prominence, such as apoptosis and strategies to identify drugs targeted to affect calcium channel function.

The broad scope of this volume serves to emphasize the swift and dynamic advances that are being made in studies of calcium signaling, a field that is central to our understanding of signal transduction mechanisms.

Joseph Eichberg, Jr., Ph.D.
Series Editor

Preface to the Second Edition

In my preface to the first edition, I reflected on the emergence of calcium signaling as a clearly identifiable field, and on the fascinating diversity of technologies that are employed. Much has happened in the calcium signaling field in the past 6 years, including advances in the available technology. This edition combines topics that have been updated from the first edition, sometimes by other authors to present a different perspective. Some topics are covered for the first time, for example, the use of high-throughput technology, and genetically tractable organisms. Some topics that have not changed significantly are not addressed in this volume, for example, subcellular fractionation.

I perhaps owe an apology to a significant number of people in the calcium signaling field who may feel that their specific area of interest is not represented. In fact, the field has grown tremendously, and as I began assembling a tentative list of topics I soon realized that it would not be possible to cover the field comprehensively. So, although this volume is titled *Calcium Signaling*, it in fact focuses strongly on the processes that generate calcium signals, rather than what calcium signals do in cells. My reasoning was that calcium "responses" often involve the same techniques used by protein chemists, molecular biologists, endocrinologists, and geneticists. Hopefully, these scientists will find their interests in other volumes in this series. Also, hopefully, they will consider expanding their experimental repertoire to include some investigations into where those calcium signals they have been so interested in actually come from.

Editor

James W. Putney, Jr. is the chief of the Calcium Regulation Section, Laboratory of Signal Transduction, at the National Institute of Environmental Health Sciences, Research Triangle Park, North Carolina. He earned his B.A. degree in chemistry from the University of Virginia in 1968, and his Ph.D. degree in pharmacology from the Medical College of Virginia in 1972. He was a postdoctoral fellow in the Department of Pharmacology at the University of Pennsylvania from 1972 to 1974, and was assistant professor and subsequently associate professor of pharmacology at Wayne State University from 1974 to 1980. From 1980 to 1986 he served as associate and then full professor of pharmacology at the Medical College of Virginia. In 1986, he moved to the National Institute of Environmental Health Sciences where he served as chief of the Laboratory of Cellular and Molecular Pharmacology from 1987 to 1996.

Dr. Putney's research interests involve various aspects of calcium signaling, especially signaling through the phospholipase C–inositol phosphate pathway. He has published over 250 research papers, reviews, and monographs on this topic and has delivered lectures at over 150 national and international conferences and symposia.

Dr. Putney is a member of the American Society for Pharmacology and Experimental Therapeutics, the American Society for Biochemistry and Molecular Biology, and the Society of General Physiologists. He is a Fellow of the American Association for the Advancement of Science. He serves, or has served in the past, on the editorial boards of *The Journal of Biological Chemistry*, *The Biochemical Journal*, *The Journal of Pharmacology and Experimental Therapeutics*, *Molecular Pharmacology*, *Pharmacology*, *The American Journal of Physiology*, and *Cell Calcium*.

Contributors

Thomas Aberle
Institute for Clinical Pharmacology and Toxicology
Saarland University
Homburg, Germany

Jens P. Anderson
Department of Physiology
University of Aarhus
Aarhus, Denmark

Michael C. Ashby
Developmental Synaptic Plasticity Unit
Porter Neuroscience Research Center
Bethesda, Maryland, U.S.A.

Richard A. Billington
Università del Piemonte Orientale
Novara, Italy

Gary St. J. Bird
National Institute of Environmental Health Sciences
Research Triangle Park, North Carolina, U.S.A.

Grant C. Churchill
Department of Pharmacology
University of Oxford
Oxford, U.K.

Kyle W. Cunningham
Department of Biology
Johns Hopkins University
Baltimore, Maryland, U.S.A.

Lianne C. Davis
Department of Pharmacology
University of Oxford
Oxford, U.K.

Clark W. Distelhorst
Departments of Medicine and Pharmacology
Case Western Reserve University
Cleveland, Ohio, U.S.A.

Leonard Dode
Department of Physiology
Catholic University of Leuven
Leuven, Belgium

Anne Dodge
Novartis Respiratory Research Centre
Horsham, West Sussex, U.K.

Nicholas J. Dolman
Synaptic Physiology Unit
Porter Neuroscience Research Center
Bethesda, Maryland, U.S.A.

Ana Y. Estevez
Department of Biology
St. Lawrence University
Canton, New York, U.S.A.

Veit Flockerzi
Institute for Clinical Pharmacology and Toxicology
Saarland University
Homburg, Germany

J. Kevin Foskett
Department of Physiology
University of Pennsylvania
Philadelphia, Pennysylvania, U.S.A.

Antony Galione
Department of Pharmacology
University of Oxford
Oxford, U.K.

Lawrence D. Gaspers
Department of Pharmacology and Physiology
New Jersey Medical School
Newark, New Jersey, U.S.A.

Armando A. Genazzani
Università del Piemonte Orientale
Novara, Italy

Martin Gosling
Novartis Respiratory Research Centre
Horsham, West Sussex, U.K.

Lucian Ionescu
Department of Physiology
University of Pennsylvania School of Medicine
Philadelphia, Pennsylvania, U.S.A.

Christine Jung
Institute for Clinical Pharmacology and Toxicology
Saarland University
Homburg, Germany

Stephan E. Lehnart
Center for Molecular Cardiology
College of Physicians and Surgeons of Columbia University
New York, New York, U.S.A.

Su Li
Novartis Respiratory Research Centre
Horsham, West Sussex, U.K.

Don-On Daniel Mak
Department of Physiology
University of Pennsylvania School of Medicine
Philadelphia, Pennsylvania, U.S.A.

Andrew R. Marks
Molecular Cardiology Program
College of Physicians & Surgeons of Columbia University
New York, New York, U.S.A.

Roser Masgrau
Institute of Neurosciences
Autonomous University of Barcelona
Bellaterra, Spain

Marcel Meissner
Institute for Clinical Pharmacology and Toxicology
Saarland University
Homburg, Germany

Walson Metzger
Department of Pharmacology and Physiology
New Jersey Medical School
Newark, New Jersey, U.S.A.

Ludwig Missiaen
Department of Physiology
Catholic University of Leuven
Leuven, Belgium

Atsushi Miyawaki
RIKEN Brain Science Institute
Wako City, Saitama, Japan

Hideaki Mizuno
RIKEN Brain Science Institute
Wako City, Saitama, Japan

Gregory R. Monteith
School of Pharmacy
University of Queensland
Brisbane, Queensland, Australia

Anthony J. Morgan
Department of Pharmacology
University of Oxford
Oxford, U.K.

Takeharu Nagai
Research Institute for Electronic Science
Hokkaido University
Sapporo, Japan

Anant B. Parekh
Department of Physiology
University of Oxford
Oxford, U.K.

Myoung K. Park
Department of Physiology
Sungkyunkwan University of School of Medicine
Suwon, Korea

Sandip Patel
Department of Physiology
University College of London
London, U.K.

Ole H. Petersen
Physiological Laboratory
University of Liverpool
Liverpool, U.K.

Stephan Philipp
Institute for Clinical Pharmacology and Toxicology
Saarland University
Homburg, Germany

Nicola Pierobon
Department of Pharmacology and Physiology
New Jersey Medical School
Newark, New Jersey, U.S.A.

Martin Poenie
Department of Zoology
University of Texas
Austin, Texas, U.S.A.

Chris T. Poll
Novartis Respiratory Research Centre
Horsham, West Sussex, U.K.

Andrea Prandini
Department of Experimental and Diagnostic Medicine
    and Interdisciplinary Center for the Study of Inflammation
University of Ferrara
Ferrara, Italy

James W. Putney, Jr.
Natonal Institute of Environmental Health Sciences
Resarch Triangle, North Carolina, U.S.A.

Luc Raeymaekers,
Department of Physiology
Catholic University of Leuven
Leuven, Belgium

Rosario Rizzuto
Department of Experimental and Diagnostic Medicine
  and Interdisciplinary Centre for the Study of Inflammation
University of Ferrara
Ferrara, Italy

Margarida Ruas
Department of Pharmacology
University of Oxford
Oxford, U.K.

Kevin Strange
Department of Anesthesiology
Vanderbilt University Medical Center
Nashville, Tennessee, U.S.A.

Alexei V. Tepikin
Physiological Laboratory
University of Liverpool
Liverpool, U.K.

Andrew P. Thomas
Department of Pharmacology and Physiology
New Jersey Medical School
Newark, New Jersey, U.S.A.

Justyn M. Thomas
Department of Pharmacology
University of Oxford
Oxford, U.K.

Bente Vilsen
Department of Physiology
University of Aarhus
Aarhus, Denmark

Xander H.T. Wehrens
Department of Molecular Physiology and Biophysics
Baylor College of Medicine
Houston, Texas, U.S.A.

Carl White
Department of Physiology
University of Pennsylvania School of Medicine
Philadelphia, Pennsylvania, U.S.A.

Ulrich Wissenbach
Institute for Clinical Pharmacology and Toxicology
Saarland University
Homburg, Germany

Frank Wuytack
Department of Physiology
Catholic University of Leuven
Leuven, Belgium

Michiko Yamasaki
Department of Pharmacology
University of Oxford
Oxford, U.K.

Table of Contents

1 Fluorescent Calcium Indicators Based on BAPTA

Martin Poenie

CONTENTS

1.1   INTRODUCTION

Researchers have been using fluorescent calcium indicators to study the regulation of intracellular calcium for over two decades. Over this time, there has also been the pursuit of ever better temporal and spatial resolution. Indeed, where once the aim of measurements was to determine the average calcium concentration in a suspension of cells over a period of minutes, the focus has now shifted to detecting subcellular calcium sparks and microdomains with dimensions measured in microns (or even nanometers) and durations measured in milliseconds [1–7]. These advances have been possible because of improvements in technology on a number of different fronts, including improved imaging devices, innovations in microscopy, and the development of new calcium indicators.

Despite the introduction of many new indicators, none of them is equally suited for all purposes, and efforts to develop better calcium indicators continue. At present, there does not appear to be an "ideal" calcium indicator with the kinetic properties, calcium affinity, and dynamic range necessary to monitor the full range of calcium excursions in the cell. The trend, therefore, has been to develop special-purpose indicators with different calcium affinities, or for measuring calcium in different cell compartments. With the proliferation of calcium indicators, investigators have found that calcium transients missed by one indicator may be detected by another [8, 9]. More often, the apparent amplitude and time course of the transient depend to a lesser or greater degree on the properties of the indicator used in the measurement [10]. Furthermore, the nature of the signal also depends on where you look in the cell, whether the nucleus versus cytoplasm, right at the membrane versus nanometers, or microns away from it, or around various organelles. Thus, it becomes critically important that the investigator understands the properties of the indicator and how $[Ca^{2+}]$ is sampled in cells.

The goal of this review is to examine the current generation of indicators with respect to their properties, intended applications, and potential problems or limitations. In discussing these indicators, there is an attempt to group them based on their most significant property or application, although some could easily fall into more than one category. It should be noted that calcium indicators will be compared primarily based on their properties in solution in our discussion. These properties can be altered substantially by the intracellular environment, a topic that is outside the scope of this review.

1.2 CALCIUM INDICATORS BASED ON BAPTA: GENERAL CONSIDERATIONS

All fluorescent calcium indicators, apart from those based on proteins, are derivatives of BAPTA [11]. The structure of BAPTA is similar to that of EGTA, but differs in the switch from the aliphatic amino groups of EGTA to the aromatic amino groups in BAPTA (Figure 1.1). This is an important feature because the aliphatic amines of EGTA have pK_a values between 8.5 and 9.5, whereas in BAPTA they are between 5.97 and 6.36 [11, 12]. The lower pK_a of the BAPTA aromatic amines leaves them largely unionized in the physiological pH range and relatively insensitive to small fluctuations in pH. Even so, some studies show that in cells, the amino groups of BAPTA-based calcium indicators are still protonated to a small extent [13, 14]. The reduced pK_a of the BAPTA amino groups has the added benefit

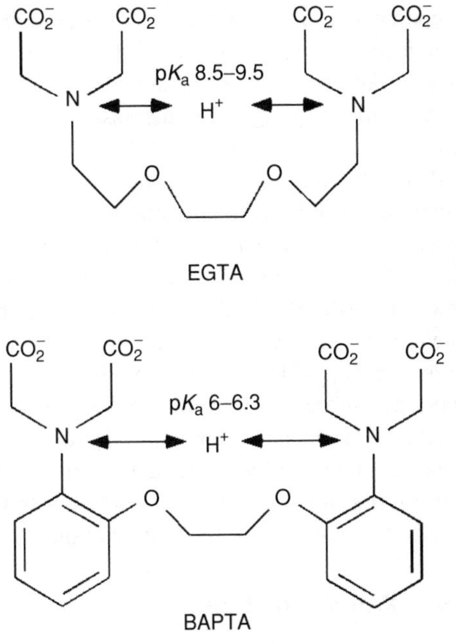

FIGURE 1.1 A comparison of EGTA and BAPTA. The key difference between EGTA and BAPTA concerns the aliphatic amino groups of EGTA as compared to the aromatic amino groups of BAPTA. The aromatic amino groups have lower pK_a values than the corresponding amino groups of EGTA. The lower pK_a of the BAPTA aromatic amines makes this chelator less sensitive to pH and kinetically faster than EGTA.

of making BAPTA much faster at binding calcium than EGTA and its relatives. With EGTA, the aromatic amino groups are largely protonated at pH 7 and these protons must dissociate before calcium can bind. The binding reaction is thus a hydrogen ion–calcium ion exchange reaction. With BAPTA, the corresponding amino groups are not protonated so there is no delay in binding calcium.

BAPTA binds to calcium with a 1:1 stoichiometry and shows high selectivity for calcium over magnesium. The stoichiometry for Mg^{2+} is also 1:1 until concentrations approach 100 mM where the stoichiometry changes to 2:1. BAPTA and its derivatives also bind tightly to a number of heavy metals including Ba^{2+}, Cd^{2+}, Hg^{2+} and Zn^{2+} and Tb^{3+}, and these can have differing effects on their spectra. Some metal ions such as Ba^{2+} and Zn^{2+} have an effect similar to calcium, whereas others such as Mn^{2+} and Ni^{2+} often quench the fluorescence of the indicator.

1.3 HIGH-AFFINITY CALCIUM INDICATORS

BAPTA itself has a high affinity for calcium ($K_d = 0.1$ μM) [11]. Fluorescent indicators derived from BAPTA have affinities for calcium that vary according to the electron-withdrawing properties of their respective fluorophores. Most BAPTA-based indicators contain a single fluorophore conjugated to the left (or A) aromatic ring of BAPTA. When calcium binds, the electronic properties of the aromatic ring and the fluorophore conjugated to it are affected. The fluorophore in turn can affect the affinity of the indicator for calcium, which depends to some degree on the nature of the fluorophore. Addition of electron donating or withdrawing substituents to the right (or B) aromatic ring can also increase or decrease its affinity for calcium. For example, a series of BAPTAs with a range of calcium K_d values have been developed by substituting various positions of the aromatic ring with groups of different electron-withdrawing potential [15, 16].

BAPTA is rarely used in cells as a calcium indicator. It is, however, often used to buffer calcium inside cells. As a buffer, it has the advantages of rapid calcium sequestration and low pH sensitivity. BAPTA also appears to be better at chelating calcium near the membranes, a feature that has been attributed to its ability to interact with membrane lipids [17, 18]. On the other hand, there are reports indicating that BAPTA may have more serious biological side effects than EGTA. One study reports that BAPTA can cause depolymerization of actin filaments and microtubules independent of its ability to chelate calcium [19]. In another study, BAPTA blocked transcription but it was not clear whether this was due to calcium chelation or a direct effect of BAPTA on the transcriptional machinery [20].

1.3.1 HIGH-AFFINITY RATIOMETRIC INDICATORS

Ratiometric indicators are those whose calcium-free and calcium-bound forms exhibit distinct and well-resolved fluorescence spectra (Figure 1.3). This shift in excitation or emission maxima offers special advantages in quantifying $[Ca^{2+}]_i$ and in imaging applications. As all the BAPTA-based indicators bind to calcium with 1:1 stoichiometry, there should be two forms of the indicator in solution, the free and calcium-bound forms, whose proportions are governed by one K_d. By monitoring the fluorescence at two excitation (or emission) wavelengths, one obtains an estimate of the concentration of the calcium-free and calcium-bound indicator. The ratio of fluorescence intensities at these two wavelengths is proportional to calcium regardless of differences in the pathlength or the absolute amount of indicator present in a particular region of the cell. One can convert this ratio to $[Ca^{2+}]_i$ by knowing the limiting fluorescence ratios, R_{min} and R_{max} rather than absolute intensity. These limiting ratios can be determined *in situ* or separately from solutions of the indicator.

Although ratiometric indicators have their advantages, they are also not without their problems. The intracellular environment can alter the spectra and dynamic range of an indicator as well as its apparent affinity for calcium. Spectral alterations make it difficult to compare the limiting fluorescence ratios between the cell and external calibration solutions. Photochemical reactions can generate degradation products that remain fluorescent but do not respond to calcium. These problems are not unique to ratiometric indicators alone but are also perhaps more apparent than when using a nonratiometric indicator.

1.3.1.1 High-Affinity Ratiometric Indicators Excited by Ultraviolet Light (Fura-2, Benzothiaza-1, Benzothiaza-2, and Indo-1)

The indicators in this group are all excited by UV light between 300 and 400 nm (see Figure 1.2 and Figure 1.3). Fura-2 has been the most popular ratiometric indicator in general and in particular for calcium-imaging using wide-field microscopy. It has not been widely used with laser scanning confocal microscopy, because it is difficult, or at least expensive, to alternate between two different UV wavelengths using lasers. Indo-1 exhibits a shift in both its excitation and emission spectra. However, there is little reason to use it for dual-excitation wavelength measurements given that its peak excitation wavelength is shorter than that of fura-2. The emission shift of indo-1 has made it very popular for flow cytometry and to a lesser extent for confocal microscopy.

Fura-2 and indo-1 have their advantages, but they are also not free from problems and limitations. The need for UV illumination is troublesome in that it causes cell damage, is poorly transmitted by glass optics, and excites endogenous fluorophores such as NADH and NADPH. The emission wavelengths of indo-1 are closer to that of the reduced pyridine nucleotides than fura-2 and thus are more affected by autofluorescence [21]. In addition, indo-1 bleaches more readily than fura-2 and seems to be rather susceptible to photochemical reactions [22, 23].

Both fura-2 and indo-1 [24] have relatively high affinity for calcium and many studies show that this can be a problem. With high-affinity indicators, large and rapid calcium transients appear to be dampened and prolonged relative to the same event measured with low-affinity indicators [10] (Figure 1.7). This has prompted a number of studies where calcium signaling events were reexamined using low-affinity indicators. One of the significant findings that emerge from these studies is that different stimuli can generate calcium transients that differ greatly in

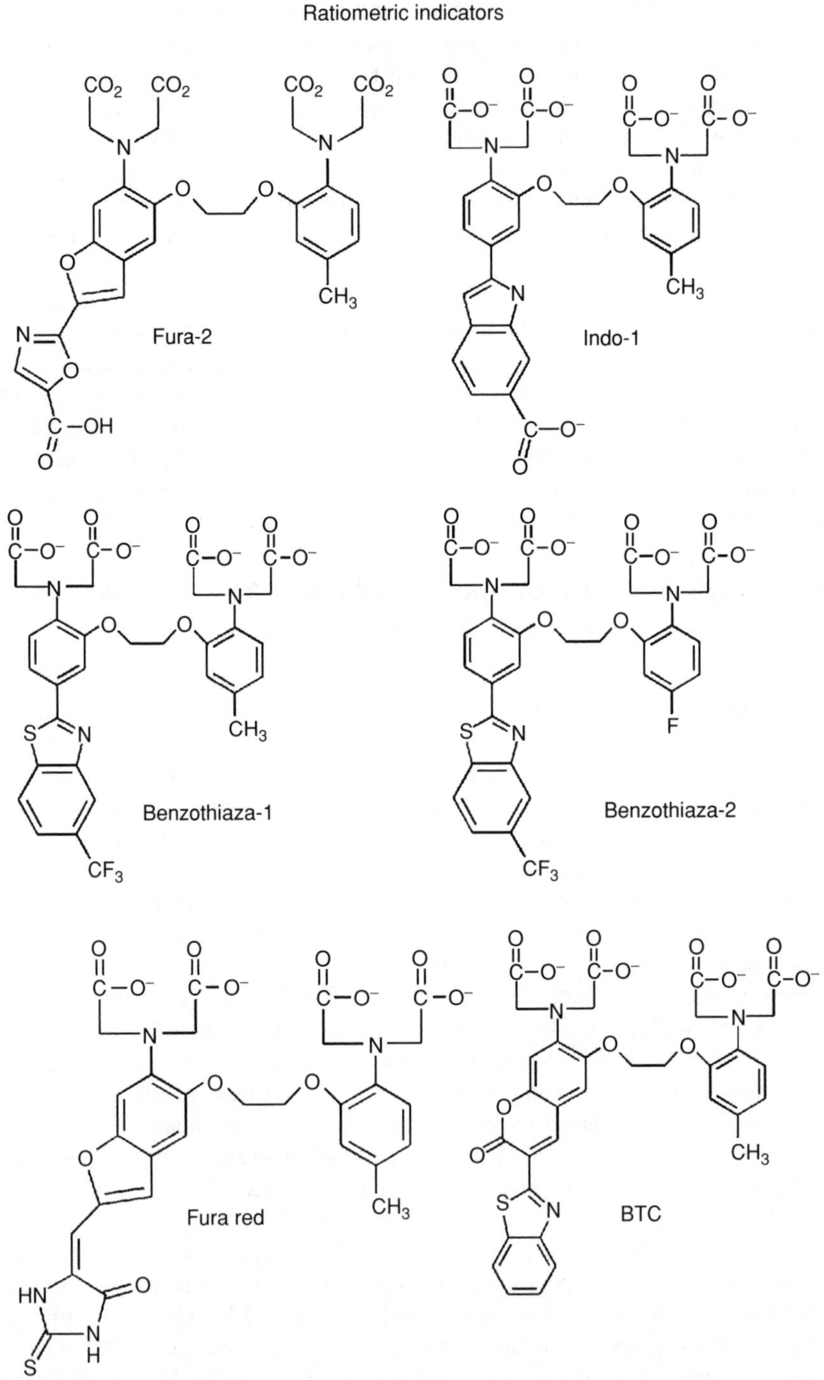

Ratiometric indicators

FIGURE 1.2 Chemical structures of ratiometric calcium indicators.

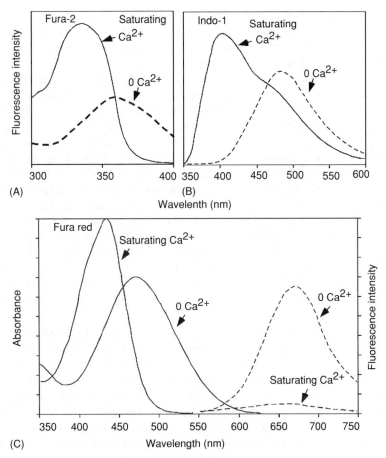

FIGURE 1.3 Spectra of the ratiometric indicators fura-2, indo-1, and Fura Red. (A) Excitation spectra of fura-2 at limiting low and high Ca^{2+} concentrations (emission 510 nm). (B) Emission spectra of Indo-1 at limiting low and high Ca^{2+} concentrations (excitation 338 nm). (C) Absorption (solid line) and fluorescence emission (dashed line) spectra of Fura Red at limiting low and high Ca^{2+} concentrations. For the fluorescence emission spectra, the excitation was set at 488 nm. (Molecular Probes, Inc., With permission.)

amplitude. Fortunately, versions of fura-2 and indo-1 with greatly reduced affinities have recently been developed [25]. These are discussed below in the section on low-affinity indicators.

1.3.1.2 High-Affinity Ratiometric Indicators Excited by Visible Light (Fura Red)

There are two ratiometric fluorescent indicators that have working wavelengths in the visible wavelength range, Fura Red and BTC [26, 27]. Due to its low affinity for calcium, BTC is discussed below in the section on low-affinity indicators. Fura Red is a ratiometric calcium indicator that can be used in a dual excitation-wavelength mode in the same manner as fura-2. However, it is rarely used this way. More often, it is loaded into cells together with fluo-3, a combination that mimics the behavior of a dual emission-wavelength indicator.

Fura Red is structurally related to fura-2 in that both contain a benzofuran fluorophore. However, the brightness and wavelength of the fluorophore can vary considerably depending on what other moiety is attached to it. In fura-2, the benzofuran is joined to an oxazole, whereas in Fura Red, the oxazole is replaced with a thiohydantoin (see Figure 1.2). This shifts the excitation and emission spectra to much longer wavelengths, but it also greatly reduces the fluorescence quantum yield. Fura Red shows a shift in excitation maxima from 474 to 437 nm when it binds calcium with a broad emission spectrum peaking at 660 nm (Figure 1.3). The quantum efficiency of the free form of the indicator is about 0.013, ten times lower than that of fura-2. The calcium-bound form has a quantum efficiency that is 40% less than that of the free indicator but the extinction coefficient is about 40% higher ([28], Iain Johnson, personal communication). Thus, the brightness of the free and bound forms of the indicator is almost the same. Practically, users find that the concentration of Fura Red in cells must be two to five times that of fluo-3 to achieve similar levels of fluorescence signal (Iain Johnson, personal communication).

The remarkably large Stokes shift of Fura Red makes it possible to simultaneously excite a mixture of fluo-3 and Fura Red at 488 nm and resolve their respective emission spectra. When a mixture of fluo-3 (or other fluorescein-based indicator) and Fura Red is titrated with calcium, the emission peak shifts from 656 nm (Fura Red) to 526 nm (fluo-3). This mimics the behavior of ratiometric indicators. Although the titrations make this method look attractive, it has problems when applied to cells. Fura Red shows differential loading in different cell types and even from cell to cell in a given culture [29]. When two indicators are loaded into cells simultaneously, differential loading results in a fluorescence ratio that varies from cell to cell. Differences in the relative amount of each indicator in the cell combined with their K_d values make it impossible to obtain a consistent calibration curve. These factors could be compounded by differences in bleaching or photo-degradation, and interactions with cellular proteins.

There are special cases where the long emission wavelength of Fura Red is an asset. For example, hemoglobin and myoglobin do not interfere with the emission spectrum of Fura Red [30], a property that is especially useful when measuring calcium in studies of the perfused heart. It also offers special benefit when trying to simultaneously monitor more than one fluorescent probe.

1.3.1.3   Summary of the Ratiometric Indicators

The properties of the ratiometric indicators are summarized in Table 1.1. At present the selection of indicators in this group is still small. In addition to fura-2, indo-1, and various spectrally similar derivatives, we now have Fura Red and BTC in this group. Fura Red and BTC are the only ratiometric indicators that have excitation

TABLE 1.1
Ratiometric Calcium Indicators

Indicator	Absorption Maxima (nm)		Emission Maxima (nm)		K_d (μM)	K_d (mM)	K_{+1} ($10^7\,M^{-1}\,s^{-1}$)	K_{-1} (s^{-1})	Reference
	Ca^{2+}-bound	Ca^{2+}-free	Ca^{2+}-bound	Ca^{2+}-free	Ca^{2+}	Mg^{2+}	Ca^{2+}	Ca^{2+}	
Benzothiaza-1	325	368	468	511	0.66				[26]
Benzothiaza-2	325	368	469	479	1.4				[26]
Fura-2	335	363	505	512	0.15–0.25		40	103	[24, 26, 52]
Indo-1	331	349	410	485	0.25				[24, 26, 52]
BisFura-2	338	366	504	511	0.37–0.47		55	257	[26, 52]
Fura Red	472	436	645	640	0.13				[26]
BTC	401	464	529	533	7–26		0.5	1000	[10, 26, 27, 71]

spectra in the visible range. These latter two indicators suffer from weaker fluorescence and in the case of BTC, a marked sensitivity to photodegradation.

1.3.2   HIGH-AFFINITY NONRATIOMETRIC CALCIUM INDICATORS

Beginning with fluo-3 and rhod-2 [31], most efforts to develop new indicators have led to indicators with longer excitation and emission wavelengths, but without the wavelength shifts characteristic of fura-2 and indo-1. These newer indicators offer a variety of different working wavelengths with brighter fluorescence and better photostability. However, these improvements sometimes come at the expense of a dynamic range or may depend on bulky fluorophores that load poorly.

With the exception of Fura Red and BTC, all long wavelength indicators use fluorescein or rhodamine derivatives as the fluorophore. The fluorescein- and rhodamine-based indicators exhibit a change in intensity rather than a spectral shift when they bind calcium. The key differences between them relate to their excitation and emission maxima, their brightness in the absence of calcium, how much their fluorescence changes upon binding calcium, their calcium affinities, and whether or not they have a tendency to accumulate in mitochondria. The fluorescein- and rhodamine-based indicators also serve as starting points for more specialized indicators including the indicator–dextran conjugates, the low-affinity indicators, and near-membrane indicators. These are discussed below in separate sections.

1.3.2.1   Indicators Based on Fluorescein Derivatives (Fluo-3, Fluo-4, Calcium Green-1, Calcium Green-2, Oregon Green 488 BAPTA)

The fluorescein-based indicators differ in whether the fluorophore is dichlorofluorescein or difluorofluorescein and in the way that the fluorophore is conjugated to the aromatic ring(s) of BAPTA (Figure 1.4 and Figure 1.5). In both fluo-3 and fluo-4, the fluorescein fluorophore is formed by directly coupling a 9-xanthone to the aromatic ring of BAPTA [31]. Thus, the left BAPTA aromatic ring becomes an integral part of the fluorescein fluorophore. The two indicators differ in that the fluorophore for fluo-3 is a dichlorofluorescein whereas that for fluo-4 is a difluorofluorescein. In Calcium Green and Oregon Green 488 BAPTA, a complete fluorescein moiety is coupled to BAPTA via an amide linkage (Figure 1.4). The distinction between Calcium Green and Oregon Green BAPTA again depends on whether dichlorofluorescein (Calcium Green) or difluorofluorescein (Oregon Green) serves as the fluorophore. The further distinction between Calcium Green-1 and -2 and Oregon Green 488 BAPTA-1 and -2 refers to whether one or two fluorophores are conjugated to BAPTA.

1.3.2.1.1   Fluo-3 and Fluo-4

Fluo-3 has two important features, its excitation maxima at 506 nm and a large increase in fluorescence upon binding calcium. Long wavelength excitation avoids problems with UV and permits excitation with an argon laser. In addition, these indicators can be used to measure changes in calcium produced by UV photolysis of caged compounds. The increase in fluorescence upon binding calcium was

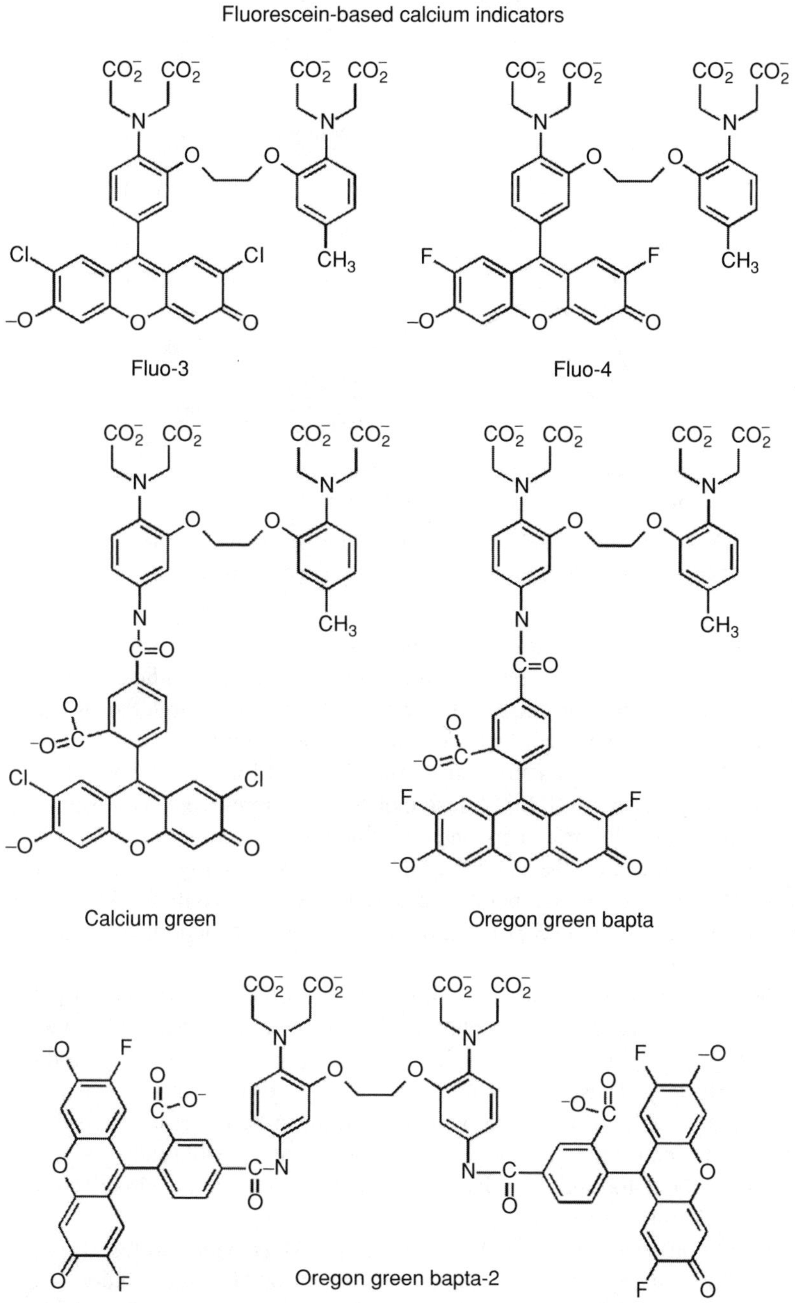

Fluorescein-based calcium indicators

FIGURE 1.4 Chemical structures of fluorescein-based indicators. Calcium Green-2 is not shown but the difference between it and Oregon Green BAPTA-2 is the same as between Calcium Green and Oregon Green 488 BAPTA. Calcium Green-2 has chlorines in place of the fluorines of Oregon Green BAPTA-2.

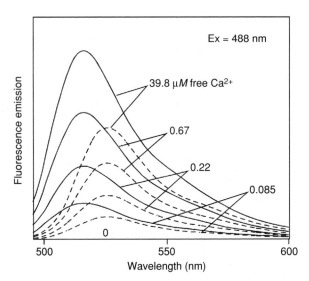

FIGURE 1.5 A spectral comparison of fluo-3 (solid line) and fluo-4 (dashed line) at various calcium concentrations. (Molecular Probes, Inc., with permission.)

originally estimated at 40-fold [31], but this was less than it should have been. Subsequent improvement in purification procedures gave an indicator with a 100-fold increase in brightness due to calcium binding [26]. This large dynamic range is important for detecting small localized changes in calcium [7].

Fluo-4 is identical in its structure to fluo-3 except for the halogen substitutions on the fluorescein ring [32]. In the design of fluo-3, fluorescein was substituted with chlorine to reduce the pK_a of its phenolic OH group from ~6.2 to ~4.5 [31]. The lower pK_a ensured that the phenolic OH was always ionized within the physiological pH range. This was important because fluorescein is only weakly fluorescent when protonated. With a pK_a of 6.4, the fluorescence of the indicators could vary with pH as well as calcium.

The substitution of fluorescein with halogens solves the problem of pH sensitivity but it also reduces its brightness. The degree of fluorescence quench by halogens generally follows the order of their appearance in the periodic table ($F < Cl < Br < I$). Thus, one would expect that difluorofluorescein would be brighter than dichlorofluorescein, which turns out to be the case. The substitution of fluorine for chlorine in fluo-4 gives an indicator that is still relatively insensitive to pH but is about 40% brighter than fluo-3 [26]. For a spectral comparison between fluo-3 and fluo-4, see Figure 1.5.

In addition to its brighter fluorescence, fluo-4 is better suited for use with confocal microscopy. With an excitation maxima at 491 nm, fluo-4 is excited much more efficiently than fluo-3 by the 488 nm argon laser line. In addition fluo-4 has greater photostability and better loading characteristics than fluo-3. Despite its improved photostability, fluo-4 has reportedly induced spurious calcium oscillations and caused increased cell death possibly due to the production of reactive oxygen compounds [33]. This is probably more generally related to the fluorescein fluorophore as increased cell die-off was also observed using calcein instead of fluo-4. The calcium affinity of fluo-4 ($K_d = 0.35$ μM) is slightly higher than that of fluo-3 ($K_d = 0.39$ μM), but this difference should not be a big factor in most studies. Where lower affinities are needed, there are a variety of derivatives not available, including fluo-5N, and those containing fluorine substitutions on the right aromatic ring of BAPTA (see the section on low-affinity indicators).

1.3.2.1.2   Calcium Green-(1 and 2) and Oregon Green 488 BAPTA-(1 and 2)

Calcium Green and Oregon Green 488 BAPTA are calcium indicators obtained by conjugating a fluorescein derivative to BAPTA through an amide linkage (Figure 1.4). They differ in that Calcium Green is based on 2,7-dichlorofluorescein whereas Oregon Green 488 BAPTA utilizes the newer 2,7-difluorofluorescein. Each of these indicators comes in two forms, one with one fluorophore and the other with two fluorophores conjugated to the BAPTA.

1.3.2.1.3   Calcium Green-1

The main improvement seen in this design is an increase in brightness. Calcium Green, for example, is five times brighter than fluo-3 [34]. On the other hand, its calcium K_d at 190 nM is less than half of that for fluo-3 and the relative increase in fluorescence upon binding calcium not nearly so large as for fluo-3 or fluo-4. The increased affinity for calcium and the increased brightness of the unbound form of Calcium Green combine to give a baseline signal in cells that is brighter than that from fluo-3 or fluo-4. However, the maximum increase in fluorescence over the baseline value would only be between two and three times that of the unstimulated level (see Figure 1.6). The difference in brightness and dynamic rage between fluo-3 and Calcium Green is compared in Figure 1.7. In addition, the excitation maximum of the Calcium Green indicators is 506 nm, which is not ideal for laser scanning confocal microscopes [26].

1.3.2.1.4   Calcium Green-2

Calcium Green-2 uses the same fluorophore as Calcium Green-1 but here there are two fluorophores, one conjugated to each of the BAPTA aromatic rings (Figure 1.4). The close proximity of two fluorophores appears to quench the fluorescence of the indicator when it is not bound to calcium. Upon binding calcium, the quenching is relieved, resulting in a large (100-fold) increase in fluorescence. In addition, Calcium Green-2 has a lower affinity for calcium ($K_d = 0.55$ μM) than Calcium Green-1. Thus, the fluorescence signal from unstimulated cells will be low and given an increase in calcium from 100 nM to 1 μM, the fluorescence would increase 10 to15-fold (Figure 1.6). This relatively large increase in signal compared to that of Calcium Green-1 is due to the reduced fluorescence of the unbound form of the indicator. While generally such an increased dynamic response is good, its significance largely depends on how well the fluorescence of the indicator can be distinguished from the background. In addition, the increase in molecular weight stemming from addition of a second fluorophore may result in poor loading using the acetoxymethyl ester form of the indicator [35].

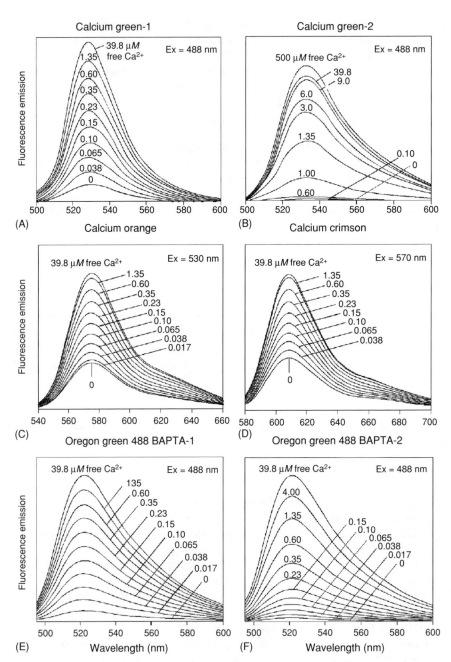

FIGURE 1.6 Fluorescence emission spectra of the "Calcium" and "Oregon Green BAPTA" series of indicators at various Ca^{2+} concentrations. (Molecular Probes, Inc., With permission.)

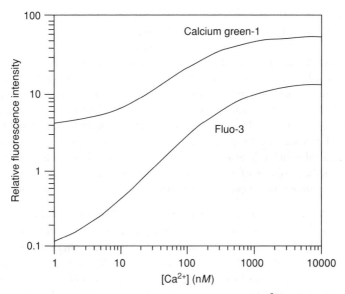

FIGURE 1.7 Comparison of fluorescence intensity responses to Ca^{2+} for fluo-3 and Calcium Green-1. Responses were calculated from the Ca^{2+} dissociation constants for the two indicators and the extinction coefficients and fluorescence quantum yields of their ion-free and ion-bound forms. They therefore represent the relative fluorescence intensities that would be obtained from equal concentrations of the two indicators excited and detected at their peak wavelengths. Note that the same trend would be observed for rhod-2 versus Calcium Orange or Calcium Crimson. (Molecular Probes, Inc., With permission.)

1.3.2.1.5 Oregon Green 488 BAPTA-1

Oregon Green 488 BAPTA is a variant of Calcium Green that uses difluorofluorescein instead of dichlorofluorescein as the fluorophore (Figure 1.4). Its chief advantages are a lower excitation maxima (494 nm), which is much closer to the 488 nm of an argon laser, and increased brightness. As with Calcium Green, the fluorescence of this indicator is relatively bright in the absence of calcium and binding to calcium gives only 14-fold increase in fluorescence intensity. In addition, the affinity for calcium ($K_d = 0.17$ μM) is even higher than that of Calcium Green-1. Assuming a 100-nM resting level, increasing calcium to 1 μM would only result in little over a twofold change in fluorescence intensity.

1.3.2.1.6 Oregon Green BAPTA-2

Oregon Green 488 BAPTA-2 is the "Oregon Green BAPTA" version of Calcium Green-2. It has two fluorophores conjugated to BAPTA and exhibits the same quenching behavior as seen with Calcium Green-2 (Figure 1.4). As a result, fluorescence in the absence of calcium is low and saturation with calcium gives a

37-fold increase in fluorescence [26] (see Figure 1.6). In addition, the affinity for calcium ($K_d = 0.58$ μM) is much reduced compared to Oregon Green BAPTA-1. This gives an indicator whose fluorescence signal will be much lower in resting cells than with Calcium Green-1. Cells that underwent an increase in calcium from 100 nM to 1 μM could show a roughly 15-fold increase in fluorescence intensity.

One of the problems with Calcium Green-2 and Oregon Green-2 is that they are relatively large molecules. Zhao and colleagues [35] have noted that loading efficiency falls off dramatically as the size of the indicator increases above 850 MW. There have been anecdotal reports that these indicators do load poorly in some cell types. A note of caution is also in order concerning the low fluorescence of these indicators in the absence of calcium due to internal quenching. This could vary significantly with environmental factors such as binding to cellular molecules.

1.3.2.1.7 Summary of Calcium Indicators Based on Fluorescein

Indicators based on fluorescein are bright but tend to bleach readily. A comparison showed the following order for indicators in terms of bleaching rate or susceptibility: Orgeon Green 488 BAPTA > fluo-3 > fluo-4 [9]. These indicators tend to load well but often show a marked accumulation in the nucleus that can complicate calcium measurements. The large dynamic range of fluo-3 and fluo-4 reportedly makes them preferable to Oregon Green 488 BAPTA in attempting to resolve elementary calcium events [9].

1.3.2.2 Indicators Based on Rhodamine (Rhod-2, X-Rhod-2, Calcium Orange, Calcium Crimson)

Since the introduction of rhod-2 [31], Molecular Probes has generated a number of new rhodamine-based calcium indicators (Figure 1.8). These indicators differ in whether the fluorophore is tetramethylrhodamine or Texas Red (sulforhodamine 101; [36]) and in the way the fluorophore is conjugated to BAPTA. In both rhod-2 and X-rhod-2 the BAPTA aromatic ring is an integral part of the rhodamine fluorophore. In Calcium Orange, a complete tetramethylrhodamine fluorophore is conjugated to BAPTA through a thiourea linkage. Calcium Crimson utilizes a complete Texas Red fluorophore coupled to BAPTA through a sulfonamide linkage.

All the rhodamine-based indicators carry a delocalized positive charge. This can present a problem when trying to load cells with the membrane-permeant acetoxymethyl ester derivative of the indicator. When esterified, the rhodamine-based indicators are hydrophobic enough to cross cell membranes despite their positive charge. The negative membrane potential across the inner mitochondrial membrane causes rhodamines, including rhodamine-based calcium indicators, to accumulate in mitochondria. Indeed, rhodamines have been used to visualize mitochondria in living cells for many years [37, 38].

1.3.2.2.1 Rhod-2 and X-Rhod-1

Rhod-2 was the first rhodamine-based indicator [31]. It had the advantage of longer excitation wavelengths than fluo-3 and was somewhat more photostable. It was not nearly as popular as fluo-3, due to a rather small increase in fluorescence upon

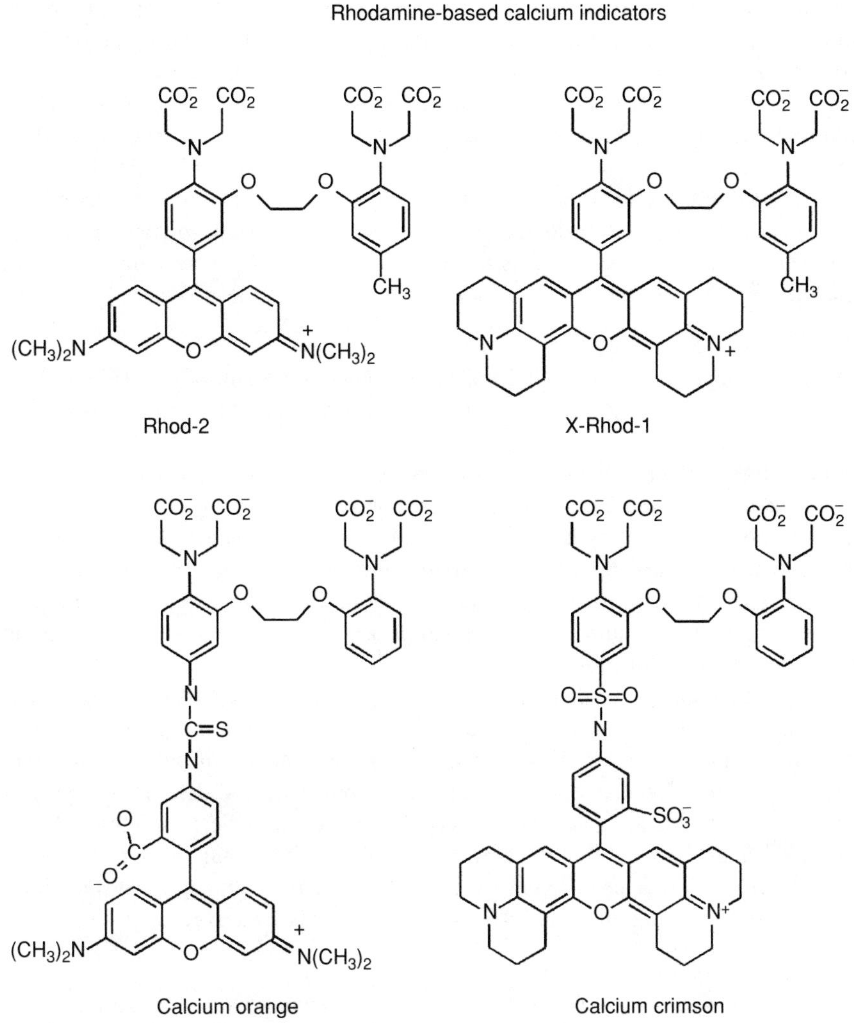

Rhodamine-based calcium indicators

FIGURE 1.8 Chemical structures of the rhodamine-based calcium indicators.

binding calcium and its tendency to accumulate in mitochondria. The small increase in fluorescence upon binding to calcium was apparently due to impurities. Molecular Probes reports that better purification procedures have yielded a preparation of rhod-2 that shows a 100-fold increase in fluorescence upon binding calcium.

Rhod-2 accumulates in mitochondria when loaded as the AM ester although the degree of accumulation seems to vary somewhat with cell type and loading conditions. The fluorescence from the residual cytosolic fraction of the indicator can be suppressed by reducing the rhodamine to dihydrorhodamine (which is nonfluorescent) prior to loading [26]. Dihydrorhodamine is oxidized back to rhodamine in the mitochondria by reduced oxygen intermediates such as peroxide or superoxide although this may also require iron or cytochromes [39].

X-rhod-1 is a variant of rhod-2 with Texas Red replacing rhodamine as the fluorophore (Figure 1.8). Like rhod-2, it shows a large (80-fold) increase in fluorescence upon binding calcium and it accumulates in mitochondria when loaded as the acetoxymethyl ester (Iain Johnson, personal communication). The distinctive advantages of X-rhod-1 are its longer excitation and emission wavelengths and its brighter fluorescence. X-rhod-2 is more bulky than rhod-2, increasing the risk of nonspecific binding and less efficient de-esterification of the AM ester.

1.3.2.2.2   Calcium Orange

Calcium Orange is the tetramethylrhodamine analog of Calcium Green (Figure 1.8). It excites maximally at 549 nm and has an affinity almost identical to that of Calcium Green-1 ($K_d = 0.19$ μM). As with other indicators in the "Calcium" series of indicators, it fluoresces rather brightly in the absence of calcium and the increase in fluorescence upon binding calcium is only about 2.5-fold (Figure 1.6). Given the high affinity and small dynamic range of this indicator, changes in $[Ca^{2+}]_i$ above baseline will produce relatively small increases in fluorescence. There are also reports that Calcium Orange can be difficult to load. For example, loading in hippocampal slices requires long incubation periods, increased temperatures, and high concentrations of Pluronic [40].

In comparison with other visible wavelength indicators, Calcium Orange exhibits the slowest photobleaching rates. As with other indicators that incorporate a rhodamine fluorophore, it shows a tendency to accumulate in mitochondria that have not been depolarized or discharged. In one study, Calcium Orange was considered to be the least useful of the visible wavelength indicators in that it failed to detect small localized calcium elevations in the cell that could be detected using other indicators [9]. This difference was attributed to the high degree of mitochondrial sequestration and the small dynamic range of this indicator.

1.3.2.2.3   Calcium Crimson

Calcium Crimson is the "Texas Red" version of Calcium Orange (Figure 1.8). Its main virtues are its excitation wavelength at 590 nm and its brighter fluorescence as compared to Calcium Orange. Like Calcium Orange it has a small dynamic range and high affinity for calcium (Figure 1.6). The Texas Red fluorophore is more bulky than rhodamine and this may cause problems in loading and nonspecific binding [35]. There are some anecdotal reports that this indicator is not very responsive to calcium after it has been loaded into cells.

1.3.2.3   Summary of Nonratiometric Indicators

The properties of these indicators are summarized in Table 1.2. Of the long wavelength ratiometric indicators, fluo-4 appears to be the most promising. A search of the literature shows that fluo-3 is by far the most popular of the current generation of long wavelength indicators and fluo-4 appears to be an improved

TABLE 1.2
Nonratiometric Calcium Indicators

	Absorption Maxima (nm)		Emission Maxima (nm)		K_d (μM)	K_d (mM)	K_{+1} ($10^7\,M^{-1}s^{-1}$)	K_{-1} (s^{-1})	Reference
	Ca^{2+}-bound	Ca^{2+}-free	Ca^{2+}-bound	Ca^{2+}-free	Ca^{2+}	Mg^{2+}	Ca^{2+}	Ca^{2+}	
Fluo-3	503	506	526	526	0.39–0.52	8.1	71	369	[26, 31, 52]
Fluo-4	494	491	516	516	0.35				[26]
Calcium Green-1	506	506	534	533	0.16–0.19		75	120	[26, 52, 34, 52]
Calcium Green-2	506	506	531	531	0.53–1.8				[26, 52, 52]
Calcium Orange	554	555	575	576	0.33–4.61				[26, 52]
Calcium Crimson	588	588	611	611	0.21–0.4				[26, 34]
Oregon Green 488 BAPTA-1	494	494	523	523	0.17				[26]
Oregon Green 488 BAPTA-2	494	494	523	523	0.38–0.58				[26, 132]
Rhod-2	552	549	581	581	0.57–2.3				[26, 31, 34]

version of fluo-3. Oregon Green 488 BAPTA bleaches more rapidly and has a lower dynamic range than other fluorescein-based indicators. Oregon Green-2 shows many favorable spectral features but problems with loading and binding to cellular components may limit its usefulness. The rhodamine-based indicators have, to a large extent, become specialized tools for studying calcium in mitochondria. Of these, rhod-2 still appears to be the most useful. They are useful as cytosolic calcium indicators when they are introduced into the cytoplasm in salt form or as dextran-conjugates.

1.4 LOW-AFFINITY CALCIUM INDICATORS FOR MEASURING LARGE OR RAPID CALCIUM TRANSIENTS

From the outset, there have been differing perspectives concerning whether it is better to measure calcium with a high-affinity or low-affinity indicator. From one perspective, it seemed better to use a high-affinity indicator where the K_d was approximately the same as the resting level of $[Ca^{2+}]_i$ because this is where the signal from the indicator would be most reliable. In addition, most cells seemed to exhibit relatively small calcium transients that seldom rose to levels greater than 1 μM. Thus, in terms of its calcium-binding characteristics, high-affinity indicators like fura-2 seemed to be well suited for most applications. A recent study by Wokosin and colleagues argued convincingly that it is better to use a moderate-affinity indicator because the signal-to-noise ratio was better over the range of calcium concentrations one is likely to encounter in cellular calcium measurements [41].

More severe problems with high affinity indicators are seen when measuring events where calcium increases to high levels or changes very quickly. Problems with the kinetics and affinity of fura-2 first became apparent when it was used for measurements in muscle [42, 43]. Muscle contraction was one of the few cases where the dynamics of $[Ca^{2+}]_i$ had been studied long before the introduction of fluorescent calcium indicators [44, 45]. Early measurements were made using metallochromic indicators (e.g., murexide, purpurate diacetic acid, antipyrlazo III, arsenazo III) or aequorin [46, 47]. Although each indicator was fraught with problems, a great deal of effort directed at overcoming their limitations led to a reasonable picture of the speed and the peak height of the muscle calcium transient. When calcium measurements were made using fura-2, the differences in kinetics and magnitude of the calcium transient were immediately evident. Attempts to resolve this disparity continue but at least two factors have emerged. One of these factors concerns the intrinsic properties of the various indicators especially with respect to their on- and off-rates for calcium binding, their dissociation constants, and their ability to buffer calcium inside the cell. Another factor concerns inter-actions between the indicator and its intracellular environment. This includes alterations in spectra, K_d, and kinetics due to binding of the indicator to proteins and the microviscosity of the environment [48, 49].

At present, it is becoming increasingly clear that cellular calcium transients can vary in their magnitude due to different stimuli and at different locations in the cell. When observed with fura-2, these distinctions tend to be blurred due to the tendency of high-affinity indicators to saturate or buffer calcium transients. The recent introduction of low-affinity indicators has led to several studies showing large differences in amplitude of calcium transients evoked by different stimuli. This may lead to renewed interest in "amplitude modulation" as an important modality in calcium signaling [50].

1.4.1 THE RELATIONSHIP BETWEEN AFFINITY AND KINETICS

High-affinity indicators present a problem in attempting to track rapid or large calcium transients. There is, for example, a big difference in the shape of a muscle calcium transient measured using low-affinity (PDAA, furaptra) and high-affinity (fura-2, fluo-3) calcium indicators. In Figure 1.9A, one sees by comparison that fura-2 gives a slower rise to peak $[Ca^{2+}]_i$ and a markedly prolonged duration of the calcium transient than seen with low-affinity indicators such as furaptra (9C) and PDAA (9D). This difference can be largely explained by the calcium-binding properties of the different indicators.

The relationship between affinity and kinetics is based on the relationship between the equilibrium binding constant (given here as K_d) and the on- and off-rate

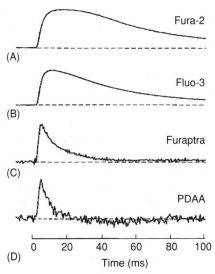

FIGURE 1.9 Comparison of the kinetics of difference calcium indicators. Recordings were collected from different experiments on intact single twitch fibers from frog muscle. Each fiber was bathed in a normal Ringer's solution at 16°C, stretched to long sarcomere length (3.5–4 μm), micro-injected with a Ca^{2+} indicator to a nonbuffering concentration, and stimulated by an action potential at zero time. The records show the fluorescence signals recorded from (A) fura-2 (B) fluo-3, and (C) furaptra. The bottom trace (D) is the delta absorbance for a muscle fiber injected with purpurate diacetic acid (PDAA). For display purposes, all signals have been scaled to have the same peak amplitude. (Data courtesy of Dr. Steve Baylor.)

constants for calcium binding. For the binding reaction between the indicator (I) and calcium (Ca):

$$K_d = [\text{Ca}] \times [\text{I}]/[\text{CaI}]$$

This reaction can be described in terms of two rate constants, the forward rate constant k_{+1} and the reverse rate constant k_{-1} such that:

$$k_{+1}([\text{Ca}] \times [\text{I}]) = V(\text{forward})$$

$$k_{-1}[\text{CaI}] = V(\text{reverse})$$

One can conceptually relate K_d and the ratio of k_{-1} and k_{+1} in terms of the equilibrium rates of association and dissociation between calcium and the indicator. For high-affinity indicators there is a huge difference between k_{-1} (a small value) and k_{+1} (a large value). Yet, by definition, at equilibrium the rate of dissociation of the calcium–indicator complex must equal the rate that free indicator associates with calcium.

Thus, at equilibrium $k_{+1}([\text{Ca}] \times [\text{I}]) = k_{-1}[\text{CaI}]$ and, therefore, $k_{-1}/k_{+1} = [\text{Ca}] \times [\text{I}]/[\text{CaI}] = K_d$.

What these equations show is that in order for the two rates to be equal, there must be a large concentration of calcium–indicator complex compared to the concentrations of free indicator and calcium. The actual ratio of free indicator and free calcium to the calcium–indicator complex is directly related to the ratio k_{-1}/k_{+1}.

Although the forward and reverse rate constants are related to how rapidly an indicator will respond to a sudden step in $[\text{Ca}^{2+}]$, the ratio of these constants (i.e., K_d) does not directly relate to the speed of the indicator. For a binding reaction between calcium and the indicator, the rate of approach to equilibrium can vary, even for indicators with the same K_d. A better description of kinetics is the relaxation time (τ), which describes how fast a reaction comes to a new equilibrium when one of the reactants is suddenly shifted away from its equilibrium concentration. When the equilibrium is suddenly shifted, there is a momentary difference between the original equilibrium concentrations and the new equilibrium. For example, given the reaction

$$[\text{A}] + [\text{B}] \longleftrightarrow [\text{C}]$$

with the original equilibrium concentrations being [A], [B], and [C], respectively, and the new equilibrium concentrations represented by [Ae], [Be], and [Ce], we can define a quantity Δx, which represents the difference between the original and new equilibrium concentrations [51]. Thus:

$$\Delta x = [\text{A}] - [\text{Ae}] = [\text{B}] - [\text{Be}] = [\text{Ce}] - [\text{C}]$$

The rate of change of Δx, given as $d(\Delta x)$, can be approximated by the equation

$$-\mathrm{d}(\Delta x)/\mathrm{d}t = [k_{+1}([\mathrm{Ae}] + [\mathrm{Be}]) + k_{-1}]\Delta x$$

From this one obtains τ given as:

$$1/\tau = [k_{+1}([\mathrm{Ae}] + [\mathrm{Be}]) + k_{-1}]$$

These equations reveal important factors that affect the speed of the indicator. On a first approximation, the equations show that, other factors being equal, τ varies with the sum of the rate constants such that the higher sum of the rate constants, the faster the reaction will reach equilibrium. For comparison, fura-2 has an on-rate of $6 \times 10^8 \, M^{-1}\mathrm{s}^{-1}$ and an off-rate of $98\,\mathrm{s}^{-1}$ whereas the corresponding values for furaptra are $7.5 \times 10^8 \, M^{-1}\mathrm{s}^{-1}$ and $26{,}760\,\mathrm{s}^{-1}$ [51, 52]. Based on these numbers, furaptra should be substantially faster than fura-2. The equations also show that both rate constants are important but their relative contribution to the overall kinetics depends on the concentration of the reactants. This is because the *contribution* of the on-rate is proportional to the concentration of the reactants, whereas the *contribution* of the off-rate is concentration independent. Thus, the speed of response partly depends on how much indicator is present. To see this, consider a typical scenario where cell is loaded with 50 μM of a high-affinity indicator such as fura-2 ($K_d \sim 250$ nM) and during some stimulus, $[\mathrm{Ca}^{2+}]_i$ rises to the micromolar range. During the course of the stimulus, as calcium combines with the indicator, the free fura-2 concentration decreases to vanishingly low concentrations. As this happens, the rate of the reaction slows down. This can be avoided if the concentration of free indicator remains relatively large compared to the concentration of calcium-bound indicator throughout the calcium transient.

These observations lead to some general conclusions regarding what constitutes a good indicator for measuring rapid calcium transients. First the sum of the on-rate and off-rate should be as large as possible. Second, the concentration of the intracellular indicator should be set to a level where it is always large relative to the equilibrium concentration of the indicator–calcium complex. Finally, in order to satisfy the previous requirements and not to drastically perturb or buffer $[\mathrm{Ca}^{2+}]_i$ levels, the affinity of the indicator needs to be relatively weak.

The thinking about indicator kinetics has mostly focused on increasing the off-rate because on-rate is essentially diffusion-limited. If that were the case, then the only way to increase the speed of an indicator is to increase its off-rate, which is tantamount to reducing its affinity for calcium. In fact, the various indicators do differ significantly in their on-rates and ideally, this parameter should be large. However, the goal is to increase the sum of the rate constants and this dictates that the off-rate must be large also. Since increasing the off-rate will lower the affinity of an indicator, a "fast" indicator will necessarily be a "low-affinity" indicator.

1.4.2 DESIGN OF LOW-AFFINITY INDICATORS

There are three main approaches to reducing the affinity of calcium indicators. One is to reduce the number of carboxyl groups in the calcium-binding pocket, a second is to conjugate a more strongly electron-withdrawing fluorophore to the left aromatic ring and the third is to add electron-withdrawing substituents to the right aromatic ring of the BAPTA core. The first approach is represented by furaptra and the various "Mag" or "Magnesium" series of indicators offered by Molecular Probes (Figure 1.10). The second approach is represented by BTC. The third approach is represented by indicators with a nitro group substituted at the 5 position or fluorines substituted at various positions of the right BAPTA ring. The 5-nitro derivatives are sold by Molecular Probes with the designation "5-N" appended to the name of the indicator (see Figure 1.10). Also in this category are the 4-, 5-, and 6- monofluoro substitutions developed by Gee and colleagues [16] as well as the 5,6-difluoro developed by London and colleagues ([25]; Figure 1.10). It should be noted that the "FF" naming convention could lead to some confusion since TEFLABS also makes the near-membrane indicator FFP18 (dubbed "fatty fura") that is not related to the "FF" 5,6-difluorinated indicators.

1.4.3 LOW-AFFINITY CALCIUM INDICATORS BASED ON APTRA

(Furaptra, Mag-fura-2, Mag-indo-1, Mag-Fluo-4, Mag-fura-5, Mag-Fura Red, Mag-Rhod-2, Mag-X-rhod-1, Magnesium Green and Magnesium Orange.)

These indicators were all developed based on the aminophenol triacetic acid (APTRA) core structure. The prototype tricarboxylate indicator is furaptra [53] that is now sold by Molecular Probes under the name Mag-fura-2 [26]. Furaptra might be considered a half-BAPTA or half-fura-2 indicator where the right side aromatic ring of BAPTA is omitted (Figure 1.10). The ethylene glycol diether connecting the two aromatic rings in fura-2 is replaced with a carboxymethyl ether group that adds a third carboxyl group to the ion-binding pocket.

All the tricarboxylate indicators show a reduced affinity for calcium and either the same or an increased affinity for magnesium over their tetracarboxylate counterparts. A number of lines of evidence point to the conclusion that the magnesium ion, which is smaller than the calcium ion, normally binds to only one side of the BAPTA molecule. Part of the evidence for this comes from the observation that when titrated over a large range of magnesium concentration, the indicator shows two distinct transitions in the spectra. Understood in this way, the reduction of the number of carboxyls from four in BAPTA to three in APTRA would have a great impact on calcium affinity, but either no effect on magnesium affinity or an increased affinity due to the incorporation of a third carboxyl into a smaller binding pocket. This expectation is borne out in the published data that show that furaptra has much reduced affinity for calcium as compared to fura-2 (>200-fold lower) whereas the magnesium affinity is 2 to 2.5 times higher than that of fura-2 [26]. It should be noted however that the published affinities for calcium and magnesium vary over a large range.

Since the introduction of furaptra, a wide variety of low-affinity tricarboxylate indicators have been developed by Molecular Probes. These are all tricarboxylate counterparts of the BAPTA-based calcium indicators. Generally, the optical properties of the tricarboxylate indicators are identical to that of the tetracarboxylate counterparts. The key differences between the various furaptra analogs lie in their

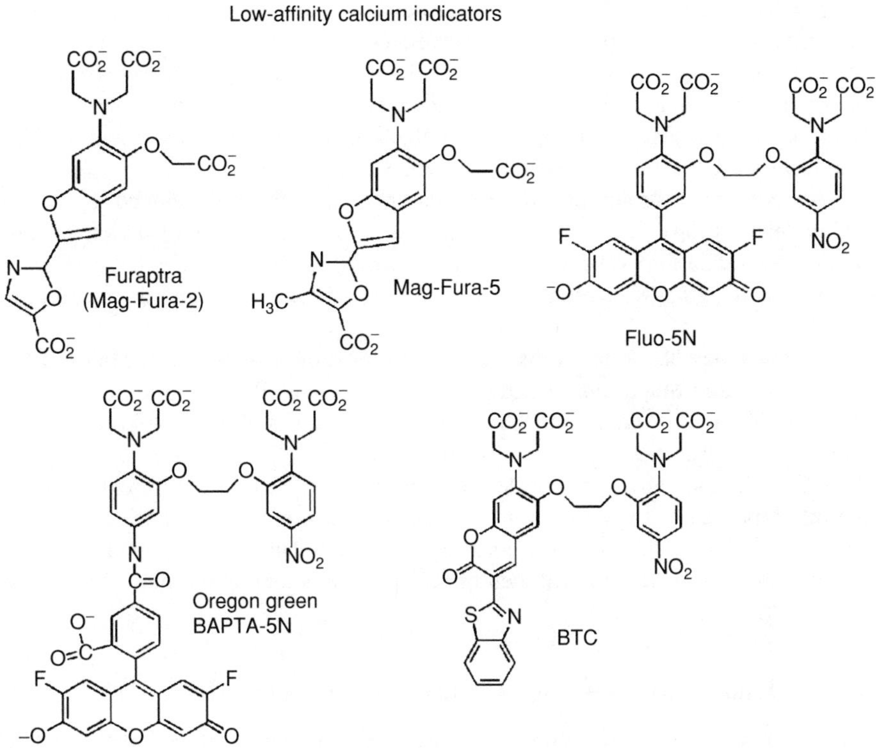

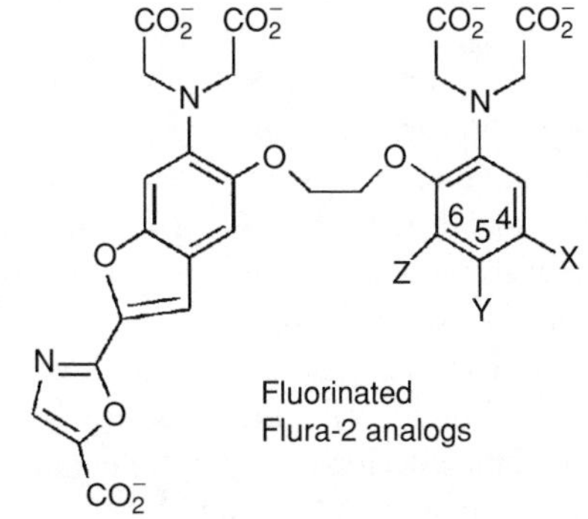

Low-affinity calcium indicators

FIGURE 1.10 Chemical structures of low-affinity calcium indicators. For the fluorinated indicators F = fluorine and H = hydrogen. It should be noted that similar fluorinated analogs can be generated for essentially all the BAPTA-based family of calcium indicators.

differing excitation and emission wavelengths, whether they exhibit shifts in their excitation or emission spectra, and in their affinities. As with their BAPTA-based counterparts, only Mag-Fura-2, Mag-Indo-1, and Mag-Fura Red exhibit spectral shifts upon binding to calcium ions. The other analogs simply exhibit a change in fluorescence intensity. However, the various single wavelength indicators in this group do offer a range of excitation wavelengths (see Table 1.3). In general, the affinities of the various single wavelength indicators vary significantly along with their working wavelengths. This is due to the electron-withdrawing property of the fluorophore such that the more strongly electron-withdrawing rhodamine analogs also have weaker affinities for both calcium and magnesium.

An additional group of APTRA variants have been developed by Otten and colleagues [54]. While these indicators are interesting from a synthetic standpoint their excitation wavelengths range from 295 to 340 nm. These short excitation wavelengths will prohibit their use as cellular calcium or magnesium indicators.

Comparisons of the tricarboxylate indicators show that furaptra, Mag-Indo-1, Mag-Fura-5, and Magnesium Orange all exhibit essentially the same fast kinetics [10]. In one study, Magnesium Green showed a slow and fast component in its response that was not seen in other indicators [10]. Mag-Fura Red showed a biphasic response during a muscle calcium transient that was different from all the other indicators and could not be explained in terms of a simple response to calcium. Mag-Fura-5, which differs from Mag-Fura-2 by a single methyl group, seems to be popular for use with neurons although it is reportedly more sensitive to pH than Mag-Fura-2 [55, 56].

1.4.4 Tetracarboxylate Low-Affinity Calcium Indicators

The major advantage of the tetracarboxylate low-affinity indicators is their relative insensitivity to magnesium. The K_d of furaptra for Mg^{2+} is reported to be in the range of 1.5 to 1.9 mM [26, 53]. Using furaptra and its relatives, many studies report $[Mg^{2+}]_i$ in the range of 0.3–1.0 mM [56–63]. Thus, a significant fraction of the indicator will be bound to magnesium in cells. The reported calcium K_d for furaptra varies over a wide range (12–60 μM) but the estimates tend to cluster in the 35–50 μM range [10, 26, 53, 62, 64–66]. Furthermore, some of these studies indicate that calcium levels can briefly reach levels in excess of 50 μM. Thus the properties of the indicator suggest the possibility of interference between calcium and magnesium. Such an interference has been reported in at least one study [67].

Interference between calcium and magnesium can take place at several levels. First, there is direct competition between Ca^{2+} and Mg^{2+} for the indicator. Second, during a large calcium transient, much of the calcium entering the cytosol becomes bound to intracellular buffers. Some of these buffers bind both Ca^{2+} and Mg^{2+} with some of the magnesium displaced as Ca^{2+} increases [68]. This presents the potential for confusion where one cannot be sure whether the observed fluorescence signal is due to calcium or magnesium.

The tetracarboxylate low-affinity indicators have substantially lower affinity for Mg^{2+} compared to their tricarboxylate counterparts. Although there are little direct titration data, the Mg^{2+} K_d for these tetracarboxylate low-affinity indicators are apparently well above 10 mM [25, 26] such that changes in Mg^{2+} within the range measured in cells should have little or no effect on indicator fluorescence [69].

TABLE 1.3
Low-Affinity Calcium Indicators

	Absorption Maxima (nm)		Emission Maxima (nm)		K_d (μM)	K_d (mM)	K_{+1} ($10^7\ M^{-1}s^{-1}$)	K_{-1} (s^{-1})	Reference
	Ca²⁺-bound	Ca²⁺-free	Ca²⁺-bound	Ca²⁺-free	Ca²⁺	Mg²⁺	Ca²⁺	Ca²⁺	
BTC	401	464[480]	529	533	7–26				[10, 26, 27, 711]
Calcium Green-5N	506	506	532	532	23–85		40	9259	[10, 26, 52, 135]
Calcium Orange-5N	549	549	582	582	20–55		21	11670	[10, 26, 27, 136]
Fluo-3FF	515	515	526	526	42				[25, 76]
Fluo-5F	493	491	520	520	2.3				[26]
Fluo-4FF			520	520	9.7				
Fluo-5N	493	491	518	518	90				[26]
Fura-4F	339	371	510	510	0.77				
Fura-5Cl	337	366	511	511	0.4				
Fura-5F	339	367	511	511	0.4				
Fura-6F	337	367	510	510	5.3				
Fura-2FF	335	363	505	505	35				[25, 73]
Furaptra (Mag-Fura-2)	335	370	491	511	20–53	1.5–5.3	75	26760	[10, 26, 52, 53]
Indo-5F	330	346	401	478	0.47				[25, 61]
Indo-1FF	330	346	401	475	26–33				[25, 61]
Mag-Indo-1	330	349	480	417	29	2.7–7.6			[10, 26]
Mag-Fura-5	332	369	505	505	31	4.3			[10, 26]
Magnesium Green	506	506	531	531	7	1.0–2.4			[10, 26]
Magnesium Orange	550	550	575	575	43	3.9–8.2			[10, 26]
Mag-Fura Red	427	483	631	659	55	2.5–9.3			[10, 26]
Mag-Fluo-4	493	490	517	517	22	4.7			[26]
Mag-Rhod-2	548	547	578	578	70				[26]
Mag-X-Rhod-1	578	575	602	602	45	10.7			[26]
Rhod-5F	578	575	602	602	1.9				[137]
Rhod-5FF	578	575	602	602	19				[137]
Rhod-5N	578	575	602	602	320				[137]
X-Rhod-FF					17				
BTC-5N					>1000				

1.4.4.1   The 5-Nitro BAPTAs (Fluo-5N, Calcium Green-5N, Calcium Orange-5N, Oregon Green 488 BAPTA-5N)

Molecular Probes has developed low-affinity analogs of fluo-4 and their Calcium Green, Calcium Orange, and Oregon Green series of indicators. This modification entails addition of a nitro group to the 5 position of the right BAPTA aromatic ring, and is designated by appending -5N to the name of the indicator (Figure 1.10). The addition of the nitro group does not change the spectral properties of these indicators but it does greatly reduce their affinity for Ca^{2+}.

In terms of their kinetics, all of the 5N series of indicators are faster than their high-affinity counterparts but none of them are as fast as furaptra. Calcium Green-5N is reportedly much slower than Calcium Orange-5N [10]. On the other hand, Calcium Orange-5N has a relatively small dynamic range. The newer Oregon Green-5N seems to exhibit better kinetics than Calcium Green and a better dynamic range than Calcium Orange. In cases where fast muscle calcium transients have been measured using both Calcium Orange-5N and Oregon Green-5N, Oregon Green-5N responded as quickly as Calcium Orange-5N, yet it gave a better signal (Julio Vergara, personal communication). Molecular Probes also offers fluo-5N, rhod-5N, and X-rhod-5N. At present, there is little information on these indicators.

1.4.4.2   Benzothiazole Derivatives (Benzothiaza-1 and -2, BTC)

Several calcium indicators have been developed based on derivatives of the benzothiazole fluorophore including the Benzothiaza-1 and -2 and BTC. Benzothiaza-1 and -2 ratiometric indicators have so far not been used for cellular calcium measurements (Figure 1.2). They show intermediate affinities for calcium with K_d values of 0.6 and 1.2 μM, respectively [26], but the excitation maxima of the calcium-bound form of the Benzothiaza indicators (325 nm) are even shorter than those for fura-2 and indo-1 [70]. Given that similar intermediate affinities can be achieved with monofluorinated versions of fura-2 and indo-1, it is unlikely that the Benzothiaza indicators will see much use.

BTC is a ratiometric indicator developed by Iatridou and colleagues [27]. It is based on the benzothiazole–coumarin fluorophore that gives a shift in excitation from 462 to 401 nm when it binds calcium (Figure 1.10 and Figure 1.11). It has a relatively weak affinity for calcium ($K_d \sim 15$ μM), which is apparently due to properties of the fluorophore. Its on- and off-rates indicate that it is slower than most of the other low-affinity indicators [10].

BTC has been used to help clarify relationships between calcium levels and cell responses that were not clear using fura-2. In studies of cultured cortical neurons by Hyrc et al. [71] BTC showed a much larger increase in calcium in response to NMDA than to AMPA. This correlated with neuronal death induced by NMDA but not by AMPA. In similar experiments using fura-2, both AMPA and NMDA gave similar increases in calcium. In other studies using mouse pancreatic acinar cells, a high calcium trigger zone was much better resolved with BTC than with fura-2 [72].

BTC has a number of favorable properties but it has a number of noticeable problems as well. Its quantum efficiency is only 25–30% that of fura-2 and it

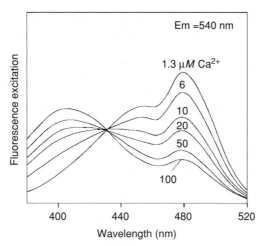

FIGURE 1.11 Excitation spectra for BTC at various Ca^{2+} concentrations. (Molecular Probes, Inc., With permission.)

photodegrades relatively rapidly to a form that is still fluorescent but insensitive to calcium [27, 71]. This is perhaps due to the same decarboxylation reaction reported for indo-1 [23]. In muscle cells, BTC gives the lowest diffusion coefficient among a number of low-affinity indicators tested, indicating a high degree of binding to myoplasmic proteins [10]. There is one report stating that BTC loads poorly compared to fura-2FF [73].

1.4.4.3 The Halogenated BAPTA Moderate- to Low-Affinity Indicators (BAPTA-4F, -5F, -5Cl, -6F, and -FF Derivatives)

A series of moderate- to low-affinity indicators have been developed where the right aromatic ring of BAPTA is substituted with fluorine or chlorine atoms at various positions on the ring. These were originally developed as derivatives of fura-2 but similar modifications can be extended to most of the BAPTA-type calcium indicators [16]. The mono-substituted derivatives place a fluorine at either the 4, 5, or 6 position on the aromatic ring. Fluorines at the ortho (6) or para (4) positions have the most dramatic effect on calcium affinity with the ortho-substitution having a much stronger effect than the para-substitution.

The "FF" series of low-affinity indicators have fluorines substituted at the 5 and 6 position on the right aromatic ring of BAPTA [74]. This derivative has a much weaker affinity for calcium than any of the mono-substituted derivatives and is comparable to that of the nitro-substituted derivatives (see Figure 1.10). This modification does not impact the fluorescence properties of the indicator, which are essentially identical to their nonfluorinated counterparts. At present, there are no data concerning the on- and off-rates for calcium binding. Judging from their structures and the data in a few published reports, it seems likely that the speed of these indicators will be similar to those of the -5N series.

One study compared calcium transients in cortical neurons measured by fura-2FF or fura-2 in response to NMDA, kainite, or high K^+ [73]. With fura-2FF, calcium transients were generally large and there was a big difference between the magnitude of calcium elevation due to high K^+ as compared to NMDA or kainate. By comparison, calcium transients measured with fura-2 were generally small and all treatments gave Ca^{2+} increases of roughly the same magnitude. Another study showed equally large Ca^{2+} increases in ATP-depleted proximal tubule cells using either Mag-Fura-2 or fura-2FF [69].

Indo-1FF has also been used to monitor changes in $[Ca^{2+}]_i$ of smooth muscle cells [75]. Here, indo-1FF showed much larger and faster changes in calcium than seen with indo-1. This study also showed that albumin and aldolase at 10 mg/ml could shift the emission spectrum of calcium-free indo-1FF toward that of the calcium-bound form with albumin having the most severe effect.

Finally, fluo-3FF has been used along with a number of other high- and low-affinity indicators to measure $[Ca^{2+}]_i$ in lizard motor nerve terminals [76]. In this study, only high-affinity calcium indicators (fluo-3, Calcium Green-2) detected an increase in calcium in response to a single action potential but when a potassium channel blocker was included, low-affinity indicators (Oregon Green BAPTA-5N, fluo-3FF) also detected a change. On the other hand, a train of depolarizing pulses could be followed accurately with the low-affinity indicators whereas the high-affinity indicators showed a decreasing response to successive pulses, probably due to saturation.

1.4.5 SUMMARY OF THE LOW-AFFINITY CALCIUM INDICATORS

The properties of the low-affinity indicators are summarized in Table 1.3. There is clearly a difference between the tricarboxylate and tetracarboxylate low-affinity indicators in their sensitivity to magnesium and in their kinetics. In general, the tricarboxylate indicators have a lower affinity for calcium, faster kinetics, and higher affinity for magnesium than the tetracarboxylate low-affinity indicators. The -5N series of indicators are a little slower than the tricarboxylate indicators with BTC substantially slower than the -5N series of indicators. Kinetics for the FF series of indicators have not been determined but they appear to be similar to those of the -5N series.

In a number of cases, the lower affinity indicators have seen differences in the amplitude of calcium signals that were not apparent when measured with fura-2. They also report fast events more faithfully than the low-affinity indicators. This includes faster time to peak and shorter durations. During a rapid train of calcium spikes, low-affinity indicators show a return of $[Ca^{2+}]_i$ to baseline whereas high-affinity indicators remain bound to calcium between spikes. On the other hand, low-affinity indicators are not suited for measuring resting calcium levels and they may obscure small differences in $[Ca^{2+}]_i$ in this range. Thus, a complete picture of $[Ca^{2+}]_i$ dynamics cells may require the use of both low- and high-affinity indicators.

1.5 INDICATORS THAT RESIST LEAKAGE AND SEQUESTRATION

Most of the BAPTA-related fluorescent calcium indicators were designed to measure the cytosolic free calcium ion concentration. Hydrophobicity of the indicator was avoided as much as possible and initially, tests were carried out to make sure that the indicator did not bind to membranes [24]. Although fura-2, with a total of five negative charges, showed little binding to membranes, it did tend to leak out of the cell and to accumulate within intracellular organelles. This phenomenon has been variously referred to as indicator leakage and compartmentalization or sequestration [77–80].

Leakage of calcium indicators creates problems in loading the indicator into cells and in obtaining an accurate measurement of $[Ca^{2+}]_i$. In extreme cases, loading of cells using the acetoxymethyl ester derivative (e.g., fura-2/AM ester) is prevented by the rapid leakage of indicator out of the cell [81, 82]. More often, one obtains a useful amount of indicator immediately after loading that appears to be evenly distributed throughout the cytosol. However, after a short time, rapid leakage out of the cytosol leaves behind only the indicator that has been sequestered in organelles. Once this happens cells show little response to changes in $[Ca^{2+}]_i$. As a consequence, measurements must be made immediately after loading and only for short periods of time.

Leakage of indicator into the extracellular media can generate artifacts when measuring $[Ca^{2+}]_i$ using cell suspensions. Baseline $[Ca^{2+}]_i$ values appear to drift upward over time as the indicator is transported from the cytosol, where calcium levels are low, to the extracellular medium where calcium levels are typically about 1 mM [83, 84] (see Figure 1.12). This problem is not so severe in imaging experiments where cells can easily be washed with fresh media and fluorescence from the cell can be distinguished from that of the surroundings.

The rapid leakage and sequestration of calcium indicators is probably due to active transporters that pump organic anions into organelles and out of the cell. One clue regarding the mechanism of this transport stems from reports that leakage is blocked by probenecid and sulfinpyrazone. These drugs block anion transport in a variety of cells [85–88]. However, probenecid blocks several different anion transporters including the hepatic anion transporter NPT1 [89], the rat kidney anion transporter ROAT1 [90], and the multidrug resistance-associated protein [89–91]. Thus probenecid sensitivity does not exactly pinpoint the transporter that exports these indicators out of the cell.

Although probenecid can prevent leakage and sequestration of calcium indicators, it seems better to avoid its use when possible. Typically, it takes large concentrations (1–5 mM) of probenecid to prevent leakage and at these concentrations it will block a number of cellular anion transporters. At a minimum, control experiments should be conducted to make sure that probenecid does not interfere with the events measured. There is at least one report showing that probenecid reduced calcium signaling [92]. Fortunately, indicators have recently been developed that resist leakage and sequestration. These include indicator–dextran conjugates and fura-PE3 and its relatives.

1.5.1   INDICATOR–DEXTRAN CONJUGATES

One solution to the problem of leakage and sequestration is to conjugate the indicator to a polymer like dextran that is retained in the cytosol. A variety of calcium indicators have been conjugated to dextran including fura-2, indo-1, rhod-2, and other rhoda-mine derivatives, fluo-4, Calcium Green-1, Oregon Green 488 BAPTA-1, and Cal-cium Crimson [26, 93, 94]. The indicators are linked to an amino dextran either through an amide linkage (fluo-4, rhod-2) or through a thiourea linkage ("Calcium" and "Oregon Green" series of indicators, Figure 1.12A). This linkage does not greatly affect the optical properties although it may cause a slight red shift in the spectrum and affinity is generally reduced compared to the parental indicator [95]. Dextran conjugates, incorporating the rhod-2 indicator and variants thereof, link the indicators to aminodextran through a carboxyl group or carboxymethyl group at-tached to the 4 position on the right BAPTA ring ([82]; Figure 1.12A). The presence of the methylene bridge greatly affects the affinity of the indicator such that when present the indicator has a relatively high affinity and when absent the affinity is much lower.

Derivatives of fluo-4 have also been developed that can also be linked to dextran [94]. These involve a strategy that is similar to that used for rhod-2 and derivatives [82]. An interesting offshoot of the fluo-4 work is the development of fluo-4 analogs linked to biotin and streptavidin as well as amine-reactive derivatives.

The synthesis of these dextrans aims for one molecule of indicator per molecule of dextran but in practice the ratio can vary. Since the indicator-to-dextran ratio can affect the affinity, K_d needs to be determined for each batch [26, 28]. The size (MW) of dextran should also be considered since entry into the nucleus can vary with molecular weight [96, 97].

Indicator–dextran conjugates are a very effective solution to the problem of leakage and sequestration. Once introduced into the cytoplasm, these dextran conjugates can remain in the cell for days [98–100]. For the most part, these indicators are spread diffusely through the cytoplasm but in neurons they can be loaded by retrograde transport [100]. The main limitation of these indicators is that they must be introduced into the cytoplasm by microinjection or other methods for breaching the cell membrane.

1.5.2   FURA-PE3 AND RELATED INDICATORS

Fura-PE3 is an alternative to indicator–dextran conjugates as a means of keeping the indicator in the cytosol. Its design was based on the assumption that anion transporters are the main cause of indicator leakage and sequestration. In addition, early on it was observed that rhod-2, when loaded into cells by iontophoresis, did not leak out of cells as rapidly as fura-2 [84]. The positive charge on rhod-2 seemed to be the major difference that might account for the difference in rates of leakage. Unfortunately, rhod-2 had the disadvantage of a rather small change in signal upon binding calcium and a tendency to enter the mitochondria when loaded into cells as the AM-ester. Thus rhod-2 did not seem to be a general solution to the problem of dye sequestration and leakage.

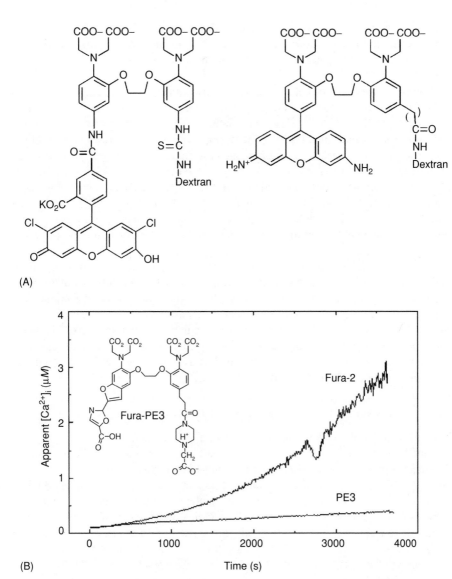

FIGURE 1.12 (A) Example structures of indicator–dextran conjugates. (B) A comparison of apparent $[Ca^{2+}]_i$ for unstimulated suspensions of 322 T lymphoma cells loaded with either fura-2 or fura-PE3. The increase in apparent $[Ca^{2+}]_i$ seen for the fura-2 loaded cell suspension is due to continuous leakage of the indicator into the extracellular media. The inset shows the structure of fura-PE3.

Fura-PE3 was designed to carry a positive charge like rhod-2 yet resist quaternization. This was achieved by adding a piperazine derivative to the right side BAPTA ring (Figure 1.12B). The piperazine derivative provided a hindered amine that is largely protonated at pH 7 but does not quaternize during subsequent steps such as in the preparation of the acetoxymethyl ester. Apart from this change, the structure and properties of fura-PE3 are essentially identical to those of fura-2. The structural features of fura-PE3 can be easily incorporated into other calcium indicators.

Fura-PE3 leaks out of cells much more slowly than fura-2, fluo-3, and related indicators (Figure 1.12B). The rate of fura-PE3 leakage is estimated at about 0.2% per minute [101], which is slow enough for cells to retain useful amounts of indicator for several hours [84]. These improved retention rates are suitable for most types of experiments and in some cases have permitted adequate loading of indicator into cells that could not be obtained with other indicators. An example of this is seen in the studies of Quigley et al. [81] who were able to obtain adequate loading of indicator into kidney proximal tubules cells using fura-PE3/AM but not with fura-2. Initial efforts to load the tubules with fura-2 showed that it was pumped out of the tubule epithelium too fast to obtain stable loading (R. Quigley, personal communication).

Initially, there was some concern that fura-PE3/AM ester, due to its positive charge, might also accumulate in mitochondria. This has not turned out to be a problem. As a rule, fura-PE3/AM loads evenly into the cytoplasm and accumulates very slowly into organelles with no special propensity to enter the mitochondria. In solution, fura-PE3/AM tends to form particulates a little more readily than fura-2/AM but these are removed with subsequent washing. To date, there have been numerous studies using fura-PE3 and these have generally confirmed its usefulness without mentioning any problems.

1.5.3 SUMMARY OF THE LEAKAGE-RESISTANT CALCIUM INDICATORS

The properties of the leakage-resistant indicators are summarized in Table 1.4. At present, there are only a few choices in this category represented by the indicator–dextran conjugates, the fura-PE3 family (fura-PE3, Indo-PE3 and fluo-LR), and perhaps rhod-2 if it is loaded as the salt. Of these, the indicator–dextrans and rhod-2 must be loaded by microinjection or other methods for breaching the membrane (scrape-loading, electroporation, etc.). The indicator–dextrans offer the greatest resistance to leakage and compartmentalization. The fura-PE3 family can be loaded as acetoxymethyl esters and although they still leak out of cells, leakage rates are slow enough for cells to be useful for at least 2 h after loading.

1.6   NEAR-MEMBRANE CALCIUM INDICATORS

There is a growing impetus to study changes in $[Ca^{2+}]$ near cell membranes both on theoretical and experimental grounds [2, 64, 102–108]. Modeling of calcium movement through channels has indicated that when calcium channels open, Ca^{2+} should rise to levels in the vicinity of 100 μM near the mouth of the channel. Experimental studies show that it takes on the order of 100 μM Ca^{2+} to achieve maximal rates of vesicle secretion in nerve terminals even though measurements of $[Ca^{2+}]_i$ in terminals never show such high levels of calcium [109]. The discrepancy is likely

TABLE 1.4
Calcium Indicators That Resist Leakage and Sequestration

| | Absorption Maxima (nm) | | Emission Maxima (nm) | | K_d (μM) | K_d (mM) | K_{+1} ($10^7 M^{-1}s^{-1}$) | K_{-1} (s^{-1}) | |
	Ca^{2+}-bound	Ca^{2+}-free	Ca^{2+}-bound	Ca^{2+}-free	Ca^{2+}	Mg^{2+}	Ca^{2+}	Ca^{2+}	Reference
Fura-dextran	337	366	494	495	0.2–0.24				[17]
Indo-dextran	341	356	408	466	0.32–0.36				[17]
Calcium green dextran	508	508	533	533	0.24–0.54				[17]
Fluo-LR	506	506	525	525	0.55				[134]
Fura-PE3	335	362	500	508	0.25	3.1–3.6			[69, 134]
Indo-PE3	330	346	408	475	0.26				[134]

due to microdomains of high calcium near the membrane, which do not extend to the entire nerve terminal.

The importance of high-calcium microdomains is not limited to nerve terminals. Other cases include calcium sparks in cardiac muscle and excitation–contraction coupling muscle in general [1, 110–114]. There are two main approaches for selectively detecting calcium levels near the membrane. One is to confine the excitation light to a small region of the cell using confocal microscopy, evanescent wave fluorescence microscopy, or multiphoton excitation [3, 111, 115–119]. Using this approach along with a calcium indicator that has a large dynamic range such as fluo-3 or fluo-4 has made it possible to image calcium entry through single open channels and in cellular microdomains [7]

The other is to use a calcium indicator that is confined to the membrane, which here is referred to as a "near-membrane" calcium indicator. Of the various methods for localizing the excitation beam, evanescent wave illumination provides the highest degree of confinement. With this approach, it is theoretically possible to restrict the effective illumination to a region less than 200 nm from the membrane [116]. The problem with this approach is getting the region of interest to lie flat up next to a coverslip. However, even with highly confined illumination, there is still a problem using diffusible calcium indicators in that they will tend to dampen or dissipate calcium gradients [120]. Also, as one confines the illumination to smaller and smaller volumes, the amount of signal is correspondingly reduced although given sufficiently high concentrations of indicator, this should not be a real limitation.

Methods for confining the illumination to a small region of the cell provide a significant improvement in resolving changes in calcium near the membrane, but one can obtain even better resolution by placing the indicator near the membrane. Accurate detection of near-membrane calcium levels is problematic using diffusible indicators, because only a small fraction of the total indicator is near the membrane. If concentrations of diffusible indicator are high, it may overwhelm the endogenous buffering and dissipate the gradient. By concentrating the indicator at the membrane, resolution is improved while retaining good signal to noise. In addition, diffusion of the indicator to adjacent regions of the membrane is limited by the relatively high viscosity of the membrane.

1.6.1   THE C_{18} NEAR-MEMBRANE CALCIUM INDICATORS (C_{18}-FURA-2, CALCIUM GREEN C_{18})

1.6.1.1 C_{18}-Fura-2 (Fura-2-C_{18})

There have been a number of attempts to develop near-membrane fluorescent calcium indicators by attaching a lipophilic group to one of the fluorescent calcium indicators such as fura-2 or Calcium Green. The first of these to be used experimentally was C_{18}-fura-2 [112]. Here, an 18-carbon aliphatic tail is attached to the right side BAPTA via an amide linkage (Figure 1.13). With this structure, the right side ring and the C_{18} aliphatic "tail" constitute one continuous hydrophobic domain. This structure is very hydrophobic and has poor water solubility. It shows a tendency to aggregate in aqueous solution and concentrate in membranes.

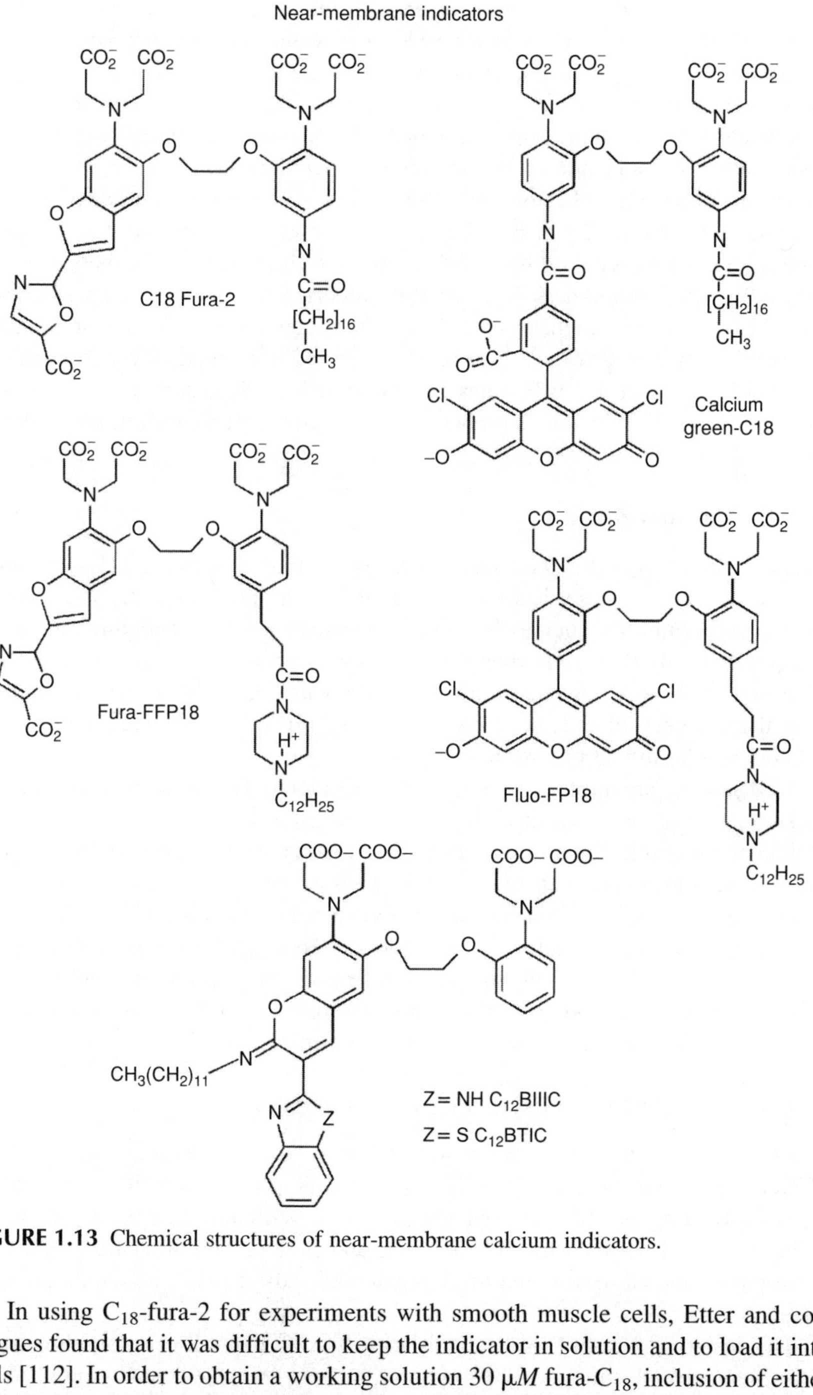

Near-membrane indicators

FIGURE 1.13 Chemical structures of near-membrane calcium indicators.

In using C_{18}-fura-2 for experiments with smooth muscle cells, Etter and colleagues found that it was difficult to keep the indicator in solution and to load it into cells [112]. In order to obtain a working solution 30 μM fura-C_{18}, inclusion of either 0.01% pluronic (a nonionic surfactant) or 50 mM EGTA was required. At these concentrations, it took 10–15 min for enough fura-C_{18} to diffuse from the patch pipette into the cell to obtain a measurable fluorescence signal. They noted that the fluorescence properties of indicator in cells were quite different from that in solution due to its tendency to aggregate and quench. Quenching is probably related to how closely the fluorophores are packed and becomes more severe as the concentration of indicator in the membrane increases [112]. Thus, it becomes important to keep the indicator concentration low enough to prevent aggregation.

Studies using fura-C_{18} or fura-2 in smooth muscle cells showed that the initial increase in calcium reported by fura-C_{18} rose 4–6 times faster than those reported by fura-2 [112]. Yet the peak $[Ca^{2+}]$ values reported by fura-C_{18} barely reached 1 μM, which is higher than that reported from fura-2 but far lower than that predicted by modeling studies. However, given that C_{18}-fura-2 has a high affinity for calcium ($K_d = 0.15$ μM), and the difficulties in quantifying calcium due to problems with aggregation, the 1 μM value is probably not a reliable estimate of the true calcium concentration.

1.6.1.2 Calcium Green C_{18}

Calcium green C_{18} is the "Calcium Green" version of C_{18}-fura-2 [1] (see Figure 1.13). It is not a ratiometric indicator but it has the advantage of an excitation maxima at 509 nm. Its calcium dissociation constant when measured in solution is 0.22 μM but only 0.06 μM when the indicator is bound to liposomes. The latter value should be more relevant to measurements with cells. There is no information about the tendency of Calcium Green C_{18} to aggregate and quench although it is probably similar to that of C_{18}-fura-2.

To date there are no studies reported where Calcium Green C_{18} has been used to measure calcium inside intact cells. It has been used with intact cells to measure calcium on the extracellular face of the plasma membrane and in permeabilized cells to detect calcium release sites. In each of these cases, the environment of the indicator can be controlled so that initial calcium levels can be kept very low [121–124]. Given its extremely low K_d, it is unlikely that Calcium Green C_{18} could be that useful for measuring near-membrane calcium inside intact cells since at baseline calcium levels the indicator would already be 75% saturated with calcium. There is the additional problem that Calcium Green has a rather low dynamic range.

1.6.1.3 Fura-FFP18 and Related Calcium Indicators

Fura-FFP18 is another near-membrane indicator based on fura-2 [84]. The key difference is in the length of the hydrophobic tail and the way it is attached to the aromatic ring of BAPTA (see Figure 1.13). With fura-FFP18, a piperazine moiety is placed between the BAPTA ring and a 12-carbon aliphatic chain. One of the piperazine amines is free to protonate and thus at pH 7, it bears a positive charge.

The design of fura-FFP18 distinguishes it from C_{18}-fura-2 in several important ways. First, the molecule is less hydrophobic and more soluble than C_{18}-fura-2. This improved solubility allows one to work with solutions of 30 μM or more FFP18 without detergents or high concentrations of EGTA [113]. It also makes it possible, even if difficult, to load the indicator into cells as an AM ester [125]. Here, there is a potential for unhydrolyzed indicator to remain bound to the outside of cells but this can be removed with fatty acid-free serum albumin. The piperazine moiety seems to reduce (but not eliminate) the problem of aggregation and quenching. This may be related to the fact that concentration of fura-FFP18 in the membrane appears to saturate at about 1×10^6 molecules per μm^2 of membrane regardless of how much free indicator is initially added to the membrane preparation [113]. With fura-C_{18}, membrane concentrations can reach levels that are tenfold higher, leading to much more severe quenching [112, 113]. Finally, the K_d of membrane-associated fura-FFP18 is 400–450 nM, several fold higher than that of C_{18}-fura-2 [84].

Fura-FFP18 is an improvement over fura-C_{18}, but it is still far from ideal as a near-membrane indicator. Its affinity is much too high to accurately track Ca^{2+} concentrations in the appropriate range (10^{-5} to 10^{-4} M). In addition, fura-FFP18 still has a tendency to aggregate and quench (Figure 1.14). It would also help to have an indicator that would load more readily as an acetoxymethyl ester than fura-FFP18.

Despite its drawbacks, FFP18 can provide new insights into the nature of calcium signaling events. In a study of porcine artery endothelial cells, fura-FFP18 detected changes in $[Ca^{2+}]$ that were not detected by cytosolic calcium indicators [8]. In another example FFP18 revealed differences in calcium levels of growth cones as they made contact with different types of substrates [126]. Recently, TEFLABS has developed near-membrane indicators based on fluo-3 (NOMO), fluo-4 (MOMO), and indo-1 (FIP18). These all involve the same structural modifications to the right BAPTA aromatic ring seen in FFP18. So far, no studies using these indicators have been published.

A new series of near-membrane calcium indicators referred to as C12-CIIC and C12 BTIC have been developed as analogs of the original BTC [27, 127]. These probes have the advantage of substantially lower affinities for calcium than previous near-membrane indicators (Figure 1.13 and Table 1.5), which should make them better for tracking calcium near the membrane. As with other near-membrane indicators, they show distortions in spectra at high calcium concentrations that may reflect aggregation of the indicator. At present, there are no published studies using these indicators for calcium measurements.

1.6.2 SUMMARY OF THE NEAR-MEMBRANE CALCIUM INDICATORS

The properties of the near-membrane calcium indicators are summarized in Table 1.5. Given that models predict that calcium near the membrane could rise to levels as high as 100 μM, clearly the current generation of near-membrane indicators is inadequate. There are good prospects for developing better near-membrane indicators but this will probably entail a major departure from current designs. Of the current generation of near-membrane indicators, the fura-FFP18 family (Fura-FFP18, FIP18, and Fluo-FP18) has lower affinities for calcium and better physical properties than C_{18}-fura-2 or Calcium Green C_{18}. Despite their weaknesses, both

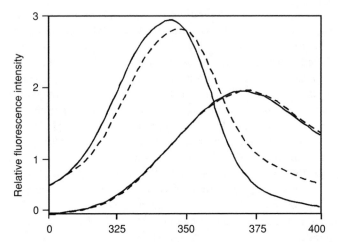

FIGURE 1.14 A spectral comparison of fura-FFP18 in free solution (solid line) or bound to liposomes (dashed line) at limiting low and high [Ca^{2+}]. For titrations in the presence of liposomes, excess fura-FFP18 was incubated with anionic liposomes that were subsequently washed by centrifugation. The excitation spectra of fura-FFP18 bound to liposomes and at saturating [Ca^{2+}] are red-shifted and quenched to a small extent compared to the indicator in free solution.

fura-FFP18 and C$_{18}$-fura-2 have generally shown that calcium levels near the membrane rise faster and to higher levels than that in the cytosol. New indicators with lowered affinity may prove more useful both in preventing aggregation and more accurately assessing near-membrane calcium concentrations.

1.7   USE OF CALCIUM INDICATORS WITH MULTIPHOTON MICROSCOPY

Multiphoton excitation fluorescence microscopy has gained in popularity as a way to obtain depth-resolved fluorescent images. The long wavelengths used in two-photon excitation give better penetration of tissues or brain slices with low background and without the collection losses inherent in confocal microscopy [128]. Frequently, these studies have relied on the same calcium indicators that are widely used for confocal microscopy. However, there are important differences between confocal and multiphoton microscopy that one should consider when evaluating calcium indicators.

With standard confocal microscopy, the expense and difficulty in generating multiple excitation wavelengths in the UV range led most users to opt for nonratiometric fluorescein-based indicators. With two-photon excitation, this limitation is generally eliminated because the excitation wavelength is now approximately doubled and the tunable lasers used here can easily accommodate wavelengths in the range of 750–800 nm that are needed for fura-2.

TABLE 1.5
Near-Membrane Calcium Indicators

	Absorption Maxima (nm)		Emission Maxima (nm)		K_d (μM) Ca^{2+}	K_d (mM) Mg^{2+}	K_{+1} ($10^7\ M^{-1}s^{-1}$) Ca^{2+}	K_{-1} (s^{-1}) Ca^{2+}	Reference
	Ca^{2+}-bound	Ca^{2+}-free	Ca^{2+}-bound	Ca^{2+}-free					
C_{18}-Fura-2	335	363	505	512	0.151				[112]
Calcium Green-C_{18}	506	506	526	526	0.069–0.23				[111, 121]
FFP18	335	364	495	502	0.33–0.51	3.3–6.8			[84]
FIP18	330	346	408	475	0.45				[134]
C12-BIIC	401	469	498	515	5.5				[127]
C12-BTIC	421	471	537	537	4.49				[127]

While two-photon excitation can easily excite both UV and visible wavelength indicators, it is still not feasible to rapidly change wavelengths so it would seem there is no advantage in using fura-2. Furthermore, various methods for estimating calcium levels from single wavelength measurements have been described [28, 65, 129, 130] and it could be argued that for the purposes of calibration there is no advantage in using ratiometric dyes. However, calibration of single wavelength indicators always requires some method for estimating the dynamic range of the indicator. In this regard, calibration procedures are simplified by using an indicator whose fluorescence decreases with increasing calcium concentrations [41, 65, 129]. Thus, it is worth considering the use of fura-2 and related indicators at longer excitation wavelengths for applications involving two-photon excitation.

A final consideration in the use of calcium indicators with two-photon excitation is the fluorescence excitation cross section that is expressed in Goppert–Mayer (GM) where $1 \text{ GM} = 10^{-50} \text{ cm}^4 \text{ s photon}^{-1}$. The fluorescence excitation cross section is in turn the product of the absorption cross section and the quantum efficiency where the absorption cross section is similar to an extinction coefficient. What is notable here is that when each is measured at the optimal wavelength, fluorescence excitation cross sections are in the range of 1–20 GM (see Figure 1.15) for all the current calcium indicators [41, 131, 132]. These values are relatively small compared to fluorophores with large two-photon cross sections that can currently reach values greater than 1000 GM [133].

1.8 CONCLUSION

There has been increasing interest in "modes" of calcium signaling and how cells interpret different patterns of calcium elevation. Cells could in principle discrimi-

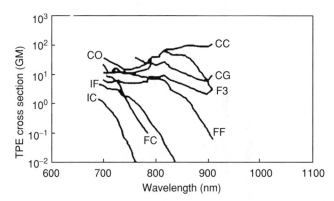

FIGURE 1.15 Two-photon excitation spectra of calcium indicators. The two-photon excitation spectra are shown for: indo-1 with Ca^{2+} (IC) and indo-1 in the absence of Ca^{2+} (IF), fura-2 with Ca^{2+} (FC) and fura-2 in the absence of Ca^{2+} (FF), calcium green-1 with Ca^{2+} (CG), calcium orange with Ca^{2+} (CO), calcium crimson with Ca^{2+} (CC), and fluo-3 with Ca^{2+}. Excitation cross sections are given in Goppert units where $1 \text{ GM} = 10^{-50} \text{ cm}^4 \text{ s photon}^{-1}$. (This figure is adapted from Xu et al. (1996) *Natl. Acad. Sci. U.S.A.* and used with permission.)

nate between calcium signals based on their amplitude or frequency as well as their spatial distribution [50, 134]. The new calcium indicators described in this review should be helpful in deciphering these signals. In general, improved indicators have extended the range of experimental systems and scenarios where one can reliably measure calcium. Specialized indicators make it possible to measure calcium in particular organelles such as the endoplasmic reticulum, mitochondria, and nucleus. Low-affinity indicators have resolved differences in the amplitude of calcium transients due to different types of stimuli applied to the same cells. These differences were not apparent when monitored with high-affinity indicators. Fast indicators offer better resolution and less perturbation of rapid calcium transients. Near-membrane indicators, while in their infancy, show differences in calcium kinetics and amplitude near the membrane even if they are not yet suitable for detecting the levels of calcium predicted by the models. Future versions of near-membrane indicators aimed at reducing their affinity and tendency to aggregate should make them more useful. The usefulness and popularity of two-photon microscopy would be aided by the development of calcium indicators with larger two-photon excitation cross sections and hopefully we should see development in this area.

ACKNOWLEDGMENTS

Figure 1.3 (Fura Red Spectra), Figure 1.5, Figure 1.6, Figure 1.7, and Figure 1.11 contain data that are the copyright of Molecular Probes, Inc., and were reproduced with permission. I wish to thank Iain Johnson and Richard Haugland of Molecular Probes for furnishing these figures and other information concerning their calcium indicators. I am in debt to Stephen Baylor both for his helpful discussions and for providing the data in Figure 1.9. I wish to thank Shanti Nulu who was a great help in gathering, analyzing, and assembling the material for this review. Finally, I would like to acknowledge the Texas Higher Education Coordinating Committee and NSF for supporting the development of fura-PE3 and fura-FFP18.

REFERENCES

1. Cheng, H., et al., Calcium sparks and $[Ca^{2+}]_i$ waves in cardiac myocytes. *Amer J Physiol*, 1996. 270(1 Pt 1): C148–59.
2. Neher, E., Vesicle pools and Ca^{2+} microdomains: new tools for understanding their roles in neurotransmitter release. *Neuron*, 1998. 20(3): 389–99.
3. Cleemann, L., G. DiMassa, and M. Morad, Ca^{2+} sparks within 200 nm of the sarcolemma of rat ventricular cells: evidence from total internal reflection fluorescence microscopy. *Adv Exp Med Biol*, 1997. 430: 57–65.
4. Macrez, N. and J. Mironneau, Local Ca^{2+} signals in cellular signalling. *Curr Mol Med*, 2004. 4(3): 263–75.
5. Niggli, E. and M. Egger, Calcium quarks. *Front Biosci*, 2002. 7: d1288–97.
6. Pacher, P., A.P. Thomas, and G. Hajnoczky, Ca^{2+} marks: miniature calcium signals in single mitochondria driven by ryanodine receptors. *Proc Natl Acad Sci USA*, 2002. 99(4): 2380–5.

7. Zou, H., et al., Imaging calcium entering the cytosol through a single opening of plasma membrane ion channels: SCCaFTs — fundamental calcium events. *Cell Calcium*, 2004. 35(6): 523–33.

8. Graier, W.F., et al., Submaximal stimulation of porcine endothelial cells causes focal Ca^{2+} elevation beneath the cell membrane. *J Physiol*, 1998. 506(Pt 1): 109–25.

9. Thomas, D., et al., A comparison of fluorescent Ca^{2+} indicator properties and their use in measuring elementary and global Ca^{2+} signals. *Cell Calcium*, 2000. 28(4): 213–23.

10. Zhao, M., S. Hollingworth, and S.M. Baylor, Properties of tri- and tetracarboxylate Ca^{2+} indicators in frog skeletal muscle fibers. *Biophys J*, 1996. 70(2): 896–916.

11. Tsien, R.Y., New calcium indicators and buffers with high selectivity against magnesium and protons: design, synthesis, and properties of prototype structures. *Biochemistry*, 1980. 19(11): 2396–404.

12. Martel, A.E. and R.M. Smith, *Critical Stability Constants*. Vol. 1. 1974, New York: Plenum Press.

13. Bancel, F., et al., Investigation of noncalcium interactions of fura-2 by classical and synchronous fluorescence spectroscopy. *Anal Biochem*, 1992. 204(2): 231–8.

14. Bancel, F., et al., Microspectrofluorometry as a tool for investigation of non-calcium interactions of Indo-1. *Cell Calcium*, 1992. 13(1): 59–68.

15. Pethig, R., et al., On the dissociation constants of BAPTA-type calcium buffers. *Cell Calcium*, 1989. 10(7): 491–8.

16. Gee, K.R., et al., New ratiometric fluorescent calcium indicators with moderately attenuated binding affinities. *BioorgMed Chem Lett*, 2000. 10: 1515–1518.

17. Alberdi, A., V. Jimenez-Ortiz, and M.A. Sosa, The calcium chelator BAPTA affects the binding of assembly protein AP-2 to membranes. *Biocell*, 2001. 25(2): 167–72.

18. Rousset, M., et al., Ca^{2+}-dependent interaction of BAPTA with phospholipids. *FEBS Lett*, 2004. 576(1–2): 41–5.

19. Saoudi, Y., et al., Calcium-independent cytoskeleton disassembly induced by BAPTA. *Eur J Biochem*, 2004. 271(15): 3255–64.

20. Shang, J. and M.A. Lehrman, Inhibition of mammalian RNA synthesis by the cytoplasmic Ca^{2+} buffer BAPTA. Analyses of [3H]uridine incorporation and stress-dependent transcription. *Biochemistry*, 2004. 43(29): 9576–82.

21. Aubin, J.E., Autofluorescence of viable cultured mammalian cells. *J Histochem Cytochem*, 1979. 27(1): 36–43.

22. Tsien, R.Y., Fluorescent indicators of ion concentrations. *Method Cell Biol*, 1989. 30: 127–56.

23. Scheenen, W.J., et al., Photodegradation of indo-1 and its effect on apparent Ca^{2+} concentrations. *Chem Biol*, 1996. 3(9): 765–74.

24. Grynkiewicz, G., M. Poenie, and R.Y. Tsien, A new generation of Ca^{2+} indicators with greatly improved fluorescence properties. *J Biol Chem*, 1985. 260(6): 3440–50.

25. London, R.E., L.A. Levy, and E. Murphy, *Fluorescent Intracellular Calcium Indicators*. 1996, US Patent 5, 516911.

26. Haugland, R.P., *Handbook of Fluorescent Probes and Research Chemicals*. 6th ed. 1996, Eugene, Oregon: Molecular Probes.

27. Iatridou, H., et al., The development of a new family of intracellular calcium probes. *Cell Calcium*, 1994. 15(2): 190–8.

28. Kao, J.P., Practical aspects of measuring [Ca^{2+}] with fluorescent indicators. *Method Cell Biol*, 1994. 40: 155–81.

29. Floto, R.A., et al., IgG-induced Ca^{2+} oscillations in differentiated U937 cells; a study using laser scanning confocal microscopy and co-loaded fluo-3 and fura-red fluorescent probes. *Cell Calcium*, 1995. 18(5): 377–89.

30. Wu, Y. and W.T. Clusin, Calcium transient alternans in blood-perfused ischemic hearts: observations with fluorescent indicator fura red. *Am J Physiol*, 1997. 273(5 Pt 2): H2161–9.

31. Minta, A., J.P. Kao, and R.Y. Tsien, Fluorescent indicators for cytosolic calcium based on rhodamine and fluorescein chromophores. *J Biol Chem*, 1989. 264(14): 8171–8.

32. Sun, W.-C., et al., Synthesis of fluorinated fluoresceins. *J Org Chem*, 1997. 62: 6469–75.

33. Knight, M.M., et al., Live cell imaging using confocal microscopy induces intracellular calcium transients and cell death. *Am J Physiol Cell Physiol*, 2003. 284(4): C1083–9.

34. Eberhard, M. and P. Erne, Calcium binding to fluorescent calcium indicators: calcium green, calcium orange and calcium crimson. *Biochem Biophys Res Commun*, 1991. 180(1): 209–15.

35. Zhao, M., S. Hollingworth, and S.M. Baylor, AM-loading of fluorescent Ca^{2+} indicators into intact single fibers of frog muscle. *Biophys J*, 1997. 72(6): 2736–47.

36. Titus, J.A., et al., Texas Red, a hydrophilic, red-emitting fluorophore for use with fluorescein in dual parameter flow microfluorometric and fluorescence microscopic studies. *J Immunol Methods*, 1982. 50(2): 193–204.

37. Johnson, L.V., M.L. Walsh, and L.B. Chen, Localization of mitochondria in living cells with rhodamine 123. *Proc Natl Acad Sci USA*, 1980. 77(2): 990–4.

38. Chen, L.B., et al., Probing mitochondria in living cells with rhodamine 123. *Cold Spring Harb Sym*, 1982. 46(Pt 1): 141–55.

39. Royall, J.A. and H. Ischiropoulos, Evaluation of $2',7'$-dichlorofluorescin and dihydrorhodamine 123 as fluorescent probes for intracellular H2O2 in cultured endothelial cells. *Arch Biochem Biophys*, 1993. 302(2): 348–55.

40. Sinha, S.R., S.S. Patel, and P. Saggau, Simultaneous optical recording of evoked and spontaneous transients of membrane potential and intracellular calcium concentration with high spatio-temporal resolution. *J Neurosci Meth*, 1995. 60(1–2): 49–60.

41. Wokosin, D.L., C.M. Loughrey, and G.L. Smith, Characterization of a range of fura dyes with two-photon excitation. *Biophys J*, 2004. 86(3): 1726–38.

42. Baylor, S.M. and S. Hollingworth, Fura-2 calcium transients in frog skeletal muscle fibres [published erratum appears in *J Physiol (Lond)*, 1988 Dec.407: p. 616]. *J Physiol*, 1988. 403: 151–92.

43. Klein, M.G., et al., Simultaneous recording of calcium transients in skeletal muscle using high- and low-affinity calcium indicators. *Biophys J*, 1988. 53(6): 971–88.

44. Ashley, C.C. and E.B. Ridgway, Simultaneous recording of membrane potential, calcium transient and tension in single muscle fibers. *Nature*, 1968. 219(159): 1168–9.

45. Miledi, R., I. Parker, and G. Schalow, Measurement of calcium transients in frog muscle by the use of arsenazo III. *Proc Roy Soc Lond — Ser B: Bio*, 1977. 198(1131): 201–10.

46. Durham, A.C. and J.M. Walton, A survey of the available colorimetric indicators for Ca^{2+} and Mg2+ ions in biological experiments. *Cell Calcium*, 1983. 4(1): 47–55.

47. Blinks, J.R., et al., Measurement of Ca^{2+} concentrations in living cells. *Prog Biophys Mol Biol*, 1982. 40(1–2): 1–114.

48. Konishi, M., et al., Myoplasmic binding of fura-2 investigated by steady-state fluorescence and absorbance measurements. *Biophys J*, 1988. 54(6): 1089–104.

49. Poenie, M., Alteration of intracellular Fura-2 fluorescence by viscosity: a simple correction. *Cell Calcium*, 1990. 11(2–3): 85–91.

50. Bootman, M.D., M.J. Berridge, and P. Lipp, Cooking with calcium: the recipes for composing global signals from elementary events. *Cell*, 1997. 91(3): 367–73.

51. Wilkinson, F., *Chemical Kinetics and Reaction Mechanisms*. 1980, New York: Van Nostrand Reinhold.

52. Naraghi, M., T-jump Study of Calcium Binding Kinetics of Calcium Chelators. *Cell Calcium*, 1997. 22: 255–268.

53. Raju, B., et al., A fluorescent indicator for measuring cytosolic free magnesium. *Am J Physiol*, 1989. 256(3 Pt 1): C540–8.

54. Otten, P.A., R.E. London, and L.A. Levy, A new approach to the synthesis of APTRA indicators. *Bioconjug Chem*, 2001. 12(1): 76–83.

55. Illner, H., J.A. McGuigan, and D. Luthi, Evaluation of mag-fura-5, the new fluorescent indicator for free magnesium measurements. *Pflugers Arch — Eur J Physiol*, 1992. 422(2): 179–84.

56. Buri, A., et al., The regulation of intracellular Mg2+ in guinea-pig heart, studied with Mg(2+)-selective microelectrodes and fluorochromes. *Exp Physiol*, 1993. 78(2): 221–33.

57. Csernoch, L., et al., Measurements of intracellular Mg2+ concentration in mouse skeletal muscle fibers with the fluorescent indicator mag-indo-1. *Biophys J*, 1998. 75(2): 957–67.

58. Dai, L.J., et al., PGE2 stimulates Mg2+ uptake in mouse distal convoluted tubule cells. *Am J Physiol*, 1998. 275(5 Pt 2): F833–9.

59. Grubbs, R.D. and A. Walter, Determination of cytosolic Mg2+ activity and buffering in BC3H-1 cells with mag-fura-2. *Mol Cell Biochem*, 1994. 136(1): 11–22.

60. Ishijima, S., T. Sonoda, and M. Tatibana, Mitogen-induced early increase in cytosolic free Mg2+ concentration in single Swiss 3T3 fibroblasts. *Am J Physiol*, 1991. 261(6 Pt 1): C1074–80.

61. Kato, H., et al., Depolarization triggers intracellular magnesium surge in cultured dorsal root ganglion neurons. *Brain Res*, 1998. 779(1–2): 329–33.

62. Murphy, E., et al., Monitoring cytosolic free magnesium in cultured chicken heart cells by use of the fluorescent indicator Furaptra. *Proc Natl Acad Sci USA*, 1989. 86(8): 2981–4.

63. Quamme, G.A. and S.W. Rabkin, Cytosolic free magnesium in cardiac myocytes: identification of a Mg2+ influx pathway. *Biochem Biophys Res Commun*, 1990. 167(3): 1406–12.

64. Berridge, M.J., Microdomains and elemental events in calcium signalling. *Cell Calcium*, 1996. 20(2): 95–6.

65. Ogden, D., et al., Analogue computation of transient changes of intracellular free Ca^{2+} concentration with the low affinity Ca^{2+} indicator furaptra during whole-cell patch-clamp recording. *Pflugers Arch — Eur J Physiol*, 1995. 429(4): 587–91.

66. Berlin, J.R. and M. Konishi, Ca^{2+} transients in cardiac myocytes measured with high and low affinity Ca^{2+} indicators. *Biophys J*, 1993. 65(4): 1632–47.

67. Hurley, T.W., M.P. Ryan, and R.W. Brinck, Changes of cytosolic Ca^{2+} interfere with measurements of cytosolic Mg2+ using mag-fura-2. *Am J Physiol*, 1992. 263(2 Pt 1): C300–7.

68. Eberhard, M. and P. Erne, Calcium and magnesium binding to rat parvalbumin. *Eur J Biochem*, 1994. 222(1): 21–6.

69. Weinberg, J.M., J.A. Davis, and M.A. Venkatachalam, Cytosolic-free calcium increases to greater than 100 micromolar in ATP-depleted proximal tubules. *J Clin Invest*, 1997. 100(3): 713–22.

70. Gee, K.R., A. Rukavishnikov, and A. Rothe, New Ca^{2+} fluoroionophores based on the BODIPY fluorophore. *Comb Chem High T Screen*, 2003. 6(4): 363–6.

71. Hyrc, K.L., J.M. Bownik, and M.P. Goldberg, Neuronal free calcium measurement using BTC/AM, a low affinity calcium indicator. *Cell Calcium*, 1998. 24(3): 165–75.

72. Ito, K., Y. Miyashita, and H. Kasai, Micromolar and submicromolar Ca^{2+} spikes regulating distinct cellular functions in pancreatic acinar cells. *EMBO J*, 1997. 16(2): 242–51.

73. Carriedo, S.G., et al., Rapid Ca^{2+} entry through Ca^{2+}-permeable AMPA/Kainate channels triggers marked intracellular Ca^{2+} rises and consequent oxygen radical production. *J Neurosci*, 1998. 18(19): 7727–38.

74. London, R.E., et al., NMR-sensitive fluorinated and fluorescent intracellular calcium ion indicators with high dissociation constants. *Am J Physiol*, 1994. 266(5 Pt 1): C1313–22.

75. Ganitkevich, V., Use of Indo-1FF for measurements of rapid micromolar cytoplasmic free Ca^{2+} increments in a single smooth muscle cell. *Cell Calcium*, 1998. 23(5): 313–22.

76. David, G., J.N. Barrett, and E.F. Barrett, Stimulation-induced changes in $[Ca^{2+}]$ in lizard motor nerve terminals. *J Physiol*, 1997. 504(Pt 1): 83–96.

77. Poenie, M., et al., Calcium rises abruptly and briefly throughout the cell at the onset of anaphase. *Science*, 1986. 233(4766): 886–9.

78. Blatter, L.A. and W.G. Wier, Intracellular diffusion, binding, and compartmentalization of the fluorescent calcium indicators indo-1 and fura-2. *Biophys J*, 1990. 58(6): 1491–9.

79. Takeuchi, K., et al., Intracellular compartmentalization of fura-2 dye demonstrated by laser-excitation fluorescence microscopy: a problem in measuring cytosolic free calcium concentration using fura-2 fluorescence in vascular smooth muscle cells. *Tohoku J Exp Med*, 1989. 159(1): 23–35.

80. Hepler, P.K. and D.A. Callaham, Free calcium increases during anaphase in stamen hair cells of Tradescantia. *J Cell Biol*, 1987. 105(5): 2137–43.

81. Quigley, R., et al., Stimulation of proximal convoluted tubule phosphate transport by epidermal growth factor: signal transduction. *Am J Physiol*, 1995. 269(3 Pt 2): F339 – 44.

82. Beierlein, M., et al., Presynaptic calcium measurements at physiological temperatures using a new class of dextran-conjugated indicators. *J Neurophysiol*, 2004. 92(1): 591–9.

83. Kermode, J.C., Q. Zheng, and E.P. Cook, Fluorescent indicators give biased estimates of intracellular free calcium change in aggregating platelets: implication for studies with human von Willebrand factor. *Blood Cell Mol Dis*, 1996. 22(3): 238–53.

84. Vorndran, C., A. Minta, and M. Poenie, New fluorescent calcium indicators designed for cytosolic retention or measuring calcium near membranes. *Biophys J*, 1995. 69(5): 2112–24.

85. Di Virgilio, F., C. Fasolato, and T.H. Steinberg, Inhibitors of membrane transport system for organic anions block fura-2 excretion from PC12 and N2A cells. *Biochem J*, 1988. 256(3): 959–63.

86. Di Virgilio, F., et al., Fura-2 secretion and sequestration in macrophages. A blocker of organic anion transport reveals that these processes occur via a membrane transport system for organic anions. *J Immunol*, 1988. 140(3): 915–20.

87. Di Virgilio, F., T.H. Steinberg, and S.C. Silverstein, Inhibition of Fura-2 sequestration and secretion with organic anion transport blockers. *Cell Calcium*, 1990. 11(2–3): 57–62.

88. Steinberg, T.H., et al., Macrophages possess probenecid-inhibitable organic anion transporters that remove fluorescent dyes from the cytoplasmic matrix. *J Cell Biol*, 1987. 105(6 Pt 1): 2695–702.

89. Yabuuchi, H., et al., Hepatic sinusoidal membrane transport of anionic drugs mediated by anion transporter Npt1. *J Pharmacol Exp Ther*, 1998. 286(3): 1391–6.

90. Sweet, D.H., N.A. Wolff, and J.B. Pritchard, Expression cloning and characterization of ROAT1. The basolateral organic anion transporter in rat kidney. *J Biol Chem*, 1997. 272(48): 30088–95.

91. Gollapudi, S., et al., Probenecid reverses multidrug resistance in multidrug resistance-associated protein-overexpressing HL60/AR and H69/AR cells but not in P-glycoprotein-overexpressing HL60/Tax and P388/ADR cells. *Cancer Chemoth Pharm*, 1997. 40(2): 150–8.

92. Packham, M.A., et al., Probenecid inhibits platelet responses to aggregating agents in vitro and has a synergistic inhibitory effect with penicillin G. *Thromb Haemostasis*, 1996. 76(2): 239–44.

93. Regehr, W.G. and P.P. Atluri, Calcium transients in cerebellar granule cell presynaptic terminals. *Biophys J*, 1995. 68(5): 2156–70.

94. Martin, V.V., et al., Novel fluo-4 analogs for fluorescent calcium measurements. *Cell Calcium*, 2004. 36(6): 509–14.

95. Konishi, M. and M. Watanabe, Resting cytoplasmic free Ca^{2+} concentration in frog skeletal muscle measured with fura-2 conjugated to high molecular weight dextran. *J Gen Physiol*, 1995. 106(6): 1123–50.

96. Stehno-Bittel, L., C. Perez-Terzic, and D.E. Clapham, Diffusion across the nuclear envelope inhibited by depletion of the nuclear Ca^{2+} store. *Science*, 1995. 270(5243): 1835–8.

97. Gerasimenko, O.V., et al., ATP-dependent accumulation and inositol trisphosphate- or cyclic ADP-ribose-mediated release of Ca^{2+} from the nuclear envelope. *Cell*, 1995. 80(3): 439–44.

98. Berger, F. and C. Brownlee, Ratio confocal imaging of free cytoplasmic calcium gradients in polarising and polarised Fucus zygotes. *Zygote*, 1993. 1(1): 9–15.

99. McClellan, A.D., D. McPherson, and M.J. O'Donovan, Combined retrograde labeling and calcium imaging in spinal cord and brainstem neurons of the lamprey. *Brain Res*, 1994. 663(1): 61–8.

100. McPherson, D.R., A.D. McClellan, and M.J. O'Donovan, Optical imaging of neuronal activity in tissue labeled by retrograde transport of Calcium Green Dextran. *Brain Res Protoc*, 1997. 1(2): 157–64.

101. Milner, E.P., Q. Zheng, and J.C. Kermode, Ristocetin-mediated interaction of human von Willebrand factor with platelet glycoprotein Ib evokes a transient calcium signal: observations with Fura-PE3. *J Lab Clin Med*, 1998. 131(1): 49–62.

102. Wu, Y.C., T. Tucker, and R. Fettiplace, A theoretical study of calcium microdomains in turtle hair cells. *Biophys J*, 1996. 71(5): 2256–75.

103. Llinas, R., M. Sugimori, and R.B. Silver, Microdomains of high calcium concentration in a presynaptic terminal. *Science*, 1992. 256(5057): 677–9.

104. Simon, S.M. and R.R. Llinas, Compartmentalization of the submembrane calcium activity during calcium influx and its significance in transmitter release. *Biophys J*, 1985. 48(3): 485–98.

105. Llinas, R., M. Sugimori, and R.B. Silver, The concept of calcium concentration micro-domains in synaptic transmission. *Neuropharmacology*, 1995. 34(11): 1443–51.

106. Naraghi, M. and E. Neher, Linearized buffered Ca^{2+} diffusion in microdomains and its implications for calculation of $[Ca^{2+}]$ at the mouth of a calcium channel. *J Neurosci*, 1997. 17(18): 6961–73.

107. Thorn, P., Spatial domains of Ca^{2+} signaling in secretory epithelial cells. *Cell Cal*, 1996. 20(2): 203–14.

108. Zucker, R.S., Calcium and transmitter release. *J Physiol*, Paris, 1993. 87(1): 25–36.

109. von Gersdorff, H. and G. Matthews, Dynamics of synaptic vesicle fusion and membrane retrieval in synaptic terminals. *Nature*, 1994. 367(6465): 735–9.

110. Lopez-Lopez, J.R., et al., Local calcium transients triggered by single L-type calcium channel currents in cardiac cells. *Science*, 1995. 268(5213): 1042–5.

111. Blatter, L.A. and E. Niggli, Confocal near-membrane detection of calcium in cardiac myocytes. *Cell Calcium*, 1998. 23(5): 269–79.

112. Etter, E.F., M.A. Kuhn, and F.S. Fay, Detection of changes in near-membrane Ca^{2+} concentration using a novel membrane-associated Ca^{2+} indicator. *J Biol Chem*, 1994. 269(13): 10141–9.

113. Etter, E.F., et al., Near-membrane $[Ca^{2+}]$ transients resolved using the Ca^{2+} indicator FFP18. *Proc Natl Acad Sci USA*, 1996. 93(11): 5368–73.

114. Marsault, R., et al., Domains of high Ca^{2+} beneath the plasma membrane of living A7r5 cells. *EMBO J*, 1997. 16(7): 1575–81.

115. Olivos Ore, L. and A.R. Artalejo, Intracellular Ca^{2+} microdomain-triggered exocytosis in neuroendocrine cells. *Trends Neurosci*, 2004. 27(3): 113–5.

116. Axelrod, D., Total internal reflection fluorescence microscopy, in *Fluorescence Microscopy of Living Cells in Culture*, D.L.T.a.Y.L. Wang, Editor. 1989, San Diego CA: Academic Press.

117. Gryczynski, I., H. Szmacinski, and J.R. Lakowicz, On the possibility of calcium imaging using Indo-1 with three-photon excitation. *Photochem Photobiol*, 1995. 62(4): 804–8.

118. Konig, K., U. Simon, and K.J. Halbhuber, 3D resolved two-photon fluorescence microscopy of living cells using a modified confocal laser scanning microscope. *Cell Mol Biol*, 1996. 42(8): 1181–94.

119. Szmacinski, H., I. Gryczynski, and J.R. Lakowicz, Three-photon induced fluorescence of the calcium probe Indo-1. *Biophys J*, 1996. 70(1): 547–55.

120. Nowycky, M.C. and M.J. Pinter, Time courses of calcium and calcium-bound buffers following calcium influx in a model cell. *Biophys J*, 1993. 64(1): 77–91.

121. Lloyd, Q.P., M.A. Kuhn, and C.V. Gay, Characterization of calcium translocation across the plasma membrane of primary osteoblasts using a lipophilic calcium-sensitive fluorescent dye, calcium green C18. *J Biol Chem*, 1995. 270(38): 22445–51.

122. Tanimura, A. and R.J. Turner, Inositol 1,4,5-trisphosphate-dependent oscillations of luminal $[Ca^{2+}]$ in permeabilized HSY cells. *J Biol Chem*, 1996. 271(48): 30904–8.

123. Tanimura, A., Y. Matsumoto, and Y. Tojyo, Polarized Ca^{2+} release in saponin-permeabilized parotid acinar cells evoked by flash photolysis of 'caged' inositol 1,4,5-trisphosphate. *Biochem J*, 1998. 332(Pt 3): 769–72.

124. Tojyo, Y., A. Tanimura, and Y. Matsumoto, Monitoring of Ca^{2+} release from intracellular stores in permeabilized rat parotid acinar cells using the fluorescent indicators Mag-fura-2 and calcium green C18. *Biochem Biophys Res Commun*, 1997. 240(1): 189–95.

125. Davies, E.V. and M.B. Hallett, Near membrane Ca^{2+} changes resulting from store release in neutrophils: detection by FFP-18. *Cell Calcium*, 1996. 19(4): 355–62.

126. Chadborn, N., et al., Direct measurement of local raised subplasmalemmal calcium concentrations in growth cones advancing on an N-cadherin substrate. *Eur J Neurosci*, 2002. 15(12): 1891–8.

127. Liepouri, F., et al., Near-membrane iminocoumarin-based low affinity fluorescent Ca(2+) indicators. *Cell Calcium*, 2002. 31(5): 221–7.

128. Zipfel, W.R., R.M. Williams, and W.W. Webb, Nonlinear magic: multiphoton microscopy in the biosciences. *Nat Biotechnol*, 2003. 21(11): 1369–77.

129. Konishi, M. and S.M. Baylor, Myoplasmic calcium transients monitored with purpurate indicator dyes injected into intact frog skeletal muscle fibers. *J Gen Physiol*, 1991. 97(2): 245–70.

130. Maravall, M., et al., Estimating intracellular calcium concentrations and buffering without wavelength ratioing. *Biophys J*, 2000. 78(5): 2655–67.

131. Xu, C., et al., Multiphoton fluorescence excitation: new spectral windows for biological nonlinear microscopy. *Proc Natl Acad Sci USA*, 1996. 93(20): 10763–8.

132. Kuba, K. and S. Nakayama, Two-photon laser-scanning microscopy: tests of objective lenses and Ca^{2+} probes. *Neurosci Res*, 1998. 32(3): 281–94.

133. Albota, M., et al., Design of organic molecules with large two-photon absorption cross sections. *Science*, 1998. 281(5383): 1653–6.

134. Thomas, A.P., et al., Spatial and temporal aspects of cellular calcium signaling. *FASEB J*, 1996. 10(13): 1505–17.

135. Oheim, M., et al., Two dye two wavelength excitation calcium imaging: results from bovine adrenal chromaffin cells. *Cell Calcium*, 1998. 24(1): 71–84.

136. Escobar, A.L., et al., Kinetic properties of DM-nitrophen and calcium indicators: rapid transient response to flash photolysis. *Pflugers Arch — Eur J Physiol*, 1997. 434(5): 615–31.

137. Vila, L., E.F. Barrett, and J.N. Barrett, Stimulation-induced mitochondrial $[Ca^{2+}]$ elevations in mouse motor terminals: comparison of wild-type with SOD1-G93A. *J Physiol*, 2003. 549(Pt 3): 719–28.

2 Fluorescent Indicators — Facts and Artifacts

Gary St. J. Bird and James W. Putney, Jr.

CONTENTS

2.1 INTRODUCTION

Elevation of intracellular Ca^{2+} is a universal signal controlling many cell functions [1]. Early years of study on calcium homeostasis relied on indirect assessments of cytoplasmic calcium by either radiocalcium measurements or monitoring Ca^{2+}-dependent ion fluxes, or the activity of Ca^{2+}-dependent enzymes [2–4]. However, the development of organic fluorescent Ca^{2+} indicators in the 1980s revolutionized the measurement of calcium ions in living cells, providing more direct methods to estimate cytoplasmic calcium and perform calcium measurements (Chapter 1). The ease with which these indicators can be introduced into intact cells and the broad choice of equipment with which to perform such measurements have combined to provide a robust and widely used technique for measuring cytoplasmic Ca^{2+} concentration ($[Ca^{2+}]_i$).

While many aspects of calcium signaling remain to be defined, our basic understanding today is that receptor-activated $[Ca^{2+}]_i$ changes are complex both spatially and temporally, involving the interplay of calcium channels and calcium pumps. The ability to use fluorescent calcium indicators to monitor, in real time, spatial and temporal changes of $[Ca^{2+}]_i$ has significantly contributed to our understanding of these fundamental homeostatic mechanisms in mammalian cells.

It has been said that cells possess a "Ca^{2+} signaling toolkit" to provide for versatility in Ca^{2+} signaling mechanisms [1]. Similarly, in this chapter, we provide a "toolkit" of protocols for dissecting the observed $[Ca^{2+}]_i$ signal and for describing the underlying calcium signaling processes. We will focus on practical approaches for manipulating $[Ca^{2+}]_i$ changes, the flexibility and limitations of using fluorescent calcium indicators, and how they can be used to reveal facts, as well as potential artifacts, about the underlying calcium signaling mechanisms.

2.1.1 Ca^{2+} SIGNALING

Many of the experimental approaches below will be described with particular emphasis on calcium signaling processes in nonexcitable cells mediated by the phosphoinositide–phospholipase C (PI-PLC) pathway [5]. However, the mechanisms and characteristics of calcium signaling events in excitable and nonexcitable cells share many common themes [6, 7], and many of these experimental approaches are applicable to a broad range of calcium signaling systems.

A general paradigm for calcium signaling is a coordinated regulation of intracellular calcium ion release and calcium ion entry across the plasma membrane of the cell. In nonexcitable cells, receptor activation of PI-PLC by neurotransmitters and hormones typically stimulates a Ca^{2+} signaling process that is biphasic, with an initial release of calcium ions to the cytoplasm from an intracellular organelle, followed by entry of calcium ions into the cytoplasm across the plasma membrane. In excitable cells, a similar scenario may occur. Alternatively, in excitable cells, coupling of release and entry can often occur in the opposite direction; i.e., entry of Ca^{2+} activates release by a process of calcium-induced calcium release [8, 9] (and see Chapter 11).

The first phase of calcium release is often attributable to inositol 1,4,5-trisphosphate (IP_3), the most widely encountered of the intracellular Ca^{2+}-release mechanisms. IP_3 acts by binding to a specific receptor (Chapter 10) on the endoplasmic reticulum (ER) (Chapter 15) or possibly to a specialized component of the endoplasmic reticulum [10]. However, there are several other potential mechanisms for controlling intracellular calcium release, including ryanodine receptors (Chapter 11), cyclic ADP ribose, and NAADP (Chapter 12). In signaling pathways involving PI-PLC, the second phase of calcium entry is most commonly attributed to capacitative calcium entry (CCE), a process of retrograde signaling such that the empty calcium-storage organelle produces a signal for calcium ion entry across the plasma membrane [11]. The nature of the retrograde signal and the target plasma membrane channel(s) underlying CCE remain poorly defined. However, channels associated with CCE have been characterized electrophysiologically (Chapter 6), and members of the TRP superfamily (Chapter 7) have been described as molecular candidates for CCE channels, in particular members of the TRPC and TRPV families [12].

Many aspects of this coordinated calcium signaling process can be described and investigated experimentally using fluorescent calcium indicators to observe $[Ca^{2+}]_i$ changes directly, especially when combined with a careful manipulation of the medium bathing cells coupled with the judicious use of pharmacological agents [13]. With a critical understanding of the basis for manipulating $[Ca^{2+}]_i$ levels and an appreciation of the flexibility and limitations of applying fluorescent calcium indicators, one can better define the underlying calcium signaling processes and minimize the potential for creating artifacts and misinterpreting measurements.

2.2 ANATOMY OF A CALCIUM SIGNAL

An anatomical picture of a calcium signal can be built by the careful application of fluorescent calcium indicators and judicious manipulation of these same signaling processes by pharmacological means. At the very least, one must consider and characterize the following parameters: (i) choice of fluorescence indicator to monitor $[Ca^{2+}]$; (ii) regulation of resting $[Ca^{2+}]$; (iii) mode of Ca^{2+}-release; (iv) source of Ca^{2+} pools; (v) mode of Ca^{2+} entry; (vi) feedback mechanisms that regulate $[Ca^{2+}]$.

2.2.1 MONITORING THE $[Ca^{2+}]_i$ SIGNAL

2.2.1.1 Fluorescent Indicator Selection

The choice of fluorescent calcium indicator is the foundation for a successful study of calcium signaling, influencing the spatial and temporal information that one can collect, and the choices one has for analyzing a response. There are several aspects of a fluorescent calcium indicator that determine its suitability for a particular study. Chapter 1 outlines the availability of an extensive selection of fluorescent Ca^{2+} indicators with varying properties that lay the fundamental groundwork for indicator selection, including wavelength selection, quantum efficiency, and ion affinity and selectivity. However, while the focus is on calcium ions, it is often the ability of these indicators to interact with other cations that provide experimental flexibility for studying calcium signaling. This is particularly interesting when considering that some of these same cations can either substitute for calcium ions in a number of biological processes, or pharmacologically interfere with these same processes (see below).

By understanding which cations will interact with a particular fluorescent calcium indicator (ion selectivity) as well as understanding the pharmacology of that cation, one can design experimental approaches involving substitution of alternative ions for calcium and thus reveal specific aspects of the underlying calcium signaling process. In this respect, we initially focus on the ratiometric indicator fura-2, which has been widely used for nearly 20 years.

2.2.1.2 Fura-2

Fura-2 provides good brightness, the ability to make ratiometric measurements, and an innovative approach for loading the indicator into intact cells [14]. Thus, it is understandable that fura-2 has for some time now been the calcium indicator of choice. It was the first widely available ratiometric indicator with an affinity and ion selectivity that made calcium signaling studies possible in intact cells [15].

Fura-2 binds certain cations in addition to Ca^{2+}, for example the divalents Sr^{2+} and Ba^{2+} and the trivalent La^{3+}, all of which render spectra similar to Ca^{2+} [16, 17], and with all the characteristics of an antiparallel intensity change at 340 and 380 nm. In contrast, Ni^{2+}, Co^{2+}, and Mn^{2+} strongly quench the fura-2 fluorescence at all wavelengths [17]. It should also be noted that Cu^{2+} can quench fura-2, which should be a concern when combining calcium studies and using $CuSO_4$ to induce protein expression [18]. While not an exhaustive list, some of the characteristics of fura-2 are compared with other indicators in Table 2.1. In addition, Table 2.2 summarizes interactions of these cations with various aspects of calcium signaling mechanisms, all of which will be discussed later in the text.

2.2.1.3 Fura-5F

Chapter 1 relates in great detail the extensive selection of fluorescent indicators available for monitoring calcium changes over a broad concentration range. At present, however, we believe that fura-5F, rather than fura-2, is a better choice for measuring $[Ca^{2+}]_i$ for most experimental situations, specifically when dealing with

TABLE 2.1
General Characteristics of Fluorescent Calcium Indicator in General Use

	Ratiometric	Ba^{2+}-Entry	Mn^{2+}-Quench	K_D (nM)
Fura-2	Yes	Yes	Yes	140
Fura-5F	Yes	Yes	Yes	400
Fluo-4	No	No	No	390
Calcium Green-1	No	Yes	No	190

TABLE 2.2
Spectral and Biological Characteristics of Ca^{2+} and Other Useful Ions for Calcium Signaling Studies

	Fura-Signal	Fura Quench	Fluo-4 Signal	Fluo-4 Quench	CaGreen-1 Signal	CaGreen-1 Quench	Block CCE	Substrate for PMCA/SERCA	Block PMCA
Ca^{2+}	Yes	No	Yes	No	Yes	No	No	Yes	No
Sr^{2+}	Yes	No	Yes	No	Yes	No	No	Yes	No
Ba^{2+}	Yes	No	Poor	No	Yes	No	No	No	No
La^{3+}	Yes	No	Yes	No	Yes	No	Yes	?	Yes
Gd^{3+}	Yes	No	Yes	No	No	Yes	Yes	?	Yes
Co^{2+}	No	Yes	Poor	No	Yes	No	No	?	?
Ni^{2+}	No	Yes	Yes	No	Yes	No	Yes	?	?
Mn^{2+}	No	Yes	Poor	No	Yes	No	No	No	?

calcium changes under certain, physiological conditions. Fura-5F, a derivative of fura-2, is handled in exactly the same way as fura-2 in terms of cell loading and recording of fluorescence signal. However, fura-5F has a higher K_D than fura-2 (400 nM vs. 140 nM, respectively), and given that most Ca^{2+} studies examining $[Ca^{2+}]_i$ in the 100 nM–1 μM range, the advantages of the higher K_D indicator are obvious. This is especially true, given the recent tendency of investigators to report ratios rather than calibrated $[Ca^{2+}]_i$ values. In such situations, the relationship between changes in ratio and $[Ca^{2+}]_i$ with the lower K_D indicator will be extremely nonlinear.

Recent experience in investigating the effects of low concentrations of the muscarinic receptor agonist, acetyl-methacholine (MCh), on calcium signaling in HEK-293 cells, highlights the potential advantages of using a higher K_D indicator [19]. It was of interest to investigate the mechanism by which low concentrations of MCh induced calcium oscillations in HEK-293 cells. Initial studies were first performed using fura-2 [20] in which we found that under normal fura-2/AM loading conditions responses to low [MCh] were rare. In fact, fura-2/AM loading conditions had to be reduced close to the limits of detection in order to see a response that was oscillatory. Under these conditions, ~50% of the cells responded to 5 μM MCh. When we subsequently used the lower affinity fura-5F, we found a substantial technical improvement in the method of investigating oscillations. One could load with sufficient fura-5F to ensure a robust fluorescence signal, and with 5 μM MCh, ~90–95% of HEK-293 cells responded with most exhibiting sustained calcium oscillations.

A spectral characterization of fura-5F is shown in Figure 2.1A and summarized in Figure 2.1D, where it appears to share many of the properties of fura-2, including a Ca^{2+}-like spectral shift with Ba^{2+}, and quenching of fluorescence by Mn^{2+}. Figure 2.1 also compares the spectral characteristics for two single wavelength calcium indicators in common usage, fluo-4 (Figure 2.1B and 2.1E) and Calcium

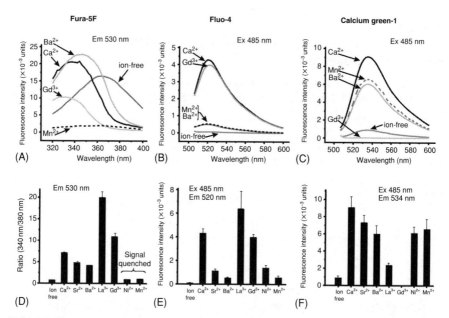

FIGURE 2.1 Spectral characteristics of the fluorescent calcium indicators fura-5F, fluo-4, and Calcium Green-1. (A–C) Free-acid forms of the fluorescent calcium indicators fura-5F, fluo-4, and Calcium Green-1 were dissolved to a final concentration of 10 μM in a buffer composed of 100 mM KCl and 20 mM HEPES (pH 7.2). Solutions devoid of cations ("Ion-free") had the buffer supplemented with 200 μM BAPTA, otherwise, the buffer contained 1 mM of each cation. All spectra were recorded on a Spectramax Gemini XS (Molecular Devices, Sunnyvale, CA). For fura-5F, emission spectra (A) were recorded at 530 nm, while scanning the sample with excitation wavelengths from 320 to 400 nm. With fluo-4 (B) and Calcium Green-1 (C), samples were excited at 485 nm, while scanning the emission spectra from 500 to 600 nm. For display purposes, only spectra for "ion-free," Ca^{2+}, Ba^{2+}, Gd^{3+}, and Mn^{2+} are shown. The data are the average for three independent determinations. (D–F) For comparison, the data for each cation are quantified by showing the 340/380 nm ratio values for fura-5F (D), and peak intensities measured at 520 nm for fluo-4 (E) and 534 nm for Calcium Green-1 (F). All data are the average $\pm$ SEM for three independent determinations.

Green-1 (Figure 2.1C and 2.1F). These characteristics differ significantly from one another and this affects the choice of calcium indicator depending on the particular calcium signaling study. For example, substituting Ba^{2+} ions for Ca^{2+} is particularly useful for monitoring the Ca^{2+}-entry process (see below) but, as indicated in Figure 2.1 and summarized in Table 2.1, only certain indicators, for example fura-5F (like fura-2) and Calcium Green-1, are suitable for such studies.

2.2.1.4 Fluorescent Indicators and Cytoplasmic Loading

The ease with which acetoxy methylester (AM) derivatives of fluorescent indicators could be introduced into cells was key to the success of this approach. In many instances, these indicators primarily concentrate and report fluorescence changes in the cytoplasm. However, the permeability of these AM-indicators is not limited to the plasma membrane. Their permeability and accumulation in intracellular organelles can present disadvantages for monitoring $[Ca^{2+}]_i$, but may sometimes offer the opportunity for monitoring calcium signaling processes within those organelles (Chapters 14 and 15).

The controllable variables that determine the extent of retention of a calcium indicator in the cytoplasm are indicator concentration, duration of incubation with the AM-indicator, and the temperature at which incubation is performed. In addition, there are variables that cannot be controlled including the activity of esterases in various intracellular compartments and the activity of plasma membrane transporters that tend to remove de-esterified indicators from the cytoplasm. Nonetheless, in many instances, optimizing the controllable parameters of incubation time and temperature will favor fluorescent indicators enriched in the cytoplasm. However, because behavior of the AM-indicators is not entirely predictable, characterization of indicator loading and distribution in the cell type of interest is strongly recommended. For example, experience with fura-2/AM loading in the rat pancreatoma cell line AR4-2J [21, 22] demonstrated that with incubations at 37°C the indicator is distributed between the cytoplasm, ER, and particularly enriched in the mitochondria. However, with AM-indicator incubations at room temperature (~20–25°C), the fura-2 signal is preferentially located in the cytoplasm [23]. In contrast, with another cell line, HEK-293, no such problems are experienced with AM-indicator loading at 37°C [19]. Of the variables that are more difficult to control, perhaps the most common is the efflux of indicators out of the cytoplasm in some cell types. The nature of this process is not fully understood, and could be due to a combination of properties of the indicator used and underlying cellular processes for transporting the indicator. Chapter 1 discusses many aspects of minimizing this problem, including the use of leakage resistant indicators, and inhibitors of anion transport inhibitors such as probenicid and sulfinpyrazone.

There are a number of ways to diagnose the problem of indicator compartmentalization. The most straightforward is to inspect the spatial distribution of the indicator, where a nonuniform, punctate fluorescence image may be indicative of compartmentalized dye. Even at low resolution, the common picture with compartmentalized dyes is an apparently dimmer nucleus. With ideal loading, the nucleus should be barely visible in the fluorescence ratio image. Permeabilization of the plasma membrane would also selectively release an indicator in the cytoplasm, leaving the indicator trapped within the organelles [21, 22]. If spatial differences cannot be observed, a combination of the following factors may indicate compartmentalization: (i) Calibrated fluorescence values of resting $[Ca^{2+}]$ are high (high ratio values are diagnostic if using ratiometric indicators). (ii) Poor or lack of agonist-induced $[Ca^{2+}]_i$ response, and a combination of agonist and a sarcoendoplasmic reticulum Ca^{2+}-ATPase (SERCA)–pump inhibitor (thapsigargin) can lead to a drop in resting $[Ca^{2+}]$ levels (see [21, 22]). If compartmentalization of the fluorescence indicator cannot be resolved by incubation conditions, then microinjection of the free acid form of the indicator directly into the cytoplasm is an option [21].

The "leakiness" of an indicator is not predictable either, with contributing factors including the nature of the indicator and cell type used, and this potential problem needs to be characterized on a case-by-case basis. For example, no problems are experienced with cytoplasmic loading of HEK-293 cells with fura-2 and fluo-4, but effective cell loading with Calcium Green-1 requires the presence of an anion transport inhibitor (probenicid or sulfinpyrazone) during AM-indicator loading and through the entirety of the experiment [24]. In contrast, CHO cells require the presence of anion transport inhibitors for both fura-2, Calcium Green-1, and fluo-3 [25–27].

2.2.1.5 Quantitative Assessment of a $[Ca^{2+}]_i$ Response

Although there are many uncertainties and errors inherent in the calibration process, it can provide a reasonable estimation of $[Ca^{2+}]_i$. Due to the nonlinear relationship between fluorescence ratio and $[Ca^{2+}]_i$, calibration is advisable for any study looking for subtle quantitative effects on calcium signaling. In addition, calibration of $[Ca^{2+}]_i$ values provides a means to compare data from one instrument with that from another, or even from one laboratory to another; ratio values on the other hand will vary widely from one system to another. However, in many instances, a general understanding of mechanisms involved in calcium signaling can be gleaned from fluorescence intensities or fluorescence ratios without attempting precise calibration of $[Ca^{2+}]_i$. In doing so, the following considerations are recommended:

1. Indicator Selection
Select a fluorescence Ca^{2+} indicator with an appropriate affinity, ideally with a K_D near or somewhat greater than the high end of the dynamic range of anticipated $[Ca^{2+}]_i$ (Chapter 1).

2. Minimize Compartmentalization
Optimize calcium dye loading conditions so that the fluorescence signal comes primarily from the compartment of interest, in most instances the cytoplasm.

3. Autofluorescence Correction
It is important to correct data for fluorescence signals not related to calcium-sensitive fluorescence changes (generally termed auto-fluorescence). Such a correction minimizes variation in data from cell-to-cell, and between experiments performed on the same or different days (assuming data recorded under the same conditions). In the case of ratiometric fura dyes, this is readily estimated by quenching fura dyes with Mn^{2+} ions (treat cells with ionomycin and $MnCl_2$ (5 μM and 20 mM, respectively, in a nominally calcium-free bathing solution). Fluorescence intensities, collected at each excitation wavelength, can then be corrected by subtracting their component autofluorescence values, and new ratio values generated. As the single wavelength indicators fluo-4 and Calcium Green-1 cannot be quenched by Mn^{2+} (Figure 2.1), this method of autofluorescence correction is not possible (but see [28, 29] for alternative calibration approaches). In this situation, nonindicator-loaded cells may provide an estimation of autofluorescence.

4. **Peaks, Areas, and Rates**

As will be described below, at any moment in time, the observed $[Ca^{2+}]_i$ signal represents a balance of many processes that coordinate to control the movement of calcium ions. During periods of cell activation, Ca^{2+}-release and Ca^{2+}-entry pathways are recruited to increase $[Ca^{2+}]_i$, and Ca^{2+}-pumps and regulatory mechanisms seek to return $[Ca^{2+}]_i$ back to resting levels. Thus, the approach one takes to evaluate the observed Ca^{2+} responses is dictated by the temporal and spatial pattern of the Ca^{2+} signal, and this then raises the issue of the suitability of using peaks, areas-under-the-curve, or initial rates to give the best indication of the magnitude of a particular response.

In general, analyzing the maximal or "Peak Response" is recommended when dealing with a response that has a rapid onset to a well-defined peak, or when dealing with a response that achieves a relatively constant steady-state level. Measurement of area-under-the-curve is not generally useful with steady-state changes, and may be too variable when describing a "peak response" as the decline of this signal is subject to regulation by various Ca^{2+}-pumps. Calculation of initial rates of rise in the fluorescence signal (or ratio) can be used, which minimizes but does not eliminate the influence of pumps and Ca^{2+} buffering. This approach is particularly useful when monitoring Ca^{2+}-entry signals with surrogate ions that are not pumped, such as Ba^{2+} and Mn^{2+} (see below).

In many instances, these approaches to evaluating the $[Ca^{2+}]_i$ signal are more reliable when dealing with responses that are spatially and temporally simple, for example when dealing with maximal receptor activation with pharmacological concentrations of agonists. However, receptor activation under more physiological conditions can lead to complex spatial and temporal $[Ca^{2+}]_i$ changes, or calcium oscillations. Under these conditions, many additional parameters beyond peak values must be considered to describe and quantify such a response [19, 30].

2.2.2   DISSECTING THE $[Ca^{2+}]_i$ SIGNAL

$[Ca^{2+}]_i$ as measured with fluorescent Ca^{2+} indicators, is a reflection of the homeostatic balance of calcium pumps and calcium channel activities at any moment in time. Any disturbance in these processes, through receptor-dependent activation of calcium channel activity for example, can lead to the rapid movement of calcium down concentration gradients until a new steady-state equilibrium is met in concert with calcium pump and feedback mechanisms. Experimental manipulations and pharmacological agents that can interfere at any point in these calcium homeostatic processes can have a profound effect on $[Ca^{2+}]$. Thus, an intimate understanding of the underlying calcium homeostatic processes and the various ways they can be manipulated, can allow those observed fluorescence signals to be interpreted in meaningful ways that speak to the underlying calcium signaling process. Some of these approaches are outlined below.

2.2.2.1 Resting $[Ca^{2+}]_i$ and the "Calcium Re-Addition" Protocol

The cytoplasm represents but one compartment for calcium ions in living cells, an intermediary space between the extracellular milieu and the various intracellular organelles key to calcium signaling, particularly the ER and mitochondria. However, enormous concentration gradients of calcium ions exist between these compartments, the highest levels in the extracellular milieu ($\sim 10^{-3} M$) and ER ($\sim 10^{-4} M$), and lowest in the cytoplasm ($\sim 10^{-7} M$) and mitochondria. Thus, under steady-state conditions when cells are unstimulated, the concentration of calcium ions in the cell cytoplasm ($\sim 10^{-7} M$) represents an exquisite balance of homeostatic mechanisms involving the tonic activity of calcium pumps and calcium channels to maintain $[Ca^{2+}]$ at these low levels. In this sense, a "resting" cell is something of a misstatement, as cells expend much energy to maintain these calcium gradients through ATP-dependent Ca^{2+} pump activity (Chapter 13). This tonic activity, however, can be revealed by rapidly disrupting the steady-state conditions and observing the resulting change in $[Ca^{2+}]_i$ as the cell seeks to reach a new steady-state $[Ca^{2+}]_i$ level.

A principal method to disrupt steady-state $[Ca^{2+}]_i$ conditions and probe Ca^{2+} pump and channel activity is by modification of the extracellular calcium conditions ($[Ca^{2+}]_o$), in most instances by restoration of extracellular Ca^{2+} to cells first stimulated in the absence of Ca^{2+} [2, 31, 32]. Basically, in this "calcium re-addition" protocol, the salt solution bathing cells is switched from one containing normal $[Ca^{2+}]_o$ (1–2 mM) to a salt solution that is depleted of $[Ca^{2+}]_o$, and then back to normal $[Ca^{2+}]_o$ conditions. The depleted $[Ca^{2+}]_o$ salt solution can be one where no $CaCl_2$ was added, which is considered "nominally Ca^{2+}-free" as (depending on the water source and purity of salts used) Ca^{2+} contaminations in this buffer can range from a few μM to as much as 100 μM. Alternatively, nominally Ca^{2+}-free solutions can be supplemented with a Ca^{2+}-chelator such as EGTA or BAPTA (100–500 μM) to ensure an essentially Ca^{2+}-free solution. While this solution changing protocol is more easily undertaken in attached cells, the principle of the protocol can be undertaken with cells in suspension.

The data shown in Figure 2.2 demonstrate this protocol in a fura-5F-loaded HEK 293 cell line that is over-expressing the V1a receptor and that responds to the PI-PLC agonist vasopressin (Vp) [33]. In these HEK 293-V1a cells, the calcium re-addition protocol (Figure 2.2A) did not result in any detectable change in the steady-state $[Ca^{2+}]_i$ under resting conditions. However, there are examples where there are more obvious effects of this calcium re-addition protocol in unstimulated cells, such as the rat pancreatoma cell line AR4-2J [23], or even in HEK 293 cells where the calcium measurement is performed by a different technique (see Figure 2.2 of Chapter 3). The introduction of a putative cation channel may also modify the characteristics of this steady-state activity; expression of constitutively active TRP channels in HEK 293 cells is a good example of this (see Figure 2.2 of Chapter 3) [24]. In all these cases, characterizing these tonic changes in $[Ca^{2+}]_i$ in resting cells is a critical control that will be important for understanding $[Ca^{2+}]_I$ in receptor-activated cells.

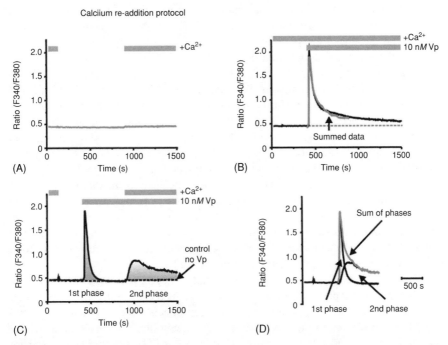

FIGURE 2.2 Biphasic calcium signaling and the calcium re-addition protocol in fura-5F-loaded HEK 293-V1a cells. HEK 293-V1a cells (HEK 293 cells expressing the V1a receptor [33]) attached to glass cover slips were loaded with calcium indicators fura-5F and cytoplasmic $[Ca^{2+}]_i$ measured at the single cell level as described by Bird and Putney [19]. In (A) and (C), cells were subject to the calcium re-addition protocol where extracellular Ca^{2+} is removed from the medium bathing the cells, and then restored at a later time point. The trace in A is for cells at rest (not stimulated), whereas in (B) the cells were activated with Vasopressin (10 nM Vp). The biphasic nature of the calcium response is illustrated in (B) with the first phase indicating intracellular Ca^{2+}-release, and the second phase Ca^{2+}-entry across the plasma membrane. The data in (B) show the response to 10 nM Vp in the continued presence of extracellular Ca^{2+}. The release and entry components in (C) were summed as shown in (D), and then plotted along with the data in (B) (gray trace). The summation of separated release and entry signals nicely recapitulates the total $[Ca^{2+}]_i$ response observed in Ca^{2+}-containing media. All data are from single observations, but each trace represents the average signal from 25 to 30 cells.

2.2.2.2 Receptor-Activated Cells and Biphasic Calcium Signaling

As described above, receptor activation of the PI-PLC pathway results in a cascade of events that leads to IP_3-mediated intracellular calcium release, and CCE across the plasma membrane. The effect is to elevate the $[Ca^{2+}]_i$ above that observed in resting cells, an effect that can be transient or sustained, but which is the result of the coordinated activity of a biphasic process of intracellular Ca^{2+}-release and Ca^{2+}-entry, and regulatory mechanisms such as Ca^{2+}-pumps. Figure 2.2B illustrates the response of fura-5F loaded HEK 293-V1a cells to Vasopressin (Vp) in the presence of extracellular Ca^{2+} ($[Ca^{2+}]_o$, 1.8 mM). On addition of Vp (10 nM), there is a rapid elevation of $[Ca^{2+}]_i$ that declines with time, but remains somewhat elevated above resting $[Ca^{2+}]_i$. The cells remain at this elevated $[Ca^{2+}]_i$ level until the stimulus is removed. The calcium re-addition protocol (Figure 2.2C) more clearly reveals the biphasic nature of the calcium signaling process. If $[Ca^{2+}]_o$ is removed prior to receptor activation (i.e., bathed in nominally Ca^{2+}-free conditions), one now sees the first phase of calcium signaling, a transient $[Ca^{2+}]_i$ response that returns to levels observed in resting cells. This first phase represents the release of calcium ions from intracellular calcium pools sensitive to IP_3 (Figure 2.2C). On adding back $[Ca^{2+}]_o$ to the bathing medium, the second phase of calcium entry is revealed resulting in a rise in $[Ca^{2+}]_i$ to an elevated and sustained phase. In Figure 2.2D, the release and entry components are summed, and these summed data are then plotted along with the data in Figure 2.2B. The summation of separated release and entry signals nicely recapitulates the total $[Ca^{2+}]_i$ response observed in Ca^{2+}-containing media.

This $[Ca^{2+}]_i$ response observed in maximally receptor-activated HEK 293-V1a cells is somewhat representative of response patterns observed in a majority of cell types. There are, however, examples where receptor activation results in a transient $[Ca^{2+}]_i$ response even in the presence of $[Ca^{2+}]_o$, with the cells reaching a steady-state $[Ca^{2+}]_i$ level close to or at resting $[Ca^{2+}]_i$ levels [34]. Alternatively, there are some cell types such as mouse lacrimal acinar cells where receptor activation results in a second phase of Ca^{2+}-entry that is substantially elevated and sustained [35].

A variation of the calcium re-addition protocol provided some of the original evidence for the CCE model in parotid acinar cells [36]. In those experiments, fura-2 loaded cells were treated with MCh in the absence of $[Ca^{2+}]_o$, which was subsequently restored after the intracellular Ca^{2+}-release transient was complete. Addition of $[Ca^{2+}]_o$ resulted in enhanced $[Ca^{2+}]_i$, compared to resting cells, indicative of increased Ca^{2+}-entry. However, if the muscarinic receptor was blocked (with atropine) after the MCh-induced Ca^{2+}-release phase was complete, an increase in $[Ca^{2+}]_i$ was still observed on addition of $[Ca^{2+}]_o$ that was, however, transient. The significance of this transient "overshoot" of Ca^{2+}-entry response was: (i) it demonstrated that intracellular Ca^{2+}-pool depletion can lead to increased permeability of the plasma membrane, (ii) this increased permeability was independent of receptor activation, and (iii) the transient nature of this response was likely due to due to refilling of intracellular Ca^{2+}-pools and subsequent inactivation of the Ca^{2+}-entry pathway. This procedure is still commonly used to detect activation of store-operated Ca^{2+} entry. However, the overshoot on Ca^{2+} addition after agonist removal is not always observed, and probably depends on the rate of Ca^{2+} entry at least transiently exceeding the rate of subsequent uptake into the endoplasmic reticulum. In these instances, the accelerated entry can be seen by use of permeant, but nonsequestered cations such as Ba^{2+} or Mn^{2+}.

2.2.3 Intracellular Ca^{2+}-Release and Ca^{2+}-Pools

As mentioned above, receptor activation in nonexcitable cells most commonly initiates a cascade involving the PI-PLC pathway, generating IP_3 that in turn interacts with its receptor (Chapter 10) on the ER and releases calcium ions into the cytoplasm. In the absence of extracellular calcium, this receptor activation leads to a transient rise in $[Ca^{2+}]_i$ emphasizing that this first phase of calcium signaling is due to mobilization of Ca^{2+} from pools sequestered within intracellular organelles. There are alternative mechanisms for regulating this Ca^{2+}-release phase, including cyclic ADP-Ribose, NAADP (Chapter 12), and activation of ryanodine receptors (Chapter 11). However, by way of example, in this section we will focus on the PI-PLC signaling pathway, and procedures that can discern Ca^{2+}-pools mobilized by agonists and a role for IP_3. In order to study calcium redistribution from intracellular calcium pools it is important to develop protocols that (i) monitor this process without interference from Ca^{2+}-entry, and (ii) utilize pharmacological agents that selectively manipulate the key players in intracellular Ca^{2+}-homeostasis. These key players include the major organelles for storing intracellular calcium, Ca^{2+}-pumps, and Ca^{2+}-release channels.

Within cells, ER represents the critical storage site for Ca^{2+} storage, release, and buffering, although there is some debate as to whether there is a specialized compartment that fulfils this role. Other specialized structures that may be involved in calcium storage include Golgi, mitochondria, granules, and nucleus. Early studies utilized permeabilized cells and subcellular fractionation techniques to identify and investigate intracellular Ca^{2+}-pools [4, 37–39]. Although such procedures can be very useful to address certain questions, they are also prone to a number of significant artifacts, including redistribution of Ca^{2+} and fragmentation of intracellular organelles. Here, we focus on protocols with intact cells that can dissect the roles of each of the major Ca^{2+} buffering organelles in calcium signaling, buffering, and storage. By focusing on the $[Ca^{2+}]_i$ transient in intact cells, one can minimize the artifacts outlined above, and manipulate this calcium release phase to describe mechanism(s) and Ca^{2+}-pools by which an agonist effects a release under more physiological conditions.

2.2.3.1 Manipulating Intracellular Ca^{2+}-Pools in Intact Cells

In developing experimental approaches for defining intracellular Ca^{2+} pools involved in intact cells, one must consider the potential organelles involved in Ca^{2+} signaling, primarily ER and mitochondria, and the pharmacological tools for probing these organelles.

ER: Within cells, ER plays several key roles in the process of regulating $[Ca^{2+}]_i$ and Ca^{2+} signaling. It is the major calcium-storing organelle and primary site of action for IP_3-induced Ca^{2+}-mobilization. It plays a critical role as a Ca^{2+}-buffer with Ca^{2+}-ATPases (SERCA pumps) that can rapidly sequester Ca^{2+} ions from the cell cytoplasm to prevent untoward changes in $[Ca^{2+}]_i$, and to replenish ER Ca^{2+} following receptor activation. In the process of Ca^{2+} signaling, it is the depletion of ER Ca^{2+} that, in some way, is coupled to an activation of Ca^{2+} entry across the plasma membrane, i.e., CCE.

In defining these roles for ER function and how they relate to agonist-activated Ca^{2+} signaling, a number of pharmacological reagents are available that target and inhibit the function of the ER SERCA Ca^{2+}-pumps. These reagents, which include thapsigargin, cyclopiazonic acid, and *tert*-butyl benzohydroquinone [13, 40–42], inhibit the SERCA Ca^{2+}-pumps responsible for storing intracellular calcium in the IP_3-sensitive pool, and make it possible to deplete this pool of its Ca^{2+} without stimulating the formation of inositol phosphates associated with agonist activation [31, 41]. As illustrated in Figure 2.3, treating intact cells with such reagents (thapsigargin in this case) leads to an elevated and sustained $[Ca^{2+}]_i$ signal (Figure 2.3A) that, like agonist activation (Figure 2.4A), is also a biphasic Ca^{2+} response involving mobilization of intracellular Ca^{2+} and Ca^{2+}-entry across the plasma membrane (Figure 2.3B).

Conveniently, these SERCA pump inhibitors are membrane permeant, providing a noninvasive technique for manipulating ER Ca^{2+}-pools in intact cells. In some cell types, the effects of thapsigargin and cyclopiazonic acid appear dose dependent [43]. In others, such as mouse lacrimal acinar cells (Figure 2.3C), concentrations of thapsigargin ranging from $20\,nM$ to $2\,\mu M$ all result in a maximally sustained and elevated $[Ca^{2+}]_i$ signal. However, the time with which this sustained level is attained is significantly delayed at the lower concentrations of thapsigargin, and likely reflects concentration dependence of the on-rate of this extremely high-affinity SERCA inhibitor.

As mentioned above, ER represents a major site of action for IP_3, and there are a variety of useful pharmacological tools for manipulating IP_3-induced calcium release [13]. The majority of these reagents have limited use for studies in intact cells as they require specialized, invasive techniques such as microinjection [35] or patch clamp (Chapter 6). There is the potential, though, to activate IP_3-receptors in intact cells using cell-permeant IP_3 analogs, which are designed along the same principles for loading calcium indicators into intact cells [44, 45]. Alternatively, IP_3 receptor function can be blocked by Xestospongin C, a compound that is membrane permeable and reasonably selective [46].

Mitochondria: In unstimulated cells, mitochondria do not appear to represent a significant Ca^{2+}-pool since they are incapable of accumulating significant amounts of Ca^{2+} under these resting conditions [4, 37]. However, once the threshold for uptake is reached ($\sim1\,\mu M$), the driving force for Ca^{2+} accumulation is substantial, and when $[Ca^{2+}]_i$ is elevated for a prolonged period of time, mitochondria can accumulate Ca^{2+} to remarkable concentrations. In addition, during relative brief periods of Ca^{2+}-release, it is now clear that mitochondria accumulate substantial Ca^{2+}, albeit transiently [47, 48]. Recent improvements in techniques using mitochondrialy targeted Ca^{2+}-sensitive proteins, based on aequorin and GFP, or with BAPTA-based dyes that accumulate in mitochondria, have provided further insight into how mitochondria accumulate Ca^{2+} rapidly and efficiently during periods of $[Ca^{2+}]_i$ elevation.

Indeed, it is suggested that mitochondria may play a significant physiological role as a Ca^{2+}-buffer through a close apposition of mitochondria with ER Ca^{2+}-release channels, and possibly Ca^{2+} channels in the plasma membrane. In these

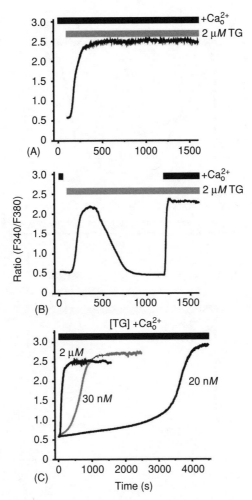

FIGURE 2.3 Thapsigargin-induced calcium signaling in mouse lacrimal acinar cells. Single mouse lacrimal cells attached to glass cover slips were loaded with calcium indicators fura-2 and cytoplasmic $[Ca^{2+}]_i$ measured at the single cell level as described by Bird et al. [35]. In (A) cells were treated with thapsigargin ($2 \mu M$) in the continued presence of extracellular Ca^{2+} (1.8 mM). In (B) a biphasic calcium response was observed in a thapsigargin-treated cell using the calcium re-addition protocol described in Figure 2.2. In (C) the effect of low concentrations of thapsigargin on the kinetics of the calcium response was monitored in the presence of extracellular Ca^{2+}. All data are representative trace observations made on single cells.

areas, local $[Ca^{2+}]_i$ could be higher than that measured in the bulk of the cytoplasm (reported by fluorescent indicators) and so above the threshold for uptake into the mitochondria [49–51]. Thus, mitochondria may play a role in shaping the spatio-temporal Ca^{2+} response during physiological receptor activation, or protecting the

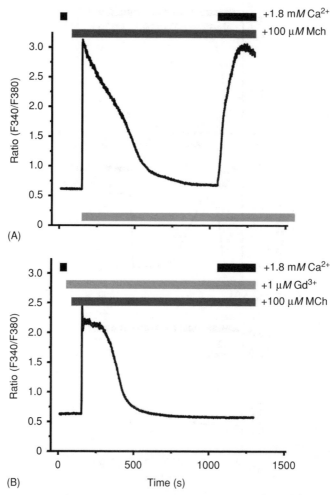

(A)

(B)

FIGURE 2.4 Agonist-induced calcium signaling in mouse lacrimal acinar cells. Single-mouse lacrimal cells attached to glass cover slips were loaded with calcium indicators fura-2 and cytoplasmic $[Ca^{2+}]_i$ measured at the single cell level as described by Bird et al. [35]. In (A) a biphasic calcium response was observed in a methacholine (MCh, $100\,\mu M$) treated cell using the calcium re-addition protocol described in Figure 2.2. (B) demonstrates the effect of low concentrations of Gd^{3+} ($1\,\mu M$) on the Ca^{2+}-entry phase. All data are representative traces observations made on single cells.

cell from the deleterious effects of prolonged $[Ca^{2+}]_i$ elevation. In addition, mitochondria function as targets of Ca^{2+} signaling because of the regulation of Ca^{2+}-sensitive enzymes in the mitochondrial matrix.

Mitochondria principally accumulate Ca^{2+} by a uniporter, and not a specific Ca^{2+}-ATPase, which can be blocked with ruthenium red [52–54]. Alternatively, as this is an electrogenic process, it can be blocked by drugs that either interfere with electron transport, such as antimycin, or collapse the proton gradient, such as protonophore uncouplers. In contrast, the release of Ca^{2+} occurs by a neutral exchange process (either a Ca^{2+} for $2H^+$ or $2Na^+$) that can be blocked by CGP37157 [55].

While in principle it should be possible to discern roles for mitochondria both as a Ca^{2+}-pool and a Ca^{2+}-buffer using the tools described above in intact cells, the Ca^{2+}-pools protocol outlined below offers the best way to define mitochondrial (and ER) Ca^{2+}-pools in an intact cell, particularly when only monitoring the $[Ca^{2+}]_i$ signal with fluorescent calcium indicator.

Ca^{2+} Ionophores: The carboxylic acid antibiotics, A23187 [56] and ionomycin [57] act as freely mobile carriers to transport divalent but not monovalent cations, and have proven most useful in manipulating the movement of Ca^{2+} ions and intracellular Ca^{2+}-pools in intact cells. These ionophores are lipid-soluble molecules that complex alkali metal cations, forming a neutral metal–antibiotic species that transports cations across a variety of membranes. With fluorescent calcium indicators excited in the UV range (namely fura indicators) a potential artifact when using A23187 is that it can contribute significant autofluorescence to the measured signal. However, this problem is avoided by using 4Br-A23187, a nonfluorescent derivative [58].

In general, ionomycin is preferred over A23187 for calcium signaling studies as it performs better in the transport of Ca^{2+} ions, and does not contribute to autofluorescence. Ionomycin is more selective for Ca^{2+} compared to A23187, and a more effective mobile ion carrier (ionomycin:Ca^{2+} stoichiometry is 1:1 compared to 2:1 for A23187). Interestingly, ionomycin does not significantly bind Sr^{2+} or Ba^{2+}, although there is some indication of La^{3+} binding [57].

Treatment of intact cells with ionomycin at high concentrations ($1–10\,\mu M$) is able to increase the permeability of the plasma membrane and other intracellular organelles, including ER and mitochondria. However, in resting cells in the absence of extracellular calcium, the effect of ionomycin on intracellular Ca^{2+}-release may reflect only the agonist-sensitive Ca^{2+}-pool, especially since the releasable Ca^{2+} content of mitochondria is generally negligible in resting cells [22]. When ionomycin is used in this way, it is important that the bathing medium contains a Ca^{2+}-chelator ($\sim100–500\,\mu M$ BAPTA), since the contaminating levels of Ca^{2+} in a nominally Ca^{2+}-free buffer would be sufficient for ionomycin to elevate $[Ca^{2+}]_i$. The rapid release of intracellular Ca^{2+}-pools by ionomycin makes it particularly suitable for getting a measure of the agonist-sensitive Ca^{2+}-pools, since the peak of the response should be minimally affected by plasma membrane pumps or release channels, especially as compared to the much more slowly acting thapsigargin [19].

In contrast, low concentrations of ionomycin appear to selectively partition into intracellular membranes and to act specifically by releasing intracellular Ca^{2+}-stores ($ED_{50} = 50\,nM$) without greatly increasing the permeability of the plasma membrane to extracellular calcium ([59]; $100\,nM$ could completely abolish TRH-induced Ca^{2+}-release).

As indicated above, ionophores can also be used to manipulate mitochondrial Ca^{2+}, although some caution is necessary. With A23187, for example, initial studies with this ionophore were performed in isolated rat liver mitochondria, where it appeared to have the additional effect of uncoupling oxidative phosphorylation and inhibiting ATP synthesis [56]. However, Abramov and Duchen [60] have characterized the effects of ionophores in intact cells indicating appropriate ways to manipulate mitochondrial Ca^{2+}, in particular using the ionophore ferutinin.

2.2.3.2 Protocol for Defining Intracellular Ca^{2+}-Pools

As outlined above, a key issue when describing the anatomy of a Ca^{2+} signaling process is defining the intracellular sources of Ca^{2+} that are mobilized during surface membrane receptor activation. Conceptually, one wants to achieve this in an intact cell to avoid any problems associated with organelle fragmentation, such as when cells are permeabilized or fractionated. In our experience, the three primary tools available to help define these Ca^{2+}-pools intact cells are the receptor agonist itself, SERCA pump inhibitors, and calcium ionophores (primarily ionomycin). By performing a series of sequential treatments with these reagents, and in the absence of $[Ca^{2+}]_o$ (to prevent complications due to Ca^{2+} entry), one can build a picture of the Ca^{2+}-pools in resting cells, and the portion of these Ca^{2+}-pools that are sensitive to mobilization by agonist activation [61, 62]. The data in Figure 2.5 illustrate this protocol with fluo-4 loaded HEK 293-V1a cells, and with all calcium measurements performed in parallel using an $FLIPR^{384}$ microplate reader (see Chapter 3).

In Figure 2.5A, the addition of ionomycin leads to a rapid and transient intracellular Ca^{2+}-release, and the subsequent treatment of the same cells with thapsigargin results in no further release of Ca^{2+}. As illustrated in Figure 2.5D, this would indicate that the intracellular Ca^{2+}-pools sensitive to thapsigargin are entirely contained within the ionomycin-sensitive pool. In Figure 2.5B, cells are first treated with thapsigargin, then a maximal agonist concentration (100 nM Vp), and finally ionomycin. In this sequence, thapsigargin causes a transient rise in $[Ca^{2+}]_i$, but no effect of the agonist is observed. However, ionomycin is still able to cause a release of intracellular Ca^{2+}. This would suggest that the agonist-sensitive pool is contained entirely within the thapsigargin-sensitive pool (i.e., ER), while the ionomycin-sensitive pool is larger, containing elements sensitive to neither thapsigargin nor agonist (Figure 2.5E). The most likely candidate for the extra ionomycin-sensitive elements is the mitochondria, although this is not rigorously established by this procedure. Finally, treating the cells first with the agonist, followed by thapsigargin and ionomycin (Figure 2.5C), shows that the agonist-sensitive Ca^{2+}-pools are a subset of the thapsigargin-sensitive Ca^{2+}-pools (compare Figure 2.5B and F to Figure 2.5C and E).

There are, however, several caveats to the data in Figure 2.5 that are subject to both the experimental conditions and technique used in the study. One is that the agonist-sensitive Ca^{2+}-pools as determined in this way do not necessarily correspond to the entire IP_3-sensitive pool. This could only be obtained if PLC-activation produced sufficient IP_3 to saturate all cellular IP_3 receptors. Second, the data shown in Figure 2.5 were performed on a population of HEK 293-V1a cells. Performing

Intracellular calcium pools: cell population measurement

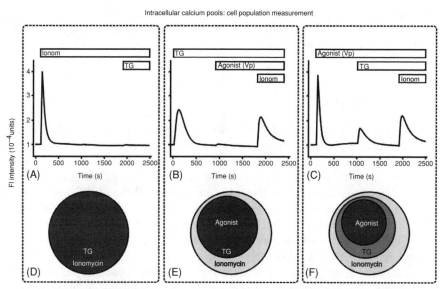

FIGURE 2.5 Examination of intracellular calcium pools in cell populations of HEK 293-V1a cells loaded with fluo-4. HEK 293-V1a cells were plated in 96-well microplates (cell density; 20,000 cells per well) 24 h before the cells were loaded with fluo-4, and calcium measurements performed with an FLIPR[384] as described in Trebak et al. [24]. Before starting the experiment, cells were bathed in medium nominally Ca^{2+}-free and supplemented with 500 μM BAPTA. In (A) cells were sequentially treated with ionomycin (Ionom, 10 μM) followed by thapsigargin (TG 2 μM). In (B) cells were sequentially treated with thapsigargin (TG 2 μM) followed by vasopressin (Vp, 100 nM), and then ionomycin (Ionom, 10 μM). In (C) cells were sequentially treated with vasopressin (Vp, 100 nM) followed by thapsigargin (TG 2 μM), and then ionomycin (Ionom, 10 μM). The figures in (D), (E), and (F) are a representation of how the intracellular Ca^{2+}-pools demonstrated in (A–C) potentially overlap (see text for description). The data are representative traces from a single experiment. Each trace is an average of data from four wells of a 96-well microplate.

this same experiment at the single cell level, the combined data from the entire population of single cells yield data similar to that shown in Figure 2.5C. However, a more detailed analysis reveals a subpopulation of cells where agonist activation appears to completely deplete the thapsigargin-sensitive Ca^{2+}-pool. Thus, the population analysis would suggest HEK 293-V1a cells contain an agonist-sensitive Ca^{2+}-pool that is a subset of the ER (thapsigargin-sensitive Ca^{2+}-pool), and that the residual ionomycin-sensitive Ca^{2+}-pool may represent an additional Ca^{2+}-pool, possibly mitochondria, which can be manipulated [61, 63]. In contrast, single cell analysis would suggest that in a proportion of the cells the entire ER Ca^{2+}-pool is agonist-sensitive (Figure 2.5E). Thus, it is possible that maximal receptor activation is unable to generate sufficient levels of second messengers, particularly IP_3, to reveal the full capacity of the thapsigargin-sensitive Ca^{2+}-pool, except in 20% of the cells. To test this idea, one could circumvent receptor activation and directly introduce IP_3 (or other second messengers) into the cytoplasm (i.e., [61]). Note that in a previous study with lacrimal acinar cells, maximal activation of PI-PLC-coupled muscarinic receptors was capable of discharging the entire thapsi-gargin-sensitive Ca^{2+}-pool [61].

2.3 Ca^{2+}-ENTRY

Following the release of store Ca^{2+} associated with activation of the PI-PLC pathway, there is generally a second, more sustained phase of Ca^{2+} signaling associated with Ca^{2+}-entry across the plasma membrane. In most, but not all instances, this second phase of calcium entry is attributed to CCE, a process not regulated by the direct actions of IP_3 on the plasma membrane but, rather, as a result of the depletion of calcium ions from an intracellular store by IP_3. That is, the empty calcium-storage organelle produces a signal for calcium ion entry across the plasma membrane. The nature of the retrograde signal and the target plasma membrane channel mediating CCE remain poorly defined. However, perhaps the clearest demonstration of the process of capacitative or store-operated entry comes from experiments utilizing thapsigargin (Figure 2.3) (a reagent that inhibits SERCA pumps responsible for storing intracellular calcium in the IP_3-sensitive pool). Thus, this reagent made it possible to deplete the agonist-sensitive intracellular Ca^{2+}-pool without stimulating receptors, G-proteins, or the formation of any inositol phosphates [31, 41].

Numerous reports have demonstrated that CCE across the plasma membrane is activated by treatment with thapsigargin, and the related agents cyclopiazonic acid and *tert*-butyl benzohydroquinone [40, 64]). However, there is the possibility in nonexcitable cells that receptor activation might activate additional pathways of Ca^{2+}-entry not related to CCE, i.e., non-CCE pathways [65–68]. Indeed, it has been suggested that receptor activation under more physiological conditions preferably activates a non-CCE pathway mediated by the actions of arachidonic acid [69]. Outlined below are ways in which Ca^{2+}-entry can be observed in intact cells using fluorescent Ca^{2+}-indicators, and pharmacological ways in which to distinguish the properties of Ca^{2+}-entry pathways that are CCE or non-CCE.

2.3.1   MONITORING Ca^{2+}-ENTRY

Since complex mechanisms of Ca^{2+} regulation can occur in response to agonist stimulation, experiments using Ca^{2+} ions to monitor Ca^{2+}-entry can lead to a number of potential artifacts. For example, activation of plasma membrane pumps by Ca^{2+} and feedback inhibition of the store-operated channels by Ca^{2+} can limit the extent of the $[Ca^{2+}]_i$ rise. Thus, a manipulation that alters steady-state Ca^{2+} accumulation might reflect changes in channel permeation, or alternatively changes in the ability of pumps to remove or sequester Ca^{2+}. The use of nontransported surrogates such as Mn^{2+} or Ba^{2+}, discussed below, greatly simplifies the analysis in such instances. Ideally, investigation of transmembrane ion fluxes can be studied under the most tightly controlled conditions by measuring membrane current with the patch-clamp technique. However, this is not always practical, and in some instances, the influx of Ca^{2+} into cells may be below the level of detection by electrophysiological techniques, yet readily detected by the more sensitive fluorescent indicators [70].

The basic approach is to separate the two phases of calcium mobilization as described in the calcium re-addition protocol. At the point of restoring $[Ca^{2+}]_o$, manipulations of the bathing solution, including cation substitutions, can enhance or distinguish Ca^{2+}-entry pathways on the basis of their pharmacological properties. As the molecular identity of the Ca^{2+}-entry channels are not known, the availability of specific pharmacological tools is somewhat limited. However, the behavior of the calcium signaling pathways and the fluorescent indicators toward various cations offers some degree of experimental flexibility, summarized in Table 2.2. More information is available for fura-2 due to its extensive use in Ca^{2+}-studies, however reference to other calcium indicators will be made when possible.

2.3.1.1　Mn^{2+}-Quench (i.e., Mn^{2+}-Entry)

Two properties have made Mn^{2+} a useful ion for studying Ca^{2+}-entry: (i) Mn^{2+} will quench fura-2; (ii) Mn appears to substitute for Ca^{2+} in a number of Ca^{2+} entry pathways, including CCE [71]. Fura-2 has high affinity for Mn^{2+} [72–74], which irreversibly binds and quenches fluorescence. Quenching of the fura-2 fluorescence can be monitored at its isobestic wavelength (~360 nm), or by an algebraic summation of fluorescence values [75, 76]. Like fura-2, Mn^{2+} will also quench fura-5F. However, the single wavelength indicators fluo-3 and fluo-4 are not quenched by Mn^{2+}. In fact, the signal of fluo-3 fully bound with Mn^{2+} ions is equivalent to the fluo-3 signal in the presence of 100 nM Ca^{2+} [28, 29]. In experiments utilizing fura-2 and Mn^{2+}, it was demonstrated that PI-PLC-linked agonists could enhance the quench of fura-2-loaded cells [77–79]. In quantifying this process, initial rates of quench are most useful, and it is also important to account for basal Mn^{2+}-quench in the absence of cell stimulation.

There are some potentially useful aspects of the Mn^{2+}-quench technique. One is the ability to monitor Ca^{2+}-release events and Mn^{2+}-quench (i.e., Mn^{2+}-entry) simultaneously; i.e., it is not necessary to separate these two phases as described by the calcium re-addition protocol. Such an approach allows a temporal comparison of intracellular Ca^{2+}-release and Ca^{2+}-entry events. For example, Shuttleworth [75] measured the coordination between Ca^{2+}-pool depletion and Ca^{2+}-entry, and observed a delay following Ca^{2+}-pool depletion before Ca^{2+}-entry (Mn^{2+}-quench). Also, Loessberg et al. [80] suggested that oscillations in Ca^{2+}-entry occurred during $[Ca^{2+}]_i$ oscillations activated by an agonist. Observations by Orrenius and colleagues [65, 81] also used the Mn^{2+}-quench to suggest that some agonists in hepatocytes used a mode of Ca^{2+}-entry distinct from CCE.

One potential artifact with Mn^{2+}-quench occurs when there is significant calcium indicator compartmentalization. Specifically, it was discovered that while Mn^{2+} was not a substrate for intracellular SERCA pumps, it could easily pass through activated IP_3 receptors in a retrograde manner [21]. Thus, with fura-2 sequestered in ER, Mn^{2+}-quench of that indicator only occurred in the presence of IP_3. An investigator may be misled, therefore, when the rate of Mn^{2+}-quench reflects movements into intracellular organelles rather than flux across the plasma membrane. Certainly, this can be avoided by minimizing indicator compartmentalization, for example, by microinjecting the free acid form of fura-2, or an indicator linked to membrane impermeant polymers (Chapter 1). On the other hand, this behavior can sometimes be used as a method for assaying the activity of IP_3-activated receptors *in situ* [82].

2.3.1.2 Sr^{2+}-Entry

Sr^{2+} can fully substitute for Ca^{2+} by rendering an excitation spectrum with fura-2, and supporting CCE and other Ca^{2+}-mediated events [17]. Unlike Ba^{2+}, Sr^{2+} is a substrate for SERCA pumps and thus is able to refill intracellular Ca^{2+}-pools [83, 84] that can be released by a subsequent receptor activation [17]. The observed K_D for Sr^{2+} (and Ba^{2+}) in supporting CCE is higher than for Ca^{2+}, although one should bear in mind that this may also reflect different affinities of these ions for fura-2 (see table in [17]). However, while this makes quantifying Sr^{2+} and Ba^{2+} levels difficult to quantitate, it does not preclude their use for qualitative analysis of underlying Ca^{2+}-homeostatic processes. In general, Ba^{2+} is more useful as a surrogate for Ca^{2+} because it is not transported into the ER and does not substitute well for Ca^{2+} in activating Ca^{2+}-activated channels (discussed below). Sr^{2+} on the other hand is pumped by both SERCA and plasma membrane pumps, and thus offers little advantage over Ca^{2+} itself. However, one group has successfully used Sr^{2+} entry to distinguish divalent cation entry through TRPC channels as compared to endogenous CCE channels [85].

2.3.1.3 Ba^{2+}-Entry

A particularly useful approach for monitoring and characterizing Ca^{2+}-entry mechanisms involves the substitution of Ba^{2+} ions for Ca^{2+}. In general, Ba^{2+} is able to substitute for Ca^{2+} in passive biological processes [86], including permeating Ca^{2+}-permeable channels, and in rendering an excitation spectrum with fura-2. On the other hand, Ba^{2+} is generally incapable of mimicking the actions of Ca^{2+} in active processes involving highly specific recognition sites. For example, Ba^{2+} does not enter ER Ca^{2+}-pools as it is a poor substrate for Ca^{2+}-pumping ATPases [16, 17, 87]. Thus, unlike Ca^{2+}, Ba^{2+} can give a measure of plasma membrane permeability without complications due to buffering and the consequences of Ca^{2+}-pool refilling, which will inactivate CCE (to some extent, these properties are true for Mn^{2+}-quench). As with Mn^{2+}-quench, quantifying initial rates of Ba^{2+} entry are recommended, as well as taking into account basal rates of entry in the absence of cell stimulation. In fact, in the case of Ba^{2+} entry, utilization of initial rates is essential because the net accumulation of Ba^{2+} seems to be limited by undefined factors unrelated to membrane permeability.

An interesting example of possible artifacts due to altered Ca^{2+} transport was encountered when studying Ca^{2+} signaling in HEK-293 cells overexpressing

TRPC3, where capacitative and noncapacitative mechanisms were operating simultaneously [24]. The problem was that in thapsigargin-treated cells, and under conditions where CCE was blocked, a transient rise in $[Ca^{2+}]_i$ was still observed in the presence of extracellular calcium that had previously been interpreted as reflecting store-operated regulation of TRPC3 channels [88]. However, expression of TRPC3 in these cells also increased the basal entry of Ca^{2+}-ions in the absence of cell stimulation, suggesting a constitutive activity for TRPC3 channels. Thus, the effect of thapsigargin described above might easily reflect the constitutive activity of TRPC3 channels, which is exaggerated by the inability (due to thapsigargin inhibition of SERCA pumps) of intracellular endoplasmic reticulum to efficiently buffer entry through these constitutive channels. However, substituting Ba^{2+} for Ca^{2+} in these experiments avoids this problem, and the Ba^{2+} entry data indicated that the apparent effect of thapsigargin in activating entry of Ca^{2+} through TRPC3 channels results from thapsigargin-induced reduction of intracellular Ca^{2+} buffering by endoplasmic reticulum rather than an actual store-dependent regulation of TRPC3 [24].

There are some potential artifacts when using Ba^{2+} as a substitute for Ca^{2+}. Ba^{2+} can significantly contribute to autofluorescence that may need to be quantitated and subtracted from component intensities [17]. Also, Ba^{2+} could potentially depolarize membrane potential through its ability to block inwardly rectifying K^+ channels (in bovine aortic endothelial cells, for example [89]), with the consequence of a reduction in driving force for cation entry. The influence of membrane potential and ways to account for or minimize its effects will be discussed later in the chapter. Briefly though, if there are concerns that membrane depolarization is influencing the observations with Ba^{2+} (or with any cation), it is recommended to remove this influence by studying the regulation of Ba^{2+}-entry in cells that are already fully depolarized [16, 67, 90], or with membrane potential controlled by patch-clamp (Chapter 6).

2.3.2   Manipulating Ca²⁺-Entry

One of the most common methods of manipulating Ca^{2+}-entry is by judicious use of pharmacological inhibitors, ideally ones with high specificity for a particular channel or class of channels. This approach has been discussed in a number of recent reviews [13, 91].

2.3.2.1   Lanthanides

Lettvin et al. predicted that despite La^{3+} and Ca^{2+} having similar hydrated radii (3. 1 ′Å vs. 2. 8 ′Å, respectively), the higher valence (charge density) of La^{3+} would cause it to bind more strongly to Ca^{2+}-binding sites, which could include membrane sites involved in the movement of Ca^{2+} ions across the plasma membrane [92]. Based on these predictions, Van Breemen et al. developed the use of lanthanides to block Ca^{2+}-entry and efflux [93]. Fortunately, the effects of lanthanides on Ca^{2+}-entry and the plasma membrane Ca^{2+}–ATPase (PMCA) can be dissociated due to the different sensitivities of these two processes. For example, Gd^{3+} is particularly suited to this task as it blocks CCE at low concentrations (~1 μM) (Figure 2.4B) [66], and only begins to block PMCA activity above 100 μM [19, 20]. With regard to Gd^{3+} block of CCE, it is suggested that this is irreversible [66], but this may vary between cell types [94]. Nonetheless, at low concentrations, Gd^{3+} inhibition is proving to be a useful pharmacological tool as it appears specific for CCE rather than non-CCE pathways [13, 19, 20, 24, 66].

At concentrations of Gd^{3+} at or above 1 mM, PMCA activity appears almost completely blocked, leading to a condition where the cell cytoplasm is isolated or "insulated" from the extracellular space [19, 93]. Since the "gadolinium insulation" blocks both the entry and efflux of Ca^{2+} ions, it is possible to investigate complex intracellular calcium signaling events independently of contributions made by constituents in the extracellular space [19].

A potential complication in using La^{3+} ions is that they are able to render an excitation spectrum with fura-2. Fortunately, in intact cells, La^{3+} poorly traverses the plasma membrane, if at all, which would also suggest that the La^{3+} block of PMCA is via an external site on the pump, not on the cytoplasmic face [17]. Also, any fura-2 signal associated with La^{3+} in unstimulated cells would be indicative of "leaky" or unhealthy cells. Care should be exercised when using La^{3+} in combination with extracellular BAPTA or EGTA, as they will bind lanthanides with high affinity [66]. However, chelation of Gd^{3+}/La^{3+} with BAPTA/EGTA may offer a convenient way to reverse their blocking effects.

2.3.2.2   2-APB (2-Aminoethoxydiphenyl Borane or 2-Aminoethyldiphenyl Borate)

2-APB, originally shown to be a membrane permeable inhibitor of IP_3 receptors [95], was subsequently shown to be a somewhat selective inhibitor for CCE over some non-CCE pathways [20, 24, 96, 97]. This appears to be an extracellular effect occurring at the plasma membrane [98–101], and independently of the IP_3 receptor [102]. Unfortunately, the effects of 2-APB are even more complex than first appreciated with observations that it can block some nonstore-operated channels (e.g., the magnesium-inhibitable cation channel [98, 103, 104]) or can activate a distinct ion channel activity depending on the concentration of 2-APB [105–107]. Thus, a careful characterization of 2-APB on Ca^{2+}-signaling is necessary.

2.3.3   REGULATION OF Ca^{2+}-ENTRY

Agonist-induced changes in $[Ca^{2+}]_i$ levels observed in cells loaded with fluorescence Ca^{2+}-indicators depend on a myriad of factors, including the underlying Ca^{2+} signaling processes and the extent of receptor activation, all of which may vary depending on the cell type used. As discussed above, stimulation of PI-PLC linked receptors activates a cascade of signaling events that elevates $[Ca^{2+}]_i$ through the coordinated activity of intracellular Ca^{2+}-release, Ca^{2+}-entry, and Ca^{2+}-pumps. Under physiological conditions, this coordinated activity results in complex changes in $[Ca^{2+}]_i$, or Ca^{2+} oscillations that result from a series of feedback mechanisms that serve to regulate intracellular Ca^{2+}-release (Chapter 11) and

Ca^{2+}-entry [108]. Fortunately, an understanding of the regulation of the Ca^{2+}-entry process has been facilitated by the electrophysiological characterization of Ca^{2+} currents associated with CCE (I_{crac}) [109, 110] (Chapter 6), and have indicated that their regulation is complex [111]. In particular, their regulation involves multiple mechanisms of both negative and positive feedback by Ca^{2+} [112–114], many of which are manifest with studies in intact cells with fluorescent Ca^{2+} indicators [115, 116]. Another complication for CCE measurements in intact cells is that elevated $[Ca^{2+}]_i$ levels can stimulate Ca^{2+}-activated channels that, in turn, influence membrane potential and thus the driving force for Ca^{2+}-entry. For example, Ca^{2+}-activated Cl^- channels can hyperpolarize cells and increase Ca^{2+} influx by providing electrical driving force [117]. In this section, we will focus on ways to monitor the CCE with fluorescent Ca^{2+} indicators that can avoid the potential impact of Ca^{2+}-dependent feedback mechanisms that otherwise might prevent or underestimate a full appreciation of its activity. In addition, we will also address the influence of membrane potential on fluorescence measurements of Ca^{2+} entry.

2.3.3.1   Feedback Inhibition of CCE by $[Ca^{2+}]_i$

The obvious way to avoid feedback regulation of CCE by $[Ca^{2+}]_i$ is to avoid or prevent a rise in $[Ca^{2+}]_i$ and still be able to monitor the Ca^{2+} entry process with a fluorescence indicator. In principle, this can be achieved by substituting surrogate ions for Ca^{2+}, primarily Mn^{2+} and Ba^{2+} (discussed above). Mn^{2+} entry is detected by quench of fluorescent Ca^{2+} indicators, specifically the fura indicators (fura-2 and fura-5F, for example). With Ba^{2+}, while it can substitute for Ca^{2+} in some Ca^{2+}-dependent processes (see above), it is useful in this context as it does not appear to negatively feedback on this type of channel [112]. Both Mn^{2+} and Ba^{2+} should provide a good measure of the unidirectional influx activity by avoiding Ca^{2+}-dependent feedback, and also because both ions are poor substrate for intracellular Ca^{2+}-ATPases.

In focusing on the regulation of CCE, the use of thapsigargin and other SERCA pump inhibitors is preferred to activate this process as this bypasses the majority of signaling events downstream of receptor activation. However, whether using thapsigargin or an agonist, each can cause a transient rise in $[Ca^{2+}]_i$ through the mobilization of Ca^{2+} from intracellular Ca^{2+}-pools. Obviously, this transient rise in $[Ca^{2+}]_i$ could influence the Ca^{2+} entry process, even if Ca^{2+}-release and Ca^{2+}-entry phases are temporally separated by the calcium readdition protocol. It is also possible that while measurements of bulk fluorescence Ca^{2+}-signals indicate $[Ca^{2+}]_i$ levels are low, standing $[Ca^{2+}]_i$ gradients exist with possibly microdomains of high $[Ca^{2+}]_i$ close to the Ca^{2+} channel of interest. One approach to minimize elevation in $[Ca^{2+}]_i$ due to release is to use low concentrations of thapsigargin. Low and high concentrations of thapsigargin result in the same extent of CCE; however, low thapsigargin takes a significantly longer period of time to achieve this (Figure 2.3C). Experiments performed in the absence of extracellular Ca^{2+} also demonstrate that these same low concentrations of thapsigargin result in the same extent of intracellular Ca^{2+}-pool depletion, but this occurs without a detectable rise in $[Ca^{2+}]_i$ [64]. Again, a combination of low thapsigargin with ion substitution should provide a measure of CCE in the absence of a global rise in $[Ca^{2+}]_i$.

2.3.3.2 Membrane Potential

Receptor activation of Ca^{2+}-entry in nonexcitable cells generally results from IP_3-mediated Ca^{2+} store depletion, or from activation of second-messenger-regulated channels rather than through voltage activation [118]. However, while membrane depolarization itself does not activate Ca^{2+}-entry in nonexcitable cells, regulation of Ca^{2+}-entry by plasma membrane potential is a widely recognized phenomenon [117]. Ca^{2+} entry through store-operated and other Ca^{2+}-permeable channels is an electrogenic process and thus is driven by a combination of chemical (concentration gradient) and electrical (membrane potential) forces. Thus, contrary to the case of voltage-activated Ca^{2+} channels, for nonvoltage-activated channels, less calcium enters upon depolarization and membrane hyperpolarization promotes calcium influx [119–122], a relationship elegantly demonstrated by Kamouchi et al. [123]. Indeed, the suggestion that membrane depolarization modulates the electrochemical gradient for Ca^{2+} is further reinforced by the ability to overcome the effects of depolarization by increasing $[Ca^{2+}]_o$ [124, 125], and Schilling et al. [16] demonstrated that the observed effect of membrane depolarization is to shift the Ca^{2+}-entry process to one that has a lower apparent affinity for Ca^{2+}. Unfortunately, the "potential dependence" of the Ca^{2+}-entry process may present a problem for Ca^{2+}-signaling studies that depend solely on fluorescent Ca^{2+}-indicators. That is, unless the membrane potential of the cells is not considered or controlled, the effect of any pharmacological agent that modulates a Ca^{2+}-entry signal must be interpreted with caution. This is especially true of drugs or biochemical manipulations that inhibit entry, as any number of toxic insults to a cell might damage the plasma membrane, causing depolarization. Furthermore, it is important to remember that while the term "nonexcitable" describes cell types that lack voltage-gated Ca^{2+} channels, this does not mean that receptor activation does not trigger membrane potential changes in these cell types [122, 123, 126, 127].

Clearly, the most straightforward way to avoid these complications is to combine fluorescence measurements of transmembrane Ca^{2+} fluxes with measurements of membrane current using the patch-clamp technique with concomitant control of membrane potential (Chapter 6). However, as argued in a recent review [70], this is not always practical, as Ca^{2+} currents, and store-operated currents in particular may, in some cell types, fall near or below the limits of detection by electrophysiological means. In such instances, it may be advisable to combine patch-clamp, to control membrane potential, with the use of fluorescent indicators to monitor cation movements. Alternatively, one can minimize these problems by carrying out experiments whereby the plasma membrane is fully depolarized [16, 67, 90], essentially a "poor man's voltage-clamp." As the driving force for Ca^{2+} entry will be reduced, smaller $[Ca^{2+}]_i$ signals can be expected, and it may be advisable to increase extracellular Ca^{2+} (or Ba^{2+}, etc.). Membrane depolarization can be achieved by replacing Na^+ with K^+ in the bathing solution [16, 121] or treating the cells with gramicidin

D (10^{-7} to $10^{-8} M$), a pore- forming ionophore that exchanges extracellular Na^+ for intracellular K^+ [122]. Pittet et al. [127] compared these approaches and recommend gramicidin D over K^+ substitution. A final note of caution: a number of studies have indicated that depolarization and/or raising extracellular K^+ concentration may have effects unrelated to the reduction of the driving force for Ca^{2+}-entry [19, 127], and may interfere with steps upstream of intracellular Ca^{2+}-release [122, 128, 129] including IP_3 production [127].

2.3.3.3 CCE and Excitable Cells

Functionally, distinguishing between excitable and nonexcitable cells can be complicated as many excitable cell types possess the receptor-regulated PI-PLC pathway and CCE-mediated Ca^{2+}-signaling. Thus, it is important to sort the contribution of various Ca^{2+}-entry pathways to the observed receptor-activated Ca^{2+}-signaling process, for example, the relative contributions of CCE, non-CCE, and voltage-dependent channels. Antagonists for L-type Ca^{2+} channels (dihydropyridines) can be useful in this approach, just as Gd^{3+} has proven useful in distinguishing CCE and non-CCE. In one example, the rat pancreatoma cell line, AR4-2J, expresses voltage-sensitive L-type Ca^{2+} channels. However, it was possible to discount a role for this channel in receptor-mediated (Substance P, Bombesin, and muscarinic activation) and thapsigargin-mediated entry as they were insensitive to inhibition by dihydropyridines [130].

2.4 CONCLUSIONS

The purpose of this chapter is to delineate useful strategies for studying Ca^{2+} signaling with fluorescent indicators. These approaches are based on the experiences of the authors in their own laboratory over the past 18 years. We have also tried to point out a number of potential artifacts and pitfalls in the use of fluorescent indicators. Most of these derive from the fact that Ca^{2+} regulation is necessarily a complex process, making it difficult to study any one facet in isolation. Our general advice is to avoid using any one technique alone if alternatives are available. This can minimize the likelihood of artifacts, as different approaches tend to have different problems. Indeed, our current understanding of the basic processes of regulated Ca^{2+}-release from intracellular stores and regulated entry across the plasma membrane is derived from results of diverse experimental approaches, including the use of fluorescent indicators, but also including the use of luminescent indicators, radioactive tracers, and electrophysiology.

REFERENCES

1. Berridge M.J., Lipp P., and Bootman M.D., The versatility and universality of calcium signalling, *Nat. Rev. Mol. Cell Biol.* 1, 11, 2000.
2. Putney J.W., Jr., Muscarinic, alpha-adrenergic and peptide receptors regulate the same calcium influx sites in the parotid gland, *J. Physiol. (Lond.)* 268, 139, 1977.

3. Laugier R., Petersen O.H., Pancreatic acinar cells: electrophysiological evidence for stimulant-evoked increase in membrane calcium permeability in the mouse, *J. Physiol. (Lond.)* 303, 61, 1980.

4. Burgess G.M., McKinney J.S., Fabiato A., Leslie B.A., and Putney J.W., Jr., Calcium pools in saponin-permeabilized guinea-pig hepatocytes, *J. Biol. Chem.* 258, 15336, 1983.

5. Bird G.S., Aziz O., Lievremont J.P., Wedel B.J., Trebak M., Vazquez G., and Putney J.W., Jr., Mechanisms of phospholipase C-regulated calcium entry, *Curr. Mol. Med.* 4, 291, 2004.

6. Putney J.W., Jr., Excitement about calcium signaling in inexcitable cells, *Science* 262, 676, 1993.

7. Putney J.W., Capacitative calcium entry in the nervous system, *Cell Calcium* 34, 339, 2003.

8. Fabiato A., Calcium-induced release of calcium from the cardiac sarcoplasmic reticulum, *Am. J. Physiol.* 245, C1, 1983.

9. Meissner G., Ryanodine receptor/Ca^{2+} release channels and their regulation by endogenous effectors, *Ann. Rev. Physiol.* 56, 485, 1994.

10. Krause K.-H., Ca^{2+}-storage organelles, *FEBS Lett.* 285, 225, 1991.

11. Putney J.W., Jr., A model for receptor-regulated calcium entry, *Cell Calcium* 7, 1, 1986.

12. Vazquez G., Wedel B.J., Aziz O., Trebak M., and Putney J.W., Jr., The mammalian TRPC cation channels, *Biochim. Biophys. Acta* 1742, 21, 2004.

13. Putney J.W., Jr., Pharmacology of capacitative calcium entry, *Mol. Interventions* 1, 84, 2001.

14. Tsien R.Y., A non-disruptive technique for loading calcium buffers and indicators into cells, *Nature* 290, 527, 1981.

15. Grynkiewicz G., Poenie M., and Tsien R.Y., A new generation of Ca^{2+} indicators with greatly improved fluorescence properties, *J. Biol. Chem.* 260, 3440, 1986.

16. Schilling W.P., Rajan L., and Strobl-Jager E., Characterization of the bradykinin-stimulated calcium influx pathway of cultured vascular endothelial cells. Saturability, selectivity, and kinetics, *J. Biol. Chem.* 264, 12838, 1989.

17. Kwan C.Y., Putney J.W., Jr., Uptake and intracellular sequestration of divalent cations in resting and methacholine-stimulated mouse lacrimal acinar cells. Dissociation by Sr^{2+} and Ba^{2+} of agonist-stimulated divalent cation entry from the refilling of the agonist-sensitive intracellular pool, *J. Biol. Chem.* 265, 678, 1990.

18. Millar N.S., Baylis H.A., Reaper C., Bunting R., Mason W.T., and Sattelle D.B., Functional expression of a cloned *Drosophila* muscarinic acetylcholine receptor in a stable Drosophila cell line, *J. Exp. Biol.* 198, 1843, 1995.

19. Bird G.S., Putney J.W., Jr., Capacitative calcium entry supports calcium oscillations in human embryonic kidney cells, *J. Physiol* 562, 697, 2005.

20. Luo D., Broad L.M., Bird G.St.J., and Putney J.W., Jr., Signaling pathways underlying muscarinic receptor-induced $[Ca^{2+}]_i$ oscillations in HEK293 cells, *J. Biol. Chem.* 276, 5613, 2001.

21. Glennon M.C., Bird G.St.J., Kwan C.-Y., and Putney J.W., Jr., Actions of vasopressin and the Ca^{2+}-ATPase inhibitor, thapsigargin, on Ca^{2+} signaling in hepatocytes, *J. Biol. Chem.* 267, 8230, 1992.

22. Glennon M.C., Bird G.S., Takemura H., Thastrup O., Leslie B.A., and Putney J.W., Jr., *In situ* imaging of agonist-sensitive calcium pools in AR4-2J pancreatoma cells. Evidence for an agonist- and inositol 1,4,5- trisphosphate-sensitive calcium pool in or closely associated with the nuclear envelope, *J Biol. Chem.* 267, 25568, 1992.

23. Bird G.S., Takemura H., Thastrup O., Putney J. W., Jr., and Menniti F. S., Mechanisms of activated Ca^{2+} entry in the rat pancreatoma cell line, AR4-2J, *Cell Calcium* 13, 49, 1992.

24. Trebak M., Bird G.St.J., McKay R.R., and Putney J.W., Jr., Comparison of human TRPC3 channels in receptor-activated and store-operated modes. Differential sensitivity to channel blockers suggests fundamental differences in channel composition, *J. Biol. Chem.* 277, 21617, 2002.

25. Edelman J.L., Kajimura M., Woldemussie E., and Sachs G., Differential effects of carbachol on calcium entry and release in CHO cells expressing the m3 muscarinic receptor, *Cell Calcium* 16, 181, 1994.

26. Lin K., Sadée W., and Quillan J.M., Rapid measurements of intracellular calcium using a fluorescence plate reader, *Biotechniques* 26, 318, 1999.

27. Yokoyama T., Kato N., and Yamada N., Development of a high-throughput bioassay to screen melatonin receptor agonists using human melatonin receptor expressing CHO cells, *Neurosci. Lett.* 344, 45, 2003.

28. Minta A., Kao J.P.Y., and Tsien R.Y., Fluorescent indicators for cytosolic calcium based on rhodamine and fluorescein chromophores, *J. Biol. Chem.* 264, 8171, 1989.

29. Kao J.P.Y., Harootunian A.T., and Tsien R.Y., Photochemically generated cytosolic calcium pulses and their detection by fluo-3, *J. Biol. Chem.* 264, 8179, 1989.

30. Uhlen P., Spectral analysis of calcium oscillations, *Sci. STKE.* 2004, 115, 2004.

31. Takemura H., Hughes A.R., Thastrup O., and Putney J.W., Jr., Activation of calcium entry by the tumor promoter, thapsigargin, in parotid acinar cells. Evidence that an intracellular calcium pool, and not an inositol phosphate, regulates calcium fluxes at the plasma membrane, *J. Biol. Chem.* 264, 12266, 1989.

32. Kwan C.Y., Takemura H., Obie J.F., Thastrup O., and Putney J.W., Jr., Effects of methacholine, thapsigargin and La^{3+} on plasmalemmal and intracellular Ca^{2+} transport in lacrimal acinar cells, *Am. J. Physiol.* 258, C1006, 1990.

33. Innamorati G., Sadeghi H., and Birnbaumer M., Transient phosphorylation of the V1a vasopressin receptor, *J. Biol. Chem.* 273, 7155, 1998.

34. Ribeiro C.M.P., Reece J., and Putney J.W., Jr., Role of the cytoskeleton in calcium signaling in NIH 3T3 cells. An intact cytoskeleton is required for agonist-induced $[Ca^{2+}]_i$ signaling, but not for capacitative calcium entry, *J. Biol. Chem.* 272, 26555, 1997.

35. Bird G.St.J., Rossier M.F., Hughes A.R., Shears S.B., Armstrong D.L., and Putney J. W., Jr., Activation of Ca^{2+} entry into acinar cells by a non-phosphorylatable inositol trisphosphate, *Nature* 352, 162, 1991.

36. Takemura H., Putney J.W., Jr., Capacitative calcium entry in parotid acinar cells, *Biochem. J.* 258, 409, 1989.

37. Blaustein M.P., Kendrick N.C., Fried R.C., and Ratzlaff R.W., Calcium metabolism at the mammalian presynaptic nerve terminal: lessons from the synaptsome. In: *Society for Neuroscience Symposia, Vol. II. Approaches to the Cell Biology of Neurons.* Cowan M.W., Ferrendelli J.A., Eds, Society for Neuroscience, Bethesda, MD, 1977, 172.

38. Streb H., Irvine R. F., Berridge M.J., and Schulz I., Release of Ca^{2+} from a nonmitochondrial store in pancreatic cells by inositol-1,4,5-trisphosphate, *Nature* 306, 67, 1983.

39. Rossier M.F., Bird G. St. J., and Putney J.W., Jr., Structural organization of the inositol 1,4,5-trisphosphate-sensitive organelle in rat liver. Evidence for linkage to the plasma membrane through actin microfilaments, *Biochem. J.* 274, 643, 1991.

40. Oldershaw K.A., Taylor C.W., 2,5-Di-(*tert*-butyl)-1,4-benzohydroquinone mobilizes inositol 1,4,5-trisphosphate-sensitive and -insensitive Ca^{2+} stores, *FEBS Lett.* 274, 214, 1990.

41. Jackson T.R., Patterson S.I., Thastrup O., and Hanley M.R., A novel tumour promoter, thapsigargin, transiently increases cytoplasmic free Ca^{2+} without generation of inositol phosphates in NG115-401L neuronal cells, *Biochem. J.* 253, 81, 1988.

42. Deng H.-W., Kwan C.-Y., Cyclopiazonic acid is a sarcoplasmic reticulum Ca^{2+}-pump inhibitor of rat aortic smooth muscle, *Acta Pharmacol. Sinica* 12, 1, 1991.

43. Dolmetsch R.E., Lewis R.S., Signaling between intracellular Ca^{2+} stores and depletion-activated Ca^{2+} channels generates $[Ca^{2+}]_i$ oscillations in T lymphocytes, *J. Gen. Physiol.* 103, 365, 1994.

44. Li W.-H., Llopis J., Whitney M., Zlokarnik G., and Tsien R.Y., Cell-permeant caged InsP$_3$ ester shows that Ca^{2+} spike frequency can optimize gene expression, *Nature* 392, 936, 1998.

45. Thomas D., Lipp P., Tovey S.C., Berridge M.J., Li W., Tsien R.Y., and Bootman M.D., Microscopic properties of elementary Ca^{2+} release sites in non-excitable cells, *Current Biol.* 10, 8, 2000.

46. Gafni J., Munsch J.A., Lam T.H., Catlin M.C., Costa L.G., Molinski T.F., and Pessah I.N., Xestospongins: potent membrane permeable blockers of the inositol 1,4,5-trisphosphate receptor, *Neuron* 19, 723, 1997.

47. Rizzuto R., Simpson A.W.M., Brini M., and Pozzan T., Rapid changes of mitochondrial Ca^{2+} revealed by specifically targeted recombinant aequorin, *Nature* 358, 325, 1992.

48. Hajnóczky G., Robb-Gaspers L.D., Seitz M.B., and Thomas A.P., Decoding of cytosolic calcium oscillations in the mitochondria, *Cell* 82, 415, 1995.

49. Rizzuto R., Intracellular Ca(2+) pools in neuronal signalling, *Curr. Opin. Neurobiol.* 11, 306, 2001.

50. Carafoli E., Historical review: mitochondria and calcium: ups and downs of an unusual relationship, *Trends Biochem. Sci.* 28, 175, 2003.

51. Parekh A.B., Store-operated Ca^{2+} entry: dynamic interplay between endoplasmic reticulum, mitochondria and plasma membrane, *J Physiol.* 547, 333, 2003.

52. González A., Schulz I., and Schmid A., Agonist-evoked mitochondrial Ca^{2+} signals in mouse pancreatic acinar cells, *J. Biol. Chem.* 275, 38680, 2000.

53. Zazueta C., Sosa-Torres M.E., Correa F., and Garza-Ortiz A., Inhibitory properties of ruthenium amine complexes on mitochondrial calcium uptake, *J. Bioenerg. Biomembr.* 31, 551, 1999.

54. Bae J.H., Park J.W., and Kwon T.K., Ruthenium red, inhibitor of mitochondrial Ca^{2+} uniporter, inhibits curcumin-induced apoptosis via the prevention of intracellular Ca^{2+} depletion and cytochrome c release, *Biochem. Biophys. Res. Commun.* 303, 1073, 2003.

55. Szalai G., Csordas G., Hantash B. M., Thomas A. P., and Hajnoczky G., Calcium signal transmission between ryanodine receptors and mitochondria, *J. Biol. Chem.* 275, 15305, 2000.

56. Reed P.W., Lardy H.A., A23187: a divalent cation ionophore, *J. Biol. Chem.* 247, 6970, 1972.

57. Liu C.-M., Herman T. E., Characterization of ionomycin as a calcium ionophore, *J. Biol. Chem.* 253, 5892, 1979.

58. Deber C.M., Tom-Kun J., Mack E., and Grinstein S., Bromo-A23187: a nonfluorescent calcium ionophore for use with fluorescent probes, *Anal. Biochem.* 146, 349, 1985.

59. Albert P.R., Tashjian A.H., Jr., Relationship of thyrotropin-releasing hormone-induced spike and plateau phases in cytosolic free Ca^{2+} concentrations to hormone secretion. Selective blockade using ionomycin and nifedipine, *J. Biol. Chem.* 259, 15350, 1984.

60. Abramov A.Y., Duchen M.R., Actions of ionomycin, 4-BrA23187 and a novel electrogenic Ca^{2+} ionophore on mitochondria in intact cells, *Cell Calcium* 33, 101, 2003.

61. Bird G.St.J., Obie J.F., and Putney J.W., Jr., Functional homogeneity of the non-mitochondrial Ca^{2+}-pool in intact mouse lacrimal acinar cells, *J. Biol. Chem.* 267, 18382, 1992.

62. Pizzo P., Fasolato C., and Pozzan T., Dynamic properties of an inositol 1,4,5-trisphosphate- and thapsigargin-insensitive calcium pool in mammalian cell lines, *J. Cell Biol.* 136, 355, 1997.

63. Fasolato C., Zottini M., Clementi E., Zachetti D., Meldolesi J., and Pozzan T., Intracellular Ca^{2+} pools in PC12 cells. Three intracellular pools are distinguished by their turnover and mechanisms of Ca^{2+} accumulation, storage and release, *J. Biol. Chem.* 266, 20159, 1991.

64. Putney J.W., Jr., Bird G.St. J., The inositol phosphate-calcium signalling system in non-excitable cells, *Endocrine Rev.* 14, 610, 1993.

65. Llopis J., Kass G.E.N., Gahm A., and Orrenius S., Evidence for two pathways of receptor-mediated Ca^{2+} entry in hepatocytes, *Biochem. J.* 284, 243, 1992.

66. Broad L.M., Cannon T. R., and Taylor C.W., A non-capacitative pathway activated by arachidonic acid is the major Ca^{2+} entry mechanism in rat A7r5 smooth muscle cells stimulated with low concentrations of vasopressin, *J. Physiol. (Lond.)* 517, 121, 1999.

67. Byron K.L., Taylor C.W., Vasopressin stimulation of Ca^{2+} mobilization, two bivalent cation entry pathways and Ca^{2+} efflux in A7r5 rat smooth muscle cells, *J. Physiol. (Lond.)* 485 (Pt. 2), 455, 1995.

68. Shuttleworth T.J., Thompson J. L., Evidence for a non-capacitative Ca^{2+} entry during $[Ca^{2+}]$ oscillations, *Biochem. J.* 316, 819, 1996.

69. Shuttleworth T.J., Arachidonic acid activates the noncapacitative entry of Ca^{2+} during $[Ca^{2+}]_i$ oscillations, *J. Biol. Chem.* 271, 21720, 1997.

70. Putney J.W., Jr., Store-operated calcium channels: how do we measure them, and why do we care? *Sci. STKE.* 2004, e37, 2004.

71. Anderson A., Mn ions pass through calcium channels, *J. Gen. Physiol.* 81, 805, 1983.

72. Hesketh T.R., Smith G.A., Moore J. P., Taylor M.V., and Metcalfe J. C., Free cytoplasmic calcium concentration and the mitogenic stimulation of lymphocytes, *J. Biol. Chem.* 258, 4876, 1983.

73. Tsien R.Y., Rink T.J., and Poenie M., Measurement of cytosolic free Ca^{2+} in individual small cells using fluorescence microscopy with dual excitation wavelengths, *Cell Calcium* 6, 145, 1985.

74. Hallam T.J., Pearson J. D., Exogenous ATP raises cytoplasmic free calcium in fura-2 loaded piglet aortic endothelial cells, *Febs Lett.* 207, 95, 1986.

75. Shuttleworth T.J., Temporal relationships between Ca^{2+} store mobilization and Ca^{2+} entry in an exocrine cell, *Cell Calcium* 15, 457, 1994.

76. Chiavaroli C., Bird G.St.J., and Putney J.W., Jr., Delayed, "All-or-none" activation of inositol 1,4,5-trisphosphate-dependent calcium signalling in single rat hepatocytes, *J. Biol. Chem.* 269, 25570, 1994.

77. Hallam T.J., Rink T.J., Agonists stimulate divalent cation channels in the plasma membrane of human platelets, *FEBS Lett.* 186, 175, 1985.

78. Sage S.O., Merritt J. E., Hallam T.J., and Rink T.J., Receptor-mediated calcium entry in fura-2-loaded human platelets stimulated with ADP and thrombin, *Biochem.J.* 258, 923, 1989.

79. Hallam T.J., Jacob R., and Merritt J. E., Influx of bivalent cations can be independent of receptor stimulation in human endothelial cells, *Biochem.J.* 259, 125, 1989.

80. Loessberg P.A., Zhao H., and Muallem S., Synchronized oscillation of Ca^{2+} entry and Ca^{2+} release in agonist-stimulated AR42J cells, *J. Biol. Chem.* 266, 1363, 1991.

81. Kass G.E.N., Chow S.C., Gahm A., Webb D.-L., Berggren P.-O., Llopis J., and Orrenius S., Two separate plasma membrane Ca^{2+} carriers participate in receptor-mediated Ca^{2+} influx in rat hepatocytes, *Biochim. Biophys. Acta* 1223, 226, 1994.

82. Hajnóczky G., Thomas A.P., The inositol trisphosphate calcium channel is inactivated by inositol trisphosphate, *Nature* 370, 474, 1994.

83. Somlyo A.V., Somlyo A.P., Strontium accumulation by sarcoplasmic reticulum and mitochondria in vascular smooth muscle, *Science* 174, 955, 1971.

84. Somlyo A.P., Somlyo A.V., Devine C.E., Peters P.D., and Hall T.A., Electron microscopy and electron probe analysis of mitochondrial cation accumulation in smooth muscle, *J. Cell Biol.* 61, 723, 1974.

85. Ma H.-T., Patterson R.L., van Rossum D.B., Birnbaumer L., Mikoshiba K., and Gill D. L., Requirement of the inositol trisphosphate receptor for activation of store-operated Ca^{2+} channels, *Science* 287, 1647, 2000.

86. Kreye V.A., Hofmann F., and Muhleisen M., Barium can replace calcium in calmodulin-dependent contractions of skinned renal arteries of the rabbit, *Pflüg. Arch.* 406, 308, 1986.

87. Broad L.M., Powis D.A., and Taylor C.W., Differentiation of BC_3H1 smooth muscle cells changes the bivalent cation selectivity of the capacitative Ca^{2+} entry pathway, *Biochem.J.* 316, 759, 1996.

88. Zhu X., Jiang M., and Birnbaumer L., Receptor-activated Ca^{2+} influx via human Trp3 stably expressed in human embryonic kidney (HEK)293 cells. Evidence for a non-capacitative calcium entry, *J. Biol. Chem.* 273, 133, 1998.

89. Franchini L., Levi G., and Visentin S., Inwardly rectifying K^+ channels influence Ca^{2+}entry due to nucleotide receptor activation in microglia, *Cell Calcium* 35, 449, 2004.

90. Lievremont J.P., Bird G.S., and Putney J.W., Jr., Canonical transient receptor potential TRPC7 can function as both a receptor- and store-operated channel in HEK-293 cells, *Am.J. Physiol. Cell Physiol.* 287, C1709, 2004.

91. Clementi E., Meldolesi J., Pharmacological and functional properties of voltage-independent Ca^{2+} channels, *Cell Calcium* 19, 269, 1996.

92. Lettvin J.Y., Pickard W.F., McCulloch W.S., and Pitts W., A theory of passive ion flux through axon membranes, *Nature* 202, 1338, 1964.

93. Van Breemen C., Farinas B., Gerba P., and McNaughton E.D., Excitation–contraction coupling in rabbit aorta studied by the lanthanum method for measuring cellular calcium influx, *Circ. Res.* 30, 44, 1972.

94. Liu X., Groschner K., and Ambudkar I.S., Distinct Ca(2+)-permeable cation currents are activated by internal Ca(2+)-store depletion in RBL-2H3 cells and human salivary gland cells, HSG and HSY, *J. Membr. Biol.* 200, 93, 2004.

95. Maruyama T., Kanaji T., Nakade S., Kanno T., and Mikoshiba K., 2APB, 2-aminoethoxydiphenyl borate, a membrane-penetrable modulator of Ins(1,4,5)P_3-induced Ca^{2+} release, *J. Biochem.* 122, 498, 1997.

96. Luo D., Broad L.M., Bird G.S., and Putney J.W., Jr., Mutual antagonism of calcium entry by capacitative and arachidonic acid-mediated calcium entry pathways, *J Biol. Chem.* 276, 20186, 2001.

97. Vazquez G., Wedel B.J., Trebak M., Bird G.St.J., and Putney J.W., Jr., Expression level of TRPC3 channel determines its mechanism of activation, *J. Biol. Chem.* 278, 21649, 2003.

98. Braun F.-J., Broad L.M., Armstrong D.L., and Putney J.W., Jr., Stable activation of single CRAC-channels in divalent cation-free solutions, *J. Biol. Chem.* 276, 1063, 2001.

99. Bakowski D., Glitsch M.D., and Parekh A.B., An examination of the secretion-like coupling model for the activation of the Ca^{2+} release-activated Ca^{2+} current Icrac in RBL-1 cells, *J. Physiol. (Lond.)* 532, 55, 2001.

100. Prakriya M., Lewis R.S., Potentiation and inhibition of Ca^{2+} release-activated Ca^{2+} channels by 2-aminoethyldiphenyl borate (2-APB) occurs independently of IP3 receptors, *J. Physiol. (Lond.)* 536, 3, 2001.

101. Iwasaki H., Mori Y., Hara Y., Uchida K., Zhou H., and Mikoshiba K., 2-Aminoethoxydiphenyl borate (2-APB) inhibits capacitative calcium entry independently of the function of inositol 1,4,5-trisphosphate receptors, *Recept. Channels* 7, 429, 2001.

102. Broad L.M., Braun F.-J., Lièvremont J.-P., Bird G. St.J., Kurosaki T., and Putney J. W., Jr., Role of the phospholipase C–inositol 1,4,5-trisphosphate pathway in calcium release-activated calcium current (I_{crac}) and capacitative calcium entry, *J. Biol. Chem.* 276, 15945, 2001.

103. Nadler M.J.S., Hermosura M.C., Inabe K., Perraud A.-L., Zhu Q., Stokes A.J., Kurosaki T., Kinet J.-P., Penner R., Scharenberg A.M., and Fleig A., LTRPC7 is a Mg-ATP-regulated divalent cation channel required for cell viability, *Nature* 411, 590595, 2001.

104. Prakriya M., Lewis R.S., Separation and characterization of currents through store-operated CRAC channels and Mg^{2+}-inhibited cation (MIC) channels, *J. Gen. Physiol.* 119, 487, 2002.

105. Braun F.J., Aziz O., and Putney J.W., Jr., 2-aminoethoxydiphenyl borane activates a novel calcium-permeable cation channel, *Mol. Pharmacol.* 63, 1304, 2003.

106. Chung M.K., Lee H., Mizuno A., Suzuki M., and Caterina M.J., 2-aminoethoxydiphenyl borate activates and sensitizes the heat-gated ion channel TRPV3, *J. Neurosci.* 24, 5177, 2004.

107. Hu H.Z., Gu Q., Wang C., Colton C.K., Tang J., Kinoshita-Kawada M., Lee L.Y., Wood J. D., and Zhu M.X., 2-aminoethoxydiphenyl borate is a common activator of TRPV1, TRPV2, and TRPV3, *J. Biol. Chem.* 279, 35741, 2004.

108. Thomas A.P., Bird G. St.J., Hajnóczky G., Robb-Gaspers L.D., and Putney J.W., Jr., Spatial and temporal aspects of cellular calcium signalling, *FASEB J.* 10, 1505, 1996.

109. Hoth M., Penner R., Depletion of intracellular calcium stores activates a calcium current in mast cells, *Nature* 355, 353, 1992.

110. Hoth M., Penner R., Calcium release-activated calcium current in rat mast cells, *J. Physiol. (Lond.)* 465, 359, 1993.

111. Putney J. W., Jr., *Capacitative Calcium Entry*, Landes Biomedical Publishing, Austin, TX,1997

112. Zweifach A., Lewis R.S., Rapid inactivation of depletion-activated calcium current (I_{CRAC}) due to local calcium feedback, *J. Gen. Physiol.* 105, 209, 1995.

113. Zweifach A., Lewis R.S., Slow calcium-dependent inactivation of depletion-activated calcium current, *J. Biol. Chem.* 270, 14445, 1995.

114. Berridge M.J., Capacitative calcium entry, *Biochem.J.* 312, 1, 1995.

115. Missiaen L., De Smedt H., Parys J.B., Oike M., and Casteels R., Kinetics of empty store-activated Ca^{2+} influx in HeLa cells, *J. Biol. Chem.* 269, 5817, 1994.

116. Louzao M.C., Ribeiro C.M. P., Bird G. St.J., and Putney J.W., Jr., Cell type-specific modes of feedback regulation of capacitative calcium entry, *J. Biol. Chem.* 271, 14807, 1996.

117. Penner R., Matthews G., and Neher E., Regulation of calcium influx by second messengers in rat mast cells, *Nature* 334, 499, 1988.

118. Bird G. St.J., Wedel B.J., Lièvremont J.-P., Trebak M., Aziz O., Vazquez G., and

Putney J.W., Jr., Mechanisms of phospholipase C-regulated calcium entry, *Curr. Mol. Med.* 4, 291, 2004.

119. Oettgen H.C., Terhorst C., Cantley L.C., and Rosoff P.M., Stimulation of the T3-T cell receptor complex induces a membrane-potential-sensitive calcium influx, *Cell* 40, 583, 1985.

120. Sage S.O., Rink T.J., Effects of ionic substitution on [Ca^{2+}]i rises evoked by thrombin and PAF in human platelets, *Eur. J. Pharmacol.* 128, 99, 1986.

121. Merritt J.E., Rink T.J., Regulation of cytosolic free calcium in fura-2-loaded rat parotid acinar cells, *J. Biol. Chem.* 262, 17362, 1987.

122. Di Virgilio F., Lew P.D., Andersson T., and Pozzan T., Plasma membrane potential modulates chemotactic peptide-stimulated cytosolic Ca^{2+} changes in human neutrophils, *J. Biol. Chem.* 262, 4574, 1987.

123. Kamouchi M., Droogmans G., and Nilius B., Membrane potential as a modulator of the free intracellular Ca^{2+} concentration in agonist-activated endothelial cells, *Gen. Physiol. Biophys.* 18, 199, 1999.

124. Mertz L.M., Baum B.J., and Ambudkar I.S., Membrane potential modulates divalent cation entry in rat parotid acini, *J. Membr. Biol.* 126, 183, 1992.

125. Mohr F.C., Fewtrell C., Depolarization of rat basophilic leukemia cells inhibits calcium uptake and exocytosis, *J. Cell Biol.* 104, 783, 1987.

126. Petersen O.H., Pedersen G.L., Membrane effects mediated by alpha- and beta-adrenoceptors in mouse parotid acinar cells, *J. Membrane Biol.* 16, 353, 1974.

127. Pittet D., Di V.F., Pozzan T., Monod A., and Lew D.P., Correlation between plasma membrane potential and second messenger generation in the promyelocytic cell line HL-60, *J. Biol. Chem.* 265, 14256, 1990.

128. Marty A., Tan Y.P., The initiation of calcium release following muscarinic stimulation in rat lacrimal glands, *J. Physiol. (Lond.)* 419, 665, 1989.

129. Zhang G.H., Melvin J.E., Membrane potential regulates Ca^{2+} uptake and inositol phosphate generation in rat sublingual mucous acini, *Cell Calcium* 14, 551, 1993.

130. Bird G.St.J., Takemura H., Thastrup O., Putney J.W., Jr., and Menniti F.S., Mechanism of activated Ca^{2+} entry in the rat pancreatoma cell line, AR4-2J, *Cell Calcium* 13, 49, 1991.

3 Fluorescence Microplate-Based Techniques for the High-Throughput Assessment of Calcium Signaling: The Highs and Lows for Calcium Researchers

Gregory R. Monteith and Gary St. J. Bird

CONTENTS

3.1 INTRODUCTION

As discussed throughout this book, the development of calcium-sensitive indicators based on the calcium chelator BAPTA revolutionized the study of cellular calcium homeostasis (Chapter 1). The first two decades of the assessment of cytoplasmic calcium ($[Ca^{2+}]_{CYT}$) using calcium-sensitive fluorescence indicators (such as fluo-3 and fura-2) were dominated by studies involving cuvette- and microscope-based equipment. Critically, the choice of this equipment dictates the spatial resolution of the calcium measurements, where large cell populations are used in cuvettes, and microscopes allow single-cell and subcellular measurements. In recent years, exciting methodology has emerged based on fluorescence microplate readers, a development spurred on by the needs of pharmaceutical companies to efficiently screen and identify bioactive molecules and therapeutic agents [1, 2]. As discussed in Chapter 17, the ubiquitous calcium signal has taken a leading role in this respect, with an array of advanced fluorescence microplate readers available that are particularly suited for the high-throughput (HT) assessment of $[Ca^{2+}]_{CYT}$.

While a useful screening tool, the ability to measure changes in $[Ca^{2+}]_{CYT}$ is key to help define and characterize underlying mechanisms of calcium signaling, and the role of $[Ca^{2+}]_{CYT}$ in cellular events ranging from contraction to synaptic transmission to gene transcription [3]. However, the adoption of advanced microplate readers for this purpose by basic research laboratories has been slow. High capital equipment costs are likely a prohibitory factor, but these concerns should be tempered by the advantages these systems can bring to the basic research environment. With increasing efficiency and flexibility for performing HT assessment of $[Ca^{2+}]_{CYT}$, the sophistication of HT-platforms to undertake complex cell signaling studies in HT mode continues to evolve. In this chapter, we will provide an overview of the opportunities available for the HT assessment of $[Ca^{2+}]_{CYT}$ using microplate readers and some of the issues and challenges related to this technology.

3.2 HT AND $[Ca^{2+}]_{CYT}$ MEASUREMENTS

Many aspects of the calcium signaling process can be described by directly observing $[Ca^{2+}]_{CYT}$ changes using fluorescent calcium indicators. Whether the equipment platform be cuvette-, microscope-, or microplate-based, it is critical to have a fundamental understanding of the basis for measuring $[Ca^{2+}]_{CYT}$ levels. This includes appreciating the flexibility and limitations of applying fluorescent calcium indicators. Chapter 2 outlines protocols that can be used to define underlying calcium signaling processes, and minimize the potential for creating artifacts or misinterpreting measurements.

The format of HT-based microplate readers dictates the limits of their ability, and flexibility, to perform $[Ca^{2+}]_{CYT}$ measurements, a central design feature being that the sample chamber is a microplate. Indeed, being microplate-based, this HT approach is somewhat of a compromise between cuvette- and microscope-based approaches. In many ways, a 96-well microplate is analogous to 96 cuvettes. Thus, when considering fluorescence-based $[Ca^{2+}]_{CYT}$ studies on a HT-platform, one must consider (i) flexibility of equipment options that influence the fluorescence-approach and (ii) methodological advantages and limitations of performing experiments with microplates.

3.3 EQUIPMENT OPTIONS FOR HT-BASED ASSESSMENT OF $[Ca^{2+}]_{CYT}$

In common with cuvette- and microscope-based systems, HT-plate readers are essentially composed of a fluoresecence light source for exciting the fluorophore, a sample compartment (the microplate well), and a detector that collects the emission fluorescence. The one additional facet important to the flexibility of these HT-platforms is the incorporation of robotics to administer solutions to each well of a microplate. Arguably FLIPR[384] (Fluorescence Imaging Plate Reader®: Molecular Devices Corporation) represents the first effective fluorescence micoplate reader for HT assessment of $[Ca^{2+}]_{CYT}$ (see schematic diagram in Figure 3.1). The key to this system's success was its ability to perform simultaneous fluorescence excitation and detection in all wells of a microplate, and perform simultaneous additions of test reagents to all wells. Technically, this was achieved by rapidly scanning (~40 ms) the entire microplate with 488-nm excitation light from an argon ion LASER, and detecting the fluorescence emission signal from the entire microplate with a camera. In addition, a programmable robotic arm carrying pipette tips

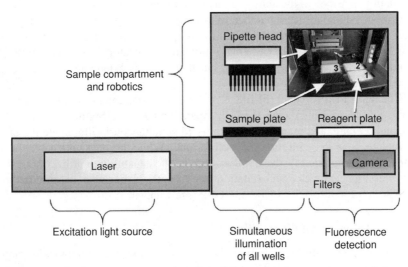

FIGURE 3.1 Schematic diagram of a HT fluorescence plate reader based on the FLIPR[384] instrument. The essential components of the HT-platform are an excitation source (LASER), a detector (camera), and a sample compartment containing areas for sample and reagent microplates and robotics for reagent addition. The inset photograph illustrates the layout of the 96-well sample microplate, three 96-well reagent plates, and the robotic head configured for carrying 96 pipette tips.

(one for each well) could collect and deliver predetermined volumes of test reagents at predetermined times. Variations on these themes are now available to researchers embarking on the HT assessment of $[Ca^{2+}]_{CYT}$, and we review equipment features in this section that impact experimental design.

3.3.1 EXCITATION LIGHT SOURCES

3.3.1.1 Lasers and LEDs

The choice of light source greatly impacts the availability of wavelength selection and thus the choice of fluorescence calcium indicator (Chapters 1 and 2). As mentioned above, the choice of an argon-ion LASER for FLIPR was a practical one dealing with the ability to deliver an adequate amount of excitation light simultaneously to all wells of a microplate. However, with only 488-nm excitation light available, the choice of fluorescent calcium indicators was limited to visible wavelength indicators such as fluo-3, fluo-4, and Calcium Green-1 [4]. An example of FLIPR-generated data using fluo-4 is shown in Figure 3.2.

While effective for $[Ca^{2+}]_{CYT}$ measurements, these single wavelength indicators are plagued with the problem of confusing calcium-dependent changes with calcium ion-independent fluorescence changes. However, the extent of this problem can be minimized on HT-platforms by three factors: (i) the ability to perform multiple simultaneous Ca^{2+} assessments under identical conditions on the same microplate, which reduces variance due to differences in Ca^{2+} indicator loading and cell plating density; (ii) the ability to incorporate and observe suitable experimental controls in parallel; (iii) a number of software corrections can be employed to "correct" fluorescence signal variations that arise from solution additions and variations in dye loading. Utilizing ratiometric indicators, such as fura-2 [5], would also minimize these problems, but this requires extending the wavelength selection into the UV range.

The latest FLIPR model (FLIPR^{TETRA}) has now replaced the argon-ion LASER with two light-emitting diodes (LEDs), expanding the wavelength selection to 408, 488, and 530 nm, and maintains the ability to simultaneously illuminate the entire microplate, be it 96-, 384-, or 1536-well formats. While this does not enable the use of the UV-ratiometric UV dyes, it does expand the capability to utilize fluorescent dyes with longer wavelength excitation (see Chapter 1), and also measure multiple fluorescent proteins such as CFP and GFP (see Chapter 4) [6].

3.3.1.2 Lamp-Based Illumination

The combination of full spectrum arc lamps (e.g., xenon-based lamps) with motorized filter-changers or monochromators offers the greatest flexibility for selecting multiple excitation wavelengths, ranging from UV to infrared. However, a significant limitation for using lamp-based illumination on a HT-platform has been the inability to deliver adequate excitation light to all wells of a microplate simultaneously. A compromise was to design microplate systems that could illuminate a single well at a time (e.g., NOVOstar, BMG LabTech; Genios Pro, Tecan), or up to

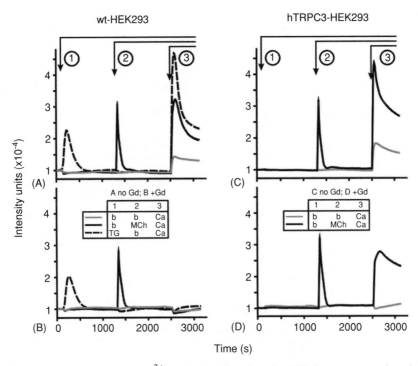

FIGURE 3.2 Assessment of $[Ca^{2+}]_{CYT}$ using fluo-4 and a HT fluorescence microplate reader. $[Ca^{2+}]_{CYT}$ responses were compared in wild type HEK293 cells (A, B) and HEK293 cells expressing hTRPC3 (C, D), and performed on a $FLIPR^{384}$ microplate reader using the fluorescence calcium indicator fluo-4. The experiment illustrates a protocol involving the addition of up to three reagents, indicated by the arrows. At the start of the experiment, cells are bathed either in a buffer nominally calcium free (A, C), or nominally calcium-free plus $10\,\mu M\ Gd^{3+}$ (B, D). All subsequent solution additions are indicated in the inset tables where b represents nominally calcium-free buffer (plus or minus $10\,\mu M\ Gd^{3+}$), MCh is $300\,\mu M$ methacholine, TG is $2\,\mu M$ thapsigargin, and Ca is $1.8\ mM\ Ca^{2+}$ (all are final concentrations in micro-wells). See Ref. [18] for a full description of $[Ca^{2+}]_{CYT}$ assay protocol. (A) and (B) characterize the MCh- and TG-induced calcium entry in wtHEK293 cells, which is fully blocked by Gd^{3+}. However, in HEK293 cells expressing hTRPC3, MCh activates a calcium entry insensitive to Gd^{3+} [18]. This experiment, performed on a 96-well microplate, involves 10 experimental conditions, each performed in duplicate, representing 20 observations (i.e., one fifth of the microplate's capacity). Not including the time for fluo-4 loading, this experiment took ~50 min which, if performed individually for each condition, would take ~17 h.

8–16 wells simultaneously (e.g., Flex Station II, Molecular Devices; VIPR, Aurora). The ability to program and automate the interchange between multiple excitation wavelengths is a clear advantage of these microplate systems, making it possible to use UV excited ratiometric calcium indicators for $[Ca^{2+}]_{CYT}$ measurements. An example of data from a microplate study using fura-2 is shown in Figure 3.3 [7].

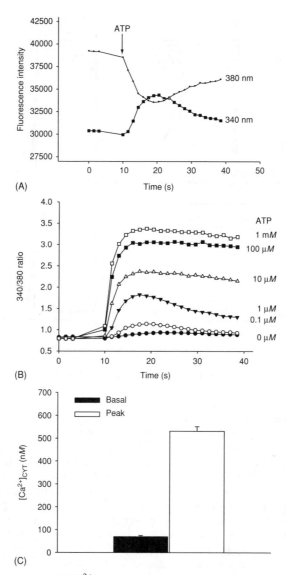

(A)

(B)

(C)

FIGURE 3.3 Assessment of $[Ca^{2+}]_{CYT}$ using fura-2 and a fluorescence microplate reader. MCF-7 cells were loaded with the calcium indicator fura-2. Activation of MCF-7 cells with ATP caused an increase in $[Ca^{2+}]_{CYT}$ as evident by a decline in fluorescence from the 380-nm excitation and increase in fluorescence from 340-nm excitation (Panel A). Panel B shows the dose-dependent increase in relative intracellular calcium. Panel C shows an example of calibrated data.

However, due to the limiting number of wells that can be observed, these microplate systems can only be considered medium throughput, not HT.

Recently, though, two HT-platforms have been developed that are equipped with Xe arc lamps and are capable of illuminating the entire microplate. The Hamamatsu FDSS 6000 combines the light from four Xe-lamps into a single-fiber optic, which is then used to illuminate either a 96- or 384-well plate (1536 currently in development) [8, 9]. In contrast, the Cell Lux (Perkin Elmer) uses a single Xe-lamp but then utilizes a bundle of 96 fiber optic light guides that delivers the excitation to each individual well. However, the limitation here is that the system is limited only to making 96 measurements simultaneously, and when using 384-well plates the measurements are performed in 96-well quadrants.

3.3.1.3 Fluorescence Emission Detection

In the three HT-platforms discussed above (FLIPR^{TETRA}, Cell Lux, and FDSS 6000), all employ CCD cameras to image the entire microplate and thus collect emission fluorescence data for each well, whether it is 96-, 384- or 1536-well format. Emission wavelengths are also selected by use of a filter, with at least two filter positions to select from. Options are also available on FDSS 6000 and Cell Lux that will allow HT-measurement of luminescence (see Chapter 4). It should be noted that the VIPR systems (medium throughput systems) use PMT detectors, providing exceptional sensitivity and temporal resolution.

3.3.1.4 Plate Formats

The key to HT is the ability to perform multiple $[Ca^{2+}]_{CYT}$ measurements in parallel, with current plate formats supporting 96, 384, or 1536 measurements. For example, if one considers a $[Ca^{2+}]_{CYT}$ measurement protocol that takes 10 min to complete in HT-mode, this same assay performed 8-wells at a time (e.g., with a Flex Station) would take 2 h to complete for the entire plate, or 16 h if performed one well at a time. The latter 8- and 16-well examples would also mean that these $[Ca^{2+}]_{CYT}$ measurements have delays built in between the first and last measurements. Unless steps are taken to maintain cells in a healthy state (e.g., minimize dye compartmentalization or leakage), such lengthy delays could potentially lead to reproducibility problems when comparing these data sets, delays that are magnified further with 384- and 1536-well formats. Hence, simultaneous excitation and detection of the entire microplate lends itself to effective HT $[Ca^{2+}]_{CYT}$ measurements.

3.3.1.5 Robotics and Solution Handling

Incorporating robotics into the FLIPR^{384} design was also a critical factor that made its introduction particularly desirable for $[Ca^{2+}]_{CYT}$ measurement studies. Specifically, it was possible to program FLIPR^{384} to perform simultaneous solution additions to all wells of a 96- or 384-well microplate, and for these solutions to be transferred from three separate reagent plates (see Figure 3.1). Thus, it was possible to employ complex protocols for defining calcium signaling processes, such as those outlined in Chapter 2, and illustrated in Figure 3.2.

Free access to the top of the microplate is essential for such solution manipulation, which is possible as fluorescence excitation and detection are performed beneath the plate in epifluorescence mode. With varying degrees of sophistication, all three HT-platforms described earlier in this chapter can transfer and manipulate solutions simultaneously in 96-well mode, with FDSS 6000 capable of controlling up to 384-wells and FLIPR^{TETRA} up to 1536-wells. All robotic manipulations are fully programmable, including pipette tip washing routines, pipette dispensing rate, and solution mixing. In a laboratory setting, any ability to automate calcium measurements can dramatically increase productivity compared to microscope- or cuvette-based protocols. Moreover, the ability to increase replicates may improve the statistical power to characterize subtle changes. Thus, it could be argued that even incorporating such fluid-handling capabilities affords advantages to fluorescence microplate readers capable of only single-, 8-, or 16-well assessment. In applications ideal for drug discovery in the pharmaceutical industry, there is extensive use of robotics for physically handling the microplate under observation, which can be exchanged for a new microplate sample held in a stacker system (see Chapter 17).

3.3.1.6   Environmental Control

In general, it is possible to measure many aspects of calcium signaling at room temperature, and in buffers that do not rely on CO_2/HCO_3^- to regulate pH. However, there may be instances where there is a requirement to maintain cells under more physiological conditions. Currently, widely used HT-platforms only permit the ability to regulate microplate temperature from ambient to ~37°C, with no capability to regulate atmospheric conditions (i.e., CO_2/air mix).

3.4   SPECIFIC ISSUES WITH [Ca^{2+}]$_{\text{CYT}}$ MEASUREMENTS IN MICROPLATES

3.4.1   Cell Plating: Adherent Cells and Cell Suspensions

Fluorescence microplate measurements of [Ca^{2+}]$_{\text{CYT}}$ can take place with cells in suspension or cells adhered to the bottom of the well. In some cases, this choice is clear, for example, hematopoietic cells (e.g., T-cells) are compatible with [Ca^{2+}]$_{\text{CYT}}$ measurements in a cell suspension, whereas primary cultured hippocampal neurons require assessment of cells adhered to the microplate surface [10]. For some cell lines, published protocols may exist where cells can be handled either in suspension or adhered to culture surfaces. In protocols where cells are required to be detached from a culture surface prior to Ca^{2+} indicator loading, the effect of the method used to detach cells on the response may need consideration. For example, care must be exercised if protease-activated receptors (e.g., the thrombin receptor) are to be characterized and the method used to detach cells involves trypsin [11].

Protocols requiring adherent cells usually require cells to be grown in the microplate at least 1 day prior to the assay. At the time of seeding the microplate with cells, it is recommended that close attention be paid to the cell number being seeded to (i) ensure uniform distribution of cell numbers in all wells and (ii) ensure conditions are optimal for cell growth and that an appropriate density of cells is present on the day of assay. Cells that are allowed to grow to confluence may cause problems, such as being easy to dislodge from the bottom of the well during solution addition, and the influence of cell density on the observed phenotypic $[Ca^{2+}]_{CYT}$ response. However, this has to be balanced with the fact that increasing cell number helps increase the fluorescence signal to be detected.

Experience with HEK293 cells indicate that optimal conditions involve seeding cells between 25,000 and 60,000 cells per well in poly-D-lysine-coated 96-well plates, 24 h prior to fluorescence indicator loading and $[Ca^{2+}]_{CYT}$ measurement. In contrast, with cells to be measured in suspension, one only needs to be concerned with cell seeding at the time of $[Ca^{2+}]_{CYT}$ measurement. For example, with Jurkat cells (loaded with the fluorescence probe in suspension just prior to the experiment) plating them at 120,000 cells per well (96-well plate) with a 10 min "settling" period is optimal. Cell suspension is not really an accurate description for the purpose of these assays, as it is important that the cells settle at the bottom of the well in order for the $[Ca^{2+}]_{CYT}$ measurement to be made.

3.4.2 FLUORESCENCE CALCIUM INDICATOR SELECTION

As discussed above, the equipment options available on the specific HT-platform will determine the variety of fluorescence calcium indicators available for use. In addition, the nature of the $[Ca^{2+}]_{CYT}$ measurement to be undertaken will also further discriminate which calcium indicator is to be used (see Chapters 1 and 2).

With lamp-based HT-systems, there are few limitations in calcium indicator selection. In general, ratiometric calcium indicators are best suited for quantifying the calcium signal, and minimizing signal artifacts due to differences in cell thickness, seeding density, and the concentration of calcium indicator. However, the nature of the $[Ca^{2+}]_{CYT}$ measurement in HT-mode makes the use of single wavelength calcium indicators such as fluo-4 a viable approach. In particular, the variability due to differences in cell loading and cell density is minimized since cells are seeded, loaded, and measured simultaneously. In addition, it is possible to dedicate a portion of the wells on a microplate to suitable control measurements, again performed in parallel, which can then be rectified by software manipulation (e.g., adjustments for variations in calcium indicator loading). It should be noted that access to fluorescence microscopy equipment allows optimization of Ca^{2+} indicator and loading method, through the assessment of indicator loading and sequestration.

3.4.3 LOADING CELLS WITH FLUORESCENT CALCIUM INDICATORS

A key part of the $[Ca^{2+}]_{CYT}$ measurement protocol is loading cells with a fluorescent calcium indicator. As described in Chapters 1 and Chapter 2, the ability of acetoxy methylester (AM) derivatives of calcium indicators to cross the plasma membrane, be de-esterified, and accumulate in the cytoplasm was paramount in the application of these indicators. But there are issues outlined in those chapters that need to be considered in the HT-environment, ranging from compartmentalization of indicators and leakage from cells, to behavior of different fluorophores within a particular cell type (e.g., HEK-293 $[Ca^{2+}]_{CYT}$ measurements employing Calcium Green-1/AM, in contrast to fluo-4/AM, require the presence of an anion transport inhibitor during indicator loading and the entirety of the experiment) [18].

In general, when loading cells with a calcium indicator, the solution bathing cells in each well are exchanged for one containing an AM-derivative of the dye for a specific period of time. Prior to the $[Ca^{2+}]_{CYT}$ measurement, the AM-derivative has to be removed, again by exchanging the buffer in each well. The main reason for this is that the AM-derivatives exhibit a contaminating fluorescence signal that can obscure the intracellular signal of interest. Cocktails of Ca^{2+}-sensitive fluorescence probes have been formulated to include extracellular dye quenchers that remove this contaminating fluorescence signal, thus eliminating the need to wash the dyes from the bathing solution before commencing the experiment [12, 13]. These so-called no-wash dyes do present a simplification of the assay protocol, particularly when considering 384- and 1536-well formats, and could minimize the potential for artifacts through physical handling. The balance between increased convenience and increased dye costs also needs to be considered. Importantly, the contents of proprietary no-wash reagents are not available, and it is therefore difficult to assess whether some components of these cocktails may alter the cellular signaling pathways being investigated. Hence, such reagents should be used with some caution in many basic research applications.

For cells adhered to the micro-well plate, exchanging bathing solutions can be achieved by using automated plate washers or manually by a researcher or technician. This choice will be dependent on access to plate washers and the robustness of the cells to automated washing. Automated plate washers are generally harsher than manual "by hand" methods, and may detach cells during washing. Different models of plate washers vary significantly in their suitability for viable cell assays. In most cases though, especially with the 96-well format, the manual technique is a suitable approach (which involves the "flicking" of an inverted plate and subsequent "blotting" to remove the solution, followed by pipetting of the new solution). However, when dealing with 384- and 1536-well formats, automation or no-wash dyes may be required for efficient and consistent loading of all wells.

3.4.4 ADDITION OF REAGENTS DURING $[Ca^{2+}]_{CYT}$ MEASUREMENTS: OPTIONS AND LIMITATIONS

Manipulating the conditions of bathing cells and observing their effects with good temporal resolution is key for an effective $[Ca^{2+}]_{CYT}$ measurement, whether it be cuvette-, microscope-, or microplate-based. Activation of cells with an agonist can generate changes in intracellular calcium that are immediate (on a second/sub-second timescale) and can continually change with time. Thus, it is imperative that a full temporal record of the calcium assay be collected, and that there be a capability to add agonist and test reagents "on-line." In general, temporal resolution does not appear to be a limiting factor for global $[Ca^{2+}]_{CYT}$ measurements on the available HT-plate readers. However, to manipulate the conditions of bathing cells, plate readers are engineered to varying degrees of complexity to handle reagent additions. The critical point here for a $[Ca^{2+}]_{CYT}$ measurement on a multi-well plate, is that plate readers are capable of making up to three separate solution additions during, and without interrupting, data collection (Figure 3.2 for example). These solutions, retrieved from reagent "reservoirs" or "reagent plates," are added to all wells simultaneously. For a 96-well plate format there is the capability to add 96 different test reagents during a single addition, and with up to three separate additions possible during a single assay.

When designing experimental protocols for HT-platforms, it is important to bear in mind that all manipulations of bathing solutions are accomplished by adding more solution to the microplate. This means that there is no way to effectively remove a reagent during the experimental protocol, such as is easily achieved through perfusion in fluorescence microscopy studies. Thus, in common with cuvette-based experiments, these "addition protocols" mean that test reagents are diluted into the solution bathing cells. This can present mixing issues (some equipments deal with this), or it may prevent the use of some reagents where their stock solutions present solubility problems and/or utilize solvents deleterious to cell viability.

A recurrent problem experienced during data collection is the appearance of "addition artifacts" that are unintended changes in the fluorescence signal during the on-line addition of reagents. Such fluorescence changes can obscure the intended calcium signal of interest, particularly when using single wavelength calcium dyes. In the main, these addition artifacts can result from (i) the physical nature of the solution addition or (ii) a nonspecific contribution of the test reagent to the fluorescence signal. There are, however, several practical approaches one can take to minimize the impact of addition artifacts.

(i) A physical artifact can be generated in a number of ways. One involves physically disturbing the cells in the microwell, a particular issue with cells in suspension, although some attached cells may also be prone to detachment, especially when confluent. Another involves the appearance of a calcium transient in cells prone to mechanical stimulation. In both cases, these effects can be minimized by (a) increasing the volume the cells are bathed in at the start of the experiment (at least 100 μl) and (b) reducing the speed with which the test reagent is injected into each well.

(ii) Some artifacts cannot be avoided, such as those created by the presence of a reagent that contributes to autofluorescence or some nonspecific interaction with the calcium dye. However, this can be accounted for by making sure that sufficient and appropriate control wells are dedicated to observing and quantifying this effect so that it can be accounted for in the test wells.

For the purpose of analysis and critical review of the experiment, any artifact in the signal not associated with an actual change in intracellular calcium can be corrected for by software manipulation. FLIPR, for example, allows the identification of "negative control wells," which the software can use to "remove" the addition artifacts and plot changes in fluorescence due to changes in calcium ions.

3.4.5 DATA ANALYSIS

As with all HT methods, data analysis is complicated by the large data sets generated. Although current microplate software is often well placed for screening applications, for many studies requiring more specific data analysis, such as characterization of calcium signaling pathways, current software is of limited use. Hence, the exporting of data to other programs is often required.

3.5 POTENTIAL APPLICATIONS FOR THE HT ASSESSMENT OF $[Ca^{2+}]_{CYT}$

The potential application of technology for the HT assessment of $[Ca^{2+}]_{CYT}$ is ever expanding. Below is a list of possible applications of the microplate readers capable of HT assessment of $[Ca^{2+}]_{CYT}$ with brief descriptions of the significance of each application.

   (i) Biomolecular screening
  (ii) Identification of functional stably transfected cell lines
 (iii) Pharmacology research, including dose–response curves
 (iv) Characterization of calcium homeostasis mechanisms
  (v) Comparison of calcium homeostasis in multiple cell lines and cells isolated from disease affected and control tissue.

(i) Biomolecular screening
The plethora of compounds in nature with biological activity and the advances in combinatorial chemistry, have led to the identification of potential drug "leads" for the treatment of a wide array of diseases. Fluorescence microplate readers and the HT assessment of $[Ca^{2+}]_{CYT}$ play a major role in these advancements, [14] see Chapter 17.

(ii) Identification of functional stably transfected cell lines
Cell lines stably expressing a specific protein are a common tool in biomedical research. Where the protein of interest is a G-protein coupled receptor involved in calcium release from intracellular stores or a Ca^{2+} channel, fluorescence microplate readers offer a rapid and efficient way to identify clones with functional expression of the protein of interest for further characterization [15].

(iii) Pharmacology research, including dose–response curves
Although the relationship between changes in $[Ca^{2+}]_{CYT}$ and the concentration of agonists or antagonists for specific receptors has been explored using fluorescence microscopy and cuvette-based studies, the advent of HT fluorescence microplate readers dramatically increases the efficiency and complexity of such studies. The use of fluorescence microplate readers for the HT assessment of $[Ca^{2+}]_{CYT}$ has been rapidly incorporated into many research programs for the characterization of receptor pharmacology, including characterization of orphan receptors [16, 17].

(iv)   Characterization of calcium homeostasis mechanisms

The approaches outlined in Chapter 2 are applicable to HT-platforms and the study of $[Ca^{2+}]_{CYT}$ regulation, including Ca-release and Ca-entry pathways. Indeed HT methods have been used to discriminate between CCE- and TRPC3-medicated calcium influx (Figure 3.2) [18]. Future studies may link HT $[Ca^{2+}]_{CYT}$ measurements with biological effects more directly through the marrying of Ca^{2+} assays with "high content screening" assays, which potentially allow HT assessment of a variety of effects including nuclear condensation, cell morphology, protein trafficking, and receptor internalization [19, 20].

(v)   Comparison of calcium homeostasis in multiple cell lines and cells isolated from disease affected and control tissue

Almost since the advent of fluorescence probes for the assessment of $[Ca^{2+}]_{CYT}$, researchers have compared calcium regulation in cells isolated from humans or animal models of disease or have compared calcium homeostasis in cell lines representing a diseased or "normal" phenotype [21]. Such studies would be strengthened in their power to link disease with alterations in calcium homeostasis if the number of comparisons was increased (e.g., assessing calcium homeostasis in a bank of diseased and normal samples). However, assessing calcium regulation in such a wide variety of samples would be extremely time consuming, and nonsimultaneous assessment may increase variability due to differences in cell loading between daily measurements. However, the development of fluorescence microplate readers should enable a wide variety of cell isolates, or cell lines representing a diseased or "normal" phenotype to be compared. It should be noted, however, that cell isolates and cell lines may differ greatly in the loading properties for Ca^{2+}-sensitive dyes. Such differences may arise from differences in dye leakage and sequestration, AM ester hydrolysis, and other intrinsic differences between cell lines such as cell thickness. Hence, the use of dyes less susceptible to these variations such as ratiometric probes (e.g., fura-2), and ratiometric probes less prone to sequestration and leakage (e.g., fura-PE3) may be better suited to such studies. The use of more advanced calibration methods, such as *in situ* assessment of R_{min}, R_{max}, and K_d may also be required (see Chapter 1). The availability of fluorescence microplate readers, capable of a wide selection of excitation wavelengths may greatly enhance the use of such technology in studies comparing calcium homeostasis in cells representing diseased and "normal" phenotypes, and may help identify diseases associated with changes in calcium homeostasis.

3.6   CHALLENGES AND FUTURE OPPORTUNITIES RELATED TO HT ASSESSMENT OF FREE CALCIUM USING FLUORESCENCE MICROPLATE READERS

One of the immediate consequences of assessing $[Ca^{2+}]_{CYT}$ using HT is the exponential increase in data, and the issues that arise regarding data analysis. Issues regarding data analysis have been faced by other technologies as they have progressed to HT protocols, such as cDNA microarrays, proteomics, and more recently high content screening. Hence, issues related to data analysis in $[Ca^{2+}]_{CYT}$ studies are easily addressed. However, since HT assessment of $[Ca^{2+}]_{CYT}$ was first embraced by pharmaceutical companies in biomolecular screening programs, software for high-end fluorescence microplate readers are often inflexible in regard to their adaptation to research that involves characterization of pathways involved in calcium homeostasis or comparison between cell lines representing diseased or "normal" phenotypes. No doubt, as the use of HT fluorescence microplate readers extends beyond screening, application software will also evolve.

The progress in the technology of instrumentation appropriate for HT assessment of $[Ca^{2+}]_{CYT}$, in particular the increase in flexibility of excitation and emission wavelengths available, gives rise to further opportunities for the possible dual assessment of multiple cellular ions, such as H^+, Na^+, and Ca^{2+}, the assessment of free Ca^{2+} in multiple cellular compartments, and the use of FRET-based probes.

3.7   SUMMARY

It is clear that the measurement of $[Ca^{2+}]_{CYT}$, like so many other biological targets, is becoming readily adaptable to the HT -environment. The application to biomolecular screening is going to be very significant, but its impact on the study of mechanisms involved in calcium homeostasis and the role of calcium signaling in disease may be even more important.

REFERENCES

1. Hodder, P., Mull, R., Cassaday, J., Berry, K., and Strulovici, B., Miniaturization of intracellular calcium functional assays to 1536-well plate format using a fluorometric imaging plate reader, *J Biomol Screen* 9 (5), 417–26, 2004.
2. Hodder, P., Cassaday, J., Peltier, R., Berry, K., Inglese, J., Feuston, B., Culberson, C., Bleicher, L., Cosford, N.D., Bayly, C., Suto, C., Varney, M., and Strulovici, B., Identification of metabotropic glutamate receptor antagonists using an automated high-throughput screening system, *Anal Biochem* 313 (2), 246–54, 2003.
3. Berridge, M.J., Bootman, M.D., and Roderick, H.L., Calcium signalling: dynamics, homeostasis and remodelling, *Nat Rev Mol Cell Biol* 4 (7), 517–29, 2003.
4. Thomas, D., Tovey, S.C., Collins, T.J., Bootman, M.D., Berridge, M.J., and Lipp, P., A comparison of fluorescent Ca^{2+} indicator properties and their use in measuring elementary and global Ca^{2+} signals, *Cell Calcium* 28 (4), 213–23, 2000.
5. Grynkiewicz, G., Poenie, M., and Tsien, R.Y., A new generation of Ca^{2+} indicators with greatly improved fluorescence properties, *J Biol Chem* 260 (6), 3440–50, 1985.
6. Falk, M.M. and Lauf, U., High resolution, fluorescence deconvolution microscopy and tagging with the autofluorescent tracers CFP, GFP, and YFP to study the structural composition of gap junctions in living cells, *Microsc Res Tech* 52 (3), 251–62, 2001.
7. Robinson, J.A., Jenkins, N.S., Holman, N.A., Roberts-Thomson, S.J., and Monteith, G.R., Ratiometric and nonratiometric Ca^{2+} indicators for the assessment of intracellular free Ca^{2+} in a breast cancer cell line using a fluorescence microplate reader, *J Biochem Biophys Methods* 58 (3), 227–37, 2004.

8. Sumichika, H., Sakata, K., Sato, N., Takeshita, S., Ishibuchi, S., Nakamura, M., Kamahori, T., Ehara, S., Itoh, K., Ohtsuka, T., Ohbora, T., Mishina, T., Komatsu, H., and Naka, Y., Identification of a potent and orally active non-peptide C5a receptor antagonist, *J Biol Chem* 277 (51), 49403–7, 2002.

9. Kawamoto, T., Kimura, H., Kusumoto, K., Fukumoto, S., Shiraishi, M., Watanabe, T., and Sawada, H., Potent and selective inhibition of the human Na+/H+ exchanger isoform NHE1 by a novel aminoguanidine derivative T-162559, *Eur J Pharmacol* 420 (1), 1–8, 2001.

10. Hemstapat, K., Smith, M.T., and Monteith, G.R., Measurement of intracellular Ca^{2+} in cultured rat embryonic hippocampal neurons using a fluorescence microplate reader: potential application to biomolecular screening, *J Pharmacol Toxicol Methods* 49 (2), 81–7, 2004.

11. Kable, E.P., Monteith, G.R., and Roufogalis, B.D., The effect of thrombin and serine proteases on intracellular Ca^{2+} in rat aortic smooth muscle cells, *Cell Signal* 7 (2), 123–9, 1995.

12. Zhang, Y., Kowal, D., Kramer, A., and Dunlop, J., Evaluation of FLIPR Calcium 3 Assay Kit — a new no-wash fluorescence calcium indicator reagent, *J Biomol Screen* 8 (5), 571–7, 2003.

13. Mehlin, C., Crittenden, C., and Andreyka, J., No-wash dyes for calcium flux measurement, *Biotechniques* 34 (1), 164–6, 2003.

14. Gribbon, P. and Sewing, A., Fluorescence readouts in HTS: no gain without pain?, *Drug Discov Today* 8 (22), 1035–43, 2003.

15. New, D.C. and Wong, Y.H., Characterization of CHO cells stably expressing a Galpha(16/z) chimera for high throughput screening of GPCRs, *Assay Drug Dev Technol* 2 (3), 269–80, 2004.

16. Witte, D.G., Cassar, S.C., Masters, J.N., Esbenshade, T., and Hancock, A.A., Use of a fluorescent imaging plate reader — based calcium assay to assess pharmacological differences between the human and rat vanilloid receptor, *J Biomol Screen* 7 (5), 466–75, 2002.

17. Shirokova, E., Schmiedeberg, K., Bedner, P., Niessen, H., Willecke, K., Raguse, J.D., Meyerhof, W., and Krautwurst, D., Identification of specific legands for orphan olfactory receptors. G protein-dependent agonism and antagonism of odorants, *J Biol Chem*, 280(12), 11807–15, 2005.

18. Trebak, M., Bird, G.S., McKay, R.R., and Putney, J.W., Jr., Comparison of human TRPC3 channels in receptor-activated and store-operated modes. Differential sensitivity to channel blockers suggests fundamental differences in channel composition., *J Biol Chem* 277 (24), 21617–23, 2002.

19. Schlag, B.D., Lou, Z., Fennell, M., and Dunlop, J., Ligand dependency of 5-hydroxytryptamine 2C receptor internalization, *J Pharmacol Exp Ther* 310 (3), 865–70, 2004.

20. Abraham, V.C., Taylor, D.L., and Haskins, J.R., High content screening applied to large-scale cell biology, *Trends Biotechnol* 22 (1), 15–22, 2004.

21. Oshima, T., Young, E.W., Bukoski, R.D., and McCarron, D.A., Rise and fall of agonist-evoked platelet Ca^{2+} in hypertensive rats, *Hypertension* 18 (6), 758–62, 1991.

4 Genetically Encoded Fluorescent Calcium Indicator Proteins

Atsushi Miyawaki, Takeharu Nagai, and Hideaki Mizuno

CONTENTS

4.1 GREEN FLUORESCENT PROTEIN-BASED FLUORESCENT INDICATORS FOR Ca^{2+}

Green fluorescent protein (GFP)-based fluorescent indicators for Ca^{2+} offer significant promise for monitoring Ca^{2+} in previously unexplored organisms, tissues, organelles, and submicroscopic environments because they are genetically encoded, function without cofactors, can be targeted to any intracellular location, and are bright enough for single-cell imaging [1–3]. These indicators may be either single GFP variants [4], circularly permuted GFP variants [5, 6], or pairs of GFP variants that permit fluorescence resonance energy transfer (FRET) [7, 8]. This review describes the design and construction of these three types of Ca^{2+} indicators, as well as their potential uses and limitations. It also focuses on recent reports on improvements and applications of Ca^{2+} indicators that have appeared in the last few years. These reports demonstrate progress toward a real understanding of the dynamic aspects of Ca^{2+} homeostasis.

4.2   Ca^{2+} INDICATORS USING SINGLE GFP VARIANTS

Our understanding of the structure–photochemistry relationship of GFP has enabled the development of genetic calcium probes based on a single GFP variant. Camgaroo-1 was constructed by inserting calmodulin (CaM) between positions 145 and 146 of YFP [4]. The Ca^{2+}-dependent conformational change of CaM in the resulting protein induces ionization of the chromophore, resulting in a fluorescence increase of up to sevenfold. The chromophore development of this probe has been improved (camgaroo-2) by an amino acid substitution, Q69M [9]. Both versions of camgaroo were expressed in *Drosophila* mushroom body neurons [10]. Experiments on the resulting brains revealed fluorescence increases upon pressure application of acetylcholine to the calyx, which delineates the dendritic field of the mushroom body neurons. These observations were the first direct evidence of the involvement of cholinergic pathways in the transmission of olfactory information from projection neurons to mushroom body neurons.

4.3   Ca^{2+} INDICATORS USING CIRCULARLY PERMUTED GFP VARIANTS

To visualize Ca^{2+}-dependent protein–protein interactions in living cells by fluorescence readout, two groups utilized circularly permuted green fluorescent proteins (cpGFPs), in which the amino and carboxyl portions had been interchanged around position 145 and reconnected by short spacers between the original termini [4]. In these studies, cpGFPs were fused to CaM and its target peptide, M13, to generate the chimeric proteins G-CaMP [5] and pericam [6], respectively. These proteins are fluorescent and their spectral properties change reversibly with Ca^{2+} concentration, probably due to the interaction between CaM and M13, leading to alteration of the environment surrounding the chromophore.

4.3.1   G-CaMP

G-CaMP is a single wavelength intensity-modulating probe for Ca^{2+}. Fast mobilization of Ca^{2+} upon depolarization was observed in myotubes expressing the G-CaMP probe [5]. G-CaMP has been used successfully to study the representation of olfactory information in *Drosophila* via a functional analysis of odor-evoked patterns of activity in neural assemblies of the *Drosophila* antennal lobe [11]. The probe was expressed in olfactory receptor neurons or projection neurons, and patterns of glomerular activity were imaged in the antennal lobe presynaptically or postsynaptically, respectively. Odors elicited specific patterns of glomerular activity that were conserved in multiple flies, and the specific responsivity of a given glomerulus was found to be a consequence of the specificity of a single odorant receptor expressed by the incoming sensory neurons. These observations are consistent with the "one neuron–one receptor" principle [12] that holds true for both vertebrate and insect olfactory systems. This type of study utilizing G-CaMP has been extended to the analysis of the mushroom body. Spatially specific odor-evoked activity was clearly observed in mushroom body neurons [13]. Recently, flies that expressed G-CaMP were successfully used to map the primary sensory neurons responsible for an innate avoidance behavior of the animal. It was determined that CO_2 activates only a single glomerulus in the antennal lobe, the V glomerulus [14]. In a different study, a transgenic mouse line expressing G-CaMP in smooth muscle was developed. Using detrusor tissue from mice, two separate types of postsynaptic Ca^{2+} signals mediated by distinct neurotransmitters were identified [15].

4.3.2 PERICAM

Three types of pericam have been obtained by mutating several amino acids adjacent to the chromophore [6]. Of these, "flash pericam" becomes bright in the presence of Ca^{2+} in a manner similar to G-CaMP, whereas "inverse pericam" displays the opposite response. A different mutant, "ratiometric pericam," undergoes changes in excitation wavelength in a Ca^{2+}-dependent manner, thereby enabling dual excitation ratiometric Ca^{2+} imaging. Their spectra in the presence and absence of Ca^{2+} are shown in Figure 4.1. Ratiometric pericam permits quantitative Ca^{2+} measurements by minimizing the effects of several artifacts that are unrelated to changes in free Ca^{2+} concentration ($[Ca^{2+}]$). It has been successfully used to monitor changes in $[Ca^{2+}]$ in mitochondria ($[Ca^{2+}]_m$) of various cell types. A study using cardiomyocytes has demonstrated that mitochondrial $[Ca^{2+}]$ oscillates synchronously with cytosolic $[Ca^{2+}]$ during the heartbeat [16].

As dual-excitation ratiometric dyes are excited alternately at two different wavelengths while the emission is collected at a single fixed wavelength, a pair of intensity measurements must be collected sequentially. Ca^{2+} imaging using ratiometric-pericam was thought to be inadequate to follow very fast Ca^{2+} dynamics or Ca^{2+} changes in highly motile cell types such as cardiomyocytes. Also, monitoring changes in $[Ca^{2+}]$ is often severely limited by poor spatiotemporal resolution of conventional wide-field microscopes. To overcome these limitations, a laser-scanning confocal microscopy (LSCM) system has been modified to obtain high-quality confocal images of Ca^{2+} using ratiometric pericam [17]. Rapid exchange between two laser beams has been achieved using acousto-optic tunable filters. Samples are scanned on each line sequentially by a violet laser diode (408 nm) and a diode-pumped solid state laser (488 nm). In this way, the ratios of the excitation peaks can be obtained at frequencies of up to 200 Hz, which enables visualization of Ca^{2+} dynamics within a motile mitochondrion.

As it has proved difficult to place cameleons into the mitochondrial matrix (see below), $[Ca^{2+}]_m$ dynamics have been explored principally by using ratiometric pericam. Results obtained using mitochondrially targeted ratiometric pericam in single cells have shown (i) facilitation of Ca^{2+} propagation within interconnected mitochondria through the formation of a mitochondrial network [18], (ii) endoplasmic reticulum-mediated transfer of Ca^{2+} from plasma membrane to mitochondria [19], (iii) contribution of mitochondria to the generation of subplasmalemmal microdomains of low Ca^{2+} that sustain the activity of capacitative Ca^{2+} entry

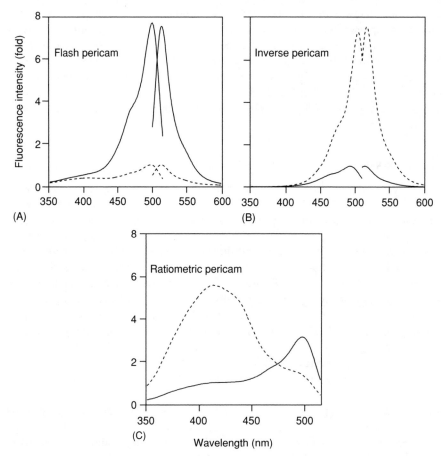

FIGURE 4.1 Ca^{2+}-dependent spectral changes of pericams. (A) Excitation and emission spectra of flash pericam and (B) inverse pericam. (C) Excitation spectra of ratiometric pericam. The data points were obtained in the presence (solid lines) or absence (dashed lines) of Ca^{2+}.

[20], and (iv) stable interactions between mitochondria and endoplasmic reticulum that lead to rapid accumulation of Ca^{2+} in a subpopulation of mitochondria [21].

Inverse pericam is the brightest of the three pericams, but is less useful because it loses its green fluorescence upon Ca^{2+} binding. To generate a bright and useful indicator for quantitative Ca^{2+} imaging, red fluorescent protein (DsRed2) was attached via a short linker, GGGSGGGS (G = glycine, S = serine), to the C terminus of inverse pericam. The resulting indicator, called "DRIP" (*D*sRed2-referenced *I*nverse *P*ericam), is a dual emission ratiometric indicator that requires two excitation wavelengths [22]. The red/green ratio of this indicator increases upon binding Ca^{2+} with the same affinity as inverse pericam. In one study, DRIP was expressed specifically in the pharyngeal muscles of *C. elegans* under the control of the myo-2 promoter, and the spatiotemporal pattern of intracellular calcium dynamics in these muscles during feeding was successfully examined [22].

4.4   Ca^{2+} INDICATORS UTILIZING PAIRS OF GFP VARIANTS THAT PERMIT FRET

FRET between two fluorophores is highly sensitive to the relative orientation and distance between the fluorophores. This technique is amenable to emission ratioing, which is more quantitative than single-wavelength monitoring, and is also an ideal readout for fast imaging by LSCM. Cameleons are chimeric proteins composed of a linear combination of a short-wavelength mutant of GFP, CaM, a glycylglycine linker, the CaM-binding peptide of myosin light chain kinase (M13), and a long-wavelength mutant of GFP [7]. Binding of Ca^{2+} to the CaM moiety of the cameleon initiates an intramolecular interaction between the CaM and M13 domains [23], causing the chimeric protein to lose its extended conformation and instead adopt a more compact one, thereby increasing the efficiency of FRET between the two constituent GFP mutants.

The original prototype cameleon was released in 1997 [7], and since then several improvements have been made to the cameleon so that it (i) can be shifted to longer wavelengths, (ii) is less sensitive to acidic pH, (iii) exhibits more efficient maturity in mammalian cells at 37°C, (iv) exhibits a significantly wider dynamic range, (v) is not perturbed by a large excess of native CaM, and (vi) has improved reaction kinetics and a K_d ideal for imaging Ca^{2+} in the ER.

Whereas the original cameleon contained blue and green mutants of GFP as donor and acceptor, respectively, CFP and YFP have been substituted to make yellow cameleons (YCs). Subsequently, red cameleons have been constructed by incorporating RFP (DsRed) as an acceptor [24]. The original version of YC (YC2.0) displayed elevated pH sensitivity because its acceptor, enhanced YFP (EYFP), is quenched by acidification with a pK_a of 7. This pH sensitivity has been markedly reduced by introducing the V68L and Q69K mutations into EYFP (EYFP.1) [25]. The resulting improved YCs, including YC2.1 and YC3.1, permit Ca^{2+} measurement without perturbation by pH changes between pH 6.5 and 8.0. Two bright versions of YFP, citrine [9] and Venus [26], which mature more efficiently at 37°C, have recently been developed and used as YC acceptors. The rapid maturation of Venus in YC2.12 or YC3.12, for example, allows for immediate detection of [Ca^{2+}] transients after gene introduction in freshly prepared brain slices [26].

Expansion of the Ca^{2+} responses of YCs has been achieved by a marriage between FRET and circular permutation techniques. To achieve a larger Ca^{2+}-dependent change in the relative orientation and distance between the fluorophores of CFP and YFP, the most recent approach [27] has been to use circularly permuted YFPs (cpYFPs), in which the amino and carboxyl portions were interchanged and reconnected by a short spacer between the original termini [4]. Circular permutation was conducted on Venus, and new termini were introduced into surface-exposed loop regions of the β-barrel. One of the variants, cp173Venus, was given a new N terminus at Asp173, which is at the opposite end of the β-barrel far removed from the original N terminus, Met1 (Figure 4.2A). YC3.60 was generated by replacing Venus in YC3.12 with cp173Venus (Figure 4.2B). Compared to YC3.12, YC3.60 is equally bright, but shows a five to sixfold larger dynamic range (Figure 4.2C). Thus,

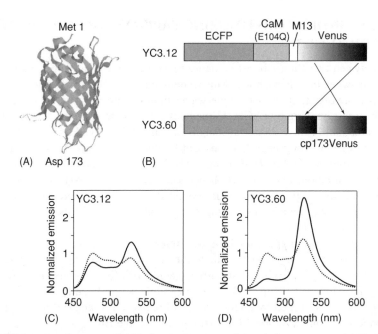

FIGURE 4.2 Development of YC3.60. (A) The three-dimensional structure of GFP with the positions of the original (Met1) and the new (Asp173) amino termini indicated. (B) Domain structures of YC3.12 and YC3.60. (CaM) *Xenopus* calmodulin; (E104Q) mutation of the conserved bidentate glutamate (E104) at position 12 of the third Ca^{2+}-binding loop to glutamine. (C) Emission spectra of YC3.12 (left) and YC3.60 (right) (excitation at 435 nm) at zero (dotted line) and saturated Ca^{2+} (solid line).

YC3.60 gives a greatly enhanced signal-to-noise ratio, thereby enabling Ca^{2+} imaging experiments that were not possible with conventional YCs. YC3.60 shows a monophasic Ca^{2+} response curve (K'_d, 0.25 μM; n, 1.7).

Additionally, the interface between CaM and the CaM-binding peptide has been redesigned to generate highly specific protein–peptide pairs that display a range of Ca^{2+} affinities and that are not perturbed by endogenous CaM and CaM-binding proteins [28]. A mutant CaM–peptide pair, D1, was cloned between CFP and citrine to yield a reengineered cameleon. This D1 cameleon is indifferent to large excess of CaM. Moreover, it displays a low Ca^{2+} affinity component with a K_d value of 60 μM, and is thus an ideal probe for measuring Ca^{2+} in the ER ($[Ca^{2+}]_{er}$). This new probe represents a dramatic improvement over YC4ER, which had previously been used to monitor $[Ca^{2+}]_{er}$ with low sensitivity. This new ER probe was used to show that, in MCF-7 cells, the antiapoptotic protein B cell lymphoma 2 (Bcl-2) lowers $[Ca^{2+}]_{er}$ by increasing Ca^{2+} leakage under resting conditions, and that inhibition of Bcl-2 by the green tea compound epigallocatechin gallate results in an increase in $[Ca^{2+}]_{er}$ [28]. Another attempt to eliminate the sensitivity of genetically encoded Ca^{2+} indicators to endogenous CaM and CaM-binding proteins involved the use of troponin C proteins from skeletal and cardiac muscle [8]. As troponin C is specific for muscle, these novel probes should not be perturbed by endogenous proteins when expressed in non-muscle tissue. A variant, TN-L15, targeted to the plasma membrane of primary hippocampal neurons, was successfully used to monitor submembrane Ca^{2+} levels that were in equilibrium with bulk cytosolic Ca^{2+} [8]. The data indicate the absence of a standing gradient from the membrane toward the cytosol.

While cameleons have proved difficult to target to the mitochondrial matrix, it has recently been shown that the use of a tandemly duplicated mitochondrial targeting sequence is capable of improving the efficacy of mitochondrial chameleon delivery [21]. YC2 was transfected into mouse hind-limb muscle *in situ,* and using a tandem repeat of the mitochondrial-targeting sequence of COX VIII was targeted successfully to the mitochondria as well as the cytosol. Ca^{2+} imaging using two-photon microscopy revealed that mitochondria in skeletal muscle take up Ca^{2+} during contraction and rapidly release it during relaxation [29].

Cameleons are advantageous for their emission of ratiometric response. They are the best choice for observing Ca^{2+} dynamics in motile animals. Calcium transients in neurons have been observed in the process of gentle touch sensation of *C. elegans* [30] or during escape behaviors of zebrafish [31].

4.4.1 PROCEDURE FOR DUAL EMISSION RATIO IMAGING OF YCs

Briefly, the procedure for time-lapse $[Ca^{2+}]_c$ imaging in cultured cells is as follows [2]:

1. Attach HeLa cells to a cover slip in a Petri dish. Transfect the cells with 1 μg of cDNA using Lipofectin (Invitrogen).
2. Between 2 and 10 days after cDNA transfection, image HeLa cells on an inverted microscope (IX70) with a cooled CCD camera (Micromax, Roper Scientific). Expose cells to reagents in HBSS containing 1.26 mM $CaCl_2$. Image acquisition and processing should be controlled by a personal computer connected to a camera and a filter wheel (Lambda 10–2, Sutter Instruments, San Rafael, CA) using the program MetaFluor (Universal Imaging, West Chester, PA). The excitation filter wheel in front of the xenon lamp and the emission filter wheel (Lambda 10–2, Sutter Instruments, San Rafael, CA) immediately below the CCD camera may also be under computer control. Whereas the excitation filter wheel may be replaced with a fixed one, an emission filter wheel is required for imaging cameleons. Excitation light from a 75 W xenon lamp should be passed through a 440DF20 (440 ± 10) excitation filter, and the light reflected onto the sample using a 455-nm long pass (455DRLP) dichroic mirror. The emitted light should be collected with a 40× (numerical aperture: 1.35) objective and passed through a 480 ± 15 or 480 ± 25 nm band pass filter (480DF30 or 480DF50, donor channel) for ECFP, and a 535 ± 12.5 nm band pass filter (535DF25, FRET channel) for YFP. Interference filters from Omega Optical or Chroma Technologies (Brattleboro, VT) are recommended.

3. Define several factors for image acquisition, including (i) excitation power, which depends on the type of light source and neutral density filter, (ii) numerical aperture of the objective, (iii) light exposure time, (iv) image acquisition interval, and (v) binning. The last three factors should be chosen based on the relative importance of temporal or spatial resolution to the experiment.

4. Choose moderately bright cells (see below). In addition, the fluorescence should be uniformly distributed in the cytosolic compartment but excluded from the nucleus, as expected for a 74-kDa protein without nuclear targeting signals. Select regions of interest so that pixel intensities are spatially averaged.

5. At the end of an experiment, convert fluorescence signals into values of $[Ca^{2+}]$. R_{max} and R_{min} can be obtained in the following manner. To saturate the intracellular indicator with Ca^{2+}, increase the extracellular $[Ca^{2+}]$ to 10–20 mM in the presence of 1–5 μM ionomycin. Wait until fluorescence intensity reaches a plateau. Then, to deplete the Ca^{2+} indicator, wash the cells with Ca^{2+}-free medium (1 μM ionomycin, 1 mM EGTA, and 5 mM MgCl$_2$ in nominally Ca^{2+}-free HBSS). The *in situ* calibration for $[Ca^{2+}]$ uses the equation $[Ca^{2+}] = K_d'[(R - R_{min})/(R_{max} - R)]^{(1/n)}$, where K_d' is the apparent dissociation constant corresponding to the Ca^{2+} concentration at which R is midway between R_{max} and R_{min}, and n is the Hill coefficient. The Ca^{2+} titration curve of YC2.1 can be fitted using a single K_d' of 0.2 μM and a single Hill coefficient of 0.62. Therefore, use $K_d' = 0.2$ μM and $n = 0.62$ for YC2.1, $K_d' = 1.5$ μM and $n = 1.1$ for YC3.1, and $K_d' = 0.2$ μM and $n = 1.7$ for YC3.60.

6. For fast and simultaneous acquisition of YFP and CFP images, a color camera (Hamamatsu Photonics, C7780–22) composed of three CCD chips (RGB: red, green, and blue) and a prism may be used. The YFP and CFP images should be captured by the G and B chips, respectively. Also, to improve spatial resolution along the z-axis, a spinning disk unit (Yokogawa, CSU21) may be placed in front of the camera. A series of confocal ratio images in pseudo-color acquired at video rate (Figure 4.3A and B) show the appearance and propagation of the $[Ca^{2+}]$ signal within the cytosol and nucleus of HeLa cells expressing YC3.60 and YC3.60$_{nu}$, respectively, after stimulation with histamine.

4.5 MONITORING RODENT NEURAL ACTIVITY WITH GFP-BASED FLUORESCENT Ca^{2+} INDICATORS

GFP-based fluorescent Ca^{2+} indicators will allow us to monitor the firing of neurons and muscles in complex tissues. To this end, it is currently thought that production of indicator transgenic animals followed by two-photon excitation microscopy-based *in vivo* analysis will produce the best results. Comparative studies on the properties of genetically encoded Ca^{2+} indicators have been performed using stable transgenic mouse lines producing inverse pericam or camgaroo-2 under the control of the tetracycline-inducible promoter [32]. A similar

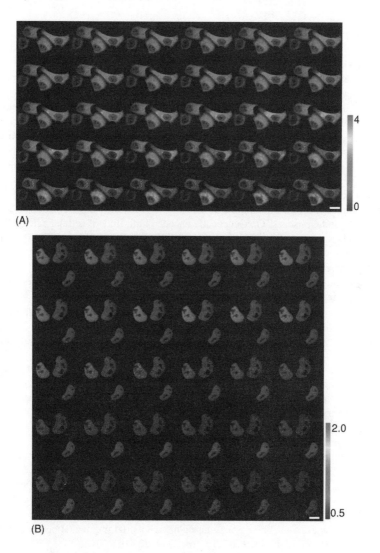

FIGURE 4.3 (See color insert following page 140.) Confocal pseudo-colored ratio images showing propagation of $[Ca^{2+}]$ in HeLa cells that expressed YC3.60 (A) and YC3.60$_{nu}$ (B).

approach was taken to characterize G-CaMP, inverse pericam, and camgaroo-2 in cultured hippocampal slices [33]. To date, the performance of these molecules for detecting neural activity varies greatly with cell types and animals used.

REFERENCES

1. Tsien, R.Y., Indicators based on fluorescence resonance energy transfer, in *Imaging in Neuroscience and Development*, Yuste, R. and Konnerth, A., Eds., Cold Spring Harbor Laboratory Press, New York, 2004.

2. Miyawaki, A., Nagai, T., and Mizuno, H., Genetic probes for calcium dynamics, in *Imaging in Neuroscience and Development*, Yuste, R. and Konnerth, A., Eds., Cold Spring Harbor Laboratory Press, New York, 2004.

3. Miyawaki, A., Visualization of the spatial and temporal dynamics of intracellular signaling, *Dev. Cell*, 4, 295, 2003.

4. Baird, G.S., Zacharias, D.A., and Tsien, R.Y., Circular permutation and receptor insertion within green fluorescent proteins, *Proc. Natl. Acad. Sci. USA*, 96, 11241, 1999.

5. Nakai, J., Ohkura, M., and Imoto, K., A high signal-to-noise Ca^{2+} probe composed of a single green fluorescent protein, *Nat. Biotechnol.*, 19, 137, 2000.

6. Nagai, T. et al., Circularly permuted green fluorescent proteins engineered to sense Ca^{2+}, *Proc. Natl. Acad. Sci. USA*, 98, 3197, 2001.

7. Miyawaki, A. et al., Fluorescent indicators for Ca^{2+} based on green fluorescent proteins and calmodulin, *Nature*, 388, 882, 1997.

8. Heim, N. and Griesbeck, O., Genetically encoded indicators of cellular calcium dynamics based on troponin C and green fluorescent protein, *J. Biol. Chem.*, 279, 14280, 2004.

9. Griesbeck, O. et al., Reducing the environmental sensitivity of yellow fluorescent protein. Mechanism and applications, *J. Biol. Chem.*, 276, 29188, 2001.

10. Yu, D. et al., Detection of calcium transients in *Drosophila* mushroom body neurons with camgaroo reporters. *J. Neurosci.*, 23, 64, 2003.

11. Wang, J.W. et al., Two-photon calcium imaging reveals an odor-evoked map of activity in the fly brain, *Cell*, 112, 271, 2003.

12. Wang, F. et al., Odorant receptors govern the formation of a precise topographic map, *Cell*, 93, 47, 1998.

13. Wang, Y. et al., Stereotyped odor-evoked activity in the mushroom body of Drosophila revealed by green fluorescent protein-based Ca^{2+} imaging, *J. Neurosci.*, 24, 6507, 2004.

14. Suh, G. S. et al., A single population of olfactory sensory neurons mediates an innate avoidance behavior in Drosophila, *Nature*, 431, 854, 2004.

15. Ji, G. et al., Ca^{2+} sensing transgenic mice: postsynaptic signaling in smooth muscle, *J. Biol. Chem.*, 279, 21461, 2004.

16. Robert, V. et al., Beat-to-beat oscillations of mitochondrial $[Ca^{2+}]$ in cardiac cells, *EMBO J.*, 20, 4998, 2001.

17. Shimozono, S. et al., Confocal imaging of subcellular Ca^{2+} concentrations using a dual-excitation ratiometric indicator based on green fluorescent protein, *Sci. STKE*, 125, PL4, 2002.

18. Frieden, M. et al., Ca^{2+} homeostasis during mitochondrial fragmentation and perinuclear clustering induced by hFis1, *J. Biol. Chem.*, 279, 22704, 2004.

19. Lalevee, N. et al., Intracellular transport of calcium from plasma membrane to mitochondria in adrenal H295R cells: implication for steroidogenesis, *Endocrinology*, 144, 4575, 2003.

20. Malli, R. et al., Sustained Ca^{2+} transfer across mitochondria is essential for mitochondrial Ca^{2+} buffering, store-operated Ca^{2+} entry, and Ca^{2+} store refilling, *J. Biol. Chem.*, 278, 44769, 2003.

21. Filippin, L. et al., Stable interactions between mitochondria and endoplasmic reticulum allow rapid accumulation of calcium in a subpopulation of mitochondria, *J. Biol. Chem.*, 278, 39224, 2003.

22. Shimozono, S. et al., Slow Ca^{2+} dynamics in pharyngeal muscles in *Caenorhabditis elegans* during fast pumping, *EMBO Rep.*, 5, 521, 2004.

23. Porumb, T. et al., A calmodulin-target peptide hybrid molecule with unique calcium-binding properties, *Protein Eng.*, 1, 109, 1994.

24. Mizuno, H. et al., Red fluorescent protein from *Discosoma* as a fusion tag and a partner for fluorescence resonance energy transfer, *Biochemistry,* 40, 2502, 2001.
25. Miyawaki, A. et al., Dynamic and quantitative Ca^{2+} measurements using improved cameleons, *Proc. Natl. Acad. Sci. USA,* 96, 2135, 1999.
26. Nagai, T. et al., A variant of yellow fluorescent protein with fast and efficient maturation for cell-biological applications, *Nat. Biotechnol.,* 20, 87, 2002.
27. Nagai, T. et al., Expanded dynamic range of fluorescent indicators for Ca^{2+} by circularly permuted yellow fluorescent proteins, *Proc. Natl. Acad. Sci. USA,* 101, 10554, 2004.
28. Palmer, A.E. et al., Bcl-2-mediated alterations in endoplasmic reticulum Ca^{2+} analyzed with an improved genetically encoded fluorescent sensor, *Proc. Natl. Acad. Sci. USA,* 101, 17404, 2004.
29. Rudolf, R. et al., *In vivo* monitoring of Ca^{2+} uptake into mitochondria of mouse skeletal muscle during contraction, *J. Cell Biol.,* 166, 527, 2004.
30. Suzuki, H. et al., *In vivo* imaging of *C. elegans* mechanosensory neurons demonstrates a specific role for the mec-4 channel in the process of gentle touch sensation, *Neuron,* 39, 1005, 2003.
31. Higashijima, S. et al., Imaging neuronal activity during zebrafish behaviour with a genetically encoded calcium indicator, *J. Neurophysiol.,* 90, 3986, 2003.
32. Hasan, M.T. et al., Functional fluorescent Ca^{2+} indicator proteins in transgenic mice under TET control, *Plos Biol.,* 2, 763, 2004.
33. Pologruto, T.A., Yasuda, R., and Svoboda, K., Monitoring neural activity and $[Ca^{2+}]$ with genetically encoded Ca^{2+} indicators, *J. Neurosci.,* 24, 9572, 2004.

5 Targeted Aequorins

Andrea Prandini and Rosario Rizzuto

CONTENTS

5.1 INTRODUCTION

Aequorin is a vintage Ca^{2+} probe through which seminal observations were made (e.g., the occurrence of agonist-dependent $[Ca^{2+}]_c$ oscillations). The advent of molecular biology techniques has recently broadened its applications, by allowing both recombinant expression (circumventing the need for microinjection of purified protein) and molecular engineering, with the development of novel Ca^{2+} probes specifically targeted to define cell locations. We will summarize here the principles of these probes and the procedures that allow their optimal utilization.

The Ca^{2+}-sensitive photoprotein aequorin derives like green fluorescent protein (GFP), the more popular acceptor of its emitted light, from the jellyfish *Aequorea victoria* [1]. Aequorin, in its active form, is composed of a 21-kDa apoprotein covalently bound to a hydrophobic prosthetic group, coelenterazine (MW~400 Da). Its polypeptide sequence includes three high-affinity Ca^{2+} binding sites (homologous to the sites present in other Ca^{2+} binding proteins, such as calmodulin). Ca^{2+} binding causes the rupture of the covalent link between the apoprotein and the prosthetic group, an irreversible reaction associated with the emission of one photon (Figure 5.1). The rate of this reaction depends on the $[Ca^{2+}]$ to which the protein is exposed. Indeed, at saturating $[Ca^{2+}]$ (>100 μM), the reaction is virtually instantaneous and the complete photoprotein pool is discharged. At lower $[Ca^{2+}]$ there is a relationship between the fractional rate of consumption and the $[Ca^{2+}]$, which has been characterized extensively for the cellular milieu (temperature, ionic condition, etc.) [2, 3]. Light emission is negligible at $[Ca^{2+}]$ lower than 10^{-7}. Then, in the physiological range of cytosolic $[Ca^{2+}]$ (10^{-7} to 10^{-5} M) the fractional rate of consumption (L/L_{max}, where L is light emission in a given instant and L_{max} is maximal light emission, i.e., the total available aequorin pool) increases steeply.

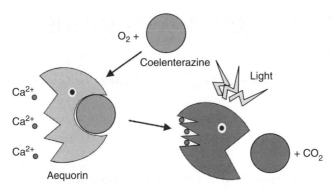

FIGURE 5.1 (See color insert following page 140.) Model of the irreversible aequorin reaction.

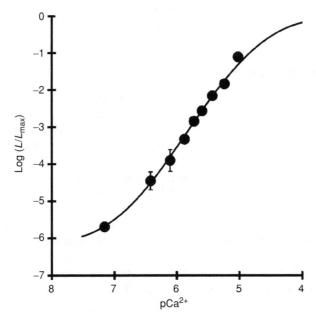

FIGURE 5.2 The $[Ca^{2+}]$ response curve of recombinant aequorin.

This relationship, shown in Figure 5.2, is the basis for the use of aequorin as a Ca^{2+} probe. Indeed, if all the light emitted by the photoprotein throughout an experiment ending with the complete discharge of the pool is collected, it is possible to estimate maximal light emission and then calculate back the $[Ca^{2+}]$ to which the photoprotein is exposed every moment.

Based on these properties, aequorin has been used extensively as a Ca^{2+} probe for the cytoplasm of intact living cells, allowing seminal observations to be made (such as, e.g., the occurrence of repetitive spiking of cytosolic $[Ca^{2+}]$ upon agonist stimulation) [4]. In these experiments, purified aequorin, extracted from the jelly-fish, was microinjected into living cells, and luminescence read in purpose-built luminometers. Due to the low light emission and the necessity of introducing it into the cytosol by traumatic maneuvers (microinjection, scrape loading), the use of the photoprotein was restricted to a few, large sessile cells and data were available only from a limited number of laboratories possessing specific knowledge and dedicated instrumentation.

The isolation of aequorin cDNA [5] elicited a renewed interest in this Ca^{2+} probe for many reasons. First, the recombinant expression of a protein from its cDNA eliminated one of the major limitations to the use of photoproteins, that is, the need of microinjecting a molecule that cannot freely cross the plasma membrane of living cells. Second, studies in populations of transfected cells allow one to obtain robust luminescent signals (and thus low-noise data), thus overcoming the intrinsic limitation of low light emission (typical of photoproteins). Most important, the protein can be molecularly engineered by appending specific targeting sequences to the Ca^{2+} measuring moiety. By this approach, novel probes can be generated that allow one to measure $[Ca^{2+}]$ with high subcellular specificity, i.e., within organelles or in defined cytosolic subcompartments. Indeed, while wild-type aequorin is exclusively cytosolic, the intracellular fate of the photoprotein can be modified by constructing chimeric cDNAs encoding a recombinant polypeptide composed of the photoprotein and a targeting signal (i.e., the amino acid sequence that directs the correct sorting of proteins to the proper intracellular localization, e.g., mitochondria, nucleus, endoplasmic reticulum, etc.). Along these lines, even more "tailored" Ca^{2+} probes can be designed: namely, by fusing aequorin with a protein of interest (e.g., an effector of the signals, such as a Ca^{2+}-dependent enzyme or channel), the Ca^{2+} microenvironment, and thus the real concentrations to which a Ca^{2+}-regulated protein is exposed can be directly assessed. Finally, a very useful application can be envisioned also for measuring $[Ca^{2+}]$ in the cytosol, i.e., the compartment in which the fluorescent dyes represent a flexible and efficient probe. This takes advantage of the procedure employed for introducing the Ca^{2+} probe, i.e., the transfection of cells with its cDNA. The aequorin cDNA can be co-transfected with a plasmid driving the expression of a protein of interest. In this way, in transient expression experiments a population of cells (20–30% of the total) is obtained that co-expresses the protein and the Ca^{2+} probe. Population measurement with cytosolic aequorin allows one to average the effect of the protein of interest on the Ca^{2+} signaling properties of a large number of cells (thus avoiding the intrinsic variability that plagues other approaches, such as the isolation of stably expressing cell clones or the single-cell analysis of positive cells, identified by co-transfecting a GFP marker).

5.2 TARGETING THE AEQUORIN cDNA TO CELLULAR COMPARTMENTS

Aequorin measurements are carried out in populations of cells (and thus the localization cannot be derived by the direct analysis of aequorin luminescence), so it is essential to demonstrate that the recombinant photoprotein is sorted to the expected cell location by immunocytochemical labeling. Given that aequorin is itself a fairly poor immunogen (and thus there are not very good antibodies available), we have appended a short (9-amino-acid) epitope tag (HA1, derived from hemagglutinin) to its protein sequence [6]. This HA1-tagged aequorin has been employed as the "building block" for all the chimaeras described here.

Cytosol: Recombinantly expressed wild-type HA1-tagged aequorin is exclusively cytosolic (cyt-AEQ) and, therefore, does not require any modification to measure [Ca^{2+}] in this compartment.

Mitochondria: Only 13 polypeptides are encoded by the mitochondrial genome. All the others are encoded by the nucleus, translated by cytosolic ribosomes, and then imported into the organelle. In most cases, the recognition as a mitochondrial protein and the import into the organelle depend on a signal sequence rich in basic residues that is located at the N-terminus of the protein, and is removed after import by proteases of the mitochondrial matrix [7]. For the mitochondrial matrix (mt-AEQ), HA1-tagged aequorin was fused to the cleavable targeting sequence (mitochondrial presequence) of a subunit (VIII) of cytochrome *c* oxidase, the terminal complex of the respiratory chain [8]. For the mitochondrial intermembrane space (mims-AEQ), aequorin was fused to the C-terminus of glycerol phosphate dehydrogenase, an integral protein of the mitochondrial inner membrane that has a large C-terminus domain protruding into the intermembrane space [9].

Nucleus: The transport of proteins to the nucleus is an active process and requires suitable nuclear localization signals (NLSs). Nuclear localization can be transient (i.e., it is physiologically regulated and varies in different cell conditions). This is achieved when specific cellular conditions (e.g., binding of hormones to the polypeptide) cause a conformational change that exposes (or masks) the NLS. NLS does not have a preferential position in the sequence and it is not removed after the import as the mitochondrial sequence [10, 11]. We developed two aequorin constructs containing an NLS. The first consists in a hybrid cDNA coding for aequorin and the NLS of the glucocorticoid receptor (excluding the hormone binding domain); the expressed protein (nu-AEQ) is constitutively located in the nucleus [12]. The second construct contains a much larger portion of the same receptor, including the hormone-binding domain, and can be directed either to the cytosol or to the nucleus based on the experimental condition (nu/cyt-AEQ). This probe is therefore cytoplasmic in the absence of glucocorticoid hormones, and nuclear in their presence. The advantage of this construct is that it allows the measurement of the cytoplasmic and nucleoplasmic [Ca^{2+}] with the very same probe and in similar experimental conditions (differing only for the presence of glucocorticoids) [13, 14].

Endoplasmic reticulum: Proteins are usually retained in this compartment via a double targeting signal. A hydrophobic sequence located at the N-terminus of the protein causes its translation on membrane-bound ribosomes and its insertion into the ER. The retention of proteins within the ER depends on a second signal that retrieves the proteins escaping into later compartments. The best characterized of these signals is the tetra-peptide KDEL, localized at the C-terminus of the protein [15]. The addition of a KDEL polypeptide to aequorin is in principle the straightforward approach to ensure a correct localization in the ER. Unfortunately, modification of the C-terminus of aequorin severely impairs its chemiluminescent properties [16]. An alternative strategy was thus designed. The aequorin–HA1 was fused to the N-terminal region of the immunoglobulin heavy chain, which includes the leader sequence (L), the VDJ, and the CH1 domains of an IgG heavy chain (er-AEQ) [17]. This chimera is retained in the ER because of the binding of the CH1 domain to the resident endogenous endoplasmic reticulum BiP [18]. This binding is displaced by the immunoglobulin light chain, so in the absence of the latter (i.e., when the heavy chain is expressed in cells other than plasma cells) the CH1 domain is selectively retained in the ER.

Other cell compartments: Alternative strategies are employed when the minimal targeting sequence responsible for the localization of the resident proteins are not fully characterized. Specifically, in the case of the sarcoplasmic reticulum (sr-AEQ) and of the Golgi apparatus (go-AEQ), aequorin targeting was achieved by fusing the photoprotein to calsequestrin, a protein confined to the terminal cisternae of striated muscle sarcoplasmic reticulum, and to the trasmembrane portion of sialyltransferase, a resident protein of the lumen of the medium-trans-Golgi, respectively [19, 20]. Finally, aequorin was directed to the subplasmalemmal rim of the cytoplasm by fusing it to SNAP-25 (25 kDa synaptosome-associated protein), a protein that is synthesized in the cytoplasm and is translocated to the plasma membrane upon posttranslational palmitoylation of cysteine residues (pm-AEQ) [21]. (Figure 5.3)

5.3   EXPRESSION AND DETECTION OF THE PROBE

The most commonly employed method to obtain expression of the recombinant protein is transfection. The transfection procedures mainly depend on the cell type employed. The calcium phosphate procedure is simple and inexpensive, and it has been successfully employed for transfecting a number of cell lines, including HeLa, L292, Cos 7, PC12, as well as primary cultures of neurons and skeletal muscle myotubes. Other transfection procedures have also been employed, such as liposomes, the gene gun, and electroporation. More recently, viral vectors have been developed. Viral constructs for some aequorins are available that can be transduced with high efficiency (80–90%) in a broad number of cells [22].

After recombinant expression the apoprotein must be reconstituted into a functional aequorin. This can be accomplished by incubating transfected cells with the chemically synthesised prosthetic group, coelenterazine. Coelenterazine is freely permeable across cell membranes and reconstitution may occur within all intracellular compartments to which the photoprotein has been targeted. During reconstitution time (1–2 h is generally sufficient for optimal results), active aequorin is partly consumed, with a rate proportional to the $[Ca^{2+}]$ of the cell domain in which the protein is located. While this does not represent a problem in compartments with low $[Ca^{2+}]$, specific experimental protocols must be employed in compartments with high $[Ca^{2+}]$ to counteract the reduction in reconstitution efficiency (see below).

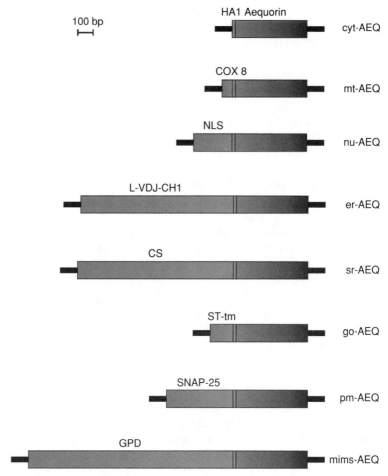

FIGURE 5.3 (See color insert following page 140.) Aequorin chimeras targeted to different intracellular compartments.

After reconstitution, the coelenterazine not incorporated in the photoprotein is washed away and luminescence measurements are carried out. The aequorin detection system is schematically shown in Figure 5.4 and consists of a perfusion chamber (where the coverslip with the cells is placed, and maintained in controlled conditions), on top of a hollow cylinder kept at a constant temperature of 37°C by water flow. Cells are continuously perfused via a peristaltic pump with thermostated saline solutions. The photomultiplier (EMI 9789 with amplifier–discriminator) is kept in a dark box and, during the experiment, the cells chamber is placed in close proximity. The output of the amplifier–discriminator is captured by an EMIC600 photon-counting board in an IBM-compatible PC. During the experiment, reagents to be tested are added to the perfusion medium and light output is continuously recorded.

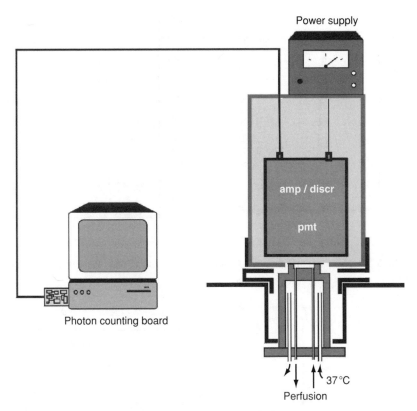

FIGURE 5.4 Schematic representation of the aequorin measuring system.

In order to calibrate the crude luminescent signal into $[Ca^{2+}]$ values an algorithm has been developed that takes into account the instant rate of photon emission and the total number of photons that can be emitted by the aequorin of the sample [6]. To estimate the latter value (i.e., the total dischargeable aequorin throughout the experiments), recording is ended after the complete lysis of the cells in a medium containing a high $[Ca^{2+}]$, thus ensuring light emission from the entire pool of reconstituted aequorin. The raw luminescence data are then calibrated using a custom-made program, and traces are obtained of the $[Ca^{2+}]$ at rest and after application of the compound(s) under investigation (representative experiments are shown in Figure 5.5 and Figure 5.6).

5.4 USE OF AEQUORIN IN CELL COMPARTMENTS WITH HIGH Ca²⁺ CONCENTRATIONS

Although aequorin is well suited for measuring $[Ca^{2+}]$ between 0.5 and 10 μM, in some intracellular compartments, the $[Ca^{2+}]$ is much higher (typically the intracellular Ca^{2+} stores). In this case, the high rate of the Ca^{2+}-dependent reaction (and light discharge) during the reconstitution phase or during the experiment reduces

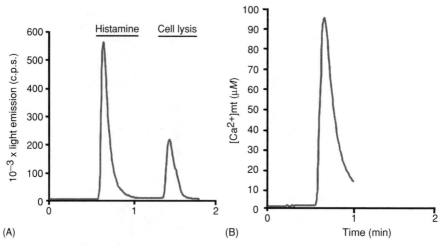

FIGURE 5.5 Calibration of light data (A) into Ca^{2+} concentration values (B) of a mitochondrial trace.

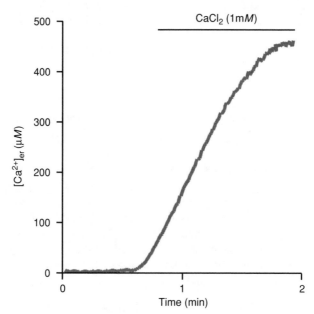

FIGURE 5.6 Representative example of a $[Ca^{2+}]_{er}$ measurement

the amount of active probe available and thus decreases the quality of the data obtained. For this purpose, it is necessary to lower the affinity of the photoprotein. There are at least three different approaches that can be combined to circumvent this experimental difficulty. The first is the mutation of the Ca^{2+}-binding sites. Indeed, all chimeras destined to compartments with high $[Ca^{2+}]$ include a mutation

(Asp-119 $\longrightarrow$ Ala) that causes an approximately 20-fold decrease in the Ca^{2+} affinity of the photoprotein [23]. The second approach is based on the use of a modified prosthetic group (coelenterazine n) that further decreases the affinity of the mutated photoprotein and gives rise to a probe with a very low rate of Ca^{2+}-dependent consumption [24]. It is also possible to follow a third approach, i.e., to use surrogate cations that elicit a slower rate of photoprotein consumption than Ca^{2+} itself, e.g., Sr^{2+} [17].

The choice of the approaches employed depend on the experimental needs, i.e., on the $[Ca^{2+}]$ range that needs to be measured. In our experience, we faced two types of experimental tasks. The first is that of measuring $[Ca^{2+}]$ within the mitochondrial matrix. In this case, $[Ca^{2+}]$ is low at rest (and this allows optimal aequorin reconstitution), but increases upon stimulation to values up to 100–200 μM (thus causing extensive consumption of wild-type aequorin in the early phases of the experiment). In this case, we obtained good results (and prolonged recordings of multiple stimulations) by using a mitochondrially targeted version of the low-affinity Asp-119 $\longrightarrow$ Ala mutant. A representative example of a mitochondrial trace obtained with mt-AEQmut is shown in Figure 5.5B.

The second case is that of the Ca^{2+} stores (ER and Golgi apparatus) that are endowed at rest with concentrations in the range of 200–500 μM. Here, two problems must be solved: aequorin reconstitution is largely inefficient with the full store and also after adequate reconstitution a probe of very low affinity should be used, to allow long-term measurements. On the former aspect, it is necessary to deplete the organelle of Ca^{2+} during reconstitution. We achieve this by switching the cells to a Ca^{2+}-free medium and applying three different protocols: (1) incubation with ionomycin (this allows a very effective emptying but requires extensive washes with a solution containing 2% BSA to remove the ionophore before beginning the physiological experiment), (2) incubation with the SERCA inhibitor tBuBHQ, and (3) repetitive challenges with IP_3-generating agonists. The two latter approaches are more effective for the ER than for the Golgi apparatus. After the Ca^{2+}-depletion protocol, coelenterazine is added (still in Ca^{2+}-free medium) and incubation is prolonged for ~1 h. Then the coverslip is transferred to the luminometer, while perfused with Ca^{2+}-free medium, and recording is started. At this point, the perfusion medium is supplemented with Ca^{2+}, and Ca^{2+} is thus accumulated into the store (thus increasing light emission). As to the optimal probe, we had successful experience in the case of both stores using the Asp-119 $\longrightarrow$ Ala mutant reconstituted with coelenterazine n. The fractional rate of consumption for $[Ca^{2+}]$ in the range of the $[Ca^{2+}]$ of these stores causes the discharge of most of the aequorin pool in ~30 min/1 h. Thus, most of the physiological experiments can be performed with this procedure: a representative example of a $[Ca^{2+}]_{er}$ measurement is shown in Figure 5.6. In most cases it is not necessary to employ the third approach, i.e., the use of the surrogate ion Sr^{2+}. This is advisable, as significant errors in the $[Ca^{2+}]_{er}$ estimates are introduced by discrepancies between the two cations (e.g., Ca^{2+} pumps have a much lower affinity for Sr^{2+}) and by artifactual mistakes in the Sr^{2+} calibration caused by traces of Ca^{2+} (that may cause substantial overestimation given the higher rate of aequorin consumption with Ca^{2+}).

5.5   TARGETING THE AEQUORIN cDNA TO SPECIFIC PROTEINS

An intriguing possibility for further expanding the uses of aequorin is to fuse the probe to specific proteins in order to monitor $[Ca^{2+}]$ to which these proteins are exposed in their specific site of actions. Specifically, this opportunity may give new information on the regulation of modulators or effectors of the Ca^{2+} signal.

In our experience a successful example of this methodology was the fusion of different PKC isoforms at the N-terminus of aequorin. These chimeras do not affect the activity of the kinases and immunocytochemical labeling, though the HA1 epitope present in the proteins confirms the correct cell localization of the chimeras both in resting state and after activation. These chimeras maintain the Ca^{2+} sensitivity and luminescence properties of aequorin. Thus, PKC–aequorin chimeras are useful biological sensors to record $[Ca^{2+}]$ to which the kinases are exposed in physiopathological conditions. Moreover, given the excellent signal-to-noise ratio of the photoprotein, they represent a simple and rapid method of pharmacological screening of new molecules acting on these important enzymes.

ACKNOWLEDGMENTS

Experimental work in the authors' laboratory was supported by grants from the Italian Ministry of Education (PRIN, FIRB, local research grants), the Emilia-Romagna region (PRRIITT), Telethon, and the Italian Association for Cancer Research.

REFERENCES

1. Shimomura O, Johnson FH, Saiga Y. Extraction, purification and properties of aequorin, a bioluminescent protein from the luminous hydromedusan, Aequorea. *J Cell Comp Physiol* 1962; 59: 223–239.
2. Moisescu DG, Ashley CC. The effect of physiologically occurring cations upon aequorin light emission. Determination of the binding constants. *Biochim Biophys Acta* 1977; 460: 189–205.
3. Cobbold PH, Rink TJ. Fluorescence and bioluminescence measurement of cytoplasmic free calcium. *Biochem J* 1987; 248: 313–328.
4. Cobbold PH. Cytoplasmic free calcium and amoeboid movement. *Nature* 1980; 285: 441–446.
5. Inouye S, Noguchi M, Sakaki Y et al. Cloning and sequence analysis of cDNA for the luminescent protein aequorin. *Proc Natl Acad Sci USA* 1985; 82: 3154–3158.
6. Brini M, Marsault R, Bastianutto C, Alvarez J, Pozzan T, Rizzuto R. Transfected aequorin in the measurement of cytosolic Ca^{2+} concentration ($[Ca^{2+}]c$). A critical evaluation. *J Biol Chem* 1995; 270: 9896–9903.
7. Hartl FU, Pfanner N, Nicholson DW, Neupert W. Mitochondrial protein import. *Biochim Biophys Acta* 1989; 988: 1–45.
8. Rizzuto R, Simpson AW, Brini M, Pozzan T. Rapid changes of mitochondrial Ca^{2+} revealed by specifically targeted recombinant aequorin. *Nature* 1992; 358: 325–327.

9. Rizzuto R, Pinton P, Carrington W et al. Close contacts with the endoplasmic reticulum as determinants of mitochondrial Ca^{2+} responses. *Science* 1998; 280: 1763–1766.

10. Kalderon D, Roberts BL, Richardson WD, Smith AE. A short amino acid sequence able to specify nuclear location. *Cell* 1984; 39: 499–509.

11. Dingwall C, Laskey RA. Nuclear targeting sequences — a consensus? *Trends Biochem Sci* 1991; 16: 478–481.

12. Brini M, Murgia M, Pasti L, Picard D, Pozzan T, Rizzuto R. Nuclear Ca^{2+} concentration measured with specifically targeted recombinant aequorin. *EMBO J* 1993; 12: 4813–4819.

13. Badminton MN, Campbell AK, Rembold CM. Differential regulation of nuclear and cytosolic Ca^{2+} in HeLa cells. *J Biol Chem* 1996; 271: 31210–31214.

14. Brini M, Marsault R, Bastianutto C, Pozzan T, Rizzuto R. Nuclear targeting of aequorin. A new approach for measuring nuclear Ca^{2+} concentration in intact cells. *Cell Calcium* 1994; 16: 259–268.

15. Munro S, Pelham HR. A C-terminal signal prevents secretion of luminal ER proteins. *Cell* 1987; 48: 899–907.

16. Nomura M, Inouye S, Ohmiya Y, Tsuji FI. A C-terminal proline is required for bioluminescence of the Ca^{2+}-binding photoprotein, aequorin. *FEBS Lett* 1991; 295: 63–66.

17. Montero M, Brini M, Marsault R et al. Monitoring dynamic changes in free Ca^{2+} concentration in the endoplasmic reticulum of intact cells. *EMBO J* 1995; 14: 5467–5475.

18. Sitia R, Meldolesi J. Endoplasmic reticulum: a dynamic patchwork of specialized subregions. *Mol Biol Cell* 1992; 3: 1067–1072.

19. Brini M, De Giorgi F, Murgia M et al. Subcellular analysis of Ca^{2+} homeostasis in primary cultures of skeletal muscle myotubes. *Mol Biol Cell* 1997; 8: 129–143.

20. Pinton P, Pozzan T, Rizzuto R. The Golgi apparatus is an inositol 1,4,5-trisphosphate-sensitive Ca^{2+} store, with functional properties distinct from those of the endoplasmic reticulum. *EMBO J* 1998; 17: 5298–5308.

21. Marsault R, Murgia M, Pozzan T, Rizzuto R. Domains of high Ca^{2+} beneath the plasma membrane of living A7r5 cells. *EMBO J* 1997; 16: 1575–1581.

22. Brini M, Bano D, Manni S, Rizzuto R, Carafoli E. Effects of PMCA and SERCA pump overexpression on the kinetics of cell Ca(2+) signalling. *EMBO J* 2000; 19: 4926–4935.

23. Kendall JM, Sala-Newby G, Ghalaut V, Dormer RL, Campbell AK. Engineering the Ca^{2+}-activated photoprotein aequorin with reduced affinity for calcium. *Biochem Biophys Res Commun* 1992; 187: 1091–1097.

24. Barrero MJ, Montero M, Alvarez J. Dynamics of $[Ca^{2+}]$ in the endoplasmic reticulum and cytoplasm of intact HeLa cells. A comparative study. *J Biol Chem* 1997; 272: 27694–27699.

6 Electrophysiological Recordings of Ca^{2+} Currents

Anant B. Parekh

CONTENTS

6.1 INTRODUCTION

The opening of plasmalemmal Ca^{2+} channels has profound effects on cell function. First and most importantly, an increase in cytoplasmic Ca^{2+} concentration is used as a key signaling messenger in virtually every cell through the phylogenetic tree, where it regulates a diverse array of fundamental physiological processes such as neurotransmitter release, muscle contraction, gene transcription, and cell growth and proliferation [1]. Second, when a Ca^{2+} channel opens, the inward flux of Ca^{2+} often depolarizes the membrane potential thereby affecting membrane excitability [2]. Finally, Ca^{2+} channels have fundamental roles in cell signaling that do not require ion permeation. In the T-tubules of skeletal muscle, the intracellular domain between loops II and III of the L-type calcium channel (Cav 1.1) can physically interact, ostensibly via the intermediary protein triadin, with ryanodine type I receptors in the sarcoplasmic reticulum. Depolarization of the T-tubules results in a conformational change in the L-type, that is, channel propagated rapidly to the ryanodine receptor, resulting in Ca^{2+} release [3].

Two major classes of Ca^{2+}-selective channel are known to exist. Voltage-gated Ca^{2+} channels are found in excitable cells like nerve and muscle, but are largely excluded from nonexcitable cells. Voltage-gated Ca^{2+} channels regulate synaptic transmission, muscle contraction and gene transcription [2]. Store-operated Ca^{2+}-permeable channels are a major source of Ca^{2+} in nonexcitable cells, where they control a diverse range of functions including secretion, enzyme activity, and cell growth and proliferation [4]. Store-operated Ca^{2+} channels are unique among known plasmalemmal ion channels in that they are specifically activated by the process of emptying the intracellular Ca^{2+} stores [4]. How this is accomplished remains contentious.

Two main methods have been utilized to measure Ca^{2+} influx namely fluorescence Ca^{2+} imaging and patch-clamp electrophysiology. These have been discussed in some depth in a recent review [4]. The most popular method is to use membrane-permeable fluorescent dyes like fura 2-AM that are easily loaded into cells in a noninvasive manner. In these experiments, an increase in cytoplasmic Ca^{2+} concentration is taken as a direct readout of plasmalemmal Ca^{2+} channel activity. Although straightforward to conduct, such experiments can be misleading. For example, the membrane potential is free to fluctuate and changes in the potential will impact upon the Ca^{2+} influx through changes in the driving force for Ca^{2+} entry. In addition, the cytoplasmic Ca^{2+} concentration at any given time reflects a balance between Ca^{2+} influx and removal. An increase in the latter would lower cytoplasmic Ca^{2+} even though Ca^{2+} channel activity might not have altered. There are a number of ways of reducing these problems when using fluorescent indicators and the reader is referred to Ref. [4] for specific details.

The flux of Ca^{2+} ions through Ca^{2+} channels generates an electric current that can be measured directly using the patch-clamp technique. Patch-clamp recordings remove the complications arising from changes in membrane potential and Ca^{2+} clearance that plague fluorescent Ca^{2+} measurements. Moreover, the whole cell configuration endows the experimenter with the ability to manipulate the intracellular composition of the cell as well as allowing membrane-impermeable drugs and proteins access to the cytoplasm. However, patch-clamp measurements are not devoid of problems. Whole cell dialysis can result in the loss of important components that regulate Ca^{2+} entry and the need to suppress other currents can result in the use of grossly nonphysiological solutions. In this chapter, I describe how store-operated and voltage-gated Ca^{2+} currents can be measured, provide electrophysiological protocols for studying some of their salient features, list the precautions needed to separate them from other currents that can be found in cells, and specify techniques available for measuring the single-channel conductance. I conclude by describing the new method of total internal reflection fluorescence microscopy that will provide further insight into Ca^{2+} channel function.

6.2 ELECTROPHYSIOLOGICAL RECORDINGS OF Ca^{2+} CURRENTS

6.2.1 Basic Considerations

A standard patch-clamp electrophysiology rig can be used to record store-operated and voltage-gated Ca^{2+} currents. It is important that the electrical noise of the recording system is minimized. The setup should be grounded well and, because store-operated currents are small, the patch-clamp amplifier is set to a high gain by choosing the high resistance setting of the feedback resistor in the amplifier headstage. For voltage-gated Ca^{2+} channels, a lower gain should be used to prevent amplifier saturation from the larger current flow.

A significant source of noise is the glass pipette itself. Hard glasses (aluminosilicate, borosilicate, or quartz) have lower dielectric loss and less noise than soft glass (soda glass, flint glass). Although soft glass can be pulled to form pipettes with lower resistance ($\sim$1MΩ) than hard glass, series resistance compensation is not a major concern in the recording of store-operated currents due to the small size of the current. Series resistance can be a problem when studying voltage-gated Ca^{2+} channels and care is therefore needed to ensure it remains as low as possible.

Noise from dielectric loss in the glass as well as the capacitance between the pipette interior and bath solution can be reduced by coating the pipette shank with silicone elastomer (Sylgard 184, Dow Corning, Midland, MI), which must then be cured (e.g., with a heat gun) before the pipette tip is fire-polished under a binocular microscope. Fire polishing smoothes the glass edges of the pipette tip (increasing the chances of seal formation) and removes contaminants left on the tip from the coating procedure. Pipettes therefore should be fire-polished after coating, not before. A simpler, albeit slightly less effective, way is to dip the pipette briefly into molten dental wax, rotate it in the wax, and then remove it and allow it to cool at room temperature. Wax clogging the pipette tip can be removed by subsequent fire-polishing. From our experience, sylgard-coated pipettes have lower electrical noise although both methods work well. To prevent dust particles in the atmosphere from settling onto the pipette tips, the pipettes should be prepared shortly before use and then stored in a clean, sealed jar or container.

Following the formation of a tight seal (typically > 10 GΩ), rupture of the patch (by suction) results in the generation of the whole cell configuration. The dominant source of noise in this configuration arises from the membrane capacitance (C) in series with the series resistance (R_s). The power spectra $S(f)$ of this combination is given by:

$$S(f) = 4kTR_s(2\pi fC)^2 / [1 + (2\pi fR_sC)^2]$$

Therefore, to minimize whole cell noise, small cells and a low series resistance are optimal. During whole cell recording, the series resistance should be monitored continuously. For the typically small store-operated currents, the voltage drop across the series resistance is typically less than 1 mV (for series resistance of <20 MΩ) and can therefore be neglected. Series resistance becomes more of an issue when store-operated currents are large, as occurs in divalent-free solution where Na$^+$ is the charge carrier or when voltage-gated Ca^{2+} channels are measured. These whole cell currents can easily exceed 1 nA in size (e.g., ventricular myocytes), resulting in substantial voltage errors unless series resistance is compensated for. In addition to voltage errors, increases in series resistance will reduce the effective rate of dialysis between the pipette and cytoplasm and hence needs to be monitored carefully.

6.2.2 HOLDING POTENTIAL

In whole cell recording, the membrane potential needs to be clamped in order to record the Ca^{2+} current. Store-operated currents are not voltage-activated and therefore cells are generally clamped at 0 mV rather than at the typical resting potential of the cell (e.g., -80 mV). At 0 mV, the electrical driving force for Ca^{2+} entry is much less than at -80 mV and this, coupled with the fact that store-operated currents like Ca^{2+} release-activated Ca^{2+} current (I_{CRAC}) show some inward rectification at hyperpolarized potentials, results in much smaller Ca^{2+} entry through the channels at 0 mV. As CRAC channels, for example, exhibit Ca^{2+}-dependent inactivation, clamping at 0 mV reduces the extent of channel inactivation thereby supporting more channel activity.

As the probability of voltage-gated Ca^{2+} channels opening is increased by depolarization, it is necessary to clamp the cells at quite hyperpolarized potentials, where channel opening is low (e.g., -80 mV). T-type (low voltage-activated) Ca^{2+} channels activate positive to -70 mV but exhibit rapid voltage-dependent inactivation. Hence, these channels will be inactivated by clamping the membrane potential at more depolarized potentials (e.g., ~-30 mV). A depolarized holding potential therefore allows for separation between T-type and the high voltage-activated channels (L, N, P/Q, R). The latter can be further separated on the basis of pharmacology, single-channel conductance, inactivation range, and kinetics of inactivation and deactivation. Differences in deactivation rates can be seen most easily by following the decay of the tail current obtained on stepping back to a hyperpolarized potential after the channels have opened.

6.2.3 VOLTAGE PULSES

Ca^{2+} currents can be recorded by applying voltage pulses. These can be either voltage steps to different potentials or voltage ramps. The advantage of ramps is that the entire current–voltage relationship can be obtained from each sweep. Ca^{2+} channels are generally very selective for Ca^{2+}, the reversal potential being very positive ($>+60$ mV). A departure from the positive reversal potential during an experiment would indicate that another current has developed (e.g., leak current) and hence the experiment may need aborting. In our laboratory, we routinely measure I_{CRAC} by applying voltage ramps that span -100 to $+100$ mV in 50 ms. We measure the amplitude of I_{CRAC} when the ramp potential is -80 mV, although any negative potential can be chosen. We choose -80 mV for two reasons. First, the current is relatively large and so analysis is simpler. Second, the transient capacity currents that arise when the membrane potential is stepped from 0 mV (holding potential) to -100 mV (start of the ramp), disappear completely by the time the ramp is at -80 mV. Whole cell currents can be affected by capacity transients if the potential chosen for measurements is set too close to the start of the ramp. This is not a problem with voltage-gated Ca^{2+} channels because the peak current tends to occur between -10 and $+10$ mV for high voltage-activated channels and -20 to 0 mV for low voltage-activated ones. For a given cell type, larger cells will tend to express more channels than smaller ones and hence the absolute magnitude of the current can vary between cells. For this reason, it is often useful to measure current density rather than absolute size. Current density can be obtained by dividing the current amplitude by membrane capacitance (which is directly proportional to the total membrane surface area).

The voltage ramp protocol provides no kinetic information and only limited insight into voltage- and Ca^{2+}-dependent inactivation of the underlying Ca^{2+} channels. Such information can be gleaned by using voltage steps. From the holding potential (e.g., 0 mV for store-operated channels and -80 mV for voltage-gated ones), hyperpolarizing (or depolarizing) pulses can be applied of varying magnitudes and durations and the kinetics of inactivation of the Ca^{2+} current during the pulse can be followed and then analyzed.

To record a Ca^{2+} current, the experimenter needs to cleanly separate the current from other membrane conductances found in the cell type used. For store-operated currents, a method is needed to deplete intracellular Ca^{2+} stores and, as store-operated currents are generally small, adopt maneuvers that enhance the size of the current. Each of these points is elaborated upon below.

6.3 STORE-OPERATED Ca^{2+} CURRENT

6.3.1 ISOLATION OF THE STORE-OPERATED CURRENT

To study a store-operated current, it is important to take precautions to ensure that other ionic currents have been suppressed. Which currents are present will depend on the cell type and this will influence the choice of experimental solution.

6.3.1.1 K$^+$ Currents

Outward K$^+$ currents are usually eliminated by using a Cs$^+$-based pipette solution. Cs$^+$ permeates most types of K$^+$ channel poorly and, in some cases, actually blocks the channels. A notable exception is the inwardly rectifying K$^+$ current that can present a problem for store-operated current measurements. Although intracellular Cs$^+$ does not permeate these channels well, it does not block them completely either. In the presence of external K$^+$, inward K$^+$ currents through the inward rectifier can arise. Following the onset of whole cell recording, cytoplasmic K$^+$ diffuses into the pipette and is replaced by Cs$^+$. Under these conditions, a small inward K$^+$ current through the inward rectifier can develop. This current has a positive reversal potential and can lead to an overestimate in the size of the store-operated one. Inwardly rectifying K$^+$ current needs to be suppressed and this can be achieved by including 5–10 mM Cs$^+$ or 0.1 mM Ba^{2+} in the extracellular solution. TEA and 4-AP, widely used general K$^+$ channel blockers, are only partially effective on inward rectifiers. As inwardly rectifying K$^+$ currents are quite widespread, it is important to ensure that they are blocked. An alternative approach is simply to use a K$^+$-based pipette solution. With 2–4 mM K$^+$ outside, the reversal potential of the current would be around −90 mV and hence only transient outward currents would be seen at more positive potentials. However, a major drawback with K$^+$-rich pipette solutions is that other K$^+$ currents may develop during the experiment, for example following receptor stimulation.

6.3.1.2 Cl$^-$ Currents

Cl$^-$ currents can be minimized by dialyzing cells with a low Cl$^-$ pipette solution, using glutamate, gluconate, or aspartate as the major anion. As the test electrode in the pipette holder is usually a silver wire coated with AgCl, at least a few mM Cl$^-$ are needed in the pipette solution to prevent voltage drift. As the coating inevitably becomes scratched during routine exchange of pipettes and is degraded with time owing to the flow of large currents, it is essential that the electrode be regularly chlorided.

6.3.1.3 Ca^{2+}-Activated Currents

Some cell types express Ca^{2+}-activated currents (potassium, chloride, nonselective) that need to be suppressed in order to record a pure store-operated current. Ca^{2+}-activated currents can be eliminated most easily by including high concentrations (several milliMolar) of Ca^{2+} chelator (EGTA or BAPTA) in the pipette solution. Because store-operated currents like I_{CRAC} are subject to Ca^{2+}-dependent inactivation mechanisms, this has the added advantage of reducing channel inactivation thereby resulting in a larger whole cell current. In those cells that are devoid of Ca^{2+}-activated currents, I_{CRAC} can be measured under conditions where no Ca^{2+} chelator is included in the pipette solution and cytoplasmic Ca^{2+} is free to fluctuate.

6.3.1.4 MagNuM/MIC

Like other Ca^{2+}-selective channels, store-operated channels like CRAC channels become permeable to Na$^+$ when divalent cations are removed from the external solution. Under these conditions, the Na$^+$ current through CRAC is approximately fivefold larger than the corresponding Ca^{2+} current [5]. However, it was reported that the Na$^+$ current could be increased further (~20-fold) by removing Mg^{2+} and Mg–ATP from the patch pipette solution [6]. Under such conditions, not only was the amplitude of the Na$^+$ current potentiated dramatically but the selectivity of the channels seemed to change too. Furthermore, single-channel currents could be observed even in the whole cell configuration and were interpreted as representing the openings of single store-operated channels [6]. However, it was subsequently found that loss of intracellular Mg^{2+}/Mg–ATP activates a nonselective current called Magnesium and Nucleotide gated inward Metal current (MagNuM), magnesium-inhibited current (MIC), or TRP-PLIK [4, 7, 8]. This current, which is not store-operated, is likely encoded by the TRPM7 gene and is widely distributed in nonexcitable cells. TRPM7 is thought to account for those earlier findings that were obtained in the absence of intracellular Mg^{2+}/Mg–ATP and which were attributed to I_{CRAC}. The TRPM7 channel conducts both Ca^{2+} and Mg^{2+} at negative potentials. At positive potentials ($>+50$ mV), a large outward current flows which is carried by Cs$^+$ (or K$^+$). This outward flux is thought to reflect relief from permeation block by divalent cations and the strong outward rectification provides a signature for identifying TRPM7 [7]. MagNuM/MIC is widespread and hence can interfere with recordings of store-operated currents. In RBL cells, dialysis with a pipette solution devoid of Mg^{2+} and Mg–ATP-activated MagNuM after a delay of ~100 s and with a half-time of 220 s [9]. Such slow kinetics is also seen for I_{CRAC} when stores are depleted slowly, for example following dialysis with strong intracellular buffer. Because it supports an inward current at negative potentials and reverses at ~+50 mV, MagNuM/MIC can overlap with store-operated currents and lead to a significant overestimation in the size of the current as well as confounding studies on permeation properties of store-operated channels [9]. It is therefore essential that MagNuM/MIC is suppressed. The simplest way to achieve this is to include Mg^{2+}/Mg–ATP in the pipette solution. The precise amount of Mg^{2+}/Mg–ATP that is required for this seems to vary between cell types. In the absence of Mg–ATP, 3 mM Mg^{2+} suppressed MagNuM in HEK-293 cells [7] whereas 1.1 mM Mg^{2+} was sufficient for RBL cells [7]. On the other hand, >6 mM Mg^{2+} has been recommended for Jurkat T lymphocytes [8]. In RBL-1 cells, we find that dialysis with 1 mM Mg^{2+} and 2 mM Mg–ATP is sufficient to suppress MagNuM fully [10]. Even after including Mg^{2+}/Mg–ATP in the pipette, it is important to check that MagNuM/MIC is not present. With Ca^{2+} as the charge carrier, I_{CRAC} is inwardly rectifying with a very positive reversal potential such that very little outward current can be detected up to +70 mV. Current through MagNuM/MIC, on the other hand, reverses at +50 mV and a large outward current is seen. Furthermore, 2-aminoethoxydiphenyl borane (2-APB) fully blocks I_{CRAC} (at concentrations >20 μM) and recovery from the block is very slow. MagNuM/MIC is only partially sensitive to similar concentrations of 2-APB and the block is readily reversible [9]. In divalent-free external solution, where Na^+ is the charge carrier, the Na^+ current through CRAC channels retains some inward rectification and reverses at ~+55 mV with PCs^+/PNa^+ of ~0.1 in divalent-free solution [8, 10]. MagNuM/MIC, on the other hand, has a linear current–voltage relationship, reverses close to 0 mV and PCs^+/PNa^+ ~1 [8]. In addition, in divalent-free external solution, spermine blocks MagNuM in a voltage-dependent manner but has no effect on the Na^+ current through CRAC channels [11]. Finally, MagNuM/MIC has a single-channel conductance of 40 pS in divalent-free external solution whereas the corresponding one for the CRAC channel is more than three orders of magnitude less (0.2 pS) [8].

6.3.1.5   TRPV6

The current through TRPV6 (initially called CaT1) channels has several features that are strikingly reminiscent of I_{CRAC}. TRPV6 currents are inwardly rectifying, highly selective for Ca^{2+} with a very positive reversal potential, exhibit a similar (but not identical) divalent cation conductivity profile, are permeable to monovalent cations in the absence of external divalents, and exhibit anomalous mole fraction effect between Ca^{2+} and Na^+. In fact, it was reported that TRPV6 comprised all or part of the CRAC channel pore [12]. Subsequent work has shown that TRPV6 is not store-operated [4, 13]. Instead, the channels are activated following a reduction in cytoplasmic Ca^{2+} concentration that alleviates a strong Ca^{2+}-dependent inactivation. As many experiments on store-operated channels involve the use of high concentrations of Ca^{2+} chelator like EGTA or BAPTA, which rapidly lower cytoplasmic Ca^{2+} levels, this raises the distinct possibility that TRPV6 could be activated under these conditions. It is therefore important to design experiments that check whether TRPV6 is contaminating the store-operated current. There are several ways whereby this can be tested. TRPV6 is largely suppressed at resting levels of cytoplasmic Ca^{2+}. Hence, dialyzing cells with a pipette solution in which Ca^{2+} is strongly buffered at levels just above resting Ca^{2+} ensures little contribution from TRPV6. For example, dialyzing RBL cells with a pipette solution containing 140 nM Ca^{2+} (10 mM total EGTA or BAPTA) fails to activate an inward current in the whole cell configuration. However, subsequent store depletion with thapsigargin or ionomycin triggers the development of I_{CRAC} [14]. Alternatively, using a pipette solution where Ca^{2+} is weakly buffered (0.1 mM EGTA) or omitting chelator from the pipette solution should not activate TRPV6 channels but will support the development of I_{CRAC}. Store-operated and TRPV6 channels also differ in their pharmacology. 2-APB fully blocks store-operated channels but has no effect on TRPV6 [13]. Hence, insensitivity of a Ca^{2+}-selective rectifying current to 2-APB would suggest that TRPV6 is operating under the conditions used. Further differences between I_{CRAC} and TRPV6 can be revealed when Na^+ is the charge carrier (divalent-free external solution). These include a voltage-dependent block/unblock of TRPV6 by intracellular Mg^{2+} that is not mirrored by CRAC and that likely accounts for an unusual slope of negative conductance in the current–voltage relationship of TRPV6 seen at quite hyperpolarized potentials, approximately four times greater Cs^+ permeability in TRPV6 that results in a less positive reversal potential, and different sensitivity to external ruthenium red [13]. In divalent-free solution, TRPV6 channels are inhibited by ruthenium red whereas CRAC channels are unaffected.

Most currents can therefore be eliminated by following standard protocols designed to suppress K^+ and Ca^{2+}-dependent conductances. However, special care is needed to block inwardly rectifying K^+ currents, MagNuM/MIC and, if present, TRPV6. These currents can be eliminated by adding 10 mM CsCl to the bath solution, and using a pipette solution that contains Mg^{2+} and Mg–ATP with Ca^{2+} either weakly buffered (although Ca^{2+}-dependent slow inactivation of store-operated channels could be a concern) or strongly buffered at slightly above resting levels.

Having designed appropriate internal and external solutions, the experimenter needs to consider liquid junctional potentials. As the patch pipette solution generally has a composition different to that of the bath solution, a junctional potential is generated across the pipette tip when it enters the bath. The size of this liquid junction potential depends on the concentrations and mobilities of the ions in both solutions. With a K^+-glutamate-based pipette solution and a standard NaCl-rich bath solution, a junction potential of $+10$ mV develops. This introduces a voltage offset such that the command potential needs to be adjusted appropriately. Electrode offsets can also arise if the Cl^- composition of the bath solution is altered during an experiment (e.g., switching from a saline ringer to one rich in sulphate), and these need compensating for if a silver chloride bath electrode is used.

6.3.2 STORE DEPLETION PROTOCOLS

There are a variety of ways to deplete stores and hence evoke store-operated Ca^{2+} entry [4]. Table 6.1 summarizes currently used methods. These can be broadly classified into two types: active and passive. Active methods increase the Ca^{2+} permeability of the stores and hence trigger a relatively rapid depletion. Such methods include an increase in cytoplasmic inositol 1,4,5-trisphosphate (InsP$_3$) levels (either by dialysis with InsP$_3$ or by stimulation of cell-surface receptors that couple to phospholipase C), dialysis with InsP$_3$ analogs, or external application of the Ca^{2+} ionophore ionomycin that passively transports Ca^{2+} across the store membrane. Slow methods rely on the passive leakage of Ca^{2+} from the stores and work by preventing the stores from refilling. Dialyzing the cytoplasm with high concentrations of the Ca^{2+} chelators EGTA or BAPTA chelate Ca^{2+} that leaks from the stores and thereby prevent store refilling. Similarly, exposure to the sarcoplasmic/endoplasmic reticulum Ca^{2+}–ATPase (SERCA) inhibitors like thapsigargin, cyclopiazonic acid, and di-*tert*-butylhydroquinone prevent the P-type ATPases from refilling the stores. In the presence of a continuous Ca^{2+} leak out, the stores are gradually depleted of Ca^{2+} and store-operated entry subsequently activates.

The method of choice will clearly depend on the nature of the specific experiment and each has advantages and disadvantages. Receptor stimulation is clearly the most relevant physiologically but is also the most complex trigger because

TABLE 6.1
Methods to Deplete Stores and Evoke Store-Operated Ca^{2+} Entry

Agent	How to Apply	Mechanism	Inhibitors	I_{CRAC} Kinetics
Receptor agonist	External	Raises $InsP_3$, which opens $InsP_3$-gated Ca^{2+} channels on the stores	Receptor antagonists, PLC blockers Heparin Xestospongin C	Fast; peaks in ~80 s
$InsP_3$	Dialysis	Opens store $InsP_3$-gated Ca^{2+} channels	Heparin, Xestospongin C	Fast; peaks in ~80 s
$InsP_3$ analogues: $Ins2,4,5P_3$; adenophostin A	Dialysis	As above	Heparin, Xestospongin C	Fast; peaks in ~80 s
Ionomycin	External	Ca^{2+} ionophore		Fast; peaks in ~80 s
Thapsigargin	External or dialysis	Prevents store refilling in presence of background Ca^{2+} leak		Slow; peaks in ~300 s
EGTA or BAPTA	Dialysis	As above		Slow; peaks in ~300 s
TPEN	External; in divalent-free solution	Chelates Ca^{2+} within the stores		Fast, peaks within a few seconds

numerous signal transduction pathways can be activated simultaneously. Hence dissecting out specific regulatory mechanisms can be difficult. Dialyzing with $InsP_3$ or its nonmetabolizable analogues like $Ins2,4,5P_3$ or adenophostin A is simpler, but one concern is that the whole cytoplasm will be flooded rapidly with $InsP_3$ and spatial gradients, which probably occur with modest levels of stimulation, will be lost. In addition, some cell types may have $InsP_3$-gated channels in the plasma membrane that will provide a nonselective current that contaminates I_{CRAC}. Ionomycin depletes stores quickly, but at the concentrations used in electrophysiological experiments (μM range), it will transport Ca^{2+} into the cell across the plasma membrane directly thereby increasing subplasmalemmal Ca^{2+} concentration. In addition, ionomycin releases Ca^{2+} from mitochondria and thus removes the organelle as a functional Ca^{2+} store. SERCA pump blockers deplete stores slowly but are very sticky and adsorb to plastics.

A novel method to empty stores is through the use of N, N, N', N'-tetrakis (2-pyridylmethyl) ethylene diamine (TPEN; [15]). This chelator, in its noncomplexed form, can cross the plasma membrane and access the stores. Once inside the stores, it chelates lumenal calcium. This results in a reduction in free Ca^{2+} within the stores and thus activation of store-operated entry. TPEN has to be applied in the absence of divalent cations to ensure it remains in its noncomplexed membrane-permeable form. TPEN has the advantage that it depletes stores without requiring a rise in cytoplasmic Ca^{2+} concentration, ruling out the possibility that a Ca^{2+}-activated pathway has been activated, rather than, or in addition to, a store-operated one. Hence, TPEN may be a good way to empty stores under physiological conditions of weak intracellular Ca^{2+} buffering. However, TPEN chelates heavy metals, many of which are important cofactors for intracellular enzymes. Moreover, the concentration of TPEN within the stores remains unknown and it presumably will access all organelles.

6.3.3 EXPERIMENTAL METHODS TO INCREASE THE SIZE OF THE STORE-OPERATED Ca^{2+} CURRENT

6.3.3.1 External Solution

As store-operated currents tend to be small, maneuvers that enhance their size will greatly facilitate the detection and analysis of the current. Depending on the cell type, the K_D for Ca^{2+} permeation through CRAC channels is between 0.8 and 2.3 mM (reviewed in Ref. [4]). With physiological concentrations of external Ca^{2+} (1–2 mM), the current amplitude would be only ~50% of maximum. However, the size of the current can be increased simply by elevating external Ca^{2+} concentration to 10–20 mM. Alternatively, external Ca^{2+} can be replaced with Ba^{2+}. Ba^{2+} permeates store-operated channels but has the additional advantage that it does not mimic Ca^{2+} in triggering Ca^{2+}-dependent inactivation pathways [4]. Ba^{2+} is toxic to cells though and as the cation is not transported by Ca^{2+} pumps, the only way it can leave the cytoplasm is via diffusion into the patch pipette. On the other hand, if the current density is low, simply raising Ca^{2+} is probably insufficient to evoke a large current. For example, the store-operated calcium current in HEK293 cells is very small. Even in 10 mM Ca^{2+} it is <-0.5 pA/pF at -80 mV. For a HEK cell of 10 pF capacitance, this corresponds to a whole cell current of just <-5 pA that is of similar size to the background leak current. Changes in leak current will therefore have a significant effect on the size of the whole cell current. A convenient way to increase the size of the current through store-operated channels several-fold especially through those that are selective for Ca^{2+} (like CRAC channels) is to exploit the fact that they lose their selectivity somewhat in the absence of divalent cations. Complete removal of external Ca^{2+} and Mg^{2+} results in a large Na^+ flux through the channels. This is a characteristic of all known Ca^{2+}-selective channels including VOCC, CRAC, and TRPV5/6. In divalent-free solution, the steady-state store-operated current in HEK293 cells increases from <-0.5 pA/pF (with Ca^{2+} as charge carrier) to ~-3 pA/pF, permitting its characterization. In divalent-free solution, it is not sufficient simply to omit external divalent cations. This is because the channels show anomalous mole fraction and the Na^+ current is blocked by Ca^{2+} and Mg^{2+}. Ca^{2+} blocks Na^+ permeation through CRAC channels with a K_D of ~10 μM. For this reason, it is important to add a chelator like EDTA that can lower both Ca^{2+} and Mg^{2+} extracellularly. Another advantage of using divalent-free solution is that small outward currents can be resolved through CRAC channels. Hence, a reversal potential for the CRAC channel can be discerned, and the permeability profile of the channels (providing clues into permeation mechanisms and pore size) can be obtained. One problem with using divalent-free solution is that channel activity is relatively transient. After initial potentiation, the current declines to a low value. The reason for this is not entirely clear but may indicate a requirement for an occupied external Ca^{2+} binding site to support channel function.

6.3.3.2 Internal Solution

CRAC channels are subject to at least three Ca^{2+}-dependent feedback inactivation mechanisms that contribute to reducing the size of the whole cell current (see Ref. [4] for a review). These mechanisms include fast inactivation that reflects local feedback by permeating Ca^{2+} ions onto a site close to the channel pore itself, and two slower processes involving Ca^{2+}-dependent store refilling and a Ca^{2+}-dependent but store-independent pathway. Although thapisgargin will prevent store refilling, it will normally increase cytoplasmic Ca^{2+} and hence could accelerate the development of the slow inactivation mechanism. An effective way to suppress these slower mechanisms is to include a high concentration of Ca^{2+} chelator (several mM EGTA) in the patch pipette. In the presence of high levels of intracellular Ca^{2+} buffering, I_{CRAC} is large. Because of its slow on-rate, EGTA has little impact on fast inactivation. However, BAPTA, which is a more rapid Ca^{2+} chelator, reduces fast inactivation. At high concentrations though (several mM), BAPTA might affect other proteins directly and therefore care is needed when using such high concentrations in a whole cell experiment [16].

6.3.4   LEAK SUBTRACTION PROTOCOLS TO ISOLATE THE STORE-OPERATED Ca^{2+} CURRENT

To isolate the pure store-operated current, it is necessary to subtract the background (leak) currents. A variety of methods have been used in the literature for leak subtraction. One method is to average a series of ramp currents that have been obtained just prior to store depletion and then subtract this from all subsequent currents. Alternatively, after the current has developed, it can be inhibited by a channel blocker (e.g., Gd^{3+} or La^{3+}) and the current that remains in the presence of the blocker is regarded as the nonstore-operated component. Finally, stores can be depleted in the absence of external Ca^{2+} and the current under these conditions is considered the background one. The store-operated current is then the current obtained in the presence of external Ca^{2+} minus the background one. Each method makes assumptions that need to be justified for the specific cell type used. Application of a channel blocker like La^{3+} (which is not specific for store-operated channels) assumes first that background conductances are unaffected and second, that the effects of the cation on surface charge can be neglected. Similar assumptions are made for the protocol in which the store-operated current is obtained by subtracting the current in the presence of Ca^{2+} from that in its absence. Both methods can be tested by examining the effects of the channel blocker or readmitting Ca^{2+} on background currents in the absence of store depletion. Subtracting currents prior to store depletion assumes the absence of time-dependent conductance changes and that adequate exchange has taken place between the pipette solution and cytoplasm. This can be tested simply by dialyzing the cell in the absence of store depletion and monitoring the extent of the background current.

6.3.5 RECORDING STORE-OPERATED Ca^{2+} CURRENT UNDER PHYSIOLOGICAL CONDITIONS

Although experiments with high intracellular Ca^{2+} buffer can be advantageous under certain conditions, it is preferable to use weak buffer when one wishes to examine how the channels are regulated physiologically. The extent of intracellular buffering (termed the Ca^{2+} binding ratio) differs between different cell types [17]. For a given injection of Ca^{2+} into the cytoplasm, the Ca^{2+} binding ratio is a measure of the number of Ca^{2+} ions that have been bound to buffer divided by the number of Ca^{2+} ions that are still free. For many cell types, the ratio is around 50–100, although pancreatic acinar cells (6000) and cerebellar Purkinje neurons (1000) are striking exceptions. A Ca^{2+} binding ratio of ~100 corresponds roughly to 20 μM free EGTA. However, when InsP$_3$ is used to deplete Ca^{2+} stores in weak buffer (80 μM free EGTA, 0.1 mM EGTA total), no I_{CRAC} can be detected [18]. This is not due to Ca^{2+}-dependent inactivation of the channels because the current is still absent when Ba^{2+} is used as the charge carrier. It turns out that if precautions are taken to ensure mitochondria remain energized in the whole cell configuration, for example by supplementing the pipette solution with a cocktail that maintains mitochondrial metabolism, then I_{CRAC} can develop in weak buffer [18]. In whole cell recording, small metabolites that are important for maintaining mitochondria in an energized state, diffuse out of the cytoplasm into the patch pipette. As the pipette has a much larger volume, this results in loss of these metabolites from the cytoplasm (washout) and mitochondrial Ca^{2+} buffering is reduced. Hence, with weak Ca^{2+} buffer in the patch pipette, it is important that intracellular Ca^{2+} buffering mechanisms are maintained. A corollary of this is that the whole cell configuration can result in the loss of diffusible factors that are important in regulating store-operated influx. However, this limitation of whole cell recording should not be over-exaggerated. The activation mechanism of I_{CRAC} does not seem to washout within the time course of a typical whole cell experiment nor do the kinetics of activation seem to be affected either [14]. Furthermore, many regulatory pathways are still functional after whole cell dialysis. Nevertheless, if washout is a concern for the type of experiment conducted, an alternative approach is to use the perforated patch configuration (see Section 6.5).

6.4 RECORDING OF VOLTAGE-GATED Ca^{2+} CURRENT

Since the discovery of voltage-gated Ca^{2+} channels by Fatt and Ginsborg in 1958 [19], voltage-gated Ca^{2+} channels have been extensively studied. Hence, we have a very good understanding of how to isolate and measure them. Here, I will summarize the main points and discuss some recent developments concerning washout of channel activity in the whole cell configuration.

As with store-operated currents, K^+ currents need to be suppressed in order to reliably measure the voltage-gated Ca^{2+} currents. As excitable cells express a variety of different K^+ channels, it is often not enough simply to use a Cs^+-based pipette solution. Instead, broad K^+ channel blockers like TEA and 4-AP are needed, depending on the complement of K^+ channels found. Ca^{2+}-activated currents can also be a problem, especially as tail currents at the end of the depolarizing pulse. Therefore, either Ca^{2+} chelators (added to the pipette solution) or selective blockers of the Ca^{2+}-dependent currents may need to be used. Voltage-gated Na^+ channels need to be blocked by tetrodotoxin, for example. Alternatively, a holding potential positive to -40 mV is often used so that the Na^+ channels inactivate. But this protocol also eliminates T-type Ca^{2+} currents. MagNuM and TRPV6 currents are unlikely to cause problems as they are very small compared with the voltage-gated currents. Certain classes of voltage-gated Ca^{2+} channel are strongly inactivated by Ca^{2+}. As with CRAC channels, Ba^{2+} does not trigger much inactivation and therefore is often used as a replacement charge carrier.

One major problem with certain voltage-gated Ca^{2+} currents is that they tend to run down quite rapidly with time. This occurs within minutes following the onset of the whole cell configuration. Channel activity in excised patches runs down even more quickly. This rundown is thought to reflect the loss of molecules that are important in maintaining channel activity in the whole cell configuration or when a patch of membrane is excised from the cell (washout). The precise mechanism of rundown is not clear and it likely reflects contributions from several mechanisms. For L-type Ca^{2+} channels, rundown can be slowed considerably by including Mg–ATP in the pipette solution [20]. In some cell types, supplementing the internal solution with cAMP or protein kinase A reverses the washout process [21]. Hence, loss of phosphorylation may be one contributing factor. Rundown of L-type Ca^{2+} channels can also be slowed by adding calpastatin to the cytosolic side of the patch [22]. Calpastatin is an endogenous inhibitor of calpain, a Ca^{2+}- activated neutral protease found in many cell types. However, synthetic inhibitors of calpain do not mimic the actions of calpastatin, leading to the suggestion that calpastatin may support Ca^{2+} channel activity by binding directly to the channels rather than through inhibition of proteolysis [23]. In addition, a cytoplasmic extract has been found to restore the L-type Ca^{2+} current after rundown in excised patches from ventricular myocytes [24]. The active ingredient had a protein-like nature in that it exhibited both heat and trypsin sensitivity. It may be related to calmodulin, which was found to re-activate L-type Ca^{2+} channels that had already rundown in excised patches [25].

P/Q-type (Cav 2.1) Ca^{2+} channels also show rundown [26]. Recombinant P/Q-type Ca^{2+} channel activity, measured with Ba^{2+} as the permeant ion, ran down with a half-time of 6 min following excision of giant membrane patches from *Xenopus* oocytes. Brief exposure of the cytosolic side of the patch to the membrane phospholipid phosphatidyl-inositol-4,5-bisphosphate $(PtdIns(4,5)P_2)$ dramatically slowed rundown. The half-time increased by almost a factor of six. $PtdIns(4,5)P_2$ was also able to reactivate channels that had already run down. Mg–ATP also slowed rundown but through an action independent of protein kinase A. Hence

PtdIns(4,5)P$_2$ is able to dramatically reduce the rate of rundown of P/Q-type channels. PtdIns(4,5)P$_2$ had an additional action on P/Q-type Ca^{2+} channels. The phospholipid exerted a voltage-dependent inhibitory effect such that weak depolarizations were unable to open the channels. Intriguingly, this inhibitory effect was prevented by protein kinase A-dependent phosphorylation. Hence the phosphorylated state of the channels may determine the efficacy of PtdIns(4,5)P$_2$ in preventing rundown [26].

In contrast to high voltage-activated Ca^{2+} channels, T-type channels are quite stable and show little rundown over a time course of several minutes.

As with store-operated Ca^{2+} current, perhaps the best way to bypass potential run-down problems is to use the perforated patch configuration. However, with the larger amplitude of the Ca^{2+} currents here and the need to suppress a multitude of K^+ currents, this is more prone to clamp artefacts and errors.

6.5 PERFORATED PATCH RECORDINGS

One way to minimize rundown of voltage-gated Ca^{2+} channels as well as the loss of important regulatory factors that support store-operated Ca^{2+} entry under physiological conditions of weak intracellular Ca^{2+} buffering is to use the perforated patch technique. With this method, the electrical connection between the cell and pipette is brought about by the insertion of pore-forming molecules that have been included in the pipette solution. If the pores are small enough, then important cytoplasmic molecules will not be able to permeate the pores and hence are restricted to the cytoplasm. In perforated patch recordings, the patch pipette tip is first dipped into the internal solution for a few seconds and then backfilled with the same solution but supplemented with nystatin (50–250 µg/ml) or amphotericin B (~30 µg/ml). These polyene antibiotics from the genus *Streptomycetes* form pores that are permeable to small ions. Following the formation of a cell-attached patch, the antibiotics form pores in the patched membrane that greatly lower the access resistance from a few GΩ to around 10–40 MΩ, providing reasonably good electrical continuity between the recording pipette and the intracellular environment. Hence, whole cell currents can now be recorded. Nystatin and amphotericin B are both selective for monovalent ions (Li^+, Na^+, K^+, Rb^+, Cs^+, and Cl^-) but are impermeable to Ca^{2+} and Mg^{2+} as well as molecules with a molecular weight > 200 (sucrose, glucose-6-phosphate, $NADP^+$, ATP). Hence, the intracellular composition of the cytoplasm is largely preserved. Conductance through the nystatin or amphotericin pores follows the profile: $Rb^+ > K^+ > Na^+ > Li^+$, which follows the order of size of the hydrated ions. Although perforated patch recordings are straightforward in principle, there are several issues the experimenter needs to pay heed to. The resistance and shape of the patch pipette as well as the time used for tip dipping can all influence the time needed to obtain an acceptable access resistance. Low resistance, bullet-shaped pipettes with a small shank are preferable to long, thin pipettes. Similarly, if seals tend to form quickly, then tip dipping should be as brief as possible. Dilute nystatin and amphotericin B solutions are sensitive to light and heat and should be kept in the dark and on ice during experiments. In addition, nystatin and amphotericin B stock solutions (stored at $<0°C$) should be added to a small tube containing internal solution and then dispersed by ultrasonication. After formation of a cell-attached patch, access resistance should fall to <40 MΩ within 5 min. Perforation is monitored by following the slow increase in the size of the transient capacity current obtained by applying small voltage steps via the patch pipette. An attractive extension of the perforated patch technique is to combine it with fluorescence measurements of cytoplasmic Ca^{2+}. The dye can easily be loaded into the cells by using the membrane-permeable ester form. In this way, cytoplasmic Ca^{2+} signals and membrane currents can be recorded simultaneously in a physiological context.

Although the perforated patch method reduces disruption to the cytoplasm, it is not without limitations. First, membrane-impermeable molecules (kinases, second messengers, various inhibitors) cannot access the cytoplasm and hence experiments are somewhat limited in scope. Second, the access resistance is typically three to five times larger than in conventional whole cell recording resulting in a poorer voltage clamp. With large currents (e.g., Na^+ current through CRAC channels in divalent-free solution), the voltage drop that occurs can be significant and series resistance compensation may be needed. Finally, nystatin and amphotericin B have a small, but finite, permeability to Cl^- ($P_{Cl}/P_{cation} \sim 0.1$). Hence, with a KCl-based pipette solution, a Donnan-type imbalance of anions between pipette solution and cytoplasm can develop, resulting in significant junction potentials and cell volume changes. For this reason, pipette solutions containing a reduced Cl^- concentration are widely used, replacing some Cl^- with sulphate or methanesulfonate.

6.6 ESTIMATE OF SINGLE Ca²⁺ CHANNEL CONDUCTANCE

A characteristic property of any ion channel is its single-channel conductance. Single-channel recording is the most direct and unequivocal way to determine this. However, if channel density is very high and single-channel conductance is very low (beyond the detection of current patch clamp amplifiers), then single-channel recordings will not be possible. Instead, channel conductance can be estimated using the technique of fluctuation analysis.

6.6.1 SINGLE-CHANNEL RECORDING

In these experiments, a high-resistance pipette (10–20 MΩ) is sealed onto the surface of the cell membrane to form a cell-attached patch. With high resistance pipettes, only a small patch of membrane is recorded from, increasing the likelihood of observing single-channel events. After a stable seal has formed, voltage-gated Ca^{2+} channels can be activated by applying depolarizing pulses whereas store-operated channels can be opened following store depletion (using receptor stimulation or SERCA pump blockers). Cell-attached patches tend to be more stable than other patch configurations and, because the cytoplasm remains unperturbed, is also the most physiological. The pipette solution is the external solution for the patch.

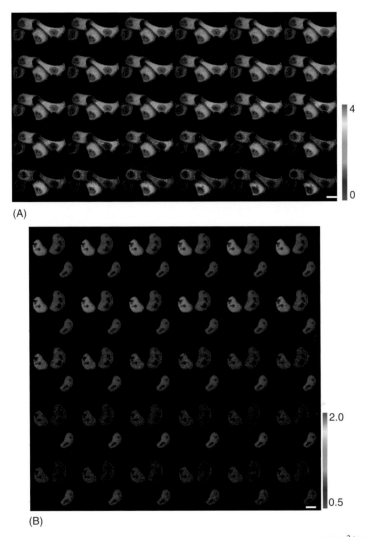

(A)

(B)

FIGURE 4.3 Confocal pseudo-colored ratio images showing propagation of $[Ca^{2+}]$ in HeLa cells that expressed YC3.60 (A) and YC3.60$_{nu}$ (B).

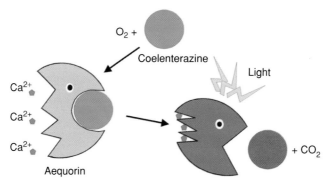

FIGURE 5.1 Model of the irreversible aequorin reaction.

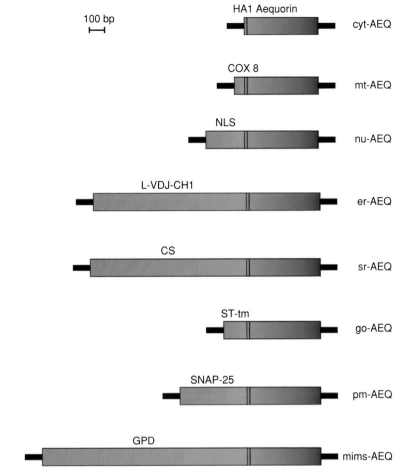

FIGURE 5.3 Aequorin chimeras targeted to different intracellular compartments.

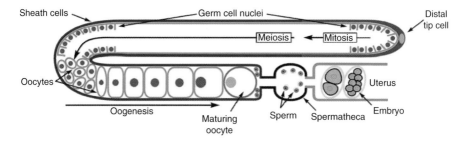

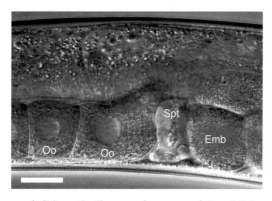

FIGURE 8.2 *Top panel:* Schematic diagram of one arm of the adult hermaphrodite gonad. *Bottom panel:* Differential interference contrast micrograph of adult hermaphrodite gonad. Oo, oocyte; Spt, spermatheca; Emb, embryo. Scale bar is 20 μm. [From Rutledge et al. [60] with permission from *Current Biology*, copyright Elsevier B.V.]

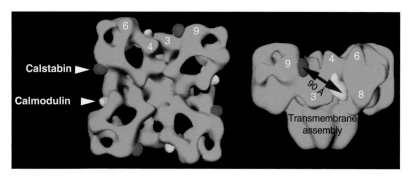

FIGURE 11.1 Three-dimensional reconstruction of ryanodine receptor in complex with calmodulin and calstabin. Solid-body representations of 3D reconstruction of RyR together with the differences attributed to calmodulin (yellow) and calstabin (pink). Numerals indicate selected RyR domains. (Modified from Wagenknecht T, Radermacher M, Grassucci R, Berbowitz J, Xin HB, Fluscher S. *J Biol Chem* 1977, 272, 32463–71.)

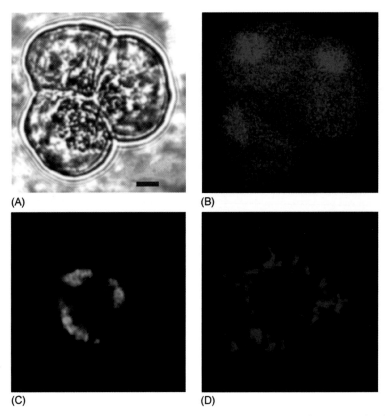

FIGURE 15.2 Measurements of cytosolic calcium and organelle location. Transmitted light image (A) of a triplet of freshly isolated pancreatic acinar cells loaded with the calcium sensitive dye Fura Red (B, note the slightly higher fluorescence in the nuclei) and NBD C_6 Ceramide (C) to visualize the Golgi. UV excitation of NADH autofluorescence reveals the location of the mitochondria (D).

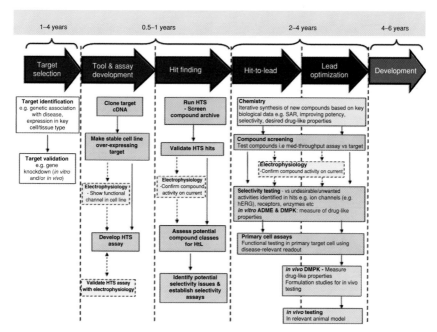

FIGURE 17.3 The main phases of drug discovery. Additional activities required for ion channel targets are indicated by boxes with dashed outlines. Data source for duration of drug discovery phases is Bannerjee, P. and Baker, A. (2001): http://www.accenture.com/xdoc/en/industries/hls/pharma/hpdd.pdf. HTS (high-throughput screen), HtL (hit to lead), ADME (absorption, distribution, metabolism, elimination), DM (drug metabolism and pharmacokinetics).

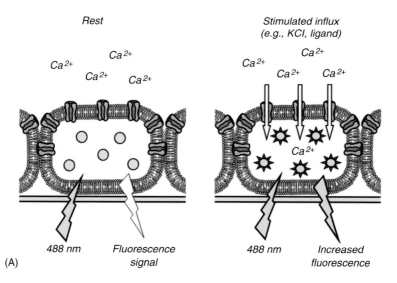

FIGURE 17.5A Ion channel assay formats commonly employed in drug discovery. Optical formats: (A) Ca^{2+} sensitive dyes such as Fura-2 and Fluo-4 allow direct real time quantitation of Ca^{2+} flux into cells and are compatible with a variety of high throughput plate readers such as FLIPR.

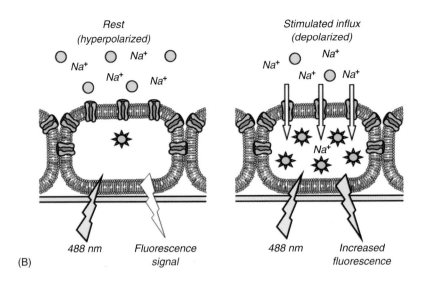

(B)

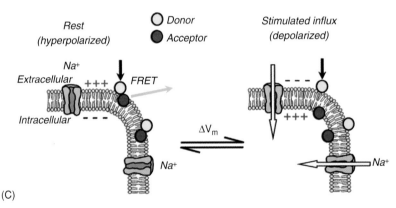

(C)

FIGURE 17.5B,C (B) Potentiometric dyes are sensitive to charges on the inner and outer leaflets of cell membranes. Redistribution dyes such as the oxonol dyes partition into the intracellular compartment upon cellular depolarization, binding to intracellular proteins and increasing their fluorescence emission. This is easily detected by standard fluorescence plate readers. (C) The FRET-based voltage-sensor probes (VSP dyes) utilize two dyes, a donor and an acceptor to produce a FRET signal. Upon cellular depolarization the FRET donor migrates to the inner leaflet of the cell membrane away from the acceptor and reduces the FRET signal. As the acceptor only has to move a short distance to elicit a change in signal, VSP dyes are very rapid to respond to changes in membrane potential.

6.6.1.1   Voltage-Gated Ca^{2+} Channels

Due to their relatively large size, single-channel recordings from voltage-gated Ca^{2+} channels have been carried out extensively. Low voltage-activated T-type channels have the smallest single-channel conductance of the voltage-gated channels but even this is nevertheless sizeable (8 pS). Because voltage-gated Ca^{2+} channels are all permeable to Ba^{2+}, a popular way to measure single-channel activity is to use a pipette solution containing isotonic $BaCl_2$ (110 mM; Ref. [27]). In addition, voltage-gated Na^+ channels are usually eliminated by including Tetrodotoxin in the pipette solution. Kinetics of single-channel openings following depolarizing pulses as well as the effects of channel block by heavy metals have provided much insight into channel gating and permeation mechanisms.

6.6.1.2   Store-Operated Ca^{2+} Channels

In some cell types like aortic smooth muscle, A431 epidermal cells and endothelia, resolvable store-operated single-channel events have been observed. In the literature, the pipette (external) is rich in its diversity, making comparisons between different published recordings difficult. In aortic myocytes, a NaCl-based solution has been used without $CaCl_2$ [28] whereas in portal vein myocytes [29] and pancreatic acinar cells [30], a standard Ringer was favored. In proliferating pulmonary myocytes [31] on the other hand, the pipette solution contained primarily Na^+-methanesulphonate. Single-channel recording in endothelia were carried out with a pipette solution containing Na_2SO_4 and 10 mM $CaCl_2$ [32]. For single-channel work on A431 epidermal cells [33] and mesangial cells [34], the pipette solution consisted of equimolar $BaCl_2$. Although cell-attached recordings are relatively straightforward, there are potential pitfalls that the experimenter needs to consider.

Because of the potential of the resting membrane, the patch potential is not the same as the pipette potential and can result in a voltage offset of up to 90 mV, depending on the resting potential. Moreover, as cell stimulation can change the membrane potential, the latter will not necessarily remain constant. It is important therefore that the membrane potential is fixed. One way to achieve this is to carry out a double patch experiment where one pipette is used to clamp the potential in the whole cell configuration and the other is used for cell-attached recording. Alternatively, the experimenter can chemically clamp the potential to 0 mV by bathing the cell in a high K^+ ringer. Unfortunately, membrane potential has not always been controlled in published recordings of single store-operated channels. A second problem concerns the nature of the channel recorded and this is a particular issue for store-operated channel recordings. The protocol is to record baseline activity (with few channel openings) for several minutes and then deplete stores. Channel activity in the patch is taken to represent store-operated channel openings. However, under these conditions, cytoplasmic Ca^{2+} is not buffered and therefore will rise following store depletion. The increased channel activity could therefore reflect a Ca^{2+}-activated conductance rather than a store-operated one. One way to test this would be to load the cytoplasm with a rapid Ca^{2+} chelator through the acetoxymethyl ester form (e.g., BAPTA-AM) and then show that a cytoplasmic Ca^{2+} increase following stimulation (by depleting stores) has been suppresed. If, after cells have been loaded with sufficient BAPTA to prevent the Ca^{2+} rise, channel activity increases after store depletion, then it is unlikely to be accounted for by a Ca^{2+}-dependent channel.

6.6.2 FLUCTUATION ANALYSIS

Consider an area of plasma membrane containing several channels. Each channel may be open or closed at any particular moment and so the total number of channels that are open will fluctuate in a random manner from one instant to the next. Fluctuation analysis is a method that extracts information from statistical descriptions of the noise. Suppose we have a cell expressing a total number N of functional store-operated channels that can be either open or closed. Let p be the probability that a particular channel is open and the current through each open channel is denoted by i.

$$\text{The whole cell current } I = Nip \tag{6.1}$$

The variance of the current Var(I) is:

$$\text{Var}(I) = Ni^2 p(1-p) = Ii(1-p) \tag{6.2}$$

or, substituting for p,

$$\text{Var}(I) = Ii - I^2/N \tag{6.3}$$

This equation is a parabola that meets the I axis at iN and has an initial slope of i.

If p is low ($\ll 1$), then it follows from Equation (6.2) that the slope of the plot of Var(I)/I yields i. One can then easily calculate the single-channel conductance. Fluctuation analysis makes certain assumptions about channel gating which need to be borne in mind. First, it assumes that the channel population is homogenous. Second, potential subconductance states of the channel are not considered. Third, open-closed transitions for each channel are assumed to be independent of the state all other channels are in. Notwithstanding these assumptions, fluctuation analysis is a powerful method to estimate single store-operated channel conductance.

For fluctuation analysis, great care is needed to ensure the background noise is kept as low as possible and that all other membrane conductances have been suppressed. This is a particularly important point for those store-operated channels, like CRAC channels, that have a very low conductance and that could easily be overestimated if another channel were to open. It is also important that the series resistance is as low as possible, otherwise this will reduce the 1-pole rolloff characteristic frequency of the membrane and overlap with the Bessel filtering applied by the experimenter. With a 15 pF cell for example, a rise in series resistance from 5 to 15 MΩ reduces the characteristic frequency from 2.1 KHz to ~700 Hz. Fluctuation analysis can easily underestimate the single-channel conductance and several control experiments are required to rule this out. First, power spectra of the current noise should be carried out to ensure that the bandwidth of the recording is sufficiently broad to capture channel gating events, including rapid transitions. Second, it is important that channel open probability is low. A high open probability will lead to a serious underestimation in i. This can be tested by measuring $Var(I)/mean(I)$ in the presence of a low concentration of channel blocker. No change in the ratio would suggest that open probability is indeed low.

Careful fluctuation analysis in Jurkat T cells with both Ca^{2+} and Na^+ as the charge carriers has revealed a tiny single-channel conductance of 10 fS with Ca^{2+} [35] and 0.2 pS with Na^+ [8]. These results have shed valuable insight into mechanisms of ion permeation through CRAC channels and also provide a useful yardstick for assessing the validity of potential candidate genes as CRAC channel clones.

6.7 TOTAL INTERNAL REFLECTION FLUORESCENCE MICROSCOPY: A NEW TOOL FOR STUDYING Ca^{2+} CHANNELS

If light traveling in a medium of high refractive index strikes a medium of low refractive index beyond a certain critical angle, then total internal reflection can take place. For cell biologists and physiologists, this can be exploited to study processes taking place close to the cell plasma membrane. Cells are grown on glass coverslips and laser light is coupled to the coverslip by the microscope objective or a prism. At a specific angle of incidence, some of the light energy penetrates the aqueous medium as an evanescent wave. The wave penetrates distances less than 100 nm from the plasma membrane. Within this local field, fluorophores can be excited, providing a means to monitor processes occurring just below the surface membrane. Using total internal reflection fluorescence microscopy (TIRFM), Demuro and Parker [36] imaged subplasmalemmal Ca^{2+} signals from several single N-type Ca^{2+} channels expressed in *Xenopus* oocytes. They found that N-type channels were relatively immobile, had a patchy distribution, and varied greatly in their gating kinetics, even for channels located close to each other.

TIRFM can also be used to track the movement of fluorescently tagged proteins, for example proteins labelled with enhanced green fluorescent protein. This approach provides new insight into protein trafficking, vesicular transport, and dynamics of signaling pathways. Clapham and colleagues recently used TIRFM to find that TRPC5 channels were stored in vesicles held just beneath the plasma membrane [37]. Epidermal growth factor activated Ca^{2+}-permeable nonselective current by promoting fusion of TRPC5-containing vesicles with the plasma membrane.

While single-channel recording remains the most direct way to study channel activity, TIRFM is a very attractive development as it is noninvasive (Ca^{2+} channels can be maintained in a physiological environment), permits the recording of numerous channels at the same time and provides spatial resolution on a submicrometer scale. TIRFM does not provide a means for voltage clamping the cell but, when combined with patch clamp electrophysiology, will be a very powerful and exciting method for studying subplasmalemmal Ca^{2+} entry and subsequent local responses.

ACKNOWLEDGMENTS

Research in the author's laboratory is supported by the Lister Institute, Medical Research Council (U.K.), and British Heart Foundation.

REFERENCES

1. Carafoli, E. Calcium signalling: a tale for all seasons. *Proc. Natl. Acad. Sci. USA*, 99, 1115–1122, 2002.
2. Hille, B. *Ionic Channels of Excitable Membranes.* Sinauer Associates, Inc., Sunderland, MA, 2002.
3. Catterall, W.A. Structure and regulation of voltage-gated Ca^{2+} channels. *Annu. Rev. Cell Dev. Biol.*, 16, 521–555, 2000.
4. Parekh, A.B. and Putney, J.W. Jr. Store-operated calcium channels. *Physiol. Rev.*, 85, 757–810, 2005.
5. Hoth, M. and Penner, R. Calcium release-activated calcium current in rat mast cells. *J. Physiol. (Lond.)*, 465, 359–386, 1993.
6. Kerschbaum, H.H. and Cahalan, M.D. Single-channel recording of a store-operated Ca^{2+} channel in jurkat T lymphocytes. *Science*, 283, 836–839, 1999.
7. Nadler, M.J. et al. LTRPC7 is a Mg-ATP-regulated divalent cation channel required for cell viability. *Nature*, 411, 590–595, 2001.
8. Prakriya, M. and Lewis, R.S. Separation and characterization of currents through store-operated CRAC channels and Mg^{2+}-inhibited cation (MIC) channels. *J. Gen. Physiol.*, 119, 487–508, 2002.
9. Hermosura, M.C. et al. Dissociation of the store-operated calcium current I_{CRAC} and the Mg^{2+}-nucleotide-regulated metal ion current MagNuM. *J. Physiol. (Lond.)*, 539, 445–458, 2002.
10. Bakowski, D. and Parekh, A.B. Permeation through store-operated CRAC channels in divalent-free solution: potential problems and implications for putative CRAC channel genes. *Cell Calcium*, 32, 379–391, 2002.
11. Kozak, J.A., Kerschbaum, H.H., and Cahalan, M.D. Distinct properties of CRAC and MIC channels in RBL cells. *J. Gen. Physiol.*, 120, 221–235, 2002.
12. Yue, L. et al. CaT1 manifests the pore properties of the calcium-release-activated calcium channel. *Nature*, 410, 705–709, 2001.
13. Voets, T. et al. CaT1 and the calcium release-activated calcium channel manifest distinct pore properties. *J. Biol. Chem.*, 276, 47767–47770, 2001.
14. Parekh, A.B. and Penner, R. Activation of store-operated calcium influx at resting $InsP_3$ levels by sensitisation of the $InsP_3$ receptor in rat basophilic leukaemia cells. *J. Physiol. (Lond.)*, 489, 377–382, 1995.
15. Hofer, A., Fasolato, C., and Pozzan, T. Capacitative Ca^{2+} entry is closely linked to the filling state of internal Ca^{2+} stores: a study using simultaneous measurements of I_{CRAC} and intraluminal $[Ca^{2+}]$. *J. Cell Biol.*, 140, 325–334, 1998.
16. Richardson, A. and Taylor, C.W. Effects of Ca^{2+} chelators on purified inositol 1,4, 5-trisphosphate $(InsP_3)$ receptors and $InsP_3$-stimulated Ca^{2+} mobilization. *J. Biol. Chem.*, 268, 11528–11533, 1993.

17. Neher, E. The use of Fura 2 for estimating Ca^{2+} buffers and Ca^{2+} fluxes. *Neurophar-macology*, 34, 1423–1442, 1995.
18. Gilabert, J.-A. and Parekh, A.B. Respiring mitochondria determine the pattern of activation and inactivation of the store-operated Ca^{2+} current I$_{CRAC}$. *EMBO J.*, 19, 6401–6407, 2000.
19. Fatt, P. and Ginsborg, B.L. The ionic requirements for the production of action potentials in crustacean muscle fibres. *J. Physiol. (Lond.)*, 142, 516–543, 1958.
20. Mcdonald T.F. et al. Regulation and modulation of Ca^{2+} channel in cardiac, skeletal, and smooth muscle cells. *Physiol. Rev.*, 74, 365–507, 1994.
21. Ono, K. and Fozzard, H.A. Phosphorylation restores activity of L-type Ca^{2+} channels after rundown in inside-out patches from rabbit cardiac cells. *J. Physiol. (Lond.)*, 454, 673–688, 1992.
22. Kameyama M, Kameyama A, Takano E, and Maki M. Rundown of the cardiac L-type Ca^{2+} channel: partial restoration of channel activity in cell-free patches by calpastatin. *Pflugers Arch.*, 435, 344–349, 1998.
23. Seydl, K. et al. Action of calpastatin in prevention of cardiac L-type Ca^{2+} channel run-down cannot be mimicked by synthetic calpain inhibitors. *Pflugers – Arch.* 429, 503–510, 1995.
24. Hao, L-Y., Kameyama, A., and Kameyama, M. A cytoplasmic factor, calpastatin and ATP together reverse run-down of Ca^{2+} channel activity in guinea-pig heart. *J. Physiol. (Lond.)*, 514, 687–699, 1999.
25. Xu, J.J. et al. Calmodulin reverses rundown of L-type Ca^{2+} channels in guinea pig ventricular myocytes. *Am. J. Physiol. Cell Physiol.*, 287, C1717–1724, 2004.
26. Wu, L. et al. Dual regulation of voltage-gated calcium channels by PtdIns(4,5)P$_2$. *Nature*, 419, 947–952, 2002.
27. Nowycky, M.C., Fox, A.P., and Tsien, R.W. Three types of neuronal calcium channel with different calcium agonist sensitivity. *Nature*, 316, 440–443, 1985.
28. Trepakova, E.S. et al. Properties of a native cation channel activated by Ca^{2+} store depletion in vascular smooth muscle cells. *J. Biol. Chem.*, 276, 7782–7790, 2001.
29. Albert, A.P. and Large, W.A. Activation of store-operated channels by noradrenaline via protein kinase C in rabbit portal vein myocytes. *J. Physiol. (Lond.)*, 544, 113–125, 2002.
30. Krause, E. et al. Depletion of intracellular calcium stores activates a calcium conducting nonselective cation current in mouse pancreatic acinar cells. *J. Biol. Chem.*, 271, 32523–32528, 1996.
31. Golovina, V.A. et al. Upregulated TRP and enhanced capacitative Ca^{2+} entry in human pulmonary artery myocytes during proliferation. *Am. J. Physiol.*, 280, H746–755, 2001.
32. Vaca, L. and Kunze, D.L. Depletion of intracellular Ca^{2+} stores activates a Ca^{2+}-selective channel in vascular endothelium. *Am. J. Physiol. (Cell Physiol.)*, 267, C920–925, 1994.
33. Zaznacheyeva, E. et al., Plasma membrane calcium channels in human carcinoma A431 cells are functionally coupled to Inositol 1,4,5-trisphosphate receptor-phosphatidylinositol 4,5-bisphosphate complexes. *J. Biol. Chem.*, 275, 4561–4564, 2000.
34. Ma, R. et al. Protein kinase C activates store-operated calcium channels in human glomerular mesangial cells. *J. Biol. Chem.*, 276, 25759–25765, 2001.
35. Zweifach, A. and Lewis, R.S. Mitogen-regulated Ca^{2+} current of T lymphocytes is activated by depletion of intracellular Ca^{2+} stores. *Proc. Natl. Acad. Sci. USA*, 90, 6295–6299, 1993.
36. Demuro, A. and Parker, I., Imaging the activity and localization of single voltage-gated Ca^{2+} channels by total internal fluorescence microscopy. *Biophys. J.*, 86, 3250–3259, 2004.

37. Bezzerides, V.J., Ramsey, I.S., Kotecha, S., Greka, A., and Clapham, D.E. Rapid vesicular translocation and insertion of TRP channels. *Nature Cell Biol.*, 6, 709–720, 2004.

7 Molecular Biology of Ca²⁺ Channels: Lessons from the TRP Superfamily

Veit Flockerzi, Thomas Aberle, Marcel Meissner, Christine Jung, Stephan Philipp, and Ulrich Wissenbach

CONTENTS

7.1 INTRODUCTION

Many cells including exocrine gland cells and vascular smooth muscle cells show a coordinated regulation of an intracellular Ca^{2+}-release mechanism and the entry of Ca^{2+} across the plasma membrane. This Ca^{2+} entry is the basis for sustained $[Ca^{2+}]_i$ elevations that are important for various cellular functions including gene expression, secretion, and cell proliferation [1]. A link between Ca^{2+} release and Ca^{2+} entry into cells was proposed by Putney [2]. His model described an initial emptying of the intracellular Ca^{2+} stores by inositol 1,4,5-trisphosphate (IP$_3$), followed by entry of Ca^{2+} into the cytosol, and refilling of the stores. This type of Ca^{2+} entry is referred to as capacitative or store-operated Ca^{2+}-entry. The first electrical measurement of current through a store-operated channel (SOC) was achieved in mast cells and this current was termed Ca^{2+}-release activated Ca^{2+} channel (CRAC) current [3]. The desire to identify the molecular nature of this SOC was and still is the motivation for many workers to characterize mammalian TRP channels. As it turned out, the activation of some of these TRP channels is linked to PLC activation; but most unanticipated is that the activation of other TRPs is due to a medley of novel regulatory mechanisms that appear to be independent of PLC signaling. These include activation by temperature, cell swelling, noxious and mechanical stimuli, and by binding of ligands such as anandamide, arachidonic acid derivates, protons, menthol, icilin, capsaicin, and phorbol esters. A number of comprehensive reviews have been published on these issues recently [4–6], and in this chapter we will discuss some of the problems that show up when cloning the cDNAs of "new" TRPs and elaborate on problems envisioned when studying TRP protein expression by antibodies.

7.2 *DROSOPHILA* TRP CHANNEL, THE FOUNDING MEMBER OF THE TRP SUPERFAMILY

TRP channels comprise a superfamily of cation permeable channels. The founding member of the TRP superfamily was identified as a *Drosophila* gene product required for visual transduction, which in the fruit fly is a phospholipase C-dependent process [7]. The name *transient receptor potential* is based on the transient rather than sustained response to light of the flies carrying a mutant in the *trp* locus [8]. Furthermore, *trp* mutants display a defect in light-induced Ca^{2+} influx. The predicted structure of TRP and the related protein, TRPL, and a variety of *in vitro* expression studies supported the proposal that TRP and TRPL [9] were cation influx channels. In addition, the studies on *Drosophila* phototransduction indicated that TRP is permeable to Ca^{2+} and is the target of a phosphoinositide cascade, leading to the suggestion that phototransduction in *Drosophila* might be analogous to the general and widespread process of phosphoinositide-mediated Ca^{2+} influx in other cells [10].

Activation of PLC could theoretically be coupled to TRP activation via production of IP_3 or diacylglycerol, metabolites of diacylglycerol, or via release of tonic inhibition of the channel due to hydrolysis of phosphatidylinositol 4,5-bisphosphate (PIP_2). According to one mechanism, referred to as store-operated Ca^{2+} entry, transient release of Ca^{2+} from internal stores induces sustained Ca^{2+} influx. The Ca^{2+} release can occur through the activation of an IP_3-gated intracellular Ca^{2+} release channel, the IP_3-receptor, which in turn results in Ca^{2+} influx. However, genetic evidence has demonstrated that activation of the *Drosophila* TRPs in photoreceptor cells is independent of the IP_3-receptor [11].

7.3 MAMMALIAN TRP-RELATED PROTEINS

On the basis of similarity to *Drosophila* TRP and TRPL protein sequences, the cDNAs of 28 mammalian TRP-related proteins have been cloned in recent years using database searches of expressed sequence tags (EST), RT-PCR, or expression-cloning strategies (see below). All TRP protein sequences comprise six predicted transmembrane domains and thereby resemble voltage-gated K^+ channels in overall transmembrane architecture (Figure 7.1). But despite these topographic similarities, TRPs show limited conservation of the S4 positive charges and P loop sequences

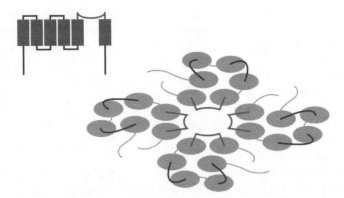

FIGURE 7.1 A TRP protein contains six domains, predicted to cross the cell membrane, and a putative pore loop within the extracellular linker separating the fifth and sixth transmembrane domains (left), but lacks the voltage-sensing element S4, present in voltage-gated channels. Four or five TRP proteins are assumed to form homooligomeric and heterooligomeric channels (right).

that are the hallmarks of the voltage-gated channel families [12] and accordingly, TRPs are only distantly related to these channels. Voltage-gated K$^+$ channels are tetramers and a similar tetrameric architecture is assumed of TRP channels. Though a good guess, direct biochemical evidence for such an architecture of TRP channels in native tissues is scarce [13–15]. In addition, it is an open issue whether TRP channels are homomeric or heteromeric complexes, or both.

As it turns out, there are not only mammalian homologs of TRP, but a panoply of TRP-related channels conserved in flies and worms (Table 7.1). These include proteins that fall into three subfamilies of channels that are the most related to TRP:

TABLE 7.1
TRP and TRP-Related Proteins

Subfamily	Worms	Flies	Mice	Humans
TRP C	3	3	7	6
TRP V	5	2	6	6
TRP M	4	1	8	8
TRP A	2	4	1	1
TRP N	1	1	—	—
TRP P	1	4	3	3
TRP ML	1	1	3	3
Total	**17**	**16**	**28**	**27**

Note: Shown are the seven subfamilies and the numbers of genes that have been identified and published. Note that the TRPC2 gene is a pseudogene in humans [48]. In mice, TRPC2 is involved in pheromone sensing [49]. No TRPN-like genes have been identified in mammals so far.

TRPC, TRPV, and TRPM. The mammalian proteins that display the greatest sequence similarity to *Drosophila* TRP belong to the TRPC-subfamily. These proteins share 30–47% amino acid homology over the ~700 to 900 amino acid residues, which encompass three to four ankyrin repeats, the six predicted transmembrane domains, and a highly conserved amino acid segment referred to as the TRP domain. This amino acid sequence motif follows immediately the predicted transmembrane domain 6 and is located within the C-terminal cytosolic tail of the protein (consensus sequence DxEWKFARxKLWxSYFxxGxTLPxPFNxxPSPK) [16]. The TRPV proteins also include at least five ankyrin repeats but lack the TRP domain, while the TRPM proteins contain a TRP domain, but no ankyrin repeats.

Two other subfamilies, TRPP and TRPML, which include the polycystic kidney disease 2 (PKD2) protein (TRPP) and mucolipin (TRPML), respectively, are also conserved throughout animal phylogeny, but are much more distantly related to TRP. A sixth subfamily, TRPA, is comprised of proteins in flies, worms, mice, and men whereas a seventh subfamily TRPN is comprised of proteins only in flies and worms such as NOMPC that have large numbers of ankyrin repeats (up to 29) and that bear some similarities to TRP.

7.4 STRATEGIES TO IDENTIFY TRP-RELATED PROTEINS

The first TRP-genes were identified by PCR-based strategies in search for homologous family members to the *Drosophila melanogaster* proteins TRP and TRP-like. Typically, degenerate primers were derived from conserved regions of the TRP proteins to amplify short cDNA fragments. Conserved regions of TRP proteins include sequences encoding the cytosolic loop between the predicted transmembrane domains 4 and 5 (consensus sequence LGPLQISLGRM), part of the predicted transmembrane domain 6 (consensus sequence VLLNMLIAM), or the TRP-motif. These fragments were then used as probes to screen cDNA libraries. This time-consuming and labor-intensive procedure was often rather ineffective because of the isolation of partial cDNAs that required repetitive rounds of screening to identify full-length clones. Furthermore, the exon trapping method [17] that utilizes RNA splicing reactions carried out in mammalian cultured cells led to the identification of an exon belonging to an additional TRP homolog [18] and to the consecutive cloning of the human TRPM2 cDNA [19], which was originally named TRPC7. In an attempt to identify proteins abundantly expressed in the retina of the squid (*Loligo forbesi*) eye, a partial protein sequence was obtained from a 92-kDa protein and a cDNA of an additional TRP homolog was isolated on the basis of this peptide sequence [20].

The ongoing genome- and expressed sequence tags (EST) projects have enormously accelerated the identification of additional TRP proteins. A breakthrough was the publication of the complete genome of the nematode *Caenorhabditis elegans* in 1998, at the time of the first sequenced multicellular organism. By comprehensive database screening using gene prediction programs, it was possible to predict a high percentage of the *C. elegans* proteom. These virtual proteins were collected in the *C. elegans* wormpep database of the Sanger Center, which was freely accessible. Using different strategies to analyze databases referred to in the corresponding chapter of the first edition of this volume [16], we were able to identify virtual proteins from the wormpep database that were similar to the known TRPC proteins by, sometime rather vague, structural criteria such as the predicted six transmembrane domains. At that time, 5 to 6 years ago, only sequences of the TRPC-subfamily and of one member of the TRPV-subfamily, TRPV1, were available. The predicted proteins T01H8.5, Co5C12.3, and F54D1.5 are now referred to as the TRP-related proteins gon2, gon2-like1, and gon2-like2, respectively. All three proteins have in common the predicted six transmembrane domain structure. The gon2 protein is required for the postembryonic mitotic cell divisions of gonadal precursor cells [21], whereas the functions of gon2-like1 and gon2-like2 remain to be elucidated. The gon2-like1 and gon2-like2 genes resemble most closely the mammalian TRPM3 gene, whereas gon2 has the highest similarity to TRPM7. The predicted proteins F28H7.10, T09A12.3, and T10B10 are now termed ocr-1, ocr-2, and ocr-3, respectively, and show highest similarity to the mammalian TRPV5/TRPV6 genes. Ocr-2 was described to be involved in the sensation of noxious chemicals [22]. The database entries C29E6.2 and M0B5.6 display a nucleotide sequence similar to the TRPA gene whereas F13D12.3 and ZK512.3 resemble more closely TRPM6 and TRPM2. The sequence of the predicted TRPprotein Y71A12B.4 is homologous to NOMPC, a gene involved in mechanosensation in insects [23]. In yeast, only one putative TRP protein was found, S61648, which is now classified as a TRP-related protein [24]. From the number of TRPs in *C. elegans*, we calculated the number of TRP-related genes probably present in humans to approximately 30, which is in good agreement with the 28 members of the TRP-gene family known today. Using these virtual protein sequences from the *C. elegans* wormpep database, we identified several homologous expressed sequence tags in human and mouse tissues. These EST clones were used as probes to isolate the full-length cDNAs of TRPV4 [25], TRPV6 [26], TRPM4 [27], and TRPM8 [28] independently from other research groups.

7.5 LIMITATIONS OF DATABASE SEARCHES OR: WHAT MAKES A TRP A TRP?

Considering ion channel function TRP-related proteins are capable of forming cation channels that carry both Na$^+$ and Ca^{2+} (TRPC-subfamily), Ca^{2+} almost exclusively (TRPV5 and TRPV6 of the TRPV-subfamily), Na$^+$ but no Ca^{2+} (TRPM4 and TRPM5 of the TRPM-subfamily), or Mg^{2+} (TRPM6 and TRPM7 of the TRPM-subfamily). As a result it is difficult to classify TRP channels according to the preferred permeant cation.

The more TRP-related proteins that were identified the more the degree of amino acid sequence homology declined. The TRPV6 protein [26], a member of the TRPV-subfamily, and Mucolipin1 [29], a protein of the TRPML-subfamily have about the same size, 725 and 580 amino acid residues, respectively, and can be aligned accurately. Their sequences share only 13% identical amino acid residues. Using the same algorithm for alignment, TRPV6 shares more identical amino acids

(13.5%) with protein kinase Cγ (GeneBank accession number NP_002730), an enzyme that is clearly not related to TRP proteins and moreover is not a transmembrane protein. Apparently, the similarity of members within the TRP superfamily can be as low as, or even lower than, their similarity to completely unrelated proteins. In comparison, the amino acid sequence identity between the voltage-dependent Ca^{2+} channel protein Ca$_v$1.2 (GeneBank accession number Q 13936), the ion-conducting pore of the cardiac L-type channel, and the Ca$_v$3.2 protein (GeneBank accession number Q 95180), the ion-conducting pore of the low voltage-activated T-type Ca^{2+} channel, is approximately 28% and thereby considerably higher than the identity found among the most distantly related TRP proteins. The calculated relative molecular weights of TRP-related proteins vary from ~ 83,000 in the case of TRPV6 to larger than ~ 233,000 in the case of TRPM6.

As has been summarized above there is common belief that the TRP-related proteins comprise a conserved structure of six transmembrane domains. However, this has been tested only in the case of the human TRPC3 protein and the results obtained are in agreement with such a topology [30]. Many programs using various algorithms predict different numbers of transmembrane domains for TRP-related proteins, leaving room for debate about the correct transmembrane topology. The TMHMM program (http://www.cbs.dtu.dk/services/TMHMM-2.0/), for example, predicts transmembrane domains for TRP-related proteins that do not fit to the assumed topology; it predicts five such domains for TRPM4, seven for TRPV5, and eight for TRPC4. Considering the absence of a signaling peptide the N-terminus of a plasmamembrane protein resides within the cytosol, whereas the number of trans-membrane domains governs the localization of its C-terminal region. A common feature among the protein sequences of most members of the TRPC-, TRPV-, TRPM-, TRPA-, and TRPN-subfamilies is that the predicted transmembrane domains are all located close together in one "cassette," independent of their numbers. In contrast, some of the members of the TRPP and TRPML-subfamilies show a somewhat different pattern with a cassette of seven predicted transmembrane regions forming the hydrophobic central core of the protein and one additional single hydrophobic domain located near the N-terminal end of the protein, thereby resulting in a 1+7 (or 8) transmembrane structure (Figure 7.2) in the case of PKD2.

A common theme connected to transmembrane topology is the plasma membrane localization. At least members of the TRPML-subfamily are supposed to reside within intracellular compartments and there is an ongoing discussion whether or whether not members of the TRPP-subfamily are localized in the plasma membrane, a feature originally believed to be common for all TRPs. The most prominent member of the TRPP-subfamily, the PKD2 protein, has been located in membranes of the endoplasmatic reticulum by antibody staining [31, 32]; it has also been reported to form plasma membrane nonselective cation channels when over-expressed in Xenopus oocytes [33]. A clue to this issue may come from recent results that suggested that at least some TRP proteins may reside within intracellular vesicles that are translocated to the plasma membrane upon appropriate stimulation [34–36].

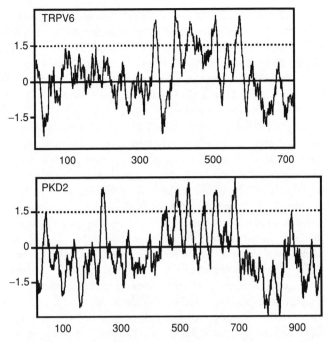

FIGURE 7.2 Hydropathy profile according to Kyte and Doolittle [47] of human TRPV6 (GeneBank accession number NP_061116), a member of the TRPV-subfamily, and of human PKD2 (GeneBank accession number NP_000288), a member of the TRPP-subfamily. Transmembrane domains were defined as regions with a hydropathy index ≥ 1.5 using a window of 19 amino acid residues. According to this analysis TRPV6 and PKD2 comprise seven and eight to nine transmembrane domains, respectively.

7.6   ANALYSIS OF TRPC EXPRESSION BY ANTIBODIES

The tissue specific expression of mammalian TRP-related gene products has been studied on the transcription level using Northern blot analysis, *in situ* hybridization, and reverse transcriptase-based polymerase chain reaction (RT-PCR), and on the protein level by antibodies. When studying the expression of a given *trp* gene the availability of TRP-gene-deficient cells or mice as controls has proven to be worthwhile.

In general, one has to consider that little data are available correlating mRNA expression of a given gene and the expression of its protein. In one of the few multigene comparison plots of mRNA versus protein abundances for cellular gene products, Anderson and Seilhamer [37] found a correlation coefficient of 0.48 between them. This result, halfway between a perfect correlation and no correlation at all, might be subject to revision as more data are collected with better measuring methods. Up to this end one has to consider that mRNA measurements cannot substitute for determining protein expression. Although there are numerous reports of the presence of mRNA encoding the TRPC proteins in animal cells and of detection of the heterologously expressed TRPC proteins by Western-blot analysis, it has proved difficult to unequivocally detect endogenous TRPC proteins. In the remainder of this chapter common problems encountered when using anti-TRPC antibodies are described. Analysis of TRPC expression by Northern blot analysis, *in situ* hybridization, and RT-PCR strategies have been covered in the corresponding chapter of the previous edition of this volume [16].

It is generally accepted that the various TRPC proteins form cation channels although it has not been demonstrated rigorously whether they are pore-forming channel subunits or auxiliary subunits like the $Ca_V\beta$-proteins of voltage-gated Ca^{2+} channels that modulate Ca^{2+} channel activity but do not contribute to the ion conducting pathway [12, 38]. In addition, the lack of specific ligands that specifically block TRPC channel activity impedes the isolation of TRPC-mediated channel activity in native tissues. In many cases, antibodies directed against a given TRPC protein are used to document expression of the TRPC protein and to correlate its expression with some electrophysiological current recordings or fura-2 measurements.

The TRPC4 protein is abundantly expressed in brain [39, 40] and there is evidence that it is also expressed in skeletal muscle [41]. We studied the expression of the TRPC4 protein using three antibodies directed against the TRPC4 protein, a commercially available antibody (Alo) and two antibodies which were prepared by us, antibody 781 and antibody 548. All three antibodies are supposed to specifically recognize the mouse TRPC4 protein. In mouse tissues, two TRPC4 variants (Gene-Bank accession numbers U50922, U50921) are expressed [40, 42], TRPC4 (relative molecular mass (M_r) ~ 102.000) and a slightly shorter version termed TRPC4$_{\Delta781-864}$ (M_r ~ 94.000), which lacks 84 amino acid residues (residues 781 to 864) due to splicing of the transcript. Both proteins are absent in TRPC4-deficient mice [40].

Figure 7.3a shows that the commercially available antibody (Alo) recognizes a single protein of 97.5 kDa both in microsomal membrane protein fractions from brain of wild type mice (lane 1) and of TRPC4-deficient mice (lane 2). In a microsomal membrane fraction of skeletal muscle from wild type mice (lane 3) and TRPC4-deficient mice (lane 4) a protein of very similar size is expressed at much lower levels. In addition, a 47-kDa and a 144 kDa protein are recognized by this antibody in the skeletal muscle of both genotypes. Apparently, the Alo antibody recognizes a protein that has a similar size as the expected TRPC4 proteins, but which is not identical with TRPC4 because expression can readily be detected in protein fractions from TRPC4-deficient mice (Figure 7.3A, lanes 2 and 4). In addition, it recognizes ~47 kDa and ~ 144kDa proteins of unknown identity in skeletal muscle. Using antibody 781 (Figure 7.3b) the two TRPC4 variants of ~102 and ~94 kDa are readily detectable in the protein fraction prepared from wild type mice (Figure 7.3b, lane 1), but are not detectable in the fractions prepared from TRPC4-deficient mice (Figure 7.3b, lane 2). In addition, no TRPC4 proteins are detected in the skeletal muscle protein fractions (Figure 7.3b, lanes 3 and 4) indicating lack of expression of TRPC4 in wild type skeletal muscle or very low protein expression levels. As a control we stripped off the antibodies and reused the blots using antibodies directed against the β1a subunit of the voltage-gated

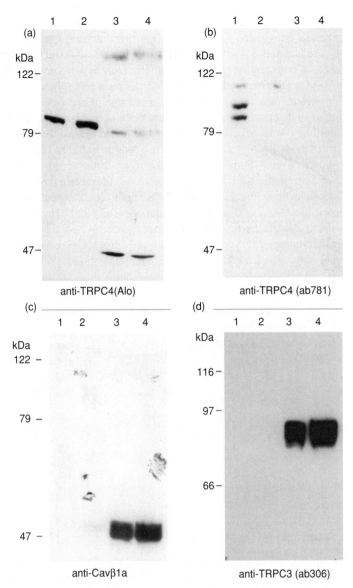

FIGURE 7.3 TRPC4 protein expression in mouse brain and skeletal muscle. 150 μg microsomal membrane proteins from brain (lanes 1 and 2) and skeletal muscle (lanes 3 and 4) of adult wild type mice (lanes 1 and 3) and adult TRPC4-deficient mice (lanes 2 and 4) were separated on 6.5% SDS polyacrylamide gels, blotted and incubated with (a) the commercially available anti-TRPC4-antibody (Alo) and affinity purified antibodies prepared by us directed against (b) TRPC4 (ab781), (c) Cavβ1a, and (d) TRPC3 (ab306). Details are described in the text.

Ca^{2+} channel/dihydropyridine receptor (CaVβ1a) [38, 43], which is abundantly expressed in skeletal muscle and with antibody 306 directed against the mouse TRPC3 protein [44]. The anti-Cavβ1a antibody readily recognizes the ~55-kDa β1a protein in skeletal muscle of both genotypes (Figure 7.3c) whereas the antibody 306 recognizes the ~81- to 90-kDa TRPC3 protein (Figure 7.3d), demonstrating abundant expression of TRPC3 in skeletal muscle, lack of TRPC3 expression in brain and, apparently, no upregulation upon deletion of the TRPC4 gene. We repeated the experiment with an additional antibody, antibody 548, which we have raised against the mouse TRPC4 protein. The antibody readily recognizes the full-length 102-kDa TRPC4 protein in brain from wild type mice (Figure 7.4a, lane 1) and an additional broad band of ~90 to ~94 kDa. A protein band of similar size is recognized in the protein fraction from brain of TRPC4-deficient mice (Figure 7.4a, lane 2), but upon closer inspection it is obvious that this protein band lacks a slower running component recognized by antibody 548 in lane 1. In order to separate the different proteins recognized by antibody 548, we blotted the proteins after running a SDS polyacrylamide gradient gel (Figure 7.4b). Under these conditions antibody 548 clearly recognizes three proteins of ~ 91, ~94, and ~102 kDa

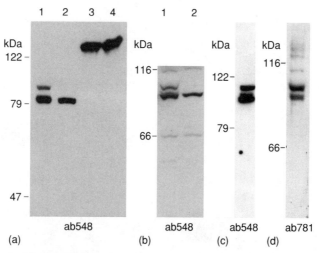

FIGURE 7.4 Specificity of anti-TRPC4 antibodies. (a) 150 μg microsomal membrane proteins from brain (lanes 1 and 2) and skeletal muscle (lanes 3 and 4) of adult wild type mice (lanes 1 and 3) and adult TRPC4-deficient mice (lanes 2 and 4) were separated on 6.5% SDS polyacrylamide gels, blotted, and incubated with the affinity purified antibody 548 (ab548) prepared by us. The identity of the ~132 kDa proteins in lanes 3 and 4 is unknown. This protein is not identical with TRPC4 because the proteins separated on lane 4 are from TRPC4-deficient mice. (b) 250 μg microsomal membrane proteins from brain of wild type mice (lane 1) and TRPC4-deficient mice (lane 2) were separated on SDS polyacrylamide gradient (7% to 4%) gels, blotted, and incubated with the affinity purified antibody 548 (ab 548) prepared by us. (c), (d) 100 μg of microsomal membrane proteins from brain of wild type mice were separated on 6.5% SDS polyacrylamide gels, blotted, and incubated with the affinity purified anti-TRPC4-antibody 548 (c) and antibody 781 (ab781, compare Figure 7.3).

(Figure 7.4b, lane 1). The ~94 and ~102 kDa represent the TRPC4 and TRPC4$_{\Delta781-864}$ proteins that are absent in protein fractions from TRPC4-deficient mice (Figure 7.4b, lane 2), whereas the ~91 kDa that is present in protein fractions of both genotypes is not related to TRPC4. In Figure 7.4c and d, additional control experiments using brain protein fractions from wild type mice are shown. Elongation of the running time of a 6.5% SDS-polyacrylamide gel allows separation of the three proteins recognized by antibody 548 (Figure 7.4c). In parallel, antibody 781 recognizes only the two TRPC4 proteins in an aliquot of the same brain protein fraction.

The data of Figure 7.3 and Figure 7.4 emphasize that results of a Western blot might lead one astray if antibodies are used that are not specific at all, as is apparently the case for the commercially available anti-TRPC4-antibody used here. To be on the safe side at least two antibodies should be employed and Western blots should be accompanied by Northern experiments, especially when control protein fractions from gene-deficient mice are not available.

A recent study [45] on TRPC1 expression encountered similar problems and emphasized the importance of establishing specificity, sensitivity, and reliability in Western blot procedures used to detect TRP-related proteins in primary cells and tissues. In the results shown in Figure 7.3, the commercial antibody or at least the batch used in these experiments does not recognize the protein it should. Accordingly, considerable caution should be exercised in the interpretation of the results obtained with poorly characterized antibodies [45].

7.7 CONCLUSIONS

Recently, the interest in TRP channels has increased considerably because these channels are emerging as essential cellular switches that allow organisms to respond to their environment. Furthermore, since mutations in several members of the TRP superfamily are critical factors in causing several diseases and TRP channels appear to be responsible for the fine-tuning of a wide variety of essential cellular processes, these channels have become attractive drug targets. At least members of the TRPC-subfamily may serve as agonist-activated channels linked to the phospholipase C metabolism and contribute to different Ca^{2+} entry pathways [46]. Together with members of other TRP-subfamilies they contribute to $[Ca^{2+}]_i$ signaling, and thus may also affect the store-operated Ca^{2+} entry process.

ACKNOWLEDGMENTS

We thank Brigitte Held for help in preparing the manuscript. This work was supported, in part, by the DFG and the Fonds der Chemischen Industrie.

REFERENCES

1. Clapham, D.E., Calcium signaling, *Cell*, 80, 259, 1995.
2. Putney, J.W., Jr., A model for receptor-regulated calcium entry, *Cell Calcium*, 7, 1, 1986.

3. Hoth, M. and Penner, R., Depletion of intracellular calcium stores activates a calcium current in mast cells, *Nature*, 355, 353, 1992.

4. Montell, C, Birnbaumer, L., Flockerzi, V., The TRP channels, a remarkably functional family, *Cell*, 108, 595, 2002.

5. Clapham, D.E., TRP channels as cellular sensors, *Nature*, 426, 517, 2003.

6. Vriens, J., Owsianik, G., Voets, T., Droogmans G, Nilius B., Invertebrate TRP proteins as functional models for mammalian channels, *Pflugers Arch.*, 449, 213, 2004.

7. Hardie, R.C. and Minke, B., The trp gene is essential for a light-activated Ca^{2+} channel in *Drosophila* photoreceptors, *Neuron*, 8, 643, 1992.

8. Montell, C. and Rubin, G.M., Molecular characterization of the *Drosophila* trp locus: a putative integral membrane protein required for phototransduction, *Neuron*, 2, 1313, 1989.

9. Phillips, A.M., Bull, A., Kelly, L.E., Identification of a *Drosophila* gene encoding a calmodulin-binding protein with homology to the *trp* phototransduction gene, *Neuron*, 8, 631, 1992.

10. Minke, B., The TRP channel and phospholipase C-mediated signaling, *Cell Mol. Neurobiol.*, 21, 629, 2002.

11. Hardie, R.C., Regulation of TRP channels via lipid second messengers, *Annu. Rev. Physiol.*, 65, 735, 2003.

12. Yu, F.H. and Catterall, W.A., The VGL-Chanome: a protein superfamily specialized for electrical signalling and ionic homeostasis, *Sci. STKE* 2004 re15.

13. Kedei, N., Szabo, T., Lile, J.D., Treanor, J.J., Olah, Z., Iadarola M.J., Blumberg P.M., Analysis of the native quaternary structure of vanilloid receptor I, *J. Biol. Chem.*, 276, 28613, 2001.

14. Hoenderop, J.G., Voets, T., Hoefs, S., Weidema, F., Prenen, J., Nilius, B., Bindels, R.J., Homo- and heterotetrameric architecture of the epithelial Ca^{2+} channels TRPV5 and TRPV6, *EMBO J.*, 22, 776, 2003.

15. Erler, I., Hirnet, D., Wissenbach, U., Flockerzi, V., Niemeyer, B.A., Ca^{2+}-selective TRPV channel architecture and function require a specific ankyrin repeat, *J. Biol. Chem.*, 279, 34456, 2004.

16. Philipp, S., Wissenbach, U., Flockerzi, V., *Molecular Biology of Calcium Channels in Calcium Signaling* (ed. Putney, J.W., Jr.), 321–342, CRC Press, Boca Raton, 2000.

17. Church, D.M., Stotler, C.J., Rutter, J.L., Murrell, J.R., Trofatter, J.A., Buckler, A.J., Isolation of genes from complex sources of mammalian genomic DNA using exon amplification, *Nat. Genet.*, 6, 98, 1994.

18. Kudoh, J., Nagamine, K., Asakawa, S., Abe, I., Kawasaki, K., Maeda, H., Tsujimoto, S., Minoshima, S., Ito, F., Shimizu, N., Localization of 16 exons to a 450-kb region involved in the autoimmune polyglandular disease type I (APECED) on human chromosome 21q22.3, *DNA Res.*, 4, 45, 1997.

19. Nagamine, K., Kudoh, J., Minoshima, S., Kawasaki, K., Asakawa, S., Ito, F., Shimizu, N., Molecular cloning of a novel putative Ca^{2+} channel protein (TRPC7) highly expressed in brain, *Genomics*, 54, 124, 1998.

20. Monk, P.D., Carne, A., Liu, S.H., Ford, J.W., Keen, J.N., Findlay, J.B., Isolation, cloning, and characterisation of a trp homologue from squid (*Loligo forbesi*) photoreceptor membranes, *J. Neurochem.*, 67, 2227, 1996.

21. West, R.J., Sun, A.Y., Church, D.L., Lambie, E.J., The *C. elegans* gon-2 gene encodes a putative TRP cation channel protein required for mitotic cell cycle progression, *Gene*, 266, 103, 2001.

22. de Bono, M., Tobin, D.M., Davis, M.W., Avery, L., Bargmann, C.I., Social feeding in *Caenorhabditis elegans* is induced by neurons that detect aversive stimuli, *Nature*, 419, 899, 2002.

23. Walker, R.G., Willingham, A.T., Zuker, C.S., A *Drosophila* mechanosensory transduction channel, *Science*, 287, 2229, 2000.

24. Palmer, C.P., Zhou, X.L., Lin, J., Loukin, S.H., Kung, C., Saimi, Y., A TRP homolog in *Saccharomyces cerevisiae* forms an intracellular Ca^{2+}-permeable channel in the yeast vacuolar membrane, *Proc. Natl. Acad. Sci. USA*, 98, 7801, 2001.

25. Wissenbach, U., Bödding, M., Freichel, M., Flockerzi, V., Trp12, a novel Trp related protein from kidney, *FEBS Lett.*, 485, 127, 2000.

26. Wissenbach, U., Niemeyer, B.A., Fixemer, T., Schneidewind, A., Trost, C., Cavalie, A., Reus, K., Meese, E., Bonkhoff, H., Flockerzi, V., Expression of CaT-like, a novel calcium-selective channel, correlates with the malignancy of prostate cancer, *J. Biol. Chem.*, 276, 9461, 2001.

27. Nilius, B., Prenen, J., Droogmans, G., Voets, T., Vennekens, R., Freichel, M., Wissenbach, U., Flockerzi, V., Voltage dependence of the Ca^{2+}-activated cation channel TRPM4, *J. Biol. Chem.*, 278, 30813, 2003.

28. Voets, T., Droogmans, G., Wissenbach, U., Janssens, A., Flockerzi, V., Nilius, B., The principle of temperature-dependent gating in cold- and heat-sensitive TRP channels, *Nature*, 430, 748, 2004.

29. Sun, M., Golden, E., Stahl, S., Falardeau, J.L., Kennedy, J.C., Acierno Jr, J.S., Bove, C., Kaneski, C.R., Nagle, J., Bromley, M.C., Colman, M., Schiffmann, R., Slangenhaupt, S.A., Mucolipidosis type IV is caused by mutations in a gene encoding a novel transient receptor potential channel, *Hum. Mol. Genet.*, 17, 2471, 2000.

30. Vannier, B., Zhu, X., Brown, D., Birnbaumer, L., The membrane topology of human transient receptor potential 3 as inferred from glycosylation-scanning mutagenesis and epitope immunocytochemistry, *J. Biol. Chem.*, 273, 8675, 1998.

31. Cai, Y., Maeda, Y., Cedzich, A., Torres, V.E., Wu, G., Hayashi, T., Mochizuki, T., Park, J.H., Witzgall, R., Somlo, S., Identification and characterization of polycystin-2, the PKD2 gene product, *J. Biol. Chem.*, 274, 28557, 1999.

32. Koulen, P., Cai, Y., Geng, L., Maeda, Y., Nishimura, S., Witzgall, R., Ehrlich, B.E., Somlo, S., Polycystin-2 is an intracellular calcium release channel, *Nat. Cell Biol.*, **4**, 191, 2002.

33. Chen, X.Z., Vassilev, P.M., Basora, N., Peng, J.B., Nomura, H., Segal, Y., Brown, E.M., Reeders, S.T., Hediger, M.A., Zhou, J., Polycystin-L is a calcium-regulated cation channel permeable to calcium ions, *Nature*, 401, 383, 1999.

34. Kanzaki, M., Zhang, Y.-Q., Mashima, H., Li, L., Shibata, H., Kojima, I., Translocation of a calcium-permeable cation channel induced by insulin-like growth factor-I, *Nat. Cell Biol.*, **1**, 165, 1999.

35. Bezzerides, V.J., Ramsey, I.S., Kotecha, S., Greka, A., Clapham, D.E., Rapid vesicular translocation and insertion of TRP channels, *Nat. Cell Biol.*, 6, 709, 2004.

36. Singh, N.B., Lockwich, T.P., Bandyopadhyay, B.C., Liu, X., Bollimuntha, S., Brazer, S., Combs, C., Das, S., Leenders, A.G.M., Sheng, Z.-H., Knepper, M.A., Ambudkar, S.V., Ambudkar, I.S., VAMP2-dependent exocytosis regulates plasma membrane insertion of TRPC3 channels and contributes to agonist-stimulated Ca^{2+} influx, *Mol. Cell*, 15, 635, 2004.

37. Anderson, L. and Seilhamer, J.A., Comparison of selected mRNA and protein abundances in human liver, *Electrophoresis*, 18, 533, 1997.

38. Berggren, P.-O., Yang, S.-N., Murakami, M., Efanov, A.M., Uhles, S., Köhler, M., Moede, T., Fernström, A., Appelskog, I.B., Aspinwall, C.A., Zaitsev, S.V., Larsson, O.,

Moitoso de Vargas, L., Fecher-Trost, C., Weißgerber, P., Ludwig, A., Leibiger, B., Juntti-Berggren, L., Barker, C.J., Gromada, J., Freichel, M., Leibiger, I.B., Flockerzi, V., Removal of Ca^{2+} channel β3 subunit enhances Ca^{2+}-oscillation frequency and insulin exocytosis, *Cell*, 119, 273, 2004.

39. Philipp, S., Cavalie, A., Freichel, M., Wissenbach, U., Zimmer, S., Trost, C., Marquart, A., Murakami, M., Flockerzi, V., A mammalian capacitative calcium entry channel homologous to Drosophila TRP and TRPL, *EMBO J.*, 15, 6166, 1996.

40. Freichel, M., Suh, S.H., Pfeifer, A., Schweig, U., Trost, C., Weißgerber, P., Biel, M., Philipp, S., Freise, D., Droogmans, G., Hofmann, F., Flockerzi, V., Nilius, B., Lack of an endothelial store-operated Ca^{2+} current impairs agonist-dependent vasorelaxation in Trp4−/− mice, *Nat. Cell Biol.*, 3, 121, 2001.

41. Vandebrouck, C., Martin, D., Coson-Van Schoor, M., Debaix, H., Gailly, P., Involvement of TRPC in the abnormal calcium influx observed in dystrophic (mdx) mouse skeletal muscle fibers, *J. Cell Biol.*, 158, 1089, 2002.

42. Roe, M.W., Worley III, J.F., Qian, F., Tamarina. N., Mittal, A.A., Dralyuk. F., Blair, N.T., Mertz, R.J., Philipson, L.H., Dukes, I.D., Characterization of a Ca^{2+} release-activatred nonselecttive cation current regulation membrane potential and $[Ca^{2+}]_i$ oscillations in transgenically derived β-cells, *J. Biol. Chem.*, 273, 10402, 1998.

43. Ruth, P., Röhrkasten, A., Biel, M., Bosse, E., Regulla, S., Meyer, H.E., Flockerzi, V., Hofmann, F., Primary structure of the β-subunit of the DHP-sensitive calcium channel from skeletal muscle, *Science*, 245, 1115, 1989.

44. Philipp, S., Strauβ, B., Hirnet, D., Wissenbach, U., Mery, L., Flockerzi, V., Hoth, M., TRPC3 mediates T cell receptor dependent calcium entry in human T lymphocytes, *J. Biol. Chem.*, 278, 26629, 2003.

45. Ong, H.-L., Chen, J., Chataway. T., Brereton, H., Zhang, L., Downs. T., Tsiokas. L., Barrit. G., Specific detection of the endogenous transient receptor potential (TRP)-1 protein in liver and airway smooth muscle cells using immunoprecipitation and Western-blot analysis, *Biochem. J.*, 364, 641, 2002.

46. Putney, J.W., Jr., The enigmatic TRPCs: multifunctional cation channels, *Trends Cell Biol.*, 14, 282, 2004.

47. Kyte, J. and Doolittle, R.F., A simple method for displaying the hydropathic character of a protein, *J. Mol. Biol.*, 157, 105, 1982.

48. Yildirim, E., Dietrich, A., Birnbaumer, L., The mouse C-type transient receptor potential 2 (TRPC2) channel: alternative splicing and calmodulin binding to its N terminus, *Proc. Natl. Acad. Sci. USA*, 100, 2220, 2003.

49. Stowers, L., Holy, T.E., Meister, M., Dulac, C., Koentges, G., Loss of sex discrimination and male-male aggression in mice deficient for TRP2, *Science*, 295, 1493, 2002.

8 Genetic and Molecular Characterization of Ca^{2+} and IP$_3$ Signaling in the Nematode *Caenorhabditis elegans*

Ana Y. Estevez and Kevin Strange

CONTENTS

8.1 INTRODUCTION

Nonmammalian model organisms such as *Escherichia coli, Saccharomyces, Caenorhabditis elegans, Drosophila*, zebrafish, and the plant *Arabidopsis* provide numerous experimental advantages for defining the molecular bases of complex physiological processes. *C. elegans* provides a particularly striking example of the experimental utility of nonmammalian model organisms (reviewed in Ref. [1]).

Worms have a short life cycle, produce large numbers of offspring by sexual reproduction, and can be cultured easily and inexpensively in the laboratory. Sexual reproduction occurs by self-fertilization in hermaphrodites or by mating with males. The reproductive and laboratory culture characteristics of *C. elegans* make it an exceptionally powerful model system for forward genetic analysis.

In addition to forward genetic tractability, *C. elegans* also has a fully sequenced and well-annotated genome. Genomic sequence and virtually all other biological data on this organism are assembled in readily accessible public databases (e.g., WormBase; http://www.wormbase.org). Numerous reagents including mutant worm strains and cosmid and YAC clones spanning the genome are freely available through public resources. Creation of transgenic worms is relatively easy, inexpensive, and rapid. *C. elegans* gene expression can be specifically and potently targeted for knockdown using RNA interference (RNAi). Finally, *C. elegans* is a highly differentiated animal but is comprised of less than 1000 somatic cells. This relatively simple anatomy greatly facilitates the study of complex physiological processes.

The focus of this chapter is to provide an overview of *C. elegans* biology and some of the powerful genetic and molecular tools available for the study of this organism. We will also describe how these tools have been used to define Ca^{2+} and IP_3 signaling in the *C. elegans* hermaphrodite gonad and intestine, Ca^{2+} signaling in excitable cells, and the role of transient receptor potential (TRP) channel-mediated Ca^{2+} signaling in sperm function.

8.2 *C. ELEGANS* BIOLOGY

8.2.1 NATURAL HISTORY AND LIFE CYCLE

C. elegans is a member of the phylum Nematoda. Nematodes, or roundworms, are some of the most numerous and widespread of all animals and are found in virtually all habitats. The phylum contains free-living species as well as species that parasitize plants and other animals. Nematodes range in size from less than 1 mm to over 35 cm in length. *C. elegans* is a free-living soil nematode about 1 mm long.

Adult *C. elegans* are predominantly hermaphroditic with males making up approximately 0.1% of wild-type populations. Self-fertilized hermaphrodites produce about 300 offspring whereas male-fertilized hermaphrodites can produce over 1000 progeny.

Under optimal laboratory conditions the average life span of *C. elegans* is 2–3 weeks. The life cycle is rapid. At 25°C, embryogenesis, the period from fertilization until hatching, occurs in 14 h. Postembryonic development occurs in four larval stages (L1–L4) that last a total of about 35 h (Figure 8.1A).

When food supply is limited, dauer larvae form after the second larval molt. Dauer larvae do not feed and have structural, metabolic, and behavioral adaptations that increase life span up to 10 times and aid in the dispersal of the animal to new habitats. Once food becomes available, dauer larvae feed and continue development to the adult stage [2, 3].

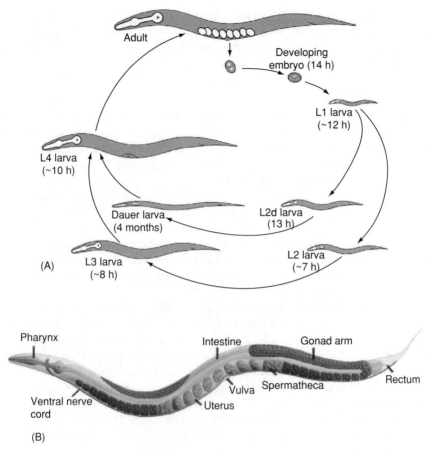

FIGURE 8.1 (A) Life cycle of *Caenorhabditis elegans* grown at 25°C on agar plates seeded with *Escherichia coli*. Eggs are laid ~5 h after fertilization and hatching occurs 9 h later. (Taken from Jorgensen F.M. and Mongo S. E., *Nat. Rev. genet*, 3, 356, 2002. With permission. Copyright Macmillan Magazines Ltd) a (B) Schematic diagram showing anatomical features of an adult *C. elegans* hermaphrodite. (From www.wormatlas.org, with permission from WormAtlas.)

8.2.2 LABORATORY CULTURE

The standard *C. elegans* laboratory strain is Bristol N2. Other strains are also used and offer certain experimental advantages. For example, the Bergerac strain exhibits a high rate of spontaneous mutation due to the activity and high copy number of the Tc1 transposon [4]. The Hawaiian strain CB4856 possesses a uniformly high density of single nucleotide polymorphisms that greatly facilitate genetic mapping [5].

Culture of *C. elegans* in the laboratory is simple and relatively inexpensive [6]. Animals are typically grown in petri dishes on agar seeded with a lawn of *E. coli* as a food source. *C. elegans* can also be grown in mass quantities using liquid culture strategies and fermentor-like devices. Worm stocks are stored frozen in liquid nitrogen indefinitely with good viability. The ability to store *C. elegans* frozen dramatically simplifies culture strategies and reduces costs associated with handling and maintaining wild-type and mutant worm strains.

8.2.3 ANATOMY

Like all nematodes, *C. elegans* has an unsegmented, cylindrical body that tapers at both ends. The body wall consists of a tough collagenous cuticle underlain by hypodermis, muscles, and nerves. A fluid-filled body cavity or pseudocoel separates the body wall from internal organs. Body shape is maintained by hydrostatic pressure in the pseudocoel (Figure 8.1B).

A newly hatched L1 larva has 558 cells. Additional divisions of somatic blast cells occur during the four larval stages eventually giving rise to 959 somatic cells in mature adult hermaphrodites and 1031 in adult males. The lineage of somatic cells in *C. elegans* is largely invariant. This invariance combined with the ability to visualize by differential interference contrast microscopy cell division and development in living embryos, larvae, and adult animals has made it possible to describe the fate map or cell lineage of the worm [7, 8].

Despite the small cell number, *C. elegans* exhibits a striking degree of differentiation. Many physiological functions found in mammals have nematode analogs. Worms possess striated and nonstriated muscle cells; a digestive tract comprised of a muscular pharynx, intestine, and rectum; and excretory and reproductive systems. The nervous system of adult hermaphrodites contains 302 neurons and 56 glial and support cells. Males have 381 neurons and 92 glial and support cells. White et al. [9] have reconstructed and mapped the connectivity of the entire hermaphrodite nervous system using serial electron microscopy. For more detailed descriptions of worm anatomy, see Refs. [1, 10] and WormAtlas (http://www.wormatlas.org/), an online resource.

8.3 FORWARD AND REVERSE GENETIC ANALYSIS

8.3.1 FORWARD GENETIC ANALYSIS

Mutagenesis and forward genetic analysis provide an unbiased approach for identifying genes that underlie complex physiological processes. In addition, analysis of gain-of-function and loss-of-function mutations provides unique molecular insights into protein structure/function and allows genes to be ordered into biological pathways or networks. A forward genetic screen involves developing an assay for a phenotype of interest, inducing random mutations in the organism's genome, isolating mutant organisms with an abnormal phenotype, and then identifying the mutated gene responsible for the abnormality.

The development of *C. elegans* as an experimental system was driven largely by the relative ease of performing forward genetic analysis. Genetic screens in worms are greatly facilitated by the animal's short life cycle, ease of laboratory culture, and fecundity. Self-fertilization in hermaphrodites allows homozygous animals to breed true and greatly facilitates the isolation and maintenance of mutant strains. It is also a useful feature if mutant worms are paralyzed or uncoordinated since reproduction does not require movement in order to find and mate with a male. Mating with males, however, is essential for moving mutations between strains.

For a detailed and illuminating discussion of forward genetic screening in *C. elegans*, the reader is referred to a recent review article by Jorgensen and Mango [11]. Briefly, once a screening assay is developed, animals are mutagenized, typically by chemical mutagens such as ethyl methane-sulfonate (EMS) or *N*-ethyl-*N*-nitro-sourea (ENU). Mutant animals are then isolated and the mutated gene identified by mapping, rescue, RNAi, and cloning strategies [5,11,12].

Mutant animals can be further mutagenized to produce double mutants. The phenotype of a double mutant may be suppressed, enhanced, or similar to the phenotype of animals with a single mutation. Mutations that enhance or suppress a phenotype may reside in genes distinct from the one mutated in the original screen. These extragenic mutations imply that the enhancer and suppressor genes interact with the first mutated gene. Genetic interactions indicate that gene products function in a common pathway. They also suggest the possibility of direct interactions between gene products (i.e., protein–protein interactions) [11].

8.3.2 Reverse Genetic Analysis

Reverse genetic methods are used to identify the function of a known gene. A typical reverse genetic strategy involves disrupting a gene's coding region or decreasing the levels of its mRNA and protein and then analyzing the resultant phenotype. One of the truly extraordinary experimental advantages of *C. elegans* is the relative ease by which gene function can be disrupted using double stranded RNA (dsRNA)-mediated gene interference (RNAi). The ability of dsRNA to readily cross *C. elegans* cell membranes is particularly advantageous for experimental disruption of gene expression. RNAi can be induced in worms by injecting them with dsRNA [13], by soaking them in dsRNA solutions [14], or by feeding them bacteria-producing dsRNA [15–17]. Worms can also be engineered with inducible transgenes that are transcribed into dsRNA [18]. When worms are fed dsRNA-producing bacteria or soaked in dsRNA solutions, the dsRNA is absorbed across the intestinal epithelium and then spreads systemically to the animal's somatic cells and germline. In cultured *C. elegans* cells, RNAi is triggered simply by adding dsRNA to the culture medium [19].

Recently, Kamath et al. [20] generated a reusable RNAi library (now available from MRC geneservice) consisting of 16,757 bacterial strains, each of which expresses a unique dsRNA. These dsRNAs correspond to ~86% of the predicted genes in the worm genome. Numerous large-scale RNAi feeding screens have been carried out in *C. elegans* making it possible to identify genes involved in a variety of basic biological processes including fat metabolism [21], ageing [22, 23], early embryonic development [24], transposon silencing [25], protection of the genome against mutations [26], osmotic stress resistance [27], and regulators of polyglutamine aggregation [28].

Gene knockout by homologous recombination is difficult and not widely used in *C. elegans*. Instead, the relative ease of culturing *C. elegans* in large numbers and the ability to store worms frozen has led to the development of so-called "target-selected gene inactivation methods." This approach involves inducing random deletion mutations in a population of worms. Mutagenized worms are subdivided into small pools, typically in 96-well culture plates. These pools are screened by PCR for a deletion mutation in a target gene. Once a pool containing the desired mutation is detected, it is subdivided and PCR screening is repeated until single worms carrying the deletion are isolated.

Jansen et al. [29] demonstrated that random deletion mutations in *C. elegans* could be generated in a large-scale manner using chemical mutagens. High through-put methods for isolation of *C. elegans* deletion mutants have been described in detail [30]. A *C. elegans* Knockout Consortium (http://elegans.bcgsc.bc.ca/knockout. shtml) is currently working to produce strains possessing deletion mutations in all identified worm genes. Deletion strains are freely available to all investigators and requests for knockout of a specific gene can be made online to the consortium (http:// elegans.bcgsc.bc.ca/cgi-bin/submit_ko.pl). A Japanese National BioResource Project is also generating and distributing *C. elegans* deletion mutants (http://shigen. lab.nig.ac.jp/nbrp/info/celegansEn.html)

Identifying where and when a gene is expressed can provide important insight into its function. Fluorescent protein transcriptional and translational reporter transgenes are widely used in *C. elegans* to identify gene expression sites (e.g., [31–33]). DNA transformation in *C. elegans* is relatively straightforward and was first reported by Stinchcomb et al. [34]. Fire [35] and Mello et al. [36] have described methods for producing and maintaining transgenic worm lines. Briefly, transforming DNA is microinjected into the distal end of the hermaphrodite gonad. Heritable DNA transformation occurs by extrachromosomal transformation, non-homologous integration, or homologous integration. Spontaneous homologous integration is extremely rare. Formation of multicopy extrachromosomal arrays is the most frequent way in which transforming DNA is inherited. Transformation by extrachromosomal arrays is often transient. Integration of transgenes and generation of stable transgenic lines is commonly carried out by gamma irradiation of trans-formed worms [37].

Recently, Praitis et al. [38] described the use of microparticle bombardment to create integrated transgenic lines in *C. elegans*. This approach is technically less demanding than microinjection and reportedly produces single- and low-copy chromosomal DNA insertions resulting in more reliable transgene expression. Low-copy integrated lines also exhibit expression of transgenes in cells of the germline. Multicopy extrachromosomal arrays are often not expressed in germ cells due to the activity of gene silencing processes (e.g., Ref. [39]).

8.4 PUBLIC RESOURCES

The "worm community" is well known for its open sharing of data and reagents. As noted earlier, numerous reagents including mutant and transgenic worm strains, cosmid and YAC clones, and EST clones are freely available from public resources (see Ref. [1] for a description of these resources and web addresses). In addition, an extraordinary wealth of data on *C. elegans* is available online. Indeed, the worm community was an early pioneer in the use of the Internet for electronic data sharing. WormBase (http://www.wormbase.org/) is a particularly noteworthy database. It provides an exhaustive catalog of worm biology including identification of all known and predicted worm genes. Gene descriptions include genome location, mutant and RNAi phenotypes, expression patterns, microarray data, gene ontology, mutant alleles, and BLAST matches.

In the following sections, we describe Ca^{2+} signaling pathways that have been studied in *C. elegans* using genetic, molecular, cell biological, and physiological tools.

8.5 CALCIUM AND IP$_3$ SIGNALING IN THE *C. ELEGANS* HERMAPHRODITE GONAD

Forward genetic screens have been invaluable for identifying the repertoire of genes involved in particular biological processes. A key consideration in developing a genetic screen is the ability to assay a phenotype of interest in a simple and straightforward manner [11]. Genes required for fertility and gonad function can be identified readily by monitoring the number of offspring produced by mutant worms.

The involvement of IP$_3$ signaling in *C. elegans* reproduction was first demonstrated by Clandinin et al. [40]. *lin-3* and *let-23* encode worm homologs of the epidermal growth factor (EGF) and the EGF receptor tyrosine kinase, respectively [41, 42]. Loss of LIN-3/LET-23-mediated signaling leads to defects in development and fertility [43, 44]. Clandinin et al. [40] mutagenized *lin-3* reduction-of-function mutants and looked for worms in which the fertility defects were "rescued" or "suppressed." From this suppressor screen, six mutants were isolated that define two genes designated *let-23* *f*ertility *e*ffector/regulator *(lfe)-1* and *lfe-2*. Genetic mapping experiments demonstrated that *lfe-1* encodes an IP$_3$ receptor (IP$_3$R). *lfe-1*, or *itr-1* (*i*nositol-1,4,5-*t*risphosphate *r*eceptor) as it is now termed, is the only IP$_3$R encoding gene in the worm genome [45]. ITR-1 functions downstream of LIN-3 and LET-23 and mutations that suppress or rescue fertility defects in *lin-3* reduction-of-function mutant worms likely result in enhanced IP$_3$R activity. For example, one suppressor mutation alters a conserved arginine residue in the IP$_3$-binding domain of ITR-1 and increases IP$_3$ binding affinity to the receptor [46, 47].

lfe-2 encodes an inositol 1,4,5-trisphosphate-3-kinase (IP$_3$K) [40]. IP$_3$ kinases phosphorylate IP$_3$ and convert it into IP$_4$ [48]. The *lfe-2* mutation that suppresses the *lin-3* fertility defect truncates the LFE-2 protein by inserting a premature stop codon into the gene. Loss-of-function mutations in *ipp-5* also suppress the fertility defects observed in *lin-3* and *let-23* mutant worms [49]. IPP-5 is an ortholog of the human type I inositol polyphosphate 5-phosphatase, which dephosphorylates IP$_3$ and converts it into IP$_2$ [50]. Loss of either LFE-2 or IPP-5 activity presumably causes elevated cytoplasmic IP_3 levels that compensate for reduced LIN-3 signaling.

The transparent cuticle and simple anatomy of *C. elegans* have allowed detailed characterization of ovulation and fertilization in living, anesthetized worms using differential interference (DIC) microscopy [51, 52]. The gonad of adult hermaphrodites consists of two identical U-shaped arms connected via spermathecae to a common uterus [53–55] (Figure 8.2). Oocytes are arranged in the proximal gonad arm in a single-file fashion and remain arrested in diakinesis of prophase I until they reach the most proximal position in. During the late stages of oogenesis, an oocyte located immediately adjacent to the spermatheca is released from cell cycle arrest and undergoes meiotic maturation. Breakdown of the nuclear envelope and a change in the shape of the oocyte due to cortical cytoskeletal rearrangements are prominent morphological features of oocyte maturation [52]. Within 5–6 min after maturation is initiated, the oocyte is ovulated into the spermatheca where it is fertilized. Upon fertilization, dilation of the spermatheca–uterine valve facilitates the entry of the embryo into the uterus. Adult hermaphrodite worms ovulate every 20–40 min [52].

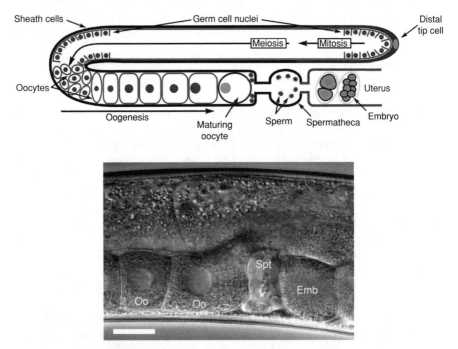

FIGURE 8.2 (See color insert following page 140.) *Top panel:* Schematic diagram of one arm of the adult hermaphrodite gonad. *Bottom panel:* Differential interference contrast micrograph of adult hermaphrodite gonad. Oo, oocyte; Spt, spermatheca; Emb, embryo. Scale bar is 20 µm. (Taken from Rutledge, E. et al. *Curr Biol.*, 11, 16, 2001. With Permission. Copyright Elsevier B. V)

GFP reporter studies demonstrated that *lfe-1/itr-1*, *lfe-2*, and *ipp-5* are expressed in the spermatheca [40, 45, 49, 56], which is also a site of *lin-3* and *let-23* expression [57]. Using DIC imaging, Clandinin et al. [40] demonstrated that the fertility defects of *lin-3* and *let-23* reduction-of-function mutants are due to failure of the spermatheca-gonad valve to dilate properly during ovulation. These defects are suppressed by mutations in *lfe-1/itr-1*, *lfe-2* [40], and *ipp-5* [49].

Spermatheca function and fertility are also severely disrupted by loss-of-function mutations or RNAi of *plc-1*, a PLC-ε homolog [58, 59]. However, in contrast to *lin-3* and *let-23* mutants, the fertility defects of *plc-1* mutants are not rescued by *itr-1* gain-of-function or *lfe-2* reduction-of-function mutations suggesting that IP$_3$ signaling is not downstream of *plc-1* in the spermatheca. *plc-1* deletion mutants are good candidates for suppressor screens aimed at isolating mutants in which fertility is restored. Such screens have the potential to identify the signaling role of PLC-1 and the signaling pathways in which it functions.

Rutledge et al. [60] postulated that IP$_3$/Ca^{2+} signaling events also regulate ovulation by regulating the contractile activity of gonad sheath cells. Sheath cells are smooth muscle-like myoepithelial cells that surround and are coupled to oocytes via gap junctions [53] (Figure 8.2). Prior to ovulation, the sheath cells contract weakly at a basal rate of ~8 contractions per minute [52, 59]. Basal contractile activity is triggered by release of major sperm protein (MSP) from sperm, which activates the VAB-1 Eph receptor tyrosine kinase [61]. Oocytes undergoing meiotic maturation secrete LIN-3, which increases both the force and rate of sheath contraction and induces dilation of the spermatheca-gonad valve [52, 59, 62]. This combination of increased sheath contraction and spermatheca dilation induces ovulation.

Yin et al. [59] utilized reverse genetic strategies and DIC imaging to test the hypothesis that IP$_3$/Ca^{2+} signaling regulates sheath cell contraction. *itr-1* reduction-of-function mutations or knockdown of ITR-1 expression by RNAi dramatically inhibit both basal and ovulatory sheath cell contractions [59]. Reduction-of-function mutations in *lin-3* or *let-23* inhibit sheath cell contraction during ovulation [59]. Gain-of-function *itr-1* mutations increase sheath contractile activity and suppress the inhibitory effects of *lin-3* and *let-23* mutations indicating that ITR-1 functions downstream of LIN-3/LET-23 signaling [59].

To identify phospholipases involved in regulating sheath contraction, Yin et al. [59] used an RNAi feeding strategy. Six *C. elegans* genes, *egl-8*, *plc-1*, *plc-2*, *plc-3*, *plc-4*, and *pll-1*, encode phospholipase C (PLC) homologs. Bacteria producing dsRNA homologous to each PLC were fed to worms. No defects in sheath cell contractions were detected in worms fed with *egl-8*, *plc-1*, *plc-2*, *pll-1*, or *plc-4* dsRNA. RNAi of *plc-3*, which encodes a PLC-γ homolog, reduced fertility, and disrupted basal and ovulatory sheath contractions. GFP reporter studies confirmed that PLC-3 is expressed in gonadal sheath cells. Gain-of-function mutations in *itr-1* prevented the inhibitory effects of *plc-3* RNAi suggesting that ITR-1 also functions downstream of PLC-3. Figure 8.3 summarizes the IP$_3$/Ca^{2+} signaling pathways that regulate sheath cell contractility and spermatheca function.

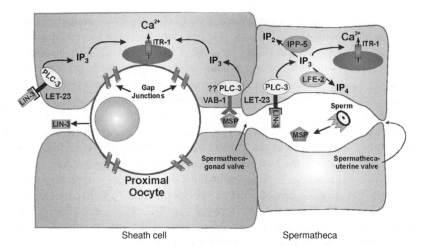

FIGURE 8.3 Schematic diagram of IP_3 signaling pathways identified in the adult hermaphrodite gonad. Major sperm protein (MSP) is released from sperm and activates the VAB-1 Eph receptor tyrosine kinase [61] on the sheath cells to regulate basal contractility. LIN-3 is an EGF-like factor secreted by maturing oocytes. Binding of LIN-3 to its cognate EGF-like receptor, LET-23, increases both the force and rate of sheath contraction and induces dilation of the spermatheca–gonad valve [52,59,62]. PLC-3 is a predicted PLC-γ homolog and generates IP_3 upon activation of LET-23 and possibly VAB-1. Binding of IP_3 to the IP_3R, ITR-1 induces Ca^{2+} release from the ER. LFE-2 is an inositol 1,4,5-trisphosphate-3-kinase (IP_3K) [40] which phosphorylates IP_3 and converts it into IP_4 [48]. IPP-5 is an ortholog of the human type I inositol polyphosphate 5-phosphatase [49], which dephosphorylates IP_3 and converts it into IP_2 [50]. See text for additional details.

8.6 OSCILLATORY Ca^{2+} SIGNALING IN THE *C. ELEGANS* INTESTINE

The *C. elegans* defecation cycle is an ultradian rhythm that consists of sequential contraction of posterior body wall muscles, anterior body wall muscles, and enteric muscles every 50 s when the animals are feeding (Figure 8.4A) [63]. Because the cycle of muscle contractions that drives defecation is a simple, invariant, and readily observable behavior, it provides a screenable phenotype for mutagenesis and forward genetic analysis.

Iwasaki et al. [64] screened for mutant worms with abnormal defecation cycles and identified 12 *dec* (*de*fecation *c*ycle defective) genes that when mutated cause the cycle to slow down or speed up. One of these mutants, *dec-4*, has prolonged or absent defecation cycles. Dal Santo et al. [56] cloned the *dec-4* gene and identified the locus as *itr-1*. They showed that reduction-of-function mutations in *itr-1* slow or eliminate the defecation cycle whereas overexpression of the receptor increases cycle frequency. Rescue of *itr-1* mutants with the wild-type gene in the intestinal cells is sufficient to restore normal cycle activity, demonstrating that ITR-1 is acting in the intestine to regulate defecation [56].

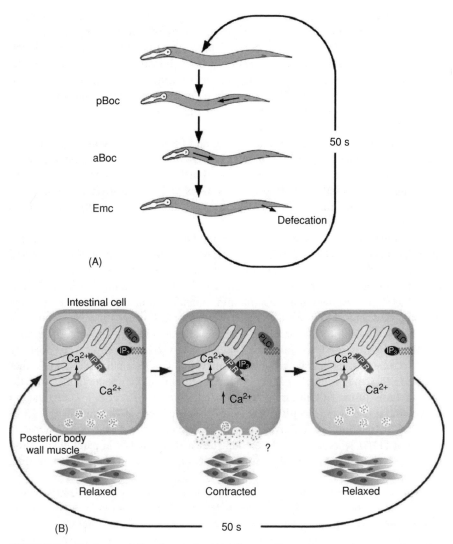

FIGURE 8.4 *C. elegans* defecation cycle. (A) Diagram illustrating muscle contractions that mediate defecation. Cycle is initiated by posterior body wall muscle contraction (pBoc). After relaxation of these muscles, the anterior body wall muscles contract (aBoc) and then expulsion occurs by enteric muscle contraction (Emc). The cycle repeats itself every 45–50 s. Figure is based on Dal Santo et al. [56]. (B) Model illustrating possible role of intracellular Ca^{2+} in regulating defecation cycle. Cyclical elevation of cytoplasmic Ca^{2+} levels driven in part by IP_3-dependent release of Ca^{2+} from intracellular stores triggers exocytic secretion from intestinal cells of an unidentified messenger that induces pBoc (see [56]). (Taken from Strange, *Physiol Rev.*, 83, 377, 2003. With permission.

To study intestinal cell Ca^{2+} dynamics during the defecation cycle, Dal Santo et al. [56], microinjected fura-2 into the intestines of live worms and observed rhythmic Ca^{2+} oscillations. Calcium levels peak immediately prior to the initiation of posterior body wall muscle contraction (pBoc). Intestinal cell Ca^{2+} oscillations are slowed or absent in *itr-1* reduction-of-function mutants. Dal Santo et al. [56] proposed that IP_3-dependent Ca^{2+} oscillations in intestinal cells control the secretion of a signal that regulates contraction of the surrounding posterior body wall muscles that drive defecation (Figure 8.4B).

We are currently using RNAi screening and fluorescent imaging techniques to identify other genes that function in intestinal oscillatory Ca^{2+} signaling. Recently, we developed an isolated intestine preparation that can be loaded with the cell permeant, acetoxymethyl (AM) ester derivatives of Ca^{2+} sensitive dyes including fluo-3, fluo-4, and fura-2. Rhythmic Ca^{2+} oscillations are readily observed in isolated intestines [65].

As noted earlier, a systemic RNAi can be induced in worms simply by feeding them bacteria producing dsRNA. We are currently performing large-scale RNAi screens to identify genes required for maintaining normal pBoc rhythm. pBoc is readily assayed simply by viewing worms through a dissecting microscope connected to a CCD camera and video monitor. Genes that disrupt pBoc timing when knocked down by RNAi are candidate oscillatory Ca^{2+} signaling genes and can then be tested using the isolated intestine preparation and Ca^{2+} imaging methods.

We recently used a combined imaging and RNAi approach to identify PLC-encoding genes that are required for intestinal Ca^{2+} signaling. RNAi of either EGL-8 (a PLC-β homolog) or PLC-3 (a PLC-γ homolog) induces dramatic arrhythmia in both pBoc and intestinal Ca^{2+} oscillations [66]. *egl-8* loss-of-function mutations also disrupt the pBoc cycle [67]. The effects of disruption of EGL-8 and PLC-3 function are additive. Genetic analysis indicated that PLC-3 acts upstream of ITR-1 whereas EGL-8 functions downstream and/or in parallel signaling pathways. An EGL-8 GFP fusion protein is localized to the apical pole of intestinal cells [68].

Mutagenesis and forward genetic screens serve not only to identify genes required for a specific biological process, but they can also provide important and novel mechanistic insights. For example, we have analyzed intestinal Ca^{2+} signaling in two IP_3R gain-of-function mutant worm strains. Both alleles are point mutations; one mutation is located in the IP_3 binding domain of the receptor and the other is located in the regulatory domain. Interestingly, these mutations have no effect on pBoc rhythm or Ca^{2+} oscillation period. However, they dramatically increase intercellular Ca^{2+} wave velocity from a mean of 22 μm/s to >130 μm/s [66]. These observations have suggested potential mechanisms of Ca^{2+} wave generation and have served as the foundation for new experiments to further define this important mode of Ca^{2+} signaling.

Patch clamp electrophysiology is an important tool for defining the role of ion channels in oscillatory Ca^{2+} signaling. We recently developed cell culture methods that allow us to culture terminally differentiated *C. elegans* somatic cells. Cell types are identified in culture by expression of cell specific GFP reporters [19, 69]. Using the whole-cell patch clamp technique we have demonstrated that cultured *C. elegans* intestinal epithelial cells express two highly Ca^{2+} selective channels [70]. One of these, ORCa (*O*utwardly *R*ectifying *Ca*lcium channel), displays strong outward rectification, inhibition by intracellular Mg^{2+}, and insensitivity to intracellular Ca^{2+} store depletion (Figure 8.5A). I_{ORCa} resembles the Mg^{2+}-inhibited current (MIC), Mg^{2+}–nucleotide regulated metal ion current (MagNUM), and TRPM7 currents studied in mammalian cells [71–74]. The other channel, (SOC) (Store-Operated Calcium Channel), is inwardly rectifying and activated by intracellular Ca^{2+} store depletion. I_{SOC} resembles the Ca^{2+}-release-activated Ca^{2+} current, I_{CRAC}, a proto-

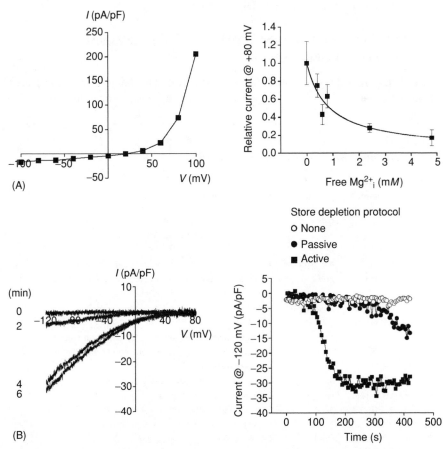

FIGURE 8.5 Calcium currents observed in cultured *C. elegans* intestinal epithelial cells. (A) I_{ORCa} is constitutively active, outwardly rectifying (left panel) and is inhibited by intracellular Mg^{2+} (right panel). (B) I_{SOC} is inwardly rectifying (left panel) and is activated by depletion of intracellular Ca^{2+} stores (right panel). Stores were passively depleted by dialyzing cells with a patch pipette solution containing 10 m*M* BAPTA or actively depleted by dialysis with 10 m*M* BAPTA, and 10 μ*M* IP_3. (Taken from Estevez, A.Y., Roberts R. K., and Strange K., *J. gen. Physiol,* 122, 207, 2003. With permission).

typical store-operated current that is expressed widely in vertebrate cells [71, 72, 75–77] (Figure 8.5B).

To determine the physiological roles of ORCa and SOC in oscillatory Ca^{2+} signaling it is important to identify the genes that encode these channels. At present, it is reasonable to postulate that the channels are encoded by one or more TRP genes [78, 79]. The *C. elegans* genome contains 18 predicted TRP-encoding genes including three TRPC, five TRPV, four TRPM, as well as six other more distantly related TRPs (Table 8. 1) [80–82]. Null or loss-of-function alleles have been identified for 13 of these genes. Patch clamp characterization of intestinal cells cultured from null or loss-of-function mutants should allow definitive testing of the hypothesis that TRP genes encode ORCa and SOC.

8.7 ROLE OF TRP CHANNEL-MEDIATED Ca^{2+} SIGNALING IN SPERM FUNCTION

Since their initial discovery in *Drosophila* [83–85], TRP ion channels have been identified in a wide range of organisms including yeast, nematodes, mice, and

---

TABLE 8.1
The *C. elegans* TRP Ion Channel Superfamily

TRP Family	Sequence Name	CGC[a] Name	General Information
TRPC	ZC21.2	*trp-1*	Expressed in motor neurons, interneurons, pharyngeal neurons, vulval, and intestinal muscle [116]
	R06B10.4	*trp-2*	
	K01A11.4	*trp-3/spe-41*	Important for conferring sperm the ability to fertilize oocytes[88]
TRPM	T01H8.5	*gon-2*	Important for postembryonic cell division of gonadal precursor cells [127]
	C05C12.3	*gtl-1/tag-33*	gon-2 *like 1*
	F54D1.5	*gtl-2*	gon-2 *like 2*
	ZK512.3	*ced-11*	When mutated results in abnormal cell death corpses
TRPV	B0212.5	*osm-9*	Important for osmosensation, chemosensation, olfactory response [116]
	F28H7.10	*ocr-1*	Expressed in a subset of *osm-9* neurons [117]
	T09A12.3	*ocr-2*	Expressed in a subset of *osm-9* neurons [117]
	T10B10.7	*ocr-3*	Expressed in a subset of *osm-9* neurons [117]
	Y40C5A.2	*ocr-4*	Expressed in a subset of *osm-9* neurons [117]
TRPN	Y71A12B.4	*trp-4*	Ce Nomp-C; mechanosensation [128]
TRPA	C29E6.2		
	M05B5.6		
TRPML	R13A5.1	*cup-5/muc-1*	Lysosome biogenesis [129]
TRPP	ZK945.9	*lov-1/pkd-1*	Location of vulva defective [130]
	Y73F8A.1	*pkd-2*	Polycystic kidney disease related [131]

[a]*Caenorhabditis* Genetics Center.

humans [81, 86, 87]. Although they are ubiquitously expressed and function in many different cell types, little is known about the physiological roles and regulation of most identified TRPs. With few exceptions, the majority of studies on TRP channels have been conducted under conditions in which they are overexpressed outside their native environment. Studies on genetically tractable organisms such as *C. elegans* and *Drosophila* can provide important insight into TRP channel physiology and the function and regulation of their mammalian counterparts [81, 82].

The *C. elegans* genome encodes three canonical TRP (TRPC) channels designated TRP-1, TRP-2, and TRP-3 (also known as SPE-41) (Table 8.1) [81, 82]. Xu and Sternberg [88] conducted targeted deletion screens to isolate TRP gene knockout animals. *trp-3* deletion mutants are nearly sterile due to a defect in oocyte and sperm fusion. Mating of *trp-3* mutant hermaphrodites with wild-type males restores normal fertility indicating that disruption of TRP-3 activity in sperm leads to the observed fertility defects. Consistent with this observation, microarray analysis as well as antibody staining studies demonstrate that TRP-3 expression is enriched in sperm [88,89].

Xu and Sternberg [88] postulated that binding of sperm to an oocyte activates TRP-3-mediated Ca^{2+} influx, which in turn triggers fusion. Fura-2 fluorescence imaging demonstrated the presence of two Ca^{2+}-permeable channels in wild-type sperm. One is a constitutively active channel (CAC) and the other is a store-operated channel (SOC) that opens upon depletion of intracellular Ca^{2+} stores with ionomycin or thapsigargin (Figure 8.6). Mature sperm from *trp-3* deletion mutants display normal CAC activity, but have severely impaired SOC activity (Figure 8.6). TRP-3 is localized to intracellular vesicles in immature sperm (spermatids) and upon sperm activation is translocated to the plasma membrane. Interestingly, mature sperm have markedly enhanced SOC activity compared to spermatids. Taken together, these data indicate that TRP-3 functions as a store-operated Ca^{2+} entry pathway that is activated by plasma membrane insertion. Importantly, the findings of Xu and Sternberg [88] establish a framework for mutagenesis and forward genetic analysis of TRP-3 function and regulation. For example, because *trp-3* knockouts are nearly sterile, suppressor screens could be conducted to isolate mutants with normal fertility. Such screens have the potential to identify novel components of the TRP-3 signaling and regulatory pathways, including mechanisms by which the channel senses Ca^{2+} store depletion.

8.8 CALCIUM SIGNALING IN *C. ELEGANS* NEURONS AND MUSCLE

Genetically encoded Ca^{2+} probes have made it possible to characterize the role of Ca^{2+} signaling in excitable cell function in *C. elegans*. In the following section, we present three examples of how a combination of *in vivo* Ca^{2+} imaging and genetic analysis has provided insights into the molecular mechanisms of muscle contraction, mechanotransduction, and neurodegeneration.

8.8.1 REGULATION OF PHARYNGEAL MUSCLE CONTRACTION

C. elegans is a filter feeder and the pharynx is a muscular organ that pumps food into the pharyngeal lumen, grinds it up, and then moves it into the intestine. To

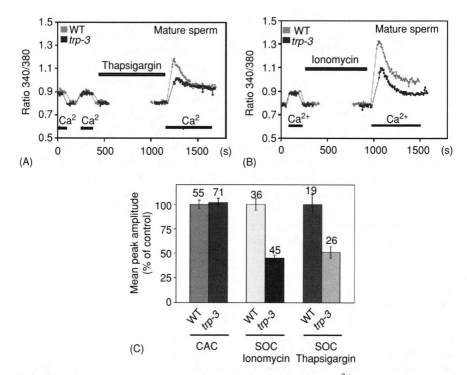

FIGURE 8.6 Constitutively active (CAC) and store-operated Ca^{2+} (SOC) channels in *C. elegans* mature sperm detected with fura-2 AM. Store depletion was induced in wild-type and *trp-3* knockout worms by incubation with thapsigargin (A) or ionomycin (B). (C) Histogram of the mean peak CAC and SOC transients in wild-type and *trp-3* knockout worms. Whereas the CAC transients are unaffected, store-operated Ca^{2+} transients are reduced ~50% in *trp-3* knockout worms compared to wild-type. (Taken from Xu, X. Z. and sternberg P. W., *Cell*, 114, 285, 2003. With Permission. Copyright Elsevier B. V.)

identify genes involved in pharyngeal function, Avery conducted a mutagenesis screen to isolate mutants with abnormal pharynx muscle contractions or "pumping" [90]. Pharyngeal pumping is essential for feeding and pumping mutants were identified by their starved appearance. This screen characterized 52 mutants including *egl-19*, which is a homolog of the α1 subunit of L-type voltage-gated Ca^{2+} channels [91].

The electrical activity of the pharynx has been characterized using an extracellular recording method [92] and microelectrode impalements [93–96]. Electrophysiological studies demonstrated that *egl-19* gain-of-function mutants have dramatically prolonged action potentials [91]. Gain-of-function mutations are localized to regions of the channel that are responsible for voltage-dependent inactivation.

Electrophysiological studies of the pharynx have been confirmed and extended using Ca^{2+} imaging methods. Cameleons are genetically encoded fusion proteins that function as ratiometric Ca^{2+} indicators (Chapter 4). Kerr et al. [97] used a cyan fluorescent protein (CFP)–yellow fluorescent protein (YFP) cameleon to study Ca^{2+} signaling in pharyngeal muscles and showed that gain-of-function mutations in *egl-19* prolong the duration of Ca^{2+} transients at the onset of pharyngeal muscle contraction. Kerr et al. [97] also demonstrated that the rate of Ca^{2+} increase during action potential firing was greatly increased in *unc-36* loss-of-function mutants. *unc-36* encodes a *C. elegans* homolog of the α2 subunit of L-type voltage-gated Ca^{2+} channels. The findings of Kerr et al. [97] suggest that UNC-36 negatively regulates Ca^{2+} influx and are in accordance with heterologous expression studies, which suggest that vertebrate α2 subunits regulate L-type Ca^{2+} channel inactivation [98–101].

Pericams are circularly permuted YFP chimeric proteins that have been engineered to sense Ca^{2+} [102]. Recently, Shimozono et al. [103] developed a ratiometric variant of inverse pericam fused with red fluorescent protein, DsRed2, at its C-terminus. This new molecule, named DsRed2-reference inverse pericam (DRIP), has also allowed the measurement of Ca^{2+} transients and dynamics in the pharynx. The studies of Shimozono et al. [103] suggest that Ca^{2+} release from intracellular stores via the ryanodine receptor UNC-68 may play a role in regulating intracellular Ca^{2+} dynamics and plasma membrane Ca^{2+} channel activity during pharyngeal contractions.

8.8.2 Mechanotransduction

C. elegans has been an invaluable model system for identifying genes involved in mechanosensory processes. Genes required for mechanosensory neuron function were identified by mutagenizing worms and assessing their ability to move away from gentle body touch [104, 105]. Touch insensitive mutant worms are mechanosensory abnormal and the genes responsible for this abnormality are termed *mec*. Approximately 18 *mec* genes have been identified to date [106]. Several of these genes are required for normal touch neuron development. At least eight *mec* genes encode proteins that have been postulated to form a mechanosensitive ion channel complex [107–109]. *mec-4* and *mec-10* are members of the DEG/ENaC cation channel superfamily and are thought to form the ion conducting pathway of the channel [110–112].

Characterizing the properties of the putative *mec* encoded mechanosensory channel complex *in vivo* by patch clamp has been difficult because of problems associated with isolating intact touch neurons (but see Ref. [111]). However, genetically encoded cameleons have allowed indirect characterization of the response of touch neurons to mechanical force. Suzuki et al. [113] expressed a CFP–YFP cameleon in the touch neurons under the control of the *mec-4* promoter. Cameleon expressing worms were glued onto agar pads and mechanically stimulated using a glass probe (Figure 8.7A). Touch activated Ca^{2+} transients are detected in wild-type animals but not in *mec-4* null mutants (Figure 8.7B). The Ca^{2+} response is also dependent on the L-type Ca^{2+} channel EGL-19. Loss-of-function mutations in *egl-19* significantly reduce the magnitude of touch evoked Ca^{2+} influx. In contrast, null mutations in the N-type voltage-gated Ca^{2+} channel

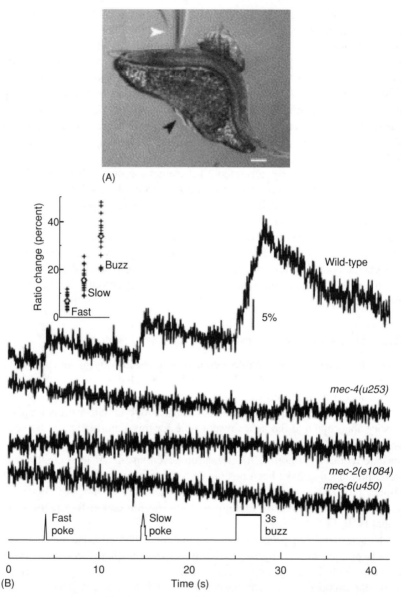

(A)

(B)

FIGURE 8.7 *In vivo* imaging of *C. elegans* mechanosensory neurons with cameleon. (A) Adult hermaphrodite worm glued to a 2% agar pad. Mechanical stimulation in the form of a fast poke, a slow poke or a buzz is applied with a probe (white arrow). Scale bar, 50 μm. (B) Calcium transients in touch neurons of wild-type or *mec* mutant worms in response to gentle stimulation. Traces represent the ratio change (percent) of YFP to CFP fluorescence intensity. Note that wild-type worms display Ca^{2+} transients in response to all three types of gentle touch stimuli whereas the responses are absent in *mec-4*, *mec-2*, and *mec-6* mutants. (Taken from Suzuki, H. et al., *Neuron*, 39, 1005, 2003. With permission. Copyright Elsevier B.V.)

gene, *unc-2* [114], have no effects on Ca^{2+} transients. Reduction-of-function alleles of the genes encoding the ryanodine receptor *unc-68*, the IP_3R *itr-1*, or the ER calcium-binding protein *calreticulin-1* (*crt-1*) also have no effects indicating that ER Ca^{2+} release does not play a role in generating the Ca^{2+} transients. Taken together, these data suggest that activation of MEC-4 upon mechanical stimulation depolarizes the touch neurons and this depolarization subsequently leads to Ca^{2+} influx via EGL-19.

Cameleons have also been used to track Ca^{2+} changes in ASH sensory neurons in response to chemical, osmotic, and mechanical stimuli [115]. Evoked ASH Ca^{2+} transients are dependent on OSM-9 (Figure 8.8), the founding member of the TRPV channel subfamily [116]. OSM-9 plays important roles in osmosensation, chemosensation, mechanosensation, and olfaction in *C. elegans* [116, 117].

8.8.3 NEURODEGENERATION

The first DEG/ENaC gene identified in *C. elegans* was *deg-1* (*deg*eneration of certain neurons). Gain-of-function mutations in *deg-1* cause various neurons to swell and degenerate [118] and the protein encoded by the gene was thus termed a "degenerin." Expression of a hyperactive form of the MEC-4 channel, termed MEC-4(d) also induces neuronal death [119, 120]. The cell death induced by MEC-4(d) and other degenerins is via necrosis rather than apoptosis [121]. To define the genes involved in necrotic cell death, Xu et al. [122] conducted a suppressor screen of MEC-4(d)-induced neurodegeneration. MEC-4(d) was expressed ectopically in the ventral nerve cord, which resulted in the necrotic death and consequent paralysis of the animals. Paralyzed worms were then mutagenized and mutants in which movement was restored were isolated. This screen identified the calreticulin encoding gene *crt-1* as a suppressor of MEC-4(d) induced

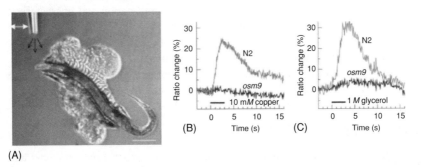

FIGURE 8.8 *In vivo* imaging of *C. elegans* ASH sensory neurons with cameleon. (A) Adult hermaphrodite worm glued to a 2% agar pad. Chemical stimuli are delivered via a glass pipette that is moved toward or away from the worm (white arrow). Scale bar, 200 μm. (B) ASH neuron Ca^{2+} transients observed in response to 10 m*M* copper or 1 *M* glycerol (C) in wild-type (N2) worms are completely absent in *osm-9* loss-of-function mutants. (Taken from Hillard, M.A. et al, *EMBO J.*, 24, 63, 2005. With permission. Copyright Macmillan Magazines Ltd.)

neurodegeneration. CRT-1 is a calcium binding protein normally expressed in the ER where it functions to modulate free-Ca^{2+} levels and also acts as a molecular chaperone [123]. The suppressing effect of loss-of-function *crt-1* mutations is dependent on Ca^{2+} release from intracellular stores and can be blocked by gain-of-function mutations in the IP_3 and ryanodine receptors ITR-1 and UNC-68, respectively, or by RNAi of another store Ca^{2+} binding protein, calnexin.

How does hyperactivation of MEC-4(d) initiate necrotic cell death? One possibility is that Na^+ influx via MEC-4(d) depolarizes the cell and activates voltage-gated Ca^{2+} channels, leading to an increase in intracellular Ca^{2+} that initiates necrosis. However, loss-of-function mutations in any of the six voltage-gated Ca^{2+} channels expressed in *C. elegans* do not inhibit MEC-4(d) induced necrosis [124]. Calcium measurements in cultured touch neurons expressing cameleons demonstrated that MEC-4(d) induces an amiloride sensitive Ca^{2+} influx. Bianchi et al. [124] have proposed that Ca^{2+} entry via the MEC-4(d) channel triggers Ca^{2+} release from intracellular stores via IP_3 and ryanodine receptors. Store Ca^{2+} release in turn activates proteases that initiate cell death [125].

8.9 CONCLUSION AND FUTURE PERSPECTIVES

C. elegans offers substantial experimental advantages for characterizing the molecular and genetic bases of biological processes common to all eukaryotes. These advantages include a fully sequenced and well-annotated genome; relative ease and economy of manipulating gene expression by RNAi, knockout and creation of transgenic animals; ready availability of numerous molecular reagents and mutant worm strains; and the ability to perform mutagenesis and forward genetic analysis. The combination of forward and reverse genetic tools with physiological methods such as patch clamp electrophysiology and intracellular Ca^{2+} imaging is a powerful approach for defining fundamental aspects of Ca^{2+} signaling in excitable and nonexcitable cells as well as for the molecular identification and characterization of plasma membrane Ca^{2+} channels. As has been observed for numerous other biological processes [1,126], genetic and molecular insights gained from an integrated understanding of Ca^{2+} signaling in *C. elegans* have and will clearly continue to provide new and important insights into vertebrate Ca^{2+} signaling mechanisms and help drive experimental studies in vertebrate cell models.

REFERENCES

1. Strange, K., From genes to integrative physiology: ion channel and transporter biology in *Caenorhabditis elegans*, *Physiol Rev.*, 83, 377, 2003.
2. Smeal, T. and Guarente L., Mechanisms of cellular senescence, *Curr. Opin. Genet. Dev.*, 7, 281, 1997.
3. Patterson, G.I. and Padgett R.W., TGF β-related pathways. Roles in *Caenorhabditis elegans* development, *Trends Genet.*, 16, 27, 2000.
4. Moerman, D.G. and Waterston R.H., Spontaneous unstable *unc-22* IV mutations in *C. elegans* var. Bergerac, *Genetics*, 108, 859, 1984.

5. Wicks, S.R. et al., Rapid gene mapping in *Caenorhabditis elegans* using a high density polymorphism map, *Nat. Genet.*, 28, 160, 2001.

6. Lewis, J.A. and Fleming J.T., Basic culture methods, In *Caenorhabditis elegans: Modern Biological Analysis of an Organism*, Epstein H.F. and Shakes D.C., Eds., Academic, New York, 1995, chap. 1.

7. Sulston, J.E. and Horvitz H.R., Post-embryonic cell lineages of the nematode, *Caenorhabditis elegans*, *Dev. Biol.*, 56, 110, 1977.

8. Sulston, J.E. et al., The embryonic cell lineage of the nematode *Caenorhabditis elegans*, *Dev. Biol.*, 100, 64, 1983.

9. White, J.G. et al., The structure of the nervous system of the nematode *C. elegans*, *Philos. Trans. R. Soc. Lond B Biol. Sci.*, 314, 1, 1986.

10. Wood, W.B., *The Nematode Caenorhabditis Elegans*, Cold Spring Harbor Laboratory Press, Cold Spring Harbor, 1988, 667 pp.

11. Jorgensen, E.M. and Mango S.E., The art and design of genetic screens: *Caenorhabditis elegans*, *Nat. Rev. Genet.*, 3, 356, 2002.

12. Swan, K.A. et al., High-throughput gene mapping in *Caenorhabditis elegans*, *Genome Res.*, 12, 1100, 2002.

13. Fire, A. et al., Potent and specific genetic interference by double-stranded RNA in *Caenorhabditis elegans*, *Nature*, 391, 806, 1998.

14. Tabara, H., Grishok A., and Mello C.C., RNAi in *C. elegans*: soaking in the genome sequence, *Science*, 282, 430, 1998.

15. Kamath, R.S. et al., Effectiveness of specific RNA-mediated interference through ingested double-stranded RNA in *Caenorhabditis elegans*, *Genome Biol.*, 2, 2.1, 2000.

16. Timmons, L., Court D. L., and Fire A., Ingestion of bacterially expressed dsRNAs can produce specific and potent genetic interference in *Caenorhabditis elegans*, *Gene*, 263, 103, 2001.

17. Timmons, L. and Fire A., Specific interference by ingested dsRNA, *Nature*, 395, 854, 1998.

18. Tavernarakis, N. et al., Heritable and inducible genetic interference by double-stranded RNA encoded by transgenes, *Nat. Genet.*, 24, 180, 2000.

19. Christensen, M. et al., A primary culture system for functional analysis of *C. elegans* neurons and muscle cells, *Neuron*, 33, 503, 2002.

20. Kamath, R.S. et al., Systematic functional analysis of the *Caenorhabditis elegans* genome using RNAi, *Nature*, 421, 231, 2003.

21. Ashrafi, K. et al., Genome-wide RNAi analysis of *Caenorhabditis elegans* fat regulatory genes, *Nature*, 421, 268, 2003.

22. Murphy, C.T. et al., Genes that act downstream of DAF-16 to influence the lifespan of *Caenorhabditis elegans*, *Nature*, 424, 277, 2003.

23. Lee, S.S. et al., A systematic RNAi screen identifies a critical role for mitochondria in *C. elegans* longevity, *Nat. Genet.*, 33, 40, 2003.

24. Zipperlen, P. et al., Roles for 147 embryonic lethal genes on *C. elegans* chromosome I identified by RNA interference and video microscopy, *EMBO J.*, 20, 3984, 2001.

25. Vastenhouw, N.L. et al., A genome-wide screen identifies 27 genes involved in transposon silencing in *C. elegans*, *Curr. Biol.*, 13, 1311, 2003.

26. Pothof, J. et al., Identification of genes that protect the *C. elegans* genome against mutations by genome-wide RNAi, *Genes Dev.*, 17, 443, 2003.

27. Lamitina, S.T. and Strange K., Transcriptional targets of the DAF-16 insulin signaling pathway protect *C. elegans* from extreme hypertonic stress, *Am. J. Physiol. Cell Physiol.*, 288, C467, 2005.

28. Nollen, E.A. et al., Genome-wide RNA interference screen identifies previously un-described regulators of polyglutamine aggregation, *Proc. Natl. Acad. Sci. USA*, 101, 6403, 2004.

29. Jansen, G. et al., Reverse genetics by chemical mutagenesis in *Caenorhabditis elegans*, *Nat. Genet.*, 17, 119, 1997.

30. Liu, L.X. et al., High-throughput isolation of *Caenorhabditis elegans* deletion mutants, *Genome Res.*, 9, 859, 1999.

31. Chalfie, M. et al., Green fluorescent protein as a marker for gene expression, *Science*, 263, 802, 1994.

32. Zhang, S., Ma C., and Chalfie M., Combinatorial marking of cells and organelles with reconstituted fluorescent proteins, *Cell*, 119, 137, 2004.

33. Hutter, H., Five-colour *in vivo* imaging of neurons in *Caenorhabditis elegans*, *J Microsc.*, 215, 213, 2004.

34. Stinchcomb, D.T. et al., Extrachromosomal DNA transformation of *Caenorhabditis elegans*, *Mol. Cell Biol.*, 5, 3484, 1985.

35. Fire, A., Integrative transformation of *Caenorhabditis elegans*, *EMBO J.*, 5, 2673, 1986.

36. Mello, C.C. et al., Efficient gene transfer in *C. elegans*: extrachromosomal maintenance and integration of transforming sequences, *EMBO J.*, 10, 3959, 1991.

37. Mello, C. and Fire A., DNA transformation, *Methods Cell. Biol.*, 48, 451, 1995.

38. Praitis, V. et al., Creation of low-copy integrated transgenic lines in *Caenorhabditis elegans*, *Genetics*, 157, 1217, 2001.

39. Kelly, W.G. and Fire A., Chromatin silencing and the maintenance of a functional germline in *Caenorhabditis elegans*, *Development*, 125, 2451, 1998.

40. Clandinin, T.R., DeModena J. A., and Sternberg P. W., Inositol trisphosphate mediates a RAS-independent response to LET-23 receptor tyrosine kinase activation in *C. elegans*, *Cell*, 92, 523, 1998.

41. Hill, R.J. and Sternberg P.W., The gene lin-3 encodes an inductive signal for vulval development in *C. elegans*, *Nature*, 358, 470, 1992.

42. Aroian, R.V. et al., The let-23 gene necessary for *Caenorhabditis elegans* vulval induction encodes a tyrosine kinase of the EGF receptor subfamily, *Nature*, 348, 693, 1990.

43. Aroian, R.V. and Sternberg P.W., Multiple functions of let-23, a *Caenorhabditis elegans* receptor tyrosine kinase gene required for vulval induction, *Genetics*, 128, 251, 1991.

44. Sternberg, P.W. et al., LET-23-mediated signal transduction during *Caenorhabditis elegans* development, *Mol. Reprod. Dev.*, 42, 523, 1995.

45. Baylis, H.A. et al., Inositol 1,4,5-trisphosphate receptors are strongly expressed in the nervous system, pharynx, intestine, gonad and excretory cell of *Caenorhabditis elegans* and are encoded by a single gene (*itr-1*), *J. Mol. Biol.*, 294, 467, 1999.

46. Walker, D.S. et al., Regulated disruption of inositol 1,4,5-trisphosphate signaling in *Caenorhabditis elegans* reveals new functions in feeding and embryogenesis, *Mol. Biol. Cell.*, 13, 1329, 2002.

47. Yoshikawa, F. et al., Mutational analysis of the ligand binding site of the inositol 1,4,5-trisphosphate receptor, *J Biol. Chem.*, 271, 18277, 1996.

48. Irvine, R.F. and Schell M. J., Back in the water: the return of the inositol phosphates, *Nat. Rev. Mol. Cell Biol.*, 2, 327, 2001.

49. Bui, Y.K. and Sternberg P.W., *Caenorhabditis elegans* inositol 5-phosphatase homolog negatively regulates inositol 1,4,5-triphosphate signaling in ovulation, *Mol. Biol. Cell.*, 13, 1641, 2002.

50. Majerus, P.W., Kisseleva M.V., and Norris F.A., The role of phosphatases in inositol signaling reactions, *J. Biol. Chem.*, 274, 10669, 1999.
51. McCarter, J. et al., Soma-germ cell interactions in *Caenorhabditis elegans*: multiple events of hermaphrodite germline development require the somatic sheath and spermathecal lineages, *Dev. Biol.*, 181, 121, 1997.
52. McCarter, J. et al., On the control of oocyte meiotic maturation and ovulation in *Caenorhabditis elegans*, *Dev. Biol.*, 205, 111, 1999.
53. Hall, D.H. et al., Ultrastructural features of the adult hermaphrodite gonad of *Caenorhabditis elegans*: relations between the germ line and soma, *Dev. Biol.*, 212, 101, 1999.
54. Hirsh, D., Oppenheim D., and Klass M., Development of the reproductive system of *Caenorhabditis elegans*, *Dev. Biol.*, 49, 200, 1976.
55. Hubbard, E.J. and Greenstein D., The *Caenorhabditis elegans* gonad: a test tube for cell and developmental biology, *Dev. Dyn.*, 218, 2, 2000.
56. Dal Santo P. et al., The inositol trisphosphate receptor regulates a 50-second behavioral rhythm in *C. elegans*, *Cell*, 98, 757, 1999.
57. Hwang, B.J. and Sternberg P. W., A cell-specific enhancer that specifies lin-3 expression in the *C. elegans* anchor cell for vulval development, *Development*, 131, 143, 2004.
58. Kariya, K. et al., Phospholipase Ce regulates ovulation in *Caenorhabditis elegans*, *Dev. Biol.*, 274, 201, 2004.
59. Yin, X. et al., Inositol 1,4,5-trisphosphate signaling regulates rhythmic contractile activity of smooth muscle-like sheath cells in the nematode *Caenorhabditis elegans*, *Mol. Biol. Cell.*, 15, 3938, 2004.
60. Rutledge, E. et al., CLH-3, a ClC-2 anion channel ortholog activated during meiotic maturation in *C. elegans* oocytes, *Curr. Biol.*, 11, 161, 2001.
61. Miller, M.A. et al., A sperm cytoskeletal protein that signals oocyte meiotic maturation and ovulation, *Science*, 291, 2144, 2001.
62. Iwasaki, K. et al., *emo-1*, a *Caenorhabditis elegans* Sec61p gamma homologue, is required for oocyte development and ovulation, *J. Cell Biol.*, 134, 699, 1996.
63. Iwasaki, K. and Thomas J. H., Genetics in rhythm, *Trends Genet.*, 13, 111, 1997.
64. Iwasaki, K., Liu D. W., and Thomas J. H., Genes that control a temperature-compensated ultradian clock in *Caenorhabditis elegans*, *Proc. Natl. Acad. Sci.*, 92, 10317, 1995.
65. Espelt, M.V., Estevez A.Y., and Strange K., Mechanisms of oscillatory Ca^{2+} signaling in the intestinal epithelium of the nematode *C. elegans*, Abstract, *FASEB J.*, 18, 459.17, 2004.
66. Espelt, M.V., Estevez A.Y., and Strange K., Oscillatory Ca^{2+} signaling in the *C. elegans* intestinal epithelium: role of the IP_3 receptor and PLC, Abstract, *FASEB J.*, 19, 383.10, 2005.
67. Lackner, M.R., Nurrish S. J., and Kaplan J. M., Facilitation of synaptic transmission by EGL-30 Gqα and EGL-8 PLCβ: DAG binding to UNC-13 is required to stimulate acetylcholine release, *Neuron*, 24, 335, 1999.
68. Miller, K.G., Emerson M.D., and Rand J. B., Goa and diacylglycerol kinase negatively regulate the Gqa pathway in *C. elegans*, *Neuron*, 24, 323, 1999.
69. Christensen, M. and Strange K., Developmental regulation of a novel outwardly rectifying mechanosensitive anion channel in *Caenorhabditis elegans*, *J. Biol. Chem.*, 276, 45024, 2001.
70. Estevez, A.Y., Roberts R.K., and Strange K., Identification of store-independent and store-operated Ca^{2+} conductances in *Caenorhabditis elegans* intestinal epithelial cells, *J. Gen. Physiol.*, 122, 207, 2003.

71. Hermosura, M.C. et al., Dissociation of the store-operated calcium current I(CRAC) and the Mg-nucleotide-regulated metal ion current MagNuM, *J. Physiol.*, 539, 445, 2002.

72. Prakriya, M. and Lewis R. S., Separation and characterization of currents through store-operated CRAC channels and Mg^{2+}-inhibited cation (MIC) channels, *J. Gen. Physiol.*, 119, 487, 2002.

73. Nadler, M.J. et al., LTRPC7 is a Mg.ATP-regulated divalent cation channel required for cell viability, *Nature*, 411, 590, 2001.

74. Kozak, J.A., Kerschbaum H.H., and Cahalan M.D., Distinct properties of CRAC and MIC channels in RBL cells, *J. Gen. Physiol.*, 120, 221, 2002.

75. Prakriya, M. and Lewis R.S., CRAC channels: activation, permeation, and the search for a molecular identity, *Cell. Calcium*, 33, 311, 2003.

76. Parekh, A.B. and Penner R., Store depletion and calcium influx, *Physiol. Rev.*, 77, 901, 1997.

77. Hoth, M. and Penner R., Depletion of intracellular calcium stores activates a calcium current in mast cells, *Nature*, 355, 353, 1992.

78. Montell, C., Physiology, phylogeny, and functions of the TRP superfamily of cation channels, *Sci STKE.*, 2001, RE1, 2001.

79. Clapham, D.E., Sorting out MIC, TRP, and CRAC ion channels, *J. Gen. Physiol.*, 120, 217, 2002.

80. Harteneck, C., Plant T.D., and Schultz G., From worm to man: three subfamilies of TRP channels, *Trends Neurosci.*, 23, 159, 2000.

81. Montell, C., The venerable inveterate invertebrate TRP channels, *Cell. Calcium*, 33, 409, 2003.

82. Vriens, J. et al., Invertebrate TRP proteins as functional models for mammalian channels, *Pflugers Arch.*, 2004.

83. Montell, C. et al., Rescue of the *Drosophila* phototransduction mutation *trp* by germline transformation, *Science*, 230, 1040, 1985.

84. Montell, C. and Rubin G.M., Molecular characterization of the *Drosophila trp* locus: a putative integral membrane protein required for phototransduction, *Neuron*, 2, 1313, 1989.

85. Hardie, R.C. and Minke B., The trp gene is essential for a light-activated Ca^{2+} channel in *Drosophila* photoreceptors, *Neuron*, 8, 643, 1992.

86. Harteneck, C., Plant T.D., and Schultz G., From worm to man: three subfamilies of TRP channels, *Trends Neurosci.*, 23, 159, 2000.

87. Clapham, D.E., Runnels L.W., and Strubing C., The TRP ion channel family, *Nat. Rev. Neurosci.*, 2, 387, 2001.

88. Xu, X.Z. and Sternberg P. W., A C. *elegans* sperm TRP protein required for sperm–egg interactions during fertilization, *Cell*, 114, 285, 2003.

89. Reinke, V. et al., A global profile of germline gene expression in *C. elegans*, *Mol. Cell.*, 6, 605, 2000.

90. Avery, L., The genetics of feeding in *Caenorhabditis elegans*, *Genetics*, 133, 897, 1993.

91. Lee, R.Y. et al., Mutations in the alpha1 subunit of an L-type voltage-activated Ca^{2+} channel cause myotonia in *Caenorhabditis elegans*, *EMBO J.*, 16, 6066, 1997.

92. Raizen, D.M. and Avery L., Electrical activity and behavior in the pharynx of *Caenorhabditis elegans*, *Neuron*, 12, 483, 1994.

93. Davis, M.W. et al., A mutation in the *C. elegans* EXP-2 potassium channel that alters feeding behavior, *Science*, 286, 2501, 1999.

94. Davis, M.W. et al., Mutations in the *Caenorhabditis elegans* Na,K-ATPase alpha-subunit gene, *eat-6*, disrupt excitable cell function, *J. Neurosci.*, 15, 8408, 1995.

95. Franks, C.J. et al., Ionic basis of the resting membrane potential and action potential in the pharyngeal muscle of *Caenorhabditis elegans*, *J. Neurophysiol.*, 87, 954, 2002.
96. Pemberton, D.J. et al., Characterization of glutamate-gated chloride channels in the pharynx of wild-type and mutant *Caenorhabditis elegans* delineates the role of the subunit GluCl-a2 in the function of the native receptor, *Mol. Pharmacol.*, 59, 1037, 2001.
97. Kerr, R. et al., Optical imaging of calcium transients in neurons and pharyngeal muscle of *C. elegans*, *Neuron*, 26, 583, 2000.
98. Singer, D. et al., The roles of the subunits in the function of the calcium channel, *Science*, 253, 1553, 1991.
99. Shirokov, R. et al., Inactivation of gating currents of L-type calcium channels. Specific role of the alpha 2 delta subunit, *J. Gen. Physiol.*, 111, 807, 1998.
100. Bangalore, R. et al., Influence of L-type Ca channel alpha 2/delta-subunit on ionic and gating current in transiently transfected HEK 293 cells, *Am. J. Physiol. Heart Circ. Physiol.*, 270, H1521, 1996.
101. De Waard, M. and Campbell K. P., Subunit regulation of the neuronal alpha 1A Ca^{2+} channel expressed in Xenopus oocytes, *J. Physiol.*, 485 (Pt 3), 619, 1995.
102. Nagai, T. et al., Circularly permuted green fluorescent proteins engineered to sense Ca^{2+}, *Proc. Natl. Acad. Sci.*, 98, 3197, 2001.
103. Shimozono, S. et al., Slow Ca^{2+} dynamics in pharyngeal muscles in *Caenorhabditis elegans* during fast pumping, *EMBO Reports*, 5, 521, 2004.
104. Chalfie, M. and Au M., Genetic control of differentiation of the *Caenorhabditis elegans* touch receptor neurons, *Science*, 243, 1027, 1989.
105. Chalfie, M. and Sulston J., Developmental genetics of the mechanosensory neurons of *Caenorhabditis elegans*, *Dev. Biol.*, 82, 358, 1981.
106. Ernstrom, G.G. and Chalfie M., Genetics of sensory mechanotransduction, *Annu. Rev. Genet.*, 36, 411, 2002.
107. Mano, I. and Driscoll M., DEG/ENaC channels: a touchy superfamily that watches its salt, *BioEssays*, 21, 568, 1999.
108. Tavernarakis, N. and Driscoll M., Molecular modeling of mechanotransduction in the nematode *Caenorhabditis elegans*, *Annu. Rev. Physiol.*, 59, 659, 1997.
109. Tavernarakis, N. and Driscoll M., Degenerins. At the core of the metazoan mechanotransducer? *Ann. N. Y. Acad. Sci.*, 940, 28, 2001.
110. Chelur, D.S. et al., The mechanosensory protein MEC-6 is a subunit of the *C. elegans* touch-cell degenerin channel, *Nature*, 420, 669, 2002.
111. O'Hagan, R., Chalfie M., and Goodman M.B., The MEC-4 DEG/ENaC channel of *Caenorhabditis elegans* touch receptor neurons transduces mechanical signals, *Nat. Neurosci.*, 8, 43, 2005.
112. Goodman, M.B. et al., MEC-2 regulates *C. elegans* DEG/ENaC channels needed for mechanosensation, *Nature*, 415, 1039, 2002.
113. Suzuki, H. et al., *In vivo* imaging of *C. elegans* mechanosensory neurons demonstrates a specific role for the MEC-4 channel in the process of gentle touch sensation, *Neuron*, 39, 1005, 2003.
114. Schafer, W.R. and Kenyon C. J., A calcium-channel homologue required for adaptation to dopamine and serotonin in *Caenorhabditis elegans*, *Nature*, 375, 73, 1995.
115. Hilliard, M.A. et al., *In vivo* imaging of *C. elegans* ASH neurons: cellular response and adaptation to chemical repellents, *EMBO J.*, 24, 63, 2005.
116. Colbert, H.A., Smith T. L., and Bargmann C. I., OSM-9, a novel protein with structural similarity to channels, is required for olfaction, mechanosensation, and olfactory adaptation in *Caenorhabditis elegans*, *J. Neurosci.*, 17, 8259, 1997.

117. Tobin, D. et al., Combinatorial expression of TRPV channel proteins defines their sensory functions and subcellular localization in *C. elegans* neurons, *Neuron*, 35, 307, 2002.

118. Chalfie, M., and Wolinsky E., The identification and suppression of inherited neuro-degeneration in *Caenorhabditis elegans*, *Nature*, 345, 410, 1990.

119. Driscoll, M., and Chalfie M., The *mec-4* gene is a member of a family of *Caenorhabditis elegans* genes that can mutate to induce neuronal degeneration, *Nature*, 349, 588, 1991.

120. Hong, K., and Driscoll M., A transmembrane domain of the putative channel subunit MEC-4 influences mechanotransduction and neurodegeneration in *C. elegans*, *Nature*, 367, 470, 1994.

121. Hall, D.H. et al., Neuropathology of degenerative cell death in *Caenorhabditis elegans*, *J. Neuroscience*, 17, 1033, 1997.

122. Xu, K., Tavernarakis N., and Driscoll M., Necrotic cell death in *C. elegans* requires the function of calreticulin and regulators of Ca^{2+} release from the endoplasmic reticulum, *Neuron*, 31, 957, 2001.

123. Park, B.J. et al., Calreticulin, a calcium-binding molecular chaperone, is required for stress response and fertility in *Caenorhabditis elegans*, *Mol. Biol. Cell.*, 12, 2835, 2001.

124. Bianchi, L. et al., The neurotoxic MEC-4(d) DEG/ENaC sodium channel conducts calcium: implications for necrosis initiation, *Nat. Neurosci.*, 7, 1337, 2004.

125. Syntichaki, P. and Tavernarakis N., Genetic models of mechanotransduction: the nematode *Caenorhabditis elegans*, *Physiol. Rev.*, 84, 1097, 2004.

126. Barr, M.M., Super models, *Physiol. Genomics*, 13, 15, 2003.

127. West, R.J. et al., The *C. elegans gon-2* gene encodes a putative TRP cation channel protein required for mitotic cell cycle progression, *Gene*, 266, 103, 2001.

128. Walker, R.G., Willingham A.T., and Zuker C.S., A *Drosophila* mechanosensory transduction channel, *Science*, 287, 2229, 2000.

129. Hersh, B.M., Hartwieg E., and Horvitz H.R., The *Caenorhabditis elegans* mucolipin-like gene cup-5 is essential for viability and regulates lysosomes in multiple cell types, *Proc. Natl. Acad. Sci.*, 99, 4355, 2002.

130. Barr, M.M. and Sternberg P. W., A polycystic kidney-disease gene homologue required for male mating behaviour in *C. elegans*, *Nature*, 401, 386, 1999.

131. Barr, M.M. et al., The *Caenorhabditis elegans* autosomal dominant polycystic kidney disease gene homologs *lov-1* and *pkd-2* act in the same pathway, *Curr. Biol.*, 11, 1341, 2001.

9 Calcium Signaling Networks in Yeast

Kyle W. Cunningham

CONTENTS

9.1 WHY STUDY YEAST?

Genome sequencing has firmly established that animals are most closely related to the fungi, less closely related to the slime molds (Mycetozoa), and distantly related to all other eukaryotes such as plants, alga, and protozoa. Eukaryotes also seem to have diverged from the Archaea more recently than the Eubacteria. This tree of life provides a useful framework to map the origin, duplication, diversification, and loss of each factor participating in calcium signaling (see Figure 9.1). Most components of the calcium signaling networks studied today in animal systems originated before or soon after the emergence of the eukaryotic domain and are retained to varying extents in diverse groups of eukaryotes. In the course of evolution, all species have lost certain various calcium signaling factors that constituted the ancestral repertoire, resulting in more streamlined and efficient signaling networks. In some lineages, particularly that which leads to mammals, certain factors have been duplicated, amplified, and diversified to yield greater complexity and its inherent advantages. The study of calcium signaling networks in nonanimal eukaryotes helps provide the proper evolutionary perspective that is key to understanding the fundamental principles that govern the operations of the system. Additionally, an understanding of calcium signaling mechanisms in diverse eukaryotes has several practical benefits. Genetic engineering of calcium signaling networks in plants

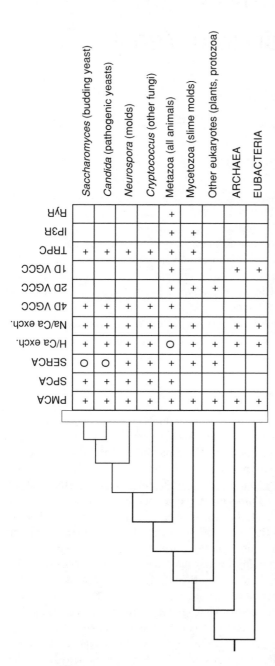

FIGURE 9.1 Evolutionary histories of enzymes that generate or dissipate calcium signals. The phylogenetic tree depicts the evolutionary relationships among four species of fungi, animals, slime molds, plants, protozoa, and prokaryotes (Archaea and Eubacteria). The presence and absence of calcium signaling factors in these species, kingdoms, and domains are indicated by "+" and "blank," respectively. "O" denotes highly probable instances of gene loss.

will potentially yield more nutritious and stress-tolerant crops. As calcium signaling networks of many pathogenic fungi provide resistance to commonly prescribed antifungal antibiotics, a better understanding of calcium signaling networks in this group may lead directly to improved human therapies.

This chapter focuses attention on calcium signaling and transport mechanisms in the budding yeast *Saccharomyces cerevisiae*, the best-characterized species in the fungal kingdom. Several technical disadvantages have limited studies of calcium signaling in yeast. For example, the very small size of yeast cells (2–3 μm in diameter) and the thick cell wall have hampered efforts to monitor Ca^{2+} channel activity and cytosolic free Ca^{2+} concentrations ($[Ca^{2+}]_{cyt}$) using the powerful electrophysiological and imaging techniques that have been so amazingly successful in animal cells. Despite these limitations, Ca^{2+} fluxes and free Ca^{2+} dynamics in various compartments can be easily measured in populations of yeast cells using isotopic tracers and targeted versions of Aequorin. Calcineurin activation is easily monitored in populations of yeast cells using a variety of independent readouts: activation of the Crz1/Tcn1 transcription factor, inhibition of Ca^{2+} transporters and Ca^{2+} channels, and others. Perhaps the most valuable tools include the arsenal of classical and molecular genetic technologies available in the species. Combinations of these approaches enable the rapid identification and accurate positioning of new components of the calcium signaling network without prior knowledge of their structure or function. The RCN regulators of calcineurin that are conserved from yeast to humans were discovered in this way. Finally, engineered yeast cells are becoming useful for identification and characterization of Ca^{2+} pumps, exchangers, channels, and regulators from plants and diverse eukaryotes. The structure, function, and regulation of the enzymes endogenous to yeast are summarized below.

9.2 Ca^{2+} PUMPS AND EXCHANGERS

9.2.1   SPCA-Type Ca^{2+} Pumps

Secretory pathway Ca^{2+} pumps/ATPases (SPCAs) are P-type ion pumps found in the Golgi complex of all fungi and animals. The founding member of the SPCA family — termed Pmr1 — was identified in yeast [1] and early genetic studies suggested it played key roles in the homeostasis of Ca^{2+} and Mn^{2+} in the secretory pathway [2–8]. Biochemical studies have shown that yeast Pmr1 and its mammalian orthologs can directly transport both Ca^{2+} and Mn^{2+} ions with high affinity [9–15]. No SPCAs exhibit sensitivity to inhibitors of SERCA-type pumps (thapsigargin, cyclopiazonic acid, 2,5-di-(*tert*-butyl)-1,4-benzohydroquinone), nor are they directly regulated by calmodulin. However, expression of the *PMR1* gene, accumulation of Pmr1 protein, and tolerance to high concentrations of Mn^{2+} in the environment are all significantly stimulated by Crz1/Tcn1 [5, 16], a transcription factor that requires activation by the Ca^{2+}/calmodulin-dependent protein phosphatase calcineurin [17, 18].

Forced overexpression of Pmr1 in yeast results in increased tolerance to elevated Mn^{2+} and Ca^{2+} in the environment [3]. Yeast mutants that lack Pmr1 exhibit a mild sensitivity to high environmental Ca^{2+} and a very severe sensitivity to high environmental Mn^{2+} [2], consistent with the presence of additional transporters in yeast that provide tolerance to Ca^{2+} but not Mn^{2+} (see below). A very interesting phenotype of yeast mutants lacking Pmr1 is their extraordinary ability, relative to wild-type yeast cells, to secrete large quantities of human proteins [1, 2, 7, 19]. This "super-secretion" phenotype of the mutants has been attributed to deficiencies in several Golgi-localized glycosyltransferases, many of which are known to be Mn^{2+} dependent. Kex2, Ca^{2+}-dependent prohormone convertase in the *trans*-Golgi network, is also severely inactivated in Pmr1-deficient mutants. All these secretory defects can be ameliorated by supplementing the culture medium with moderate levels of Mn^{2+} or Ca^{2+} and worsened by addition of chelators. These early studies highlight a major role for SPCAs in supplying sufficient quantities of Ca^{2+} and Mn^{2+} to the central compartments of the secretory pathway.

SPCA-type Ca^{2+} pumps probably diverged from SERCA-type Ca^{2+} pumps shortly before the divergence of animals and fungi. SERCA-type Ca^{2+} pumps are not found in yeast or its closest relatives but nevertheless are found in genomes of more distantly related ascomycetes and basidiomycetes groups of fungi [20]. Does the endoplasmic reticulum of yeast function without Ca^{2+} or does Pmr1 fulfill the roles of both SERCA and SPCA in yeast? Several lines of investigation suggest the latter explanation. First, Pmr1-deficient mutants maintain $[Ca^{2+}]_{ER}$ at lower levels than wild-type cells [8]. Second, Pmr1-deficient mutants exhibit high levels of unfolded or misfolded proteins in the ER, as judged by significant activation of the "unfolded protein response" or UPR signaling pathway, which again can be remedied by supplementation of moderate Ca^{2+} or Mn^{2+} [21]. Though molecular chaperones in the yeast ER are not known to be dependent on Ca^{2+} [22], their orthologs in mammalian ER are well known to be Ca^{2+}-dependent enzymes. Third, Pmr1-deficient cells exhibit high levels of Ca^{2+} influx akin to that of thapsigargin-treated mammalian cells [23]. Both situations stimulate the influx of Ca^{2+} through channels in the plasma membrane, elevate $[Ca^{2+}]_{cyt}$, activate calcineurin, and cause the accumulation of an unidentified small molecule termed CIF that may participate in "store-operated" or "capacitative" Ca^{2+} entry [24]. As described below in more detail, the elevated Ca^{2+} influx observed in Pmr1-deficient mutants may be mechanistically distinct from capacitative Ca^{2+} entry in mammals and instead caused by the activation of a MAP-kinase cascade that is sensitive to stress in the ER, Golgi complex, and/or plasma membrane [25]. Fourth, these phenotypes can be suppressed by expression of mammalian SERCA in the ER of Pmr1-deficient yeast [7, 26].

Pmr1-deficient mutants exhibit several phenotypes attributable to accumulation of excess Mn^{2+} in the cytoplasm. Loss of Mn^{2+} transport by Pmr1 reverses the sensitivity of *sod1* mutants (cells lacking cytosolic superoxide dismutase) to oxygen and chemicals that induce reactive oxygen species [4], and also it strongly diminishes retrotransposition events by endogenous retrotransposons [27]. Apparently, elevated cytosolic Mn^{2+} can directly scavenge free radicals and directly inhibit the activity of reverse transcriptase in these situations.

Mutations in the *ATP2C1* gene encoding SPCA1 in humans result in a rare skin blistering disorder known as Hailey–Hailey disease [28, 29]. Judging from the spectrum of mutations in the human *ATP2C1* gene, Hailey–Hailey disease appears to be caused by haplo insufficiency, which partially decreases SPCA1 activity in keratinocytes [30]. It is not yet clear whether the disorder can be attributed to elevated cytosolic Ca^{2+} or Mn^{2+}, depleted secretory Ca^{2+} or Mn^{2+}, membrane stress, or some other effect contributing to the disease pathology that has not been fully elucidated. Sufferers of Hailey–Hailey disease retain one functional copy of *ATP2C1* and therefore may respond to therapies that topically induce SPCA1 expression in the affected regions of the skin.

9.2.2 PMCA-Type Ca^{2+} Pumps

Members of the plasma membrane calcium ATPase (PMCA) superfamily of Ca^{2+} pumps are ubiquitous among eukaryotes and even observed in certain archaea and eubacteria. The first nonanimal PMCA — Pmc1 — was identified in yeast and partially characterized [3]. Several surprises ensued, the first of which was the localization of Pmc1 not to the plasma membrane but to the limiting membrane of the vacuole, a lysosome-like organelle in fungi. Also, Pmc1 lacks the C-terminal Ca^{2+}/calmodulin-binding domain and seems not to be a direct target of calmodulin *in vivo*. Finally, expression of Pmc1 is massively increased upon activation of the same Ca^{2+}, calmodulin, calcineurin, Crz1/Tcn1 signaling pathway that also induces Pmr1 expression [5, 16]. Isolated vacuoles from Pmc1-deficient mutants lack an ATP-dependent high-affinity Ca^{2+} transporter, but the ion selectivity and catalytic properties of Pmc1 have not been carefully studied. Pmc1 has been purified in association with Nyv1, a v-SNARE protein involved in fusion of vacuole membranes, and seems to be inhibited *in vivo* through dose-dependent associations with Nyv1 [31], but the physiological role of this interaction has not been firmly established.

Pmc1-deficient mutants are completely viable in standard conditions, but they accumulated fivefold lower Ca^{2+} in a stable "non-exchangeable" pool (probably the vacuole, which contains large amounts of inorganic polyphosphate as a Ca^{2+} sink) without altering the smaller pool of "exchangeable" Ca^{2+} (probably representing the cytoplasm and secretory compartments) [3]. As expected, Pmc1-deficient mutants are much more sensitive to high environmental Ca^{2+} than wild type cells in growth assays, calcineurin activity assays, and in aequorin luminescence assays that directly monitor $[Ca^{2+}]_{cyt}$ [32]. These effects can be partially suppressed by overexpressing Pmr1. Interestingly, Pmc1 is strongly partially mislocalized to the ER and Golgi complex and essential for viability when the SPCA Pmr1 has been mutated [3, 33]. Therefore, the primary role of Pmc1 is to promote tolerance to high Ca^{2+} conditions and the primary role of Pmr1 is to promote secretory functions in low Ca^{2+} conditions, though each Ca^{2+} pump can partially compensate for the loss of the other.

9.2.3 H^+/Ca^{2+} Exchangers and Related Enzymes

Classical genetic approaches in yeast also revealed the first eukaryotic H^+/Ca^{2+} exchanger — Vcx1 (also called Hum1) [5, 34]. Close homologs of Vcx1 are found in the genomes of many bacteria, archaea, and eukaryotes, but they are not found in any animals. Vcx1 and its orthologs are distantly related to the superfamily of Na^+/Ca^{2+} exchangers that are equally widespread in nature. Yeast retains two homologs of Na^+/Ca^{2+} exchangers (*YJR106w* and *YDL206w*) and another putative H^+/Ca^{2+} exchanger (*YNL321w*) that are all conserved in other fungi [35] but not yet characterized in any system. Vcx1 localizes to the limiting membrane of the vacuole and absolutely depends on the large pH gradient generated by the vacuolar H^+-ATPase to concentrate Ca^{2+} in the lumen [5, 34]. Vcx1 has low affinity but very high selectivity for Ca^{2+} [36, 37]; ordinarily it does not transport Mn^{2+}, but its overexpression weakly increases Mn^{2+} tolerance [34] and several mutant derivatives are capable of transporting Mn^{2+} and providing strong Mn^{2+} tolerance [38].

With its high level of expression and large capacity relative to Pmc1, Vcx1 was predicted to have a major impact on the shape and magnitude of Ca^{2+} signals. Surprisingly, the study of Vcx1-deficient mutants demonstrate a much smaller role of Vcx1 (relative to Pmc1) in the available assays of $^{45}Ca^{2+}$ exchangeability, $[Ca^{2+}]_{cyt}$, calcineurin activation, and growth in high Ca^{2+} environments [5]. However, the activity of Vcx1 in these assays increased dramatically when calcineurin was inactivated with either drugs or mutations. Heterologously expressed Vcx1-like proteins from plants usually do not respond to calcineurin [39, 40]. Additionally, several mutant derivatives of Vcx1 that appear to be resistant to the negative effects of calcineurin have been isolated but the amino acid substitutions fail to provide insights into the mechanism of regulation by calcineurin [5, 37]. At the present time, calcineurin has no strong effects on Vcx1 expression, localization, or mobility on SDS-PAGE and only seems to affect Vcx1 activity *in vivo*. These findings suggest that calcineurin potently inhibits Vcx1 function *in vivo* by an unknown (potentially indirect) posttranslational mechanism.

Why does calcineurin strongly increase the function of one vacuolar Ca^{2+} transporter (Pmc1) and potently inhibit the other (Vcx1)? Questions like these typically bring important insights, but at present the answers are not fully elucidated. Hyperactive calcineurin-insensitive variants of Vcx1 appear to compete with Pmr1 for substrates (cytosolic Ca^{2+} and potentially Mn^{2+}), leading to membrane stress and activation of compensatory Ca^{2+} influx channels, elevation of $[Ca^{2+}]_{cyt}$, and activation of calcineurin [23]. A throttle on Vcx1 may therefore prevent this potent transporter from competing with Pmr1, which provides necessary Ca^{2+} and Mn^{2+} ions to the secretory pathway. Another possible benefit of the calcineurin–Vcx1 interaction is amplification of calcium signals arising from a double-negative feedback loop (calcineurin inhibition of Vcx1, Vcx1 depletion of cytosolic Ca^{2+} necessary for calcineurin activity). In this case, small rises in $[Ca^{2+}]_{cyt}$ may weakly activate calcineurin, which then inhibits Vcx1 resulting in greater elevation of $[Ca^{2+}]_{cyt}$. Basal activities of Pmr1 and Pmc1 would prevent runaway elevation of $[Ca^{2+}]_{cyt}$ in most conditions. However, neither of these possible rationales can explain why Vcx1 exists in yeast. With all its regulatory devices, Vcx1 is not available to sequester Ca^{2+} in times of Ca^{2+} overload and is not necessary — perhaps even harmful — for Ca^{2+} homeostasis in low Ca^{2+} conditions. Vcx1 exhibits strong effects only when it is overexpressed, mutated to alter ion selectivity and calcineurin sensitivity, or disregulated by the loss of calcineurin. Additional studies on the molecular mechanism of Vcx1 regulation by calcineurin may shed light on this perplexing question and provide insights into similar mechanisms that may operate in other fungi and possibly even animals.

9.2.4 MISSING AND NOVEL Ca^{2+} TRANSPORTERS?

SERCA-type Ca^{2+} pumps clearly operate in most fungi but are lacking in all budding yeasts sequenced to date. Several studies have reported some functional redundancy between Pmr1 and Cod1/Spf1 [41, 42], a P-type ATPase of the poorly characterized P5 subfamily that is conserved in all fungi, animals, and many other eukaryotic kingdoms [43]. There is no direct evidence that Cod1/Spf1 can transport Ca^{2+}: its autophosphorylation and ATPase activities were not stimulated by Ca^{2+} and its overexpression in yeast does not confer Ca^{2+} tolerance like Pmr1, Pmc1, and Vcx1. The functional redundancy with Pmr1 has been attributed to roles in secretory protein modification and folding in the ER [42]. Mutants lacking Cod1/Spf1 exhibit profound stimulation of the UPR signaling pathway and activation of Ca^{2+} influx channels possibly by the same stress-sensitive pathway that operates in Pmr1-deficient mutants [25]. No substrates of P5-type ATPases have been identified, so the possibility that they directly modulate Ca^{2+} signaling cannot be ruled out completely. Other than the uncharacterized homologs of H^+/Ca^{2+} and Na^+/Ca^{2+} described above, no other factors in yeast have emerged as possible Ca^{2+} transporters.

9.3 Ca^{2+} CHANNELS

9.3.1 TRPC-TYPE Ca^{2+} CHANNELS

Homology searches of the yeast genome revealed a distant homolog of mammalian TRPC-type Ca^{2+} channels — termed Yvc1 — localizing unexpectedly to the limiting membrane of the vacuole rather than the plasma membrane. Electrophysiological characterization of Yvc1 in isolated vacuoles revealed a large-conductance nonselective cation channel that was activated by membrane stretch but insensitive to inositol 1,4,5-trisphosphate (IP3) and other small molecules [44]. Ca^{2+} permeation and mechanosensitivity were also demonstrated *in vivo* using cytoplasmic aequorin as a probe [45]. Within a few seconds of hypertonic shock, yeast cells induced a dramatic elevation of $[Ca^{2+}]_{cyt}$ that peaked at ~1 min and declined slowly. This calcium signal did not require extracellular Ca^{2+} or the plasma membrane Ca^{2+} channels described below and instead completely required Yvc1 (as well as vacuolar Ca^{2+} transporter — Pmc1 or Vcx1 — to generate the mobile pool of vacuolar Ca^{2+}). How Yvc1 responds to hypertonic shock and membrane stretch remains a fascinating question. The physiological role of this Ca^{2+} release channel in yeast also remains to be established.

Release of Ca^{2+} from the vacuole has been postulated as an important step in the process by which vacuoles undergo homotypic fusion [46]. Surprisingly, vacuole fusion *in vitro* and *in vivo* is not dependent on Yvc1, Pmc1, and Vcx1 [47, 48].

Release of Ca^{2+} from vacuoles *in vitro* requires formation of trans-SNARE complexes, a crucial intermediate step in the homotypic fusion pathway [49]. This vacuolar Ca^{2+} release therefore involves an unknown Ca^{2+} channel that is intimately associated with the fusion machinery.

9.3.2 VGCC-Type Ca^{2+} Channels

All fungal genomes sequenced to date contain a single homolog of the classic four-domain voltage-gated Ca^{2+} channel (VGCC) catalytic subunit that has been studied thoroughly in mammals. All animals sequenced to date express at least four different VGCC catalytic subunits falling into the L-, N/P/Q-, and T-type subclasses. Alignment of the fungal and animal VGCC catalytic subunits suggests a single ancestral gene was duplicated and diverged early in the animal lineages to yield the various subclasses. Fungal VGCCs have not been characterized using electrophysiological techniques because most fungal cells synthesize a tough cell wall and express numerous other ion channels and transporters in their plasma membrane that overwhelm the VGCC signal. Nevertheless, $^{45}Ca^{2+}$ influx assays, aequorin luminescence assays, and calcineurin activation assays have all been used to study VGCC function and regulation in yeast.

Genetic screens have yielded two classes of yeast mutants that specifically lack VGCC function [23, 50–52]. The first class carries mutations in the gene encoding Cch1, the homolog of mammalian VGCC catalytic or α-subunits. The second class carries mutations in the gene encoding Mid1, which is not obviously related to any proteins in animals despite reports to the contrary [53]. All fungal genomes sequenced to date contain a single homolog of Mid1 and multiple sequence alignments indicate a transmembrane topology and composition that is strikingly similar to the uncleaved VGCC regulatory subunits or α2δ-subunits of animals. Cch1 and Mid1 co-localize to the plasma membrane and co-purify after solubilization with mild detergents [23]. Gross overexpression of either Cch1 or Mid1 did not increase VGCC activity in yeast, but the forced overexpression of both proteins led to modest increase in activity [54], possibly due to limiting expression of a third subunit. Indeed, a recent genetic screen has revealed a novel four-transmembrane protein that is required for VGCC activity in yeast and is homologous to the γ-subunits of mammalian VGCCs (our unpublished observations). Thus, like mammalian VGCCs, the yeast VGCC is composed of Cch1, Mid1, and probably other proteins.

Does the yeast VGCC respond to membrane depolarization? The four voltage-sensing S4 domains of animal α-subunits contain four to five positively charged residues arranged along one face of a hydrophobic α-helix. Cch1 and its orthologs in other fungi all contain about half this number of positively charged residues and therefore might not respond to depolarization in the same range as that of mammalian cells. The transmembrane potential of yeast and most other fungi has been estimated to be much more negative than that of mammals, typically around −180 mV, so it is possible that the fewer positive charges in the S4 domains tune these channels to a more electronegative range. On the other hand, action potentials are not known to be generated in fungi and efforts to alter membrane potential with ionic imbalances have not yielded evidence for voltage sensitivity, except for one recent case. Changing the environmental pH from acid to alkaline results in a rapid and transient elevation of $[Ca^{2+}]_{cyt}$ and activation of calcineurin in a fashion that completely depends on Cch1 and Mid1 [55]. Interestingly, the yeast VGCC is also rapidly activated by rising levels of glucose-1-phosphate (relative to glucose-6-phosphate) that occur naturally as a consequence of energy source availability [56, 57]. Because the major electrogenic pump of the plasma membrane — the Pma1 proton pump — responds to glucose metabolism [58, 59], yeast VGCC activation by glucose metabolism and alkaline shock may be mechanistically related.

Other stimuli are well known to stimulate Ca^{2+} influx activity of the yeast VGCC, albeit with much slower kinetics. Haploid yeast cells begin to generate Cch1- and Mid1-dependent Ca^{2+} signals after ~40 min of treatment with mating pheromones [23, 51, 52, 60–64]. The pheromone-responsive transcription factor Ste12 seems to be necessary and sufficient for this phenomenon, but the targets of Ste12 that participate in the phenomenon have not been identified [64]. An unidentified protein kinase may also be involved because the calcineurin protein phosphatase appears to exert a strong negative feedback on VGCC activity in these conditions [64].

As mentioned earlier, Ca^{2+} influx via the yeast VGCC is also stimulated in mutants lacking Pmr1, Cod1/Spf1, Cdc1, and many other mutants that affect secretory protein biogenesis and trafficking [23, 41, 42, 51]. Drugs that block N-glycosylation or protein folding in the ER also stimulate Cch1- and Mid1-dependent Ca^{2+} influx after ~40 min of treatment [21]. In all these situations, calcineurin appears to negatively regulate VGCC activity by a feedback mechanism. Recent findings suggest that calcineurin directly dephosphorylates Cch1 *in vitro* and *in vivo* and that the Slt2/Mpk1 stress-activated MAP kinase is likely to phosphorylate Cch1 [25]. Slt2/Mpk1 is necessary for VGCC activation in response to several if not all of these membrane stresses but is not necessary for VGCC activation in response to mating pheromones when the Fus3 MAP kinase is also activated.

One of the most exciting observations regarding the VGCC in yeast is its activation by all major classes of antifungal drugs in therapeutic use today. Fungal infections are discomforting to healthy individuals and frequently life-threatening to immunodeficient or immunosuppressed patients. The major classes of antifungal drugs block sterol biosynthesis in the ER (Terbinafine and azole-class antibiotics such as Fluconazole, Miconazole, etc.) and sterol function in the plasma membrane (Amphotericin B). None of these antibiotics kill fungal cells; they simply slow growth and proliferation. Remarkably, all of these antifungal compounds activate VGCC and calcineurin signaling in yeast and it is the activation of the calcium signaling pathway that prevents cell death in yeast and diverse fungal pathogens [65–71]. The combination of these antifungal drugs with inhibitors of calcineurin (FK506 or Cyclosporin A) results in a potent fungicidal activity that is not observed with either medication alone. What limits the use of FK506 and Cyclosporin A in antifungal therapies is their immunosuppressive side effects. The development of new drugs that specifically inactivate fungal VGCCs may dramatically improve the effectiveness of existing antifungal therapies.

9.3.3 OTHER Ca²⁺ CHANNELS?

Heterologous expression of Mid1 in Cos-7 cells caused the appearance of a stretch-activated nonselective cation channel [53], suggesting that Mid1 may function alone as a mechanosensitive Ca^{2+} channel. In yeast, however, every example of Mid1-dependent Ca^{2+} influx and signaling is also dependent on Cch1 and there is little evidence that Mid1 can function as a Ca^{2+}-permeable channel in the absence of Cch1. Confusingly, the phenotypes of *mid1* and *cch1* knockout mutants are sometimes different from the *cch1 mid1* double mutant. For example, *cch1* mutants and *cch1 mid1* double mutants exhibit significant activation of the UPR signaling pathway (indicative of ER stress) whereas *mid1* mutants and wild-type cells do not (our unpublished data). Small differences in growth rates also suggest that Cch1 retains unknown functions that are independent of Mid1 and Ca^{2+} influx [72]. The mounting evidence from budding and pathogenic yeasts [73] all support the notion that Mid1 and Cch1 function together as regulatory and catalytic subunits of a VGCC-like Ca^{2+} channel.

The response to mating pheromones triggers Ca^{2+} influx by two separate pathways, a high-affinity Ca^{2+} influx system (HACS) that depends on Cch1 and Mid1 and a low-affinity Ca^{2+} influx system (LACS) that depends on Figure 9.1 [74]. Figure 9.1 is a four-transmembrane protein possibly related to Stargazin (our unpublished observations), a regulatory subunit of AMPA- and NMDA-receptor ion channels in mammals. Homologs of these glutamate-sensitive ion channels are not recognizable in the genomes of fungi. Figure 9.1 is probably not a catalytic subunit of a fungal ion channel because its transmembrane segments seem to evolve much more rapidly than those of catalytic subunits such as Yvc1 and Cch1 and its expression is massively induced by the response to mating pheromones [75]. Obviously more work is warranted to identify its other subunits, to determine its modes of regulation, and to determine what if anything may respond to LACS activation.

Could other classes of Ca^{2+}-permeable ion channels operate in fungi? No fungal genomes sequenced to date contain homologs of mammalian IP3-receptors or Ryanodine-receptors. This is surprising because the more distant slime mold *Dictyostelium* expresses an IP3-responsive Ca^{2+} channel that is obviously related to mammalian IP3-receptors [76]. There is even one report of IP3-triggered Ca^{2+} release from purified yeast vacuoles [77]. Additionally, homotypic vacuole fusion seems to involve release of Ca^{2+} from the vacuole lumen by an unknown Yvc1-independent channel [49]. Identification of the factors required for these activities may therefore elucidate new classes of ion channels, which may be conserved in animals or other eukaryotes.

9.4 SYNOPSIS AND FUTURE DIRECTIONS

All components of the calcium signaling network in yeast (except for the vacuolar H^+/Ca^{2+} exchanger) are retained in animals and seem only to differ in their modes of regulation. This level of understanding has arisen primarily through the use of powerful biochemical, genetic, and genomic technologies available in the model system. These core technologies of the system almost certainly will reveal new components of the network that have eluded detection in all systems. The main technical disadvantages include the great difficulty with imaging $[Ca^{2+}]_{cyt}$ in live yeast cells and with electrophysiological recordings of ion channels in the plasma membrane of live yeast cells. If some of these limitations can be overcome, a complete map of the calcium signaling network and all its dynamics might become possible for this simple organism at all stages of its life cycle. Such advances could help pave the way toward a general understanding of calcium signaling in all organisms.

REFERENCES

1. Rudolph, H.K., Antebi, A., Fink, G.R., Buckley, C.M., Dorman, T.E., LeVitre, J., Davidow, L.S., Mao, J.I., and Moir, D.T. The yeast secretory pathway is perturbed by mutations in *PMR1*, a member of a Ca^{2+} ATPase family. *Cell*, 58, 133, 1989.
2. Antebi, A. and Fink, G.R. The yeast Ca^{2+}-ATPase homologue, *PMR1*, is required for normal Golgi function and localizes in a novel Golgi-like distribution. *Mol Biol Cell*, 3, 633, 1992.
3. Cunningham, K.W. and Fink, G.R. Calcineurin-dependent growth control in *Saccharomyces cerevisiae* mutants lacking *PMC1*, a homolog of plasma membrane Ca^{2+} ATPases. *J Cell Biol*, 124, 351, 1994.
4. Lapinskas, P.J., Cunningham, K.W., Liu, X.F., Fink, G.R., and Culotta, V.C. Mutations in *PMR1* suppress oxidative damage in yeast lacking superoxide dismutase. *Mol Cell Biol*, 15, 1382, 1995.
5. Cunningham, K.W. and Fink, G.R. Calcineurin inhibits *VCX1*-dependent H^+/Ca^{2+} exchange and induces Ca^{2+} ATPases in yeast. *Mol Cell Biol*, 16, 2226, 1996.
6. Halachmi, D. and Eilam, Y. Elevated cytosolic free Ca^{2+} concentrations and massive Ca^{2+} accumulation within vacuoles, in yeast mutant lacking *PMR1*, a homolog of Ca^{2+}-ATPase. *FEBS Lett*, 392, 194, 1996.
7. Dürr, G., Strayle, J., Plemper, R., Elbs, S., Klee, S.K., Catty, P., Wolf, D.H., and Rudolph, H.K. The medial-Golgi ion pump Pmr1 supplies the yeast secretory pathway with Ca^{2+} and Mn^{2+} required for glycosylation, sorting, and endoplasmic reticulum-associated protein degradation. *Mol Biol Cell*, 9, 1149, 1998.
8. Strayle, J., Pozzan, T., and Rudolph, H.K. Steady-state free Ca^{2+} in the yeast endoplasmic reticulum reaches only 10 ÌM and is mainly controlled by the secretory pathway pump Pmr1. *EMBO J*, 18, 4733, 1999.
9. Sorin, A., Rosas, G., and Rao, R. PMR1, a Ca^{2+}-ATPase in yeast Golgi, has properties distinct from sarco/endoplasmic reticulum and plasma membrane calcium pumps. *J Biol Chem*, 272, 9895, 1997.
10. Mandal, D., Woolf, T.B., and Rao, R. Manganese selectivity of pmr1, the yeast secretory pathway ion pump, is defined by residue gln783 in transmembrane segment 6. Residue Asp778 is essential for cation transport. *J Biol Chem*, 275, 23933, 2000.
11. Wei, Y., Chen, J., Rosas, G., Tompkins, D.A., Holt, P.A., and Rao, R. Phenotypic screening of mutations in Pmr1, the yeast secretory pathway Ca^{2+}/Mn^{2+}-ATPase, reveals residues critical for ion selectivity and transport. *J Biol Chem*, 275, 23927, 2000.

12. Mandal, D., Rulli, S.J., and Rao, R. Packing interactions between transmembrane helices alter ion selectivity of the yeast Golgi Ca^{2+}/Mn^{2+}-ATPase PMR1. *J Biol Chem*, 278, 35292, 2003.

13. Fairclough, R.J., Dode, L., Vanoevelen, J., Andersen, J.P., Missiaen, L., Raeymaekers, L., Wuytack, F., and Hovnanian, A. Effect of Hailey–Hailey disease mutations on the function of a new variant of human secretory pathway Ca^{2+}/Mn^{2+}-ATPase (hSPCA1). *J Biol Chem*, 278, 24721, 2003.

14. Ton, V.K., Mandal, D., Vahadji, C., and Rao, R. Functional expression in yeast of the human secretory pathway Ca^{2+}, Mn^{2+}-ATPase defective in Hailey–Hailey disease. *J Biol Chem*, 277, 6422, 2002.

15. Van Baelen, K., Vanoevelen, J., Missiaen, L., Raeymaekers, L., and Wuytack, F. The Golgi PMR1 P-type ATPase of *Caenorhabditis elegans*. Identification of the gene and demonstration of calcium and manganese transport. *J Biol Chem*, 276, 10683, 2001.

16. Matheos, D.P., Kingsbury, T.J., Ahsan, U.S., and Cunningham, K.W. Tcn1p/Crz1p, a calcineurin-dependent transcription factor that differentially regulates gene expression in *Saccharomyces cerevisiae*. *Genes Dev*, 11, 3445, 1997.

17. Stathopoulos, A.M. and Cyert, M.S. Calcineurin acts through the *CRZ1/TCN1* encoded transcription factor to regulate gene expression in yeast. *Genes Dev*, 11, 3432, 1997.

18. Stathopoulos-Gerontides, A., Guo, J.J. and Cyert, M.S. Yeast calcineurin regulates nuclear localization of the Crz1p transcription factor through dephosphorylation. *Genes Dev*, 13, 798, 1999.

19. Moir, D.T. and Davidow, L.S. Production of proteins by secretion from yeast. *Methods Enzymol*, 194, 491, 1991.

20. Benito, B., Garciadeblas, B., and Rodriguez-Navarro, A. Molecular cloning of the calcium and sodium ATPases in *Neurospora crassa*. *Mol Microbiol*, 35, 1079, 2000.

21. Bonilla, M., Nastase, K.K., and Cunningham, K.W. Essential role of calcineurin in response to endoplasmic reticulum stress. *EMBO J*, 21, 2343, 2002.

22. Parlati, F., Dominguez, M., Bergeron, J.J., and Thomas, D.Y. *Saccharomyces cerevisiae CNE1* encodes an endoplasmic reticulum (ER) membrane protein with sequence similarity to calnexin and calreticulin and functions as a constituent of the ER quality control apparatus. *J Biol Chem*, 270, 244, 1995.

23. Locke, E.G., Bonilla, M., Liang, L., Takita, Y., and Cunningham, K.W. A homolog of voltage-gated Ca^{2+} channels stimulated by depletion of secretory Ca^{2+} in yeast. *Mol Cell Biol*, 20, 6686, 2000.

24. Csutora, P., Su, Z., Kim, H.Y., Bugrim, A., Cunningham, K.W., Nuccitelli, R., Keizer, J.E., Hanley, M.R., Blalock, J.E., and Marchase, R.B. Calcium influx factor is synthesized by yeast and mammalian cells depleted of organellar calcium stores. *Proc Natl Acad Sci USA*, 96, 121, 1999.

25. Bonilla, M. and Cunningham, K.W. MAP kinase stimulation of Ca^{2+} signaling is required for survival of endoplasmic reticulum stress in yeast. *Mol Biol Cell*, 14, 4296, 2003.

26. Degand, I., Catty, P., Talla, E., Thinès-Sempoux, D., De Kerchove D'Exaerde, A., Goffeau, A., and Ghislain, M. Rabbit sarcoplasmic reticulum Ca^{2+}-ATPase replaces yeast PMC1 and PMR1 Ca^{2+}-ATPases for cell viability and calcineurin-dependent regulation of calcium tolerance. *Mol Microbiol*, 31, 545, 1999.

27. Bolton, E.C., Mildvan, A.S., and Boeke, J.D. Inhibition of reverse transcription *in vivo* by elevated manganese ion concentration. *Mol Cell*, 9, 879, 2002.

28. Hu, Z., Bonifas, J.M., Beech, J., Bench, G., Shigihara, T., Ogawa, H., Ikeda, S., Mauro, T., and Epstein, E.H., Jr. Mutations in ATP2C1, encoding a calcium pump, cause Hailey–Hailey disease. *Nat Genet*, 24, 61, 2000.

29. Sudbrak, R., Brown, J., Dobson-Stone, C., Carter, S., Ramser, J., White, J., Healy, E., Dissanayake, M., Larregue, M., Perrussel, M., Lehrach, H., Munro, C.S., Strachan, T., Burge, S., Hovnanian, A., and Monaco, A.P. Hailey–Hailey disease is caused by mutations in ATP2C1 encoding a novel Ca^{2+} pump. *Hum Mol Genet*, 9, 1131, 2000.

30. Missiaen, L., Raeymaekers, L., Dode, L., Vanoevelen, J., Van Baelen, K., Parys, J.B., Callewaert, G., De Smedt, H., Segaert, S., and Wuytack, F. SPCA1 pumps and Hailey–Hailey disease. *Biochem Biophys Res Commun*, 322, 1204, 2004.

31. Takita, Y., Engstrom, L., Ungermann, C., and Cunningham, K.W. Inhibition of the Ca^{2+}-ATPase Pmc1p by the v-SNARE protein Nyv1p. *J Biol Chem*, 276, 6200, 2001.

32. Miseta, A., Kellermayer, R., Aiello, D.P., Fu, L., and Bedwell, D.M. The vacuolar Ca^{2+}/H^+ exchanger Vcx1p/Hum1p tightly controls cytosolic Ca^{2+} levels in *S. cerevisiae*. *FEBS Lett*, 451, 132, 1999.

33. Marchi, V., Sorin, A., Wei, Y., and Rao, R. Induction of vacuolar Ca^{2+}-ATPase and H^+/Ca^{2+} exchange activity in yeast mutants lacking Pmr1, the Golgi Ca^{2+}-ATPase. *FEBS Lett*, 454, 181, 1999.

34. Pozos, T.C., Sekler, I., and Cyert, M.S. The product of *HUM1*, a novel yeast gene, is required for vacuolar Ca^{2+}/H^+ exchange and is related to mammalian Na^+/Ca^{2+} exchangers. *Mol Cell Biol*, 16, 3730, 1996.

35. Zelter, A., Bencina, M., Bowman, B.J., Yarden, O., and Read, N.D. A comparative genomic analysis of the calcium signaling machinery in *Neurospora crassa, Magnaporthe grisea*, and *Saccharomyces cerevisiae*. *Fungal Genet Biol*, 41, 827, 2004.

36. Dunn, T., Gable, K., and Beeler, T. Regulation of cellular Ca^{2+} by yeast vacuoles. *J Biol Chem*, 269, 7273, 1994.

37. Pittman, J.K., Cheng, N.H., Shigaki, T., Kunta, M., and Hirschi, K.D. Functional dependence on calcineurin by variants of the *Saccharomyces cerevisiae* vacuolar Ca^{2+}/H^+ exchanger Vcx1p. *Mol Microbiol*, 54, 1104, 2004.

38. del Pozo, L., Osaba, L., Corchero, J., and Jimenez, A. A single nucleotide change in the MNR1 (VCX1/HUM1) gene determines resistance to manganese in *Saccharomyces cerevisiae*. *Yeast*, 15, 371, 1999.

39. Hirschi, K.D., Zhen, R.G., Cunningham, K.W., Rea, P.A., and Fink, G.R. CAX1, an H^+/Ca^{2+} antiporter from *Arabidopsis*. *Proc Natl Acad Sci USA*, 93, 8782, 1996.

40. Cheng, N.H., Pittman, J.K., Shigaki, T., and Hirschi, K.D. Characterization of CAX4, an Arabidopsis H^+/cation antiporter. *Plant Physiol*, 128, 1245, 2002.

41. Cronin, S.R., Rao, R., and Hampton, R.Y. Cod1p/Spf1p is a P-type ATPase involved in ER function and Ca^{2+} homeostasis. *J Cell Biol*, 157, 1017, 2002.

42. Vashist, S., Frank, C.G., Jakob, C.A., and Ng, D.T. Two distinctly localized p-type ATPases collaborate to maintain organelle homeostasis required for glycoprotein processing and quality control. *Mol Biol Cell*, 13, 3955, 2002.

43. Schultheis, P.J., Hagen, T.T., O'Toole, K.K., Tachibana, A., Burke, C.R., McGill, D.L., Okunade, G.W., and Shull, G.E. Characterization of the P5 subfamily of P-type transport ATPases in mice. *Biochem Biophys Res Commun*, 323, 731, 2004.

44. Palmer, C.P., Zhou, X.L., Lin, J., Loukin, S.H., Kung, C., and Saimi, Y. A TRP homolog in *Saccharomyces cerevisiae* forms an intracellular Ca^{2+}-permeable channel in the yeast vacuolar membrane. *Proc Natl Acad Sci USA*, 98, 7801, 2001.

45. Denis, V. and Cyert, M.S. Internal Ca^{2+} release in yeast is triggered by hypertonic shock and mediated by a TRP channel homologue. *J Cell Biol*, 156, 29, 2002.

46. Peters, C. and Mayer, A. Ca^{2+}/calmodulin signals the completion of docking and triggers a late step of vacuole fusion. *Nature*, 396, 575, 1998.

47. Ungermann, C., Wickner, W., and Xu, Z. Vacuole acidification is required for trans-SNARE pairing, LMA1 release, and homotypic fusion. *Proc Natl Acad Sci USA*, 96, 11194, 1999.

48. Bayer, M.J., Reese, C., Buhler, S., Peters, C., and Mayer, A. Vacuole membrane fusion: V0 functions after trans-SNARE pairing and is coupled to the Ca^{2+}-releasing channel. *J Cell Biol*, 162, 211, 2003.

49. Merz, A.J. and Wickner, W.T. trans-SNARE interactions elicit Ca^{2+} efflux from the yeast vacuole lumen. *J Cell Biol*, 164, 195, 2004.

50. Iida, H., Nakamura, H., Ono, T., Okumura, M.S., and Anraku, Y. *MID1*, a novel *Saccharomyces cerevisiae* gene encoding a plasma membrane protein, is required for Ca^{2+} influx and mating. *Mol Cell Biol*, 14, 8259, 1994.

51. Paidhungat, M. and Garrett, S. A homolog of mammalian, voltage-gated calcium channels mediates yeast pheromone-stimulated Ca^{2+} uptake and exacerbates the *cdc1*(Ts) growth defect. *Mol Cell Biol*, 17, 6339, 1997.

52. Fischer, M., Schnell, N., Chattaway, J., Davies, P., Dixon, G., and Sanders, D. The *Saccharomyces cerevisiae CCH1* gene is involved in calcium influx and mating. *FEBS Lett*, 419, 259, 1997.

53. Kanzaki, M., Nagasawa, M., Kojima, I., Sato, C., Naruse, K., Sokabe, M., and Iida, H. Molecular identification of a eukaryotic, stretch-activated nonselective cation channel. *Science*, 285, 882, 1999.

54. Iida, K., Tada, T., and Iida, H. Molecular cloning in yeast by *in vivo* homologous recombination of the yeast putative alpha1 subunit of the voltage-gated calcium channel. *FEBS Lett*, 576, 291, 2004.

55. Viladevall, L., Serrano, R., Ruiz, A., Domenech, G., Giraldo, J., Barcelo, A., and Arino, J. Characterization of the calcium-mediated response to alkaline stress in *Saccharomyces cerevisiae*. *J Biol Chem*, 279, 43614, 2004.

56. Fu, L., Miseta, A., Hunton, D., Marchase, R.B., and Bedwell, D.M. Loss of the major isoform of phosphoglucomutase results in altered calcium homeostasis in *Saccharomyces cerevisiae*. *J Biol Chem*, 275, 5431, 2000.

57. Tokes-Fuzesi, M., Bedwell, D.M., Repa, I., Sipos, K., Sumegi, B., Rab, A., and Miseta, A. Hexose phosphorylation and the putative calcium channel component Mid1p are required for the hexose-induced transient elevation of cytosolic calcium response in *Saccharomyces cerevisiae*. *Mol Microbiol*, 44, 1299, 2002.

58. Eilam, Y. and Othman, M. Activation of Ca^{2+} influx by metabolic substrates in *Saccharomyces cerevisiae*: role of membrane potential and cellular ATP levels. *J Gen Microbiol*, 136, 1990.

59. Eilam, Y., Othman, M., and Halachmi, D. Transient increase in Ca^{2+} influx in *Saccharomyces cerevisiae* in response to glucose: effects of intracellular acidification and cAMP levels. *J Gen Microbiol*, 136, 2537, 1990.

60. Ohsumi, Y. and Anraku, Y. Specific induction of Ca^{2+} transport activity in *MATa* cells of *Saccharomyces cerevisiae* by a mating pheromone, alpha factor. *J Biol Chem*, 260, 10482, 1985.

61. Iida, H., Yagawa, Y., and Anraku, Y. Essential role for induced Ca^{2+} influx followed by $[Ca^{2+}]i$ rise in maintaining viability of yeast cells late in the mating pheromone response pathway. A study of $[Ca^{2+}]i$ in single *Saccharomyces cerevisiae* cells with imaging of fura-2. *J Biol Chem*, 265, 13391, 1990.

62. Nakajima-Shimada, J., Iida, H., Tsuji, F.I., and Anraku, Y. Monitoring of intracellular calcium in *Saccharomyces cerevisiae* with an apoaequorin cDNA expression system. *Proc Natl Acad Sci USA*, 88, 6878, 1991.

63. Prasad, K.R. and Rosoff, P.M. Characterization of the energy-dependent, mating factor-activated Ca^{2+} influx in *Saccharomyces cerevisiae*. *Cell Calcium*, 13, 615, 1992.
64. Muller, E.M., Locke, E.G., and Cunningham, K.W. Differential regulation of two Ca^{2+} influx systems by pheromone signaling in *Saccharomyces cerevisiae*. *Genetics*, 159, 1527, 2001.
65. Marchetti, O., Moreillon, P., Glauser, M.P., Bille, J., and Sanglard, D. Potent synergism of the combination of fluconazole and cyclosporine in *Candida albicans*. *Antimicrob Agents Chemother*, 44, 2373, 2000.
66. Bonilla, M. and Cunningham, K.W. Calcium release and influx in yeast: TRPC and VGCC rule another kingdom. *Sci STKE*, 2002, PE17, 2002.
67. Del Poeta, M., Cruz, M.C., Cardenas, M.E., Perfect, J.R., and Heitman, J. Synergistic antifungal activities of bafilomycin A1, fluconazole, and the pneumocandin MK-0991/caspofungin acetate (L-743,873) with calcineurin inhibitors FK506 and L-685,818 against *Cryptococcus neoformans*. *Antimicrob Agents Chemother*, 44, 739, 2000.
68. Edlind, T., Smith, L., Henry, K., Katiyar, S., and Nickels, J. Antifungal activity in *Saccharomyces cerevisiae* is modulated by calcium signalling. *Mol Microbiol*, 46, 257, 2002.
69. Cruz, M.C., Goldstein, A.L., Blankenship, J.R., Del Poeta, M., Davis, D., Cardenas, M.E., Perfect, J.R., McCusker, J.H., and Heitman, J. Calcineurin is essential for survival during membrane stress in *Candida albicans*. *EMBO J*, 21, 546, 2002.
70. Blankenship, J.R., Steinbach, W.J., Perfect, J.R., and Heitman, J. Teaching old drugs new tricks: reincarnating immunosuppressants as antifungal drugs. *Curr Opin Investig Drugs*, 4, 192, 2003.
71. Onyewu, C., Blankenship, J.R., Del Poeta, M., and Heitman, J. Ergosterol biosynthesis inhibitors become fungicidal when combined with calcineurin inhibitors against *Candida albicans*, *Candida glabrata*, and *Candida krusei*. *Antimicrob Agents Chemother*, 47, 956, 2003.
72. Courchesne, W.E., and Ozturk, S. Amiodarone induces a caffeine-inhibited, MID1-depedent rise in free cytoplasmic calcium in *Saccharomyces cerevisiae*. *Mol Microbiol*, 47, 223, 2003.
73. Kaur, R., Castano, I., and Cormack, B.P. Functional genomic analysis of fluconazole susceptibility in the pathogenic yeast *Candida glabrata*: roles of calcium signaling and mitochondria. *Antimicrob Agents Chemother*, 48, 1600, 2004.
74. Muller, E.M., Mackin, N.A., Erdman, S.E., and Cunningham, K.W. Figure 9.1p facilitates Ca^{2+} influx and cell fusion during mating of *Saccharomyces cerevisiae*. *J Biol Chem*, 278, 38461, 2003.
75. Erdman, S., Lin, L., Malczynski, M., and Snyder, M. Pheromone-regulated genes required for yeast mating differentiation. *J Cell Biol*, 140, 461, 1998.
76. Traynor, D., Milne, J.L., Insall, R.H., and Kay, R.R. Ca^{2+} signalling is not required for chemotaxis in Dictyostelium. *EMBO J*, 19, 4846, 2000.
77. Belde, P.J., Vossen, J.H., Borst-Pauwels, G.W., and Theuvenet, A.P. Inositol 1,4,5-trisphosphate releases Ca^{2+} from vacuolar membrane vesicles of *Saccharomyces cerevisiae*. *FEBS Lett*, 323, 113, 1993.

10 Nuclear Patch Clamp Electrophysiology of Inositol Trisphosphate Receptor Ca^{2+} Release Channels

Don-On Daniel Mak, Carl White, Lucian Ionescu, and J. Kevin Foskett

CONTENTS

10.1   INTRODUCTION

Modulation of cytoplasmic free-calcium concentration ($[Ca^{2+}]_i$) is a ubiquitous signaling system involved in the regulation of numerous processes, including transepithelial transport, learning and memory, muscle contraction, synaptic transmission, secretion, motility, membrane trafficking, excitability, gene expression, and cell division. Activation of phospholipases Cβ and Cγ (PLC) by ligand interaction with G-protein- or tyrosine kinase-linked receptors, respectively, results in hydrolysis of phosphatidylinositol-4,5-bisphosphate, generating inositol 1,4,5-trisphosphate ($InsP_3$). $InsP_3$ binds to its receptor ($InsP_3R$), a ligand-gated Ca^{2+} release channel [1, 2] localized in the endoplasmic reticulum (ER) [3]. Analyses of $InsP_3$-mediated $[Ca^{2+}]_i$ signals in single cells have revealed them to be unexpectedly complex. In the temporal domain, this complexity is manifested as repetitive spikes or oscillations, with frequencies often tuned to levels of stimulation, suggesting that $[Ca^{2+}]_i$ signals may be transduced by frequency encoding as well as amplitude. In the spatial domain, $[Ca^{2+}]_i$ signals may initiate at specific locations and remain highly localized or propagate as waves [4–7]. Thus, $InsP_3$-mediated $[Ca^{2+}]_i$ signals are often organized to provide different signals to discrete parts of the cell.

10.2   MOLECULAR IDENTIFICATION OF THE InsP$_3$R

Three genes [8–14] and alternatively spliced forms [15–17] encode for a family of $InsP_3Rs$ in mammalian cells, including humans. The three full-length sequences are 60–80% homologous [18]. The $InsP_3R$ is ubiquitously expressed, perhaps in all cell types [19–23]. The three isoforms have distinct and overlapping patterns of expression with most cells expressing more than one type [17, 21, 23–26], and expression levels can be modified during differentiation [17, 27, 28] and by use-dependent degradation [29–34]. This diversity of expression is impressive and suggests that cells require distinct $InsP_3Rs$ to regulate specific functions. Nevertheless, the functional implications of this diversity, both at the single channel as well as cellular levels, remain largely unexplored. Gaining insights into the functional correlates and physiological implications of this diversity requires the use and development of novel expression and assay systems, a goal that constituted part of the rationale for studying the detailed properties of the mammalian $InsP_3R$ in its native ER membrane environment, discussed below.

The $InsP_3Rs$ are ~2700–2800 amino acid intracellular membrane proteins that exist as homo- or hetero-tetramers [18, 35–39]. Structurally, each monomer in the tetramer contains a cytoplasmic amino-terminus comprising ~85% of the protein, a hydrophobic region predicted to contain six membrane-spanning helices that contribute to the ion-conducting pore of the $InsP_3R$ channel, and a relatively short cytoplasmic carboxy-terminus [37, 40] (Figure 10.1). Functionally, the amino-

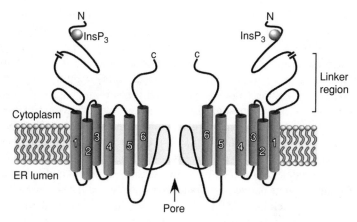

FIGURE 10.1 Topology of an InsP$_3$R channel in the ER membrane. Schematic diagram showing the topology of two InsP$_3$R molecules in a tetrameric InsP$_3$R channel in the ER membrane. InsP$_3$ binds in a cytoplasmic-oriented region near the amino-terminus of each monomer. The ion-permeable pore is formed by equal contributions from each monomer of membrane helices and a luminal loop sequence that contains the ion selectivity filter sequence.

terminal region can be divided into a proximal InsP$_3$-binding domain and a more distal "regulatory"/"linker" domain. InsP$_3$ binding to the InsP$_3$R is stoichiometric and localized by mutagenesis and an x-ray crystal structure to a region within residues 226–578 [41–43]. Binding of InsP$_3$ gates the channel open, mediated possibly by the linker region, because ~2000 amino acids separate the InsP$_3$-binding domain from the pore. As this region contains consensus sequences for phosphorylation and binding by calmodulin (CaM) and ATP, it may also integrate other signaling pathways or metabolic states with the function of the InsP$_3$R. Physiologically relevant concentrations of ATP strongly influence the Ca^{2+} activation properties of the channel [44, 45]. ATP also modulates the effects of the ligand adenophostin on channel kinetics as well as its binding affinity for the channel [46]. This latter effect revealed an unexpected role of the linker region in regulation of the properties of the ligand-binding domain.

10.3   DEVELOPMENT OF NUCLEAR PATCH CLAMP ELECTROPHYSIOLOGY TO STUDY SINGLE InsP$_3$R CHANNELS

The InsP$_3$R is itself a ligand-gated ion channel [1, 2]. The most powerful approach to study the detailed permeation and gating properties of single ion channels is the patch clamp technique [47]. However, because intracellular membranes are not accessible to patch pipettes (although see Ref. [48]), single intracellular-membrane localized ion channels have been studied traditionally following their isolation and reconstitution into artificial lipid bilayer membranes. The first electrophysiological recordings of single InsP$_3$R channels were made using this approach [1, 49–51], and some laboratories continue to use this technique to study both endogenous as well as recombinant InsP$_3$R channel [52–54]. As in all *in vitro* reconstitution systems, there is a concern that the isolation, purification, and reconstitution protocols may disrupt normal functional properties of the channels. Furthermore, the *in vitro* environment used for channel current recordings, including the bilayer lipid composition and composition of the bathing solutions, may alter the normal channel permeation or gating properties. We therefore sought to develop a system that would enable InsP$_3$R channels to be recorded in their native ER membrane environment in the presence of physiologically relevant solutions. As the ER is continuous with the outer membrane of the nuclear envelope [55], we reasoned that ER-localized ion channels would also be present in the nuclear envelope outer membrane, and that single-channel properties of the InsP$_3$R in its native membrane environment could be examined by patch clamping isolated nuclei. As detailed below, we have employed this approach successfully to record single InsP$_3$R channels using nuclei isolated from a variety of cells. Initially, we used nuclei isolated from *Xenopus* oocytes to study its endogenous InsP$_3$R [56]. Surprisingly, formation of seals with gigaohm resistances, a prerequisite for studying ion channel activity, was achieved with very high frequency despite the high density of nuclear pores in the nuclear envelope [57]. Only one patch out of many thousands obtained thus far has had an nS channel-like conductance that might be compatible with the large size of nuclear pores [58]. Although the tips of the patch pipettes are small, it is inconceivable that all of the patches were devoid of nuclear pores by chance. We assume, without understanding the mechanisms involved, that the nuclear pores are all closed and impermeable in our nuclear membrane patches. Our studies over the past several years [44–46, 59–62, 63–68] have shown that InsP$_3$R channels recorded in their native ER membranes behave far more robustly and very differently from those observed in reconstitution studies, particularly the responses to ligands and kinetic features, validating our hypothesis that recording these channels in their native ER membrane is the preferred approach for single-channel investigations of the InsP$_3$R.

Similar success in reliably obtaining gigaohm seals on outer nuclear membranes has also been achieved using nuclei isolated from CHO, Cos-7, DT40, HeLa, and insect Sf9 cells, suggesting that nuclear patch clamp electrophysiology is a general approach that can be applied to many cell types. In some of these systems we have successfully recorded either endogenous or recombinant InsP$_3$R single-channel activities. In the rest of this chapter, we provide detailed procedures for recording InsP$_3$R channels by nuclear patch clamp electrophysiology, describing the three cellular systems with which we have had most experience: *Xenopus* oocytes, insect Sf9 cells, and avian DT40 cells.

10.3.1 *Xenopus* Oocytes

10.3.1.1 Frog Husbandry

Female *Xenopus laevis* frogs raised specifically for laboratory use are purchased from suppliers (suppliers we have used include: NASCO, 901 Janesville Avenue, P.O. Box 901, Fort Atkinson, WI 53538-0901, (800) 558-9595, Cat # LM00531M;

Xenopus One, 5654 Merkle Road, Dexter, MI 48130, (734) 426-2083, Cat. # 4216; Xenopus Express, 6008 Glen Harwell Rd, Plant City, FL 33566-9784, (800) 936-6787, Cat. # OP IMP FM). There is no strong correlation between the general health of the frogs and the suppliers, so the supplier used is usually dictated by the availability of the frogs. The quantity and quality of oocytes vary greatly from frog to frog even among frogs ordered at the same time from the same supplier. To avoid possible seasonal variability, the frogs are exposed to a constant light and dark cycle of 12 h each at a constant temperature of about 16°C in a temperature-controlled room. This is critical to the quality of the oocytes that can be obtained from the frogs. Frogs kept in temperatures higher than 20°C yield oocytes that die sooner after isolation and from which nuclei are more difficult to isolate.

10.3.1.2 Solutions for Oocytes

We use essentially only two solutions for handling oocytes: the standard oocyte solution (SOS) that contains (in mM) 100 NaCl, 2 KCl, 1.8 CaCl$_2$, 1 MgCl$_2$, 5 HEPES (pH 7.6 with NaOH); and the Ca^{2+}-free SOS that has the same composition as SOS but with no added CaCl$_2$. Ca^{2+}-free SOS is used in situations when presence of Ca^{2+} may be harmful to the oocytes, whereas SOS is used in all other occasions. Extracellular Ca^{2+} is necessary for the long-term health of the oocytes, so they should not be kept in Ca^{2+}-free medium for extensive periods. All solutions used are filter-sterilized with 2-μm filters and stored in sterilized containers.

10.3.1.3 Surgical Extraction, Defolliculation, and Maintenance of Oocytes

Procedures used for surgical ovary extraction, oocyte defolliculation, oocyte mRNA injection, and maintenance of oocytes are slight variations on those that have been described elsewhere [69–72]. Since a mature female *Xenopus laevis* frog can literally have thousands of oocytes, a frog can be used for multiple oocyte extractions. Extracted pieces of ovary are kept in SOS solution, and individual oocytes are isolated as soon as possible. Although individual oocytes can be isolated from pieces of ovary manually, it is tedious and not practical if any significant quantity of oocytes is needed. We therefore isolate oocytes using a collagenase treatment. Only mature, healthy stage V and VI oocytes are selected for our experiments [73, 74]. These oocytes are about 1–1.5 mm in diameter and show clear separation of pigment in the darker animal and lighter vegetal hemispheres (Figure 10.2). The pigmentation in the animal hemisphere of these oocytes is distinctly lighter than that in the earlier stages. Stage VI oocytes can be distinguished by an unpigmented equatorial band between the two hemispheres. The most visible indication of the state of health of an oocyte is the maintenance of pigmentation in the two hemispheres (Figure 10.2). Healthy oocytes have an even, dark pigmentation over all the animal hemisphere. For long-term maintenance (3–7 days), defolliculated oocytes (injected or uninjected) are kept individually in 96-well plates, with each well containing 200 μl of SOS solution. Isolation of individual oocytes prevents cell contents released from dying oocytes from harming other oocytes. The incubating SOS solution contains 50 μg/ml of the antibiotic gentamicin (Gibco, Cat. #

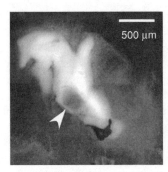
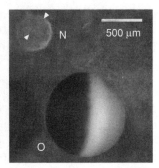

FIGURE 10.2 Isolation of *Xenopus oocyte* nucleus. Left: a nucleus is isolated from an oocyte that has been teased open. The arrowhead marks the translucent nucleus clearly visible among the cytoplasmic debris. Right: a completely isolated, translucent nucleus (N) is placed next to an intact stage VI oocyte (O) for comparison. Small arrowheads mark blebs visible on the nuclear envelope.

15750–060). The health of the oocytes can be seriously compromised by prolonged exposure to high temperature (>20°C), which should be avoided as much as possible. All oocytes are kept in an incubator at 16°C.

10.3.1.4 Oocyte Nucleus Isolation

An oocyte is transferred to a nuclear isolation chamber (we use the cover of a 35-mm Petri dish because of its lower rim and larger area) containing the experimental bath solution. Two pairs of fine forceps are used to gently tease open the oocyte. Since the nucleus is usually located just beneath the animal pole of the oocyte, we start at the vegetal pole to avoid accidentally damaging the nucleus before we can visually locate it in the cytoplasm. Basically, the oocyte is torn into two pieces with the forceps. Then the piece that is more likely to contain the nucleus (usually the larger piece or the one with more of the animal hemisphere) is further torn into two and so on until the nucleus, a translucent ovoid, separates out from the surrounding cytoplasm, a milky white opaque substance (Figure 10.2). An isolated oocyte nucleus is about 300–500 μm in diameter (Figure 10.2). As the nucleus tends to stick to the bottom of the isolation chamber, it should be transferred onto a Petri dish containing the experimental bath solution as soon as possible using a simple nucleus transfer device (Figure 10.3B). The Petri dish with the isolated nucleus is then placed on the stage of an inverted microscope where the patch clamp experiment is performed.

Manual isolation of an oocyte nucleus is a delicate procedure during which only a minimum of mechanical trauma is experienced by the nucleus, including the nuclear envelope. As the ability to obtain gigaohm seals during patch clamp experiments using the isolated nucleus depends critically on the integrity of the nucleus structure, care should be taken to avoid damaging the nucleus during isolation and transfer. A damaged nucleus can be identified by leakage of nucleoplasm from it. Forceps with well-maintained fine tips, much practice, and healthy

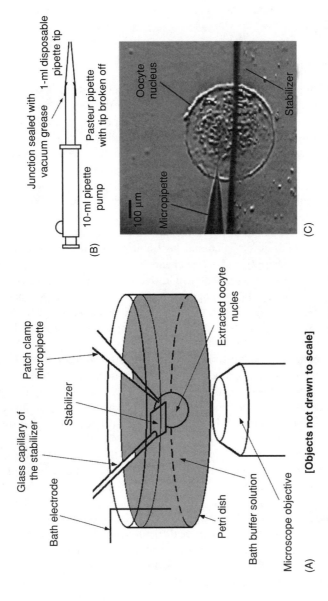

FIGURE 10.3 Patch clamping an oocyte nucleus. (A) Schematic diagram of the experimental setup for nuclear patch clamping using isolated oocyte nucleus. (B) Schematic diagram of nucleus transfer apparatus. (C) Photo-micrograph of an isolated oocyte nucleus patch clamped.

oocytes can improve the success rate of isolating intact nuclei. Unhealthy oocytes and oocytes exposed to high temperature (as little as 20 min at room temperature) have a sticky cytoplasm from which the nucleus cannot be isolated easily. Such cytoplasm will stick to the nucleus and prevent gigaohm seal formation even if the nucleus is isolated.

10.3.1.5   Oocyte Nucleus Patch Clamping

The experimental setup used for nuclear patch clamp experiments to obtain single-channel current records of $InsP_3R$ channel is similar to the regular patch clamp setup [47]. As the oocyte nucleus is large (300–500 μm), and it is the nuclear membrane on the top of the nucleus that is patch clamped (Figure 10.3A), a low-power objective (10 × magnification) with great depth of field (able to focus at least 500 μm deep into the experimental chamber) is required. As in all patch clamp setups, a microscope with a long working distance condenser is good to have but not necessary. To restrict the movement of the nucleus, a stabilizer fashioned out of a piece of cover slip attached to a glass capillary using Sylgard 184 (Dow Corning, Midland MI) is positioned on top of the isolated nucleus using a micromanipulator (Figure 10.3). The distribution of functional $InsP_3R$ in the outer nuclear envelope appears to be heterogeneous [62], and active channels are more frequently detected at the bases of bleb-like structures (~30–50 μm in size) on the surface of the isolated nucleus (Figure 10.2). With video imaging and a stabilized nucleus, multiple membrane patches can be obtained from the area identified to have a higher density of $InsP_3R$ to increase the probability (P_d) of obtaining a membrane patch that exhibits $InsP_3R$ channel activity [62].

Patch clamp pipettes are pulled from borosilicate glass capillaries (WPI, Sarasota, FL, Cat. # 1B150F-4) using a horizontal pipette puller (Sutter Instruments, Novato, CA, Cat. # P-97). Pipette resistance is about 5–20 MΩ when filled with a 140 mM KCl solution. Using a micromanipulator, the patch pipette is positioned so its tip is just above the membrane of the isolated nucleus (Figure 10.3). A positive pressure (~5 mmHg) is maintained during this time inside the patch pipette to minimize the contamination of the pipette solution by the bath solution due to diffusion through the pore of the pipette. Then the positive pressure is relieved and the pipette lowered so its tip comes into contact with the nuclear envelope of the isolated nucleus. Care is taken not to puncture the nucleus because when the integrity of the nuclear envelope is compromised, nucleoplasm will leak out and the nucleus becomes deflated. It is much more difficult to obtain gigaohm seal from such a damaged nucleus. Gigaohm seal is obtained usually within a few seconds after negative pressure (~10–20 mmHg) is applied inside the pipette.

It is well known that there is a high density of large nuclear pores present in the nuclear envelope. In the membrane isolated at the tip of the patch pipette (~0.5–1 μm diameter), there should be tens of nuclear pores. Remarkably, gigaohm seals form regularly in nuclear patch clamp experiments (~80%). This indicates that the vast majority of the nuclear pores are closed in an isolated nucleus. Nevertheless, it appears that the solution in the perinuclear space or lumen of the nuclear envelope (the region between the outer and inner nuclear membranes) is in ionic equilibrium with the bath solution, because no junction potential difference has been detected between the two solutions in all our patch clamp experiments.

When an isolated nucleus is patch clamped, the cytoplasmic side of any channel located in the outer nuclear membrane is exposed to the solution inside the patch pipette (cytoplasmic-side in). Thus, if the pipette solution contains InsP$_3$ and a suitable concentration (between 40 nM and 80 μM) of free Ca^{2+}, then InsP$_3$R channels present in the membrane patch isolated at the tip of the patch pipette will be activated and transmembrane channel currents can be detected as the InsP$_3$R channels gate (see Section 4).

As the InsP$_3$R channels (endogenous *Xenopus* type 1 and recombinant rat type 3) observed in isolated oocyte nuclei inactivate irreversibly after a relatively short time (~30–60 s) [56, 62, 64], channel currents are recorded as soon as seal resistance increases above 200 MΩ without excising the isolated membrane patch, to maximize data obtained. Thus, the majority of our data are obtained in the on-nucleus configuration, although the excised patch configuration can be easily and reliably obtained by raising the micropipette, when needed.

The InsP$_3$R is a relatively large conductance channel (channel conductance of the *Xenopus* type 1 and recombinant rat type 3 InsP$_3$R in symmetric 140 mM KCl solution is 360 pS) so that under normal patch clamp recording conditions, the signal-to-noise ratio in the recording is not a significant problem. Thus, extraordinary measures to reduce noise, like coating patch pipette tip with Sylgard or using quartz glass for patch pipettes, are not necessary. Current records are filtered between 1 and 5 kHz.

One advantage of using *Xenopus* oocyte nucleus to study the InsP$_3$R is that the type 1 InsP$_3$R is the only intracellular Ca^{2+} release channel detectably expressed in the *Xenopus* oocytes [75–77]. Thus, any InsP$_3$-stimulated channel observed in the outer nuclear membrane of a nucleus isolated from an uninjected oocyte must be the endogenous type 1 InsP$_3$R.

The major problem with studying the InsP$_3$R channel by nuclear patch clamping of isolated *Xenopus* oocyte nuclei is the probability (P_d) of detecting InsP$_3$R channel activity. From our experience, P_d of oocytes obtained from the same frog at the same time is very consistent and does not change with time after the oocytes are defolliculated. P_d of different batches of oocytes obtained from different frogs at the same time appears to be, for some unknown reason, somewhat similar, even if the frogs donating the oocytes are maintained in different locations. Infrequently, for several weeks, the P_d can be very high (~0.6 to 0.8) and most membrane patches obtained exhibit multiple channels. During other periods, the P_d for the endogenous channel is so low (<0.05) that the oocytes can be used for expression of recombinant exogenous InsP$_3$R by cRNA injection (below). Unfortunately, even if optimal ligand conditions are used, the P_d in uninjected oocytes for the majority of the time is between 0.1 and 0.2, too high for studying recombinant InsP$_3$R but low enough to slow progress in investigating endogenous InsP$_3$R substantially. We have tried altering the feeding pattern and light cycle the frogs are kept under, using oocytes obtained from frogs injected with human chorionic gonadotropin that induces maturation of the oocytes [69], using tap water (allowed to stand for at least 48 h to remove harmful gas used in water treatment) instead of deionized water, and using frogs from different suppliers, but we have found no reliable means to increase the P_d.

During the periods when the P_d for the endogenous channel is <0.05, the oocytes can be used for expression of recombinant exogenous InsP$_3$R by cRNA injection. For our studies of the recombinant rat type 3 InsP$_3$R [45, 61, 64, 65], 23 nl of 1 µg/µl cRNA solution was normally injected. Type 3 InsP$_3$R expression was usually detectable by Western blot analysis 72 h after cRNA injection, and functionally in nuclear patch clamp experiments 96 h after injection. The P_d at this time increased to ~0.4, or 20-fold higher than before injection. In control experiments, the endogenous channel P_d does not increase over this period. Accordingly, the InsP$_3$R channels detected in membrane patches obtained from nuclei isolated from cRNA-injected oocytes are overwhelmingly (>95%) homotetrameric recombinant channels [45, 64, 65].

10.3.1.6   Solutions Used in Patch Clamp Experiments

Because the InsP$_3$R is an intracellular channel, most of the solutions used in our experiments contain high K^+ concentration (140 mM KCl), with pH buffered to 7.3 with 10 mM HEPES. The InsP$_3$R channel has a relatively high permeability for monovalent cations [56, 65], so K^+ ions are the major charge carriers in our experiments. This is important as it enables the free Ca^{2+} concentrations experienced by the InsP$_3$R to be tightly controlled. This is critical for studies of the gating properties of the InsP$_3$R as, in addition to being a permeant ion, Ca^{2+} also acts as a ligand that regulates channel gating.

To ensure proper buffering of free $[Ca^{2+}]$, it is important to use different Ca^{2+} chelators with appropriate Ca^{2+} affinities for different ranges of $[Ca^{2+}]$. For 20 nM < $[Ca^{2+}]_i$ < 1 µM, BAPTA (1,2-bis(O-aminophenoxy) ethane-N,N,N',N'-tetraacetic acid, from Molecular Probes, Eugene, OR, Cat # B1204) is used. BAPTA is used instead of EGTA because of its faster Ca^{2+} binding kinetics and lower pH dependence. Also, our patch clamp experiments do not require a large quantity of buffer solutions, so the higher cost of BAPTA is not a major concern. For 800 nM < $[Ca^{2+}]_i$ < 5 µM, 5,5'-dibromo BAPTA (Molecular Probes, Cat # D1211) is used. For 3 µM < $[Ca^{2+}]_i$ < 30 µM, HEDTA (N-(2-Hydroxyethyl) ethylenediamine-N,N',N'-triacetic acid from Sigma-Aldrich, Cat # H2378) is used. Generally, 500 µM of Ca^{2+} chelator is used. For $[Ca^{2+}]_i$ < 10 nM, 1 mM BAPTA is used for extra buffering capacity. We found that chelators at these concentrations have no significant effects on InsP$_3$R channel gating [78]. For $[Ca^{2+}]_i$ > 30 µM, no Ca^{2+} chelator other than Na$_2$ATP (when Na$_2$ATP is called for in our experiments) is used to buffer $[Ca^{2+}]_i$. It is a common mistake to use the same Ca^{2+} chelator for all $[Ca^{2+}]_i$, from nanomolar to micromolar range. This will cause great inaccuracy in evaluation of $[Ca^{2+}]_i$. We have found that the Ca^{2+} chelators in solution lose chelating capacity over time. Furthermore, their purity may vary from batch to batch, and many chelators can adsorb moisture. Thus, computer software like Maxchelator (C. Patton, Stanford University, CA) can only serve as guides for making experimental solutions with desired $[Ca^{2+}]_i$. $[Ca^{2+}]_i$ of the solutions used in our patch clamp experiments are directly measured using either Ca^{2+}-selective minielectrodes [79–81] or by fluorimetry with Ca^{2+}-sensitive dyes. All solutions are used within 3 months after they are made.

In general, the bath solutions in most of our experiments contained no ATP, whereas the pipette solutions contained various concentrations of Na$_2$ATP.

10.3.2 Sf9 CELLS

Sf9 cells express the PLC–InsP$_3$ system [82] that is coupled to endogenous as well as expressed recombinant G-protein-coupled plasma membrane receptors [83, 84]. Although the sequence of Sf9 cell InsP$_3$R has not been determined, all other invertebrates appear to possess only a single InsP$_3$R isoform, corresponding to the type 1 channel. Thus, the Sf9 cell is similar to the *Xenopus* oocyte in expressing homotetramers of the type 1 InsP$_3$R. However, in comparison with the oocyte system for patch clamp electrophysiology of the InsP$_3$R in native ER membranes, the Sf9 cell system offers distinct advantages. Foremost among them is the high probability of detecting active InsP$_3$R channels in patched nuclear membranes. In the oocyte, as well as CHO and Cos-7 cells, there is a relatively low and variable probability of endogenous InsP$_3$R channel detection. Under optimal stimulatory conditions for the oocyte system, patch clamp of the *Xenopus* InsP$_3$R usually detects channels in 10–20% of patches [62]. In contrast, we observe InsP$_3$R channel activity in ~80% of the patches in Sf9 cell nuclei, with most patches containing multiple channels consistently among different batches of cells. A second advantage concerns the duration of channel activity. Previous work on the endogenous *Xenopus* type 1 and recombinant rat type 3 InsP$_3$Rs recorded in *Xenopus* nuclei showed that in response to continuous stimulation by InsP$_3$, channel activity usually inactivated relatively quickly, with a time constant ~30 s [62, 64], making it difficult to observe channel behaviors over long periods and limiting the number of open–closed transitions available for analyses. In contrast, channels observed in the Sf9 nuclear patches last substantially longer (see Section 10.4.2), with many channels lasting for very long times (>10 min). This longer activity provides more event transitions for gating analyses, which has enabled observations of novel gating behavior; and has enabled studies into the regulation of channel inactivation. A disadvantage of the Sf9 cell system is that the consistently high rate of endogenous InsP$_3$R expression precludes its use for studies of expressed recombinant channels.

10.3.2.1 Sf9 Cell Culture and Nucleus Isolation

Spodoptera frugiperda (Sf9) cells (Invitrogen, Cat. # 11496–015) are grown and maintained in SF-900II serum-free media (GIBCO) in suspension culture according to the manufacturer's protocols. The cells are seeded at $0.5–0.7 \times 10^6$ viable cells/ml and subcultured when they reach a density of $\sim 2.5 \times 10^6$ cells/ml, approximately every 3–4 days, to avoid cell clumping. For maximum detection of functional InsP$_3$R channels in patch clamp experiments, each freshly thawed batch of cells should be subcultured three to four times before it is used for electrophysiology. The cells can then be propagated and used for nucleus isolation for up to 7–8 weeks in culture before a new lot is thawed and expanded.

In order to prepare the nuclear homogenization mixture, cells are moved from the suspension culture to a T-25 flask, allowed to attach to the bottom of the flask for 1 h, and then washed twice with Ca^{2+}, Mg^{2+}-free PBS. The ice-cold nuclear isolation solution, containing (in mM): 140 KCl, 250 sucrose, 1.5 β-mercaptoethanol, 10 Tris-HCl (pH to 7.4), with complete protease inhibitor cocktail (Roche Molecular Biochemicals, Indianapolis, Indiana) and 0.05 mM phenylmethylsulfonylfluoride (PMSF) is added to the flask and the cells are then detached by gentle scraping. It is critical to add the protease inhibitor cocktail before detaching the cells from the flask, because we have observed that the normally high rate of success of detecting InsP$_3$R channels by patch clamping of the isolated nuclei is dependent on the presence of the protease inhibitor cocktail in the homogenization mixture before intact cells are disrupted mechanically. About 1–2 ml of the cell mixture is then rapidly transferred to an ice-cold Dounce homogenizer. Homogenization of the mixture is performed using two to four strokes of the pestle. Then 20–30 μl of the homogenized mixture is added to 1 ml standard bath solution (in mM: 140 KCl, 10 HEPES, 0.5 BAPTA, pH 7.3, and free $[Ca^{2+}]$ adjusted to ~300 nM) in an experimental chamber on the stage of an inverted microscope. Isolated nuclei are 5–10 μM in diameter, and can be distinguished from intact cells based on their unique morphology (Figure 10.4). To ensure a high level of integrity of nuclei, fresh homogenizations should be performed every 2 h.

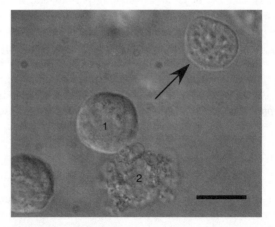

FIGURE 10.4 Isolated nucleus from Sf9 cells. Arrow indicates an intact nucleus isolated from an Sf9 cell suitable for nuclear patch clamp experiments. Intact cell (1) and a partially damaged cell (2) also present in field. Scale bar is 10 μm.

10.3.2.2   Sf9 Cell Nucleus Patch Clamping

A patch pipette with resistance of 20–25 MΩ is positioned so that its tip comes into contact with the outer membrane of the nucleus. Gentle suction is then applied (10–20 mmHg) and gigaohm seals are obtained in ~80% of the attempts. In order to maximize the duration of channel activity recorded, current recording is started as soon as seal resistance exceeds 200 MΩ. Bath and pipette solutions with carefully buffered [Ca^{2+}], the same as those described above for *Xenopus* protocols, were used. In most of the multichannel patches the amplitudes of the individual channels detected in the same patch were similar (Figure 10.6B). However, the endogenous channel occasionally opened to subconductance levels, with dwell times lasting from less than 10 ms to >100 ms. The substates accounted for less than 0.1% of all openings. The most predominant subconductance state had amplitude half of the main conductance state.

10.3.3   DT40 CELLS

Electrophysiological characterization of single recombinant InsP$_3$R is necessary to provide insights into the differences in channel gating and permeation properties of different channel isoforms, as well as into the roles for channel function of specific sequences in the primary structures of the channels. However, studies of recombinant InsP$_3$R permeation and gating in native ER membranes by nuclear patch clamping in oocyte and Sf9 cell systems are complicated by the presence of endogenous InsP$_3$R channels. Native channels may not be easily distinguished from the expressed recombinant ones by either their gating or conductance characteristics; and they may interact with the expressed channels to form heteroligomeric channels with variable subunit stoichiometries. Whereas the occurrence of batches of *Xenopus* oocytes that lack endogenous channel activities in nuclear patching has provided a system for functional studies of single recombinant rat type 3 InsP$_3$R [45, 64, 65], a more ideal system would be one in which recombinant channels could be introduced into a cell that lacked endogenous InsP$_3$R channels in a true InsP$_3$R-null background. At present, the only such cell type available is a genetically engineered DT40 cell line. The DT40 is a B cell line initially derived from an avian leucosis virus-induced lymphoma in white leghorn chicken [85]. It is unique among vertebrate cells in that it displays a remarkably high frequency of homologous gene targeting [86], which has been exploited to develop a mutant DT40 cell line (InsP$_3$R-KO) deficient in the expression of all three isoforms of the InsP$_3$R [87]. We have subsequently introduced recombinant rat types 1 and 3 InsP$_3$R into the InsP$_3$R-KO cells and have successfully recorded single-channel activity from isolated nuclei.

10.3.3.1   DT40 Cell Culture

Both wild-type and InsP$_3$R-KO DT40 lines are available from Riken Bioresource Center, Japan (cell bank numbers RCB1464 and RCB1467, respectively). Cells are grown in suspension culture at 39°C in RPMI 1640 media (Gibco) and supplemented with 10% (v/v) FBS, 2 mM L-glutamine, 1% chicken serum, 50 μM

β-mercaptoethanol, 100 U/ml penicillin, and 100 μg/ml streptomycin in a humidi-
fied 95/5% air/CO_2 atmosphere. It is acceptable to culture at 37°C without any loss
of cell viability, the only effect being an increase in the doubling time from about 10
to 18 h. It is of primary importance, however, that the culture be maintained in
continuous logarithmic growth at $(5–10) \times 10^5$ cells/ml. Care should be taken not to
grow cells beyond a density of 2.5×10^6 cells/ml as this may decrease viability;
similarly, cells should not be split to a density below 2×10^5 cells/ml. Cells
suspended at a density of 2×10^6 cells/ml in normal culture media with the addition
of 10% DMSO can be frozen and stored in liquid nitrogen.

10.3.3.2   Generation of Recombinant InsP₃R-Expressing DT40-KO Cells

Because of the generally poor transfection efficiency of DT40 cells, we produce
stable DT40 cell lines expressing recombinant InsP₃R to ensure that every nucleus
patched is derived from an expressing cell. Electroporation is the most widely used
technique for the introduction of foreign DNA into DT40 cells, but the efficiency is
generally low. The use of linearized DNA is widely reported in the literature, and it
is suggested that this may increase transfection efficiency and likelihood of integration when generating stable transfectants. Using the following protocol we
routinely achieve a transfection efficiency of 5–10% using eGFP (circular) as a
reporter: 10^7 cells are suspended in 800 μl PBS in a 4 mm gap electroporation
cuvette with 10–40 μg DNA. After 15 min incubation on ice, the preparation is
electroporated once at 550 V and 25 μF using a Gene Pulser II (BioRad). The
cuvette is left on ice for a further 30 min before the cell suspension is added to 20 ml
warm culture media. eGFP can be visualized after 18–24 h. We have had much
greater success using a Nucleofector Device from Amaxa Biosystems. Using their
transfection kits and parameter recommendations, we achieve transfection efficien-
cies of 40% and have subsequently used this approach to express InsP₃R and
generate stable clones. The InsP₃R rat type 1 or type 3 cDNA was cloned into the
pIRES2-EGFP vector (Clontech). The pIRES-EGFP expression system permits
InsP₃R and eGFP to be translated from a single RNA, enabling identification of
transfected cells by fluorescence. Immediately after Amaxa electroporation, cells
are suspended in 15 ml media and cultured for 24–36 h. GFP-expressing cells are
then concentrated to >95% using a fluorescence-activated cell sorter (Becton
Dickinson) and re-suspended in media containing the selection reagent G418
(2 mg/ml; Gibco). Cells are maintained in G418-containing media for 2 weeks,
after which they are subcloned by a second round of cell sorting configured to
deposit a single GFP-positive cell into each well of a 96-well plate, with each well
containing 200 μl of 20% conditioned media with G418. Individual colonies should
be visible after 7 days of undisturbed culture at 37°C. The InsP₃R expression level
can then be confirmed with Western blot. It is important to select and amplify at
least six clones and monitor expression levels over the course of several weeks, as
we have noticed a loss of InsP₃R expression in some clones.

We have also successfully employed a pantropic retroviral expression system to
introduce recombinant DNA into DT40 cells (Clontech, Cat. # 631512). Virus is

generated by cotransfecting a retroviral expression vector with the envelope glyco-protein from the vesicular stomatitis virus (VSV-G) into a packaging cell line that stably expresses the viral gag and pol genes. The expression vector, a kind gift from Dr Masamitsu Iino, University of Tokyo, contained the rat type 1 InsP$_3$R-pIRES2-EGFP sequence cloned into the retroviral expression vector pMX. We generated viral particles as described [88] and infected DT40-KO cells. Although the expression efficiency for this technique was low (2–5%) we were able to success-fully isolate stable InsP$_3$R-expressing clones using the fluorescence subcloning method outlined above.

10.3.3.3 DT40 Cell Nuclear Membrane Patch Clamp Electrophysiology

Cells in log phase growth are harvested and centrifuged at 100g for 5 min, washed once with ice-cold PBS and once in cold isolating solution containing (mM): 150 KCl, 250 sucrose, 1.5 β-mercaptoethanol, 10 Tris–HCl, pH 7.5. The cells are then re-suspended in isolating solution containing complete protease inhibitor cocktail, one tablet per 40 ml isolating solution (Roche Molecular Biochemicals) and 0.05 mM phenylmethylsulphonyl fluoride (PMSF) and stored on ice. To isolate nuclei, 1 ml of the preparation is homogenized (one to two strokes) using a Dounce glass homogenizer that yields about 2–5% isolated nuclei. An aliquot of the homogenized preparation (25 μl) is plated onto a 1 ml glass-bottomed bath contain-ing 2 ml of standard bath solution and allowed to adhere for 5 min. It is often difficult to cleanly isolate nuclei from DT40 cells, so usually some plasma mem-brane and cytoplasmic debris remain attached. However, this distinct morphology can be used to quickly identify disrupted cells (Figure 10.5). Indeed, it is quite easy to patch onto the cleanly exposed nuclear surface of a nucleus still partially contained within a broken cell. The patch clamp setup and experimental consider-ations are essentially the same as described for oocyte and Sf9 systems. We routinely acquire stable (10–60 min) seals (20–100 GΩ) in 70–80% of the attempts. Rat type-1 channels recorded in DT40-KO cell nuclei have a similar conductance (~380 pS) to endogenous and recombinant channels recorded in oocyte nuclei (Figure 10.6C). However, we have noted some differences in gating behavior. The maximum P_o is only 0.15 under optimal conditions (1 μM Ca^{2+}, 10 μM IP$_3$) and the mean open time is about 1 ms. The rapid gating observed in DT40 cell nuclei therefore necessitates sampling at a higher frequency of 20 kHz with filter frequency of 10 kHz.

10.4 ION CHANNEL PROPERTIES OF THE InsP$_3$R IN NATIVE ER MEMBRANES BY NUCLEAR PATCH CLAMPING

10.4.1 Basic Conductance Properties of the InsP$_3$R

The conductance properties of both endogenous *Xenopus* type 1 and recombinant rat type 3 InsP$_3$R in the outer nuclear membrane of the *Xenopus* oocyte nuclei are very similar. In the presence of symmetric 140 mM KCl and absence of significant concentrations of divalent cations, the most frequently observed open conductance

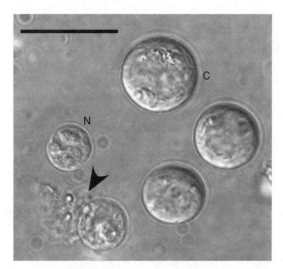

FIGURE 10.5 Isolated nucleus from DT40 cells. An intact nucleus (N) isolated from a DT40 cell next to a partially isolated nucleus with cytoplasmic debris (arrowhead) still attached. Partially isolated nuclei are preferentially selected for nuclear patch clamp experiments. Also present in cell homogenate are intact cells (C). Scale bar is 15 μm.

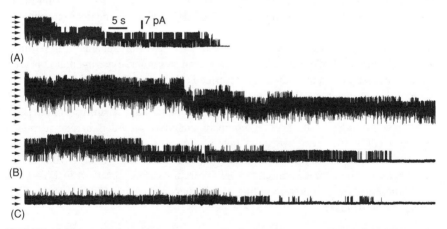

FIGURE 10.6 InsP$_3$R channel current traces. Typical current traces showing multiple active InsP$_3$R channels recorded in ER membranes by nuclear patch clamp electrophysiology. (A) Endogenous *Xenopus* type 1 InsP$_3$R channels. (B) Endogenous Sf9–InsP$_3$R channels. (C) Recombinant rat type 1 InsP$_3$R channels expressed in a DT40-KO cell line. Each trace shows a continuous current record obtained in one nuclear patch clamp experiment using standard pipette and bath solutions with the pipette solution containing 1 μ*M* [Ca^{2+}]$_i$ and 10 μ*M* InsP$_3$ plotted on the same scale. Arrows indicate current levels corresponding to different number of open channels, with lowest arrow indicating closed channel current level.

state of InsP$_3$R channel exhibits linear (Ohmic) channel current vs. transmembrane potential (I–V) relation with a slope conductance of 360 ± 10 pS [63, 64] (Figure 10.6A). A similar conductance is observed for the recombinant rat type 1 channel expressed in Cos-7 cell [66, 67] and DT40-KO nuclei (Figure 10.6C). In contrast, the conductance of the Sf9-InsP$_3$R is ~480 pS (Figure 10.6B). Using reversal potentials observed under asymmetric ionic conditions, and employing the Goldman–Hodgkin–Katz equations [90], the oocyte InsP$_3$R is found to be a divalent selective cation channel with relative permeabilities of P_{Ca}: P_{Ba}: P_{Mg}: P_K: $P_{Cl} = 5.5 : 4.3 : 3.2 : 1 : 0.23$ [56, 62, 63, 64, 66]. Divalent cations can act as permeant ion blockers, reducing the K^+ conductance of InsP$_3$R and causing rectification of the I–V relation of the channel [63, 64]. In symmetric 140 mM KCl solution containing 3 mM MgCl$_2$, the slope conductance of the InsP$_3$R expressed in the outer *Xenopus* oocyte nuclear membrane is 120 pS at 0 mV and 360 pS at 60 mV [63, 64]. In addition to the main conductance state, the InsP$_3$R channel also exhibits conductance substates occasionally, some of which (flicker mode) have gating kinetics markedly different from that of the main conductance state. These conductance substates occur <1% of the time the InsP$_3$R channel is open and therefore are not studied in detail [56, 62, 64].

A basic six transmembrane-domain topology of IP$_3$Rs has been proposed that is shared with other cation channels [40] (Figure 10.1). By analogy, putative transmembrane helices 5 and 6 and the intervening intraluminal loop would constitute the permeation pathway [40], a proposal that was verified by deletion mutant analysis [90]. The identification of the pore in the InsP$_3$R was established by mutating residues in the putative 5/6 loop and recording single recombinant channels in a mammalian cell nuclear membrane [66]. Two residues of the r-InsP$_3$R-1 channel were identified that mediate ion selectivity and conductance [66]. A molecular mechanism for divalent cation selectivity in InsP$_3$Rs was proposed involving a ring of four aspartic acid residues at position 2550 that form a Ca^{2+} binding site within the ion conduction pathway, with ion conductance determined by valine at position 2548.

In addition to the InsP$_3$R channel, other ion channels are detected in our nuclear patch clamp experiments from time to time. These can be readily distinguished from the InsP$_3$R by their conductance (no non-InsP$_3$R channel with conductances between 250 and 400 pS in symmetric 140 mM KCl solutions has been detected in all our nuclear patch clamp experiments using isolated oocyte nuclei), insensitivity to InsP$_3$ stimulation (these channels can be observed in both the presence and absence of InsP$_3$), and insensitivity to competitive inhibition by heparin (InsP$_3$R channel is inhibited by 100 μg/ml heparin (Sigma-Aldrich Cat. # H3400)).

10.4.2 GATING REGULATION OF THE InsP$_3$R

The InsP$_3$R channel is normally closed until it is gated open by the binding of InsP$_3$. The crystal structure of the core binding-region (residues 225–604) showed that InsP$_3$ binds near a cleft formed at the interface of a beta-sheet rich β-trefoil domain and an α-helix-rich armadillo repeat [43]. The activity of the InsP$_3$R is sensitive to both Ca^{2+} and $InsP_3$. Indeed, $InsP_3R$ channel regulation by $[Ca^{2+}]_i$ is a fundamental component of most models of oscillations and wave propagation. The appreciable K^+ permeability of the $InsP_3R$ channel has enabled its gating to be studied in native ER membranes by nuclear patch clamp electrophysiology under nearly physiological conditions with $[Ca^{2+}]$ highly controlled experimentally. Our electrophysiological studies have revealed that Ca^{2+} is the fundamental ligand of the $InsP_3R$ [60, 61, 65, 78]. We elucidated the mechanism of $InsP_3$ channel activation by examining the $InsP_3$ dependence of Ca^{2+} regulation [78]. Whereas previous studies of reconstituted channels in bilayers suggested that the channel had a narrow bell-shaped $[Ca^{2+}]_i$-response (maximum open probability P_o was ~0.05) centered at ~300 nM, we found that the *Xenopus* type 1 channel in native ER membrane had high P_o (~0.8, as much as 20-fold higher than observed in bilayer studies) that remained elevated in the presence of saturating (10 µM) $InsP_3$, even as $[Ca^{2+}]_i$ was raised to very high concentrations (Figure 10.7), displaying two functional Ca^{2+}binding sites: activating, with half-maximal activating $[Ca^{2+}]_i$, K_{act}, of 210 nM and Hill coefficient H_{act} ~ 2; and inhibitory, with half-maximal inhibitory $[Ca^{2+}]_i$, K_{inh}, of 54 µM and Hill coefficient H_{inh} ~ 4. We have observed similar biphasic $[Ca^{2+}]_i$ regulation with comparable Hill parameter values for the recombinant rat type 3 channel expressed in oocytes [65], recombinant rat type 1 channel expressed in Cos-7 cells [68] as well as for the endogenous insect Sf9 $InsP_3R$. Activation by Ca^{2+} enables the channel to participate in Ca^{2+}-induced Ca^{2+} release (CICR), which is important for propagating Ca^{2+} release. Inhibition by high $[Ca^{2+}]_i$ may be important in terminating Ca^{2+} release [92–97]. Lowering $[InsP_3]$ into the more physiological range was without effect on Ca^{2+} activation parameters or H_{inh}, but decreased K_{inh} with a half-maximal activating $[InsP_3]$, K_{IP3}, of 50 nM, and Hill coefficient H_{IP3} of 4 for $InsP_3$ (Figure 10.7). Accordingly, cytoplasmic Ca^{2+} at low concentrations and $InsP_3$ both activate the channel, but they affect it in fundamentally different ways. Ca^{2+} binding directly activates it, like a conventional agonist. In contrast, $InsP_3$ binding activates it indirectly, solely by decreasing the affinity of the Ca^{2+} inhibitory site, with no direct stimulatory effect itself. Under conditions of low $[InsP_3]$, Ca^{2+} preferentially binds to the inhibitory site due to its higher affinity ($K_{inh} < K_{act}$), causing the channel to be inactive. Under conditions of stimulation associated with higher $[InsP_3]$, K_{inh} becomes $>K_{act}$, so Ca^{2+} binds preferentially to the Ca^{2+} activation site, activating the channel. These results demonstrated that Ca^{2+} is a true agonist, whereas the sole function of $InsP_3$ is to relieve Ca^{2+} inhibition of channel gating. Importantly, we showed that the recombinant rat type 3 (r-$InsP_3R$-3) channel is regulated by $InsP_3$ by an analogous mechanism [66], and the endogenous Sf9 channel also appears to be regulated analogously. Thus, this mechanism likely accounts for how all $InsP_3R$ are gated by $InsP_3$. Furthermore, adenophostin also regulates the channel by this mechanism [46]. Allosteric tuning of Ca^{2+} inhibition by $InsP_3$ enables individual $InsP_3R$ channels to respond in a graded fashion, with implications for localized and global Ca^{2+} signaling.

Whereas $InsP_3$ allosterically regulates the Ca^{2+} inhibition binding sites, we discovered a complementary allosteric regulation of the Ca^{2+} activation sites of

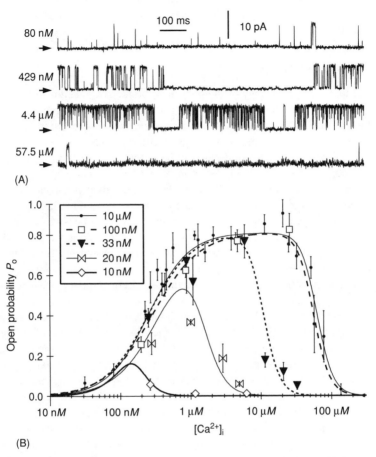

FIGURE 10.7 Ligand regulation of InsP$_3$R channel gating. (A) Typical current traces obtained for *Xenopus* type 1 InsP$_3$R by patch clamping isolated oocyte nuclei with standard pipette and bath solutions. The pipette solutions contained 10 μM InsP$_3$ and $[Ca^{2+}]_i$ as tabulated. Arrows indicate the closed channel current levels. (B) The dependence of InsP$_3$R channel open probability (P_o) on $[Ca^{2+}]_i$ in the presence of various [InsP$_3$] as tabulated. (Figure is modified from Figure 10.1 in [78].)

both types 1 [44] and 3 [45] channels by ATP. Together, therefore, these two regulators shape the $[Ca^{2+}]$ responses of the gating of the channel.

The major regulatory influences on InsP$_3$R gating identified to date impinge on the Ca^{2+} regulation of the channel, emphasizing the importance of the mechanisms by which Ca^{2+} both activates as well as inhibits the channel. Nevertheless, the nature of the Ca^{2+} binding sites is unknown. The InsP$_3$R binds Ca^{2+} [97–99] and CaM [100–101], so these modulators might be directly involved. CaM has been proposed to mediate Ca^{2+} inhibition [101, 105–111] although our results [60] as well as others [104, 112] provide no support for this hypothesis. Although it remains possible that other proteins either mediate or modulate the effects of Ca^{2+}, we believe that Ca^{2+} exerts its profound effects on channel gating by directly binding to it at multiple sites.

10.5 CONCLUSIONS

Patch clamp electrophysiology of the outer membrane of isolated nuclei is a robust methodology for studying permeation and gating of single endogenous as well as recombinant InsP$_3$R channels in their native ER membrane environment. This approach preserves the normal physiological channel milieu and enables the use of physiologically relevant solutions for recording. Furthermore, the approach appears to be one that is generally applicable to many cell types. Our studies indicate that different cell types may provide distinct advantages depending upon the experimental goals. *Xenopus* oocytes express the type 1 InsP$_3$R as their only intracellular Ca^{2+} release channel. Their nuclei are large and easy to isolate. Many fluorescence-imaging studies have explored the contributions of the InsP$_3$R to microscopic and global $[Ca^{2+}]$ signals in the oocyte cytoplasm. Thus, the oocyte system is ideal for efforts to integrate single-channel properties into models of $[Ca^{2+}]_i$ signals in cells. Furthermore, oocytes are a convenient system for recombinant InsP$_3$R channel expression. A disadvantage of *Xenopus* oocyte nuclei has been the relatively low probability and variability of detecting functional InsP$_3$R channel activities in patched membranes. Furthermore, the presence of endogenous InsP$_3$R complicates, although does not preclude, studies of recombinant InsP$_3$R channels because of the lack of a null background. Insect Sf9 cells express the sole invertebrate InsP$_3$R isoform. They consistently express the highest density of functional InsP$_3$R channels on their outer nuclear membrane of any system we have studied to date. This has greatly facilitated detailed studies that had been limited previously by the relative infrequency in detecting InsP$_3$R channels in other systems. Furthermore, long channel current recordings can be obtained that facilitate data acquisition and observations of different gating modes (Figure 10.6). A disadvantage of this system is that the amino acid sequence of the Sf9 channel, as well as much of its genome, is not yet known. Furthermore, the high endogenous InsP$_3$R channel detection precludes use of this system for recombinant InsP$_3$R channel studies. Finally, chicken DT40 pre-B cell lines exist with any or all of the three isoforms of the InsP$_3$R genetically deleted. A significant advantage of this system is that recombinant InsP$_3$R channels can be expressed and studied in a completely InsP$_3$R-null background. Thus, this cell system will be particularly useful in the future for comparing the electrophysiological properties of the different mammalian channel isoforms and their splice variants. It will also be valuable in studies of engineered mutant InsP$_3$R channels in attempts to define specific sequence determinants of channel permeation and gating properties and of protein interactions that have consequences for channel function. Finally, because attainment of gigaohm seals has been possible on nuclei isolated from several different cell types, it is likely that other cell types will be discovered that present unique advantages in patch clamp electrophysiology of ER-localized InsP$_3$R in their isolated nuclei.

ACKNOWLEDGMENTS

We thank coworkers who have assisted in our nuclear patch clamp studies, including Sean McBride, Nataliya Petrenko, Darren Boehning, Alexander Semenyutin, and Rob Lee. Our studies have been supported by grants and fellowships from the National Institutes of Health and the American Heart Foundation.

REFERENCES

1. Maeda, N., Kawasaki, T., Nakade, S., Yokota, N., Taguchi, T., Kasai, M., and Mikoshiba, K., Structural and functional characterization of inositol 1,4, 5-trisphosphate receptor channel from mouse cerebellum, *J Biol Chem* 266 (2), 1109–16, 1991.
2. Ferris, C.D., Huganir, R.L., Supattapone, S., and Snyder, S.H., Purified inositol 1,4, 5-trisphosphate receptor mediates calcium flux in reconstituted lipid vesicles, *Nature* 342 (6245), 87–9, 1989.
3. Ferreri-Jacobia, M., Mak, D.-O.D., and Foskett, J.K., Translational mobility of the type 3 inositol 1,4,5-trisphosphate receptor Ca^{2+} release channel in endoplasmic reticulum membrane, *J Biol Chem* 280 (5), 3824–31, 2005.
4. Berridge, M.J., Bootman, M.D., and Lipp, P., Calcium — a life and death signal, *Nature* 395 (6703), 645–8, 1998.
5. Berridge, M.J., Inositol trisphosphate and calcium signalling, *Nature* 361 (6410), 315–25, 1993.
6. Clapham, D.E., Calcium signaling, *Cell* 80 (2), 259–68, 1995.
7. Thomas, A.P., Bird, G.S., Hajnoczky, G., Robb-Gaspers, L.D., and Putney, J.W., Jr., Spatial and temporal aspects of cellular calcium signaling, *FASEB J* 10 (13), 1505–17, 1996.
8. Blondel, O., Takeda, J., Janssen, H., Seino, S., and Bell, G.I., Sequence and functional characterization of a third inositol trisphosphate receptor subtype, IP$_3$R-3, expressed in pancreatic islets, kidney, gastrointestinal tract, and other tissues, *J Biol Chem* 268 (15), 11356–63, 1993.
9. De Smedt, F., Verjans, B., Mailleux, P., and Erneux, C., Cloning and expression of human brain type I inositol 1,4,5-trisphosphate 5-phosphatase. High levels of mRNA in cerebellar Purkinje cells, *FEBS Lett* 347 (1), 69–72, 1994.
10. Furuichi, T., Yoshikawa, S., Miyawaki, A., Wada, K., Maeda, N., and Mikoshiba, K., Primary structure and functional expression of the inositol 1,4,5-trisphosphate-binding protein P400, *Nature* 342 (6245), 32–8, 1989.
11. Maranto, A.R., Primary structure, ligand binding, and localization of the human type 3 inositol 1,4,5-trisphosphate receptor expressed in intestinal epithelium, *J Biol Chem* 269 (2), 1222–30, 1994.
12. Mignery, G.A., Sudhof, T.C., Takei, K., and De Camilli, P., Putative receptor for inositol 1,4,5-trisphosphate similar to ryanodine receptor, *Nature* 342 (6246), 192–5, 1989.
13. Ross, C.A., Danoff, S.K., Schell, M.J., Snyder, S.H., and Ullrich, A., Three additional inositol 1,4,5-trisphosphate receptors: molecular cloning and differential localization in brain and peripheral tissues, *Proc Natl Acad Sci USA* 89 (10), 4265–9, 1992.
14. Sudhof, T.C., Newton, C.L., Archer, B.T., 3rd, Ushkaryov, Y.A., and Mignery, G.A., Structure of a novel InsP$_3$ receptor, *EMBO J* 10 (11), 3199–206, 1991.

15. Danoff, S.K., Ferris, C.D., Donath, C., Fischer, G.A., Munemitsu, S., Ullrich, A., Snyder, S.H., and Ross, C.A., Inositol 1,4,5-trisphosphate receptors: distinct neuronal and nonneuronal forms derived by alternative splicing differ in phosphorylation, *Proc Natl Acad Sci USA* 88 (7), 2951–5, 1991.

16. Ferris, C.D. and Snyder, S.H., IP$_3$ receptors. Ligand-activated calcium channels in multiple forms, *Adv Second Messenger Phosphoprotein Res* 26, 95–107, 1992.

17. Nakagawa, T., Shiota, C., Okano, H., and Mikoshiba, K., Differential localization of alternative spliced transcripts encoding inositol 1,4,5-trisphosphate receptors in mouse cerebellum and hippocampus: *in situ* hybridization study, *J Neurochem* 57 (5), 1807–10, 1991.

18. Patel, S., Joseph, S.K., and Thomas, A.P., Molecular properties of inositol 1,4, 5-trisphosphate receptors, *Cell Calcium* 25 (3), 247–64, 1999.

19. Fujino, I., Yamada, N., Miyawaki, A., Hasegawa, M., Furuichi, T., and Mikoshiba, K., Differential expression of type 2 and type 3 inositol 1,4,5-trisphosphate receptor mRNAs in various mouse tissues: *in situ* hybridization study, *Cell Tissue Res* 280 (2), 201–10, 1995.

20. Furuichi, T., Simon-Chazottes, D., Fujino, I., Yamada, N., Hasegawa, M., Miyawaki, A., Yoshikawa, S., Guenet, J. L., and Mikoshiba, K., Widespread expression of inositol 1,4,5-trisphosphate receptor type 1 gene (Insp3r1) in the mouse central nervous system, *Receptors Channels* 1 (1), 11–24, 1993.

21. De Smedt, H., Missiaen, L., Parys, J. B., Bootman, M.D., Mertens, L., Van Den Bosch, L., and Casteels, R., Determination of relative amounts of inositol trisphosphate receptor mRNA isoforms by ratio polymerase chain reaction, *J Biol Chem* 269 (34), 21691–8, 1994.

22. Sharp, A.H., Nucifora, F.C., Jr., Blondel, O., Sheppard, C.A., Zhang, C., Snyder, S.H., Russell, J.T., Ryugo, D.K., and Ross, C.A., Differential cellular expression of isoforms of inositol 1,4,5-triphosphate receptors in neurons and glia in brain, *J Comp Neurol* 406 (2), 207–20, 1999.

23. Taylor, C.W., Genazzani, A.A., and Morris, S.A., Expression of inositol trisphosphate receptors, *Cell Calcium* 26 (6), 237–51, 1999.

24. Newton, C.L., Mignery, G.A., and Sudhof, T.C., Co-expression in vertebrate tissues and cell lines of multiple inositol 1,4,5-trisphosphate (InsP$_3$) receptors with distinct affinities for InsP$_3$, *J Biol Chem* 269 (46), 28613–9, 1994.

25. De Smedt, H., Missiaen, L., Parys, J.B., Henning, R.H., Sienaert, I., Vanlingen, S., Gijsens, A., Himpens, B., and Casteels, R., Isoform diversity of the inositol trisphosphate receptor in cell types of mouse origin, *Biochem J* 322 (Pt 2), 575–83, 1997.

26. Bush, K.T., Stuart, R.O., Li, S.H., Moura, L.A., Sharp, A.H., Ross, C.A., and Nigam, S.K., Epithelial inositol 1,4,5-trisphosphate receptors. Multiplicity of localization, solubility, and isoforms, *J Biol Chem* 269 (38), 23694–9, 1994.

27. Kume, S., Muto, A., Okano, H., and Mikoshiba, K., Developmental expression of the inositol 1,4,5-trisphosphate receptor and localization of inositol 1,4,5-trisphosphate during early embryogenesis in *Xenopus laevis*, *Mech Dev* 66 (1–2), 157–68, 1997.

28. Shiraishi, K., Okada, A., Shirakawa, H., Nakanishi, S., Mikoshiba, K., and Miyazaki, S., Developmental changes in the distribution of the endoplasmic reticulum and inositol 1,4,5-trisphosphate receptors and the spatial pattern of Ca^{2+} release during maturation of hamster oocytes, *Dev Biol* 170 (2), 594–606, 1995.

29. Wojcikiewicz, R.J., Furuichi, T., Nakade, S., Mikoshiba, K., and Nahorski, S.R., Muscarinic receptor activation down-regulates the type I inositol 1,4,5-trisphosphate receptor by accelerating its degradation, *J Biol Chem* 269 (11), 7963–9, 1994.

30. Magnusson, A., Haug, L.S., Walaas, S.I., and Ostvold, A.C., Calcium-induced degradation of the inositol (1,4,5)-trisphosphate receptor/Ca^{2+}-channel, *FEBS Lett* 323 (3), 229–32, 1993.

31. Lee, B., Gai, W., and Laychock, S.G., Proteasomal activation mediates down-regulation of inositol 1,4,5-trisphosphate receptor and calcium mobilization in rat pancreatic islets, *Endocrinology* 142 (5), 1744–51, 2001.

32. Bokkala, S. and Joseph, S.K., Angiotensin II-induced down-regulation of inositol trisphosphate receptors in WB rat liver epithelial cells. Evidence for involvement of the proteasome pathway, *J Biol Chem* 272 (19), 12454–61, 1997.

33. Wojcikiewicz, R.J., Type I, II, and III inositol 1,4,5-trisphosphate receptors are unequally susceptible to down-regulation and are expressed in markedly different proportions in different cell types, *J Biol Chem* 270 (19), 11678–83, 1995.

34. Zhu, C.C., Furuichi, T., Mikoshiba, K., and Wojcikiewicz, R.J., Inositol 1,4,5-trisphosphate receptor down-regulation is activated directly by inositol 1,4,5-trisphosphate binding. Studies with binding-defective mutant receptors, *J Biol Chem* 274 (6), 3476–84, 1999.

35. Yoshida, Y. and Imai, S., Structure and function of inositol 1,4,5-trisphosphate receptor, *Jpn J Pharmacol* 74 (2), 125–37, 1997.

36. Monkawa, T., Miyawaki, A., Sugiyama, T., Yoneshima, H., Yamamoto-Hino, M., Furuichi, T., Saruta, T., Hasegawa, M., and Mikoshiba, K., Heterotetrameric complex formation of inositol 1,4,5-trisphosphate receptor subunits, *J Biol Chem* 270 (24), 14700–4, 1995.

37. Joseph, S.K., Boehning, D., Pierson, S., and Nicchitta, C.V., Membrane insertion, glycosylation, and oligomerization of inositol trisphosphate receptors in a cell-free translation system, *J Biol Chem* 272 (3), 1579–88, 1997.

38. Joseph, S.K., Bokkala, S., Boehning, D., and Zeigler, S., Factors determining the composition of inositol trisphosphate receptor hetero-oligomers expressed in COS cells, *J Biol Chem* 275 (21), 16084–90, 2000.

39. Joseph, S.K., Lin, C., Pierson, S., Thomas, A.P., and Maranto, A.R., Heteroligomers of type-I and type-III inositol trisphosphate receptors in WB rat liver epithelial cells, *J Biol Chem* 270 (40), 23310–6, 1995.

40. Michikawa, T., Hamanaka, H., Otsu, H., Yamamoto, A., Miyawaki, A., Furuichi, T., Tashiro, Y., and Mikoshiba, K., Transmembrane topology and sites of N-glycosylation of inositol 1,4,5-trisphosphate receptor, *J Biol Chem* 269 (12), 9184–9, 1994.

41. Yoshikawa, F., Morita, M., Monkawa, T., Michikawa, T., Furuichi, T., and Mikoshiba, K., Mutational analysis of the ligand binding site of the inositol 1,4,5-trisphosphate receptor, *J Biol Chem* 271 (30), 18277–84, 1996.

42. Yoshikawa, F., Iwasaki, H., Michikawa, T., Furuichi, T., and Mikoshiba, K., Cooperative formation of the ligand-binding site of the inositol 1,4, 5-trisphosphate receptor by two separable domains, *J Biol Chem* 274 (1), 328–34, 1999.

43. Bosanac, I., Alattia, J.R., Mal, T.K., Chan, J., Talarico, S., Tong, F.K., Tong, K.I., Yoshikawa, F., Furuichi, T., Iwai, M., Michikawa, T., Mikoshiba, K., and Ikura, M., Structure of the inositol 1,4,5-trisphosphate receptor binding core in complex with its ligand, *Nature* 420 (6916), 696–700, 2002.

44. Mak, D.-O.D., McBride, S., and Foskett, J.K., ATP regulation of type 1 inositol 1,4, 5-trisphosphate receptor channel gating by allosteric tuning of Ca^{2+} activation, *J Biol Chem* 274 (32), 22231–7, 1999.

45. Mak, D.-O.D., McBride, S., and Foskett, J. K., ATP regulation of recombinant type 3 inositol 1,4,5-trisphosphate receptor gating, *J Gen Physiol* 117 (5), 447–56, 2001.

46. Mak, D.-O.D., McBride, S., and Foskett, J.K., ATP-dependent adenophostin activation of inositol 1,4,5-trisphosphate receptor channel gating: kinetic implications for the durations of calcium puffs in cells, *J Gen Physiol* 117 (4), 299–314, 2001.

47. Penner, R., A practical guide to patch clamping, in *Single-Channel Recording*, 2nd Edition, eds., Sakmann, B. and Neher, E. Plenum Press, New York, 1995, 3–30.

48. Jonas, E.A., Knox, R.J., and Kaczmarek, L.K., Giga-ohm seals on intracellular membranes: a technique for studying intracellular ion channels in intact cells, *Neuron* 19 (1), 7–13, 1997.

49. Bezprozvanny, I., Watras, J., and Ehrlich, B.E., Bell-shaped calcium-response curves of Ins(1,4,5)P$_3$ and calcium-gated channels from endoplasmic reticulum of cerebellum, *Nature* 351 (6329), 751–4, 1991.

50. Watras, J., Bezprozvanny, I., and Ehrlich, B.E., Inositol 1,4,5-trisphosphate-gated channels in cerebellum: presence of multiple conductance states, *J Neurosci* 11 (10), 3239–45, 1991.

51. Ehrlich, B.E. and Watras, J., Inositol 1,4,5-trisphosphate activates a channel from smooth muscle sarcoplasmic reticulum, *Nature* 336 (6199), 583–6, 1988.

52. Thrower, E.C., Park, H.Y., So, S.H., Yoo, S.H., and Ehrlich, B.E., Activation of the inositol 1,4,5-trisphosphate receptor by the calcium storage protein chromogranin A, *J Biol Chem* 277 (18), 15801–6, 2002.

53. Ramos, J., Jung, W., Ramos-Franco, J., Mignery, G.A., and Fill, M., Single channel function of inositol 1,4,5-trisphosphate receptor type-1 and -2 isoform domain-swap chimeras, *J Gen Physiol* 121 (5), 399–411, 2003.

54. Tu, H., Wang, Z., Nosyreva, E., De Smedt, H., and Bezprozvanny, I., Functional characterization of mammalian inositol 1,4,5-trisphosphate receptor isoforms, *Biophys J* 88 (2), 1046–55, 2005.

55. Dingwall, C. and Laskey, R., The nuclear membrane, *Science* 258 (5084), 942–7, 1992.

56. Mak, D.-O.D. and Foskett, J.K., Single-channel inositol 1,4,5-trisphosphate receptor currents revealed by patch clamp of isolated *Xenopus* oocyte nuclei, *J Biol Chem* 269 (47), 29375–8, 1994.

57. Goldberg, M.W. and Allen, T.D., The nuclear pore complex: three-dimensional surface structure revealed by field emission, in-lens scanning electron microscopy, with underlying structure uncovered by proteolysis, *J Cell Sci* 106 (Pt 1), 261–74, 1993.

58. Mazzanti, M., DeFelice, L.J., Cohn, J., and Malter, H., Ion channels in the nuclear envelope, *Nature* 343 (6260), 764–7, 1990.

59. Zeng, W.Z., Mak, D.-O.D., Li, O., Shin, D.M., Foskett, J.K., and Muallem, S., A new mode of Ca^{2+} signaling by G protein-coupled receptors: gating of IP$_3$ receptor Ca^{2+} release channels by G beta gamma, *Curr Biol* 13 (10), 872–76, 2003.

60. Mak, D.-O.D., McBride, S.M., Petrenko, N.B., and Foskett, J.K., Novel regulation of calcium inhibition of the inositol 1,4,5-trisphosphate receptor calcium-release channel, *J Gen Physiol* 122 (5), 569–81, 2003.

61. Mak, D.-O.D., McBride, S.M., and Foskett, J.K., Spontaneous channel activity of the inositol 1,4,5-trisphosphate (InsP$_3$) receptor (InsP$_3$R). Application of allosteric modeling to calcium and InsP$_3$ regulation of InsP$_3$R single-channel gating, *J Gen Physiol* 122 (5), 583–603, 2003.

62. Mak, D.-O.D., and Foskett, J.K., Single-channel kinetics, inactivation, and spatial distribution of inositol trisphosphate (IP$_3$) receptors in *Xenopus* oocyte nucleus, *J Gen Physiol* 109 (5), 571–87, 1997.

63. Mak, D.-O.D., and Foskett, J.K., Effects of divalent cations on single-channel conduction properties of *Xenopus* IP$_3$ receptor, *Am J Physiol* 275 (1 Pt 1), C179–88, 1998.

64. Mak, D.-O.D., McBride, S., Raghuram, V., Yue, Y., Joseph, S.K., and Foskett, J.K., Single-channel properties in endoplasmic reticulum membrane of recombinant type 3 inositol trisphosphate receptor, *J Gen Physiol* 115 (3), 241–56, 2000.

65. Mak, D.-O.D., McBride, S., and Foskett, J.K., Regulation by Ca^{2+} and inositol 1,4, 5-trisphosphate (InsP₃) of single recombinant type 3 InsP₃ receptor channels. Ca^{2+} activation uniquely distinguishes types 1 and 3 Ins P₃ receptors, *J Gen Physiol* 117 (5), 435–46, 2001.

66. Boehning, D., Mak, D.-O.D., Foskett, J.K., and Joseph, S.K., Molecular determinants of ion permeation and selectivity in inositol 1,4,5-trisphosphate receptor Ca^{2+} channels, *J Biol Chem* 276 (17), 13509–12, 2001.

67. Boehning, D., Joseph, S.K., Mak, D.-O.D., and Foskett, J.K., Single-channel recordings of recombinant inositol trisphosphate receptors in mammalian nuclear envelope, *Biophys J* 81 (1), 117–24, 2001.

68. Yang, J., McBride, S., Mak, D.-O.D., Vardi, N., Palczewski, K., Haeseleer, F., and Foskett, J.K., Identification of a family of calcium sensors as protein ligands of inositol trisphosphate receptor Ca^{2+} release channels, *Proc Natl Acad Sci U.S.A* 99 (11), 7711–6, 2002.

69. Soreq, H. and Seidman, S., *Xenopus* oocyte microinjection: from gene to protein, *Methods Enzymol* 207, 225–65, 1992.

70. Peng, H.B., *Xenopus laevis*: practical uses in cell and molecular biology. Solutions and protocols, *Methods Cell Biol* 36, 657–62, 1991.

71. Camacho, P. and Lechleiter, J.D., *Xenopus* oocytes as a tool in calcium signaling research, in *Calcium Signaling*, 1st edition, Putney, J.W., Jr. Ed., CRC Press, Boca Raton, 2000, 157–181.

72. Stühmer, W. and Parekh, A.B., Electrophysiological recordings from *Xenopus* oocytes, in *Single-Channel Recording*, 2nd edition, Sakmann, B. and Neher, E. Eds., Plenum Press, New York, 1995, 341–356.

73. Dumont, J.N., Oogenesis in *Xenopus laevis* (Daudin). I. Stages of oocyte development in laboratory maintained animals, *J Morphol* 136 (2), 153–79, 1972.

74. Goldin, A.L., Maintenance of *Xenopus laevis* and oocyte injection, *Methods Enzymol* 207, 266–79, 1992.

75. Parys, J.B., Sernett, S.W., DeLisle, S., Snyder, P.M., Welsh, M.J., and Campbell, K.P., Isolation, characterization, and localization of the inositol 1,4, 5-trisphosphate receptor protein in *Xenopus laevis* oocytes, *J Biol Chem* 267 (26), 18776–82, 1992.

76. Kume, S., Muto, A., Aruga, J., Nakagawa, T., Michikawa, T., Furuichi, T., Nakade, S., Okano, H., and Mikoshiba, K., The *Xenopus* IP₃ receptor: structure, function, and localization in oocytes and eggs, *Cell* 73 (3), 555–70, 1993.

77. Callamaras, N. and Parker, I., Inositol 1,4,5-trisphosphate receptors in *Xenopus laevis* oocytes: localization and modulation by Ca^{2+}, *Cell Calcium* 15 (1), 66–78, 1994.

78. Mak, D.-O.D., McBride, S., and Foskett, J.K., Inositol 1,4,5-trisphosphate activation of inositol trisphosphate receptor Ca^{2+} channel by ligand tuning of Ca^{2+} inhibition, *Proc Natl Acad Sci USA* 95 (26), 15821–5, 1998.

79. Baudet, S., Hove-Madsen, L., and Bers, D.M., How to make and use calcium-specific mini- and microelectrodes, *Methods Cell Biol* 40, 93–113, 1994.

80. Sigel, E. and Affolter, H., Preparation and utilization of an ion-specific calcium minielectrode, *Methods Enzymol* 141, 25–36, 1987.

81. Schefer, U., Ammann, D., Pretsch, E., Oesch, U., and Simon, W., Neutral carrier based Ca^{2+}-selective electrode with detection limit in the sub-nanomolar range, *Anal Chem* 58, 2282–5, 1986.

82. Raghu, P., Habib, S., Hasnain, S.E., and Hasan, G., Development of a functional assay for Ca^{2+} release activity of IP_3R and expression of an IP_3R gene fragment in the baculovirus-insect cell system, *Gene* 190 (1), 151–6, 1997.

83. Knight, P.J., Pfeifer, T.A., and Grigliatti, T.A., A functional assay for G-protein-coupled receptors using stably transformed insect tissue culture cell lines, *Anal Biochem* 320 (1), 88–103, 2003.

84. Schreurs, J., Yamamoto, R., Lyons, J., Munemitsu, S., Conroy, L., Clark, R., Takeda, Y., Krause, J.E., and Innis, M., Functional wild-type and carboxyl-terminal-tagged rat substance P receptors expressed in baculovirus-infected insect Sf9 cells, *J Neurochem* 64 (4), 1622–31, 1995.

85. Baba, T.W., Giroir, B.P., and Humphries, E.H., Cell lines derived from avian lymphomas exhibit two distinct phenotypes, *Virology* 144 (1), 139–51, 1985.

86. Buerstedde, J.M. and Takeda, S., Increased ratio of targeted to random integration after transfection of chicken B cell lines, *Cell* 67 (1), 179–88, 1991.

87. Sugawara, H., Kurosaki, M., Takata, M., and Kurosaki, T., Genetic evidence for involvement of type 1, type 2 and type 3 inositol 1,4,5-trisphosphate receptors in signal transduction through the B-cell antigen receptor, *EMBO J* 16 (11), 3078–88, 1997.

88. Miyakawa, T., Mizushima, A., Hirose, K., Yamazawa, T., Bezprozvanny, I., Kurosaki, T., and Iino, M., Ca^{2+}-sensor region of IP_3 receptor controls intracellular Ca^{2+} signaling, *EMBO J* 20 (7), 1674–80, 2001.

89. Hille, B., *Ionic Channels of Excitable Membranes*, 2nd ed. Sinauer Associates, Sunderland, MA, 1995.

90. Ramos-Franco, J., Galvan, D., Mignery, G.A., and Fill, M., Location of the permeation pathway in the recombinant type 1 inositol 1,4,5-trisphosphate receptor, *J Gen Physiol* 114 (2), 243–50, 1999.

91. Combettes, L., Claret, M., and Champeil, P., Calcium control of $InsP_3$-induced discharge of calcium from permeabilised hepatocyte pools, *Cell Calcium* 14 (4), 279–92, 1993.

92. Finch, E.A., Turner, T.J., and Goldin, S.M., Calcium as a coagonist of inositol 1,4,5-trisphosphate-induced calcium release, *Science* 252 (5004), 443–6, 1991.

93. Iino, M. and Tsukioka, M., Feedback control of inositol trisphosphate signalling by calcium, *Mol Cell Endocrinol* 98 (2), 141–6, 1994.

94. Ilyin, V. and Parker, I., Role of cytosolic Ca^{2+} in inhibition of $InsP_3$-evoked Ca^{2+} release in *Xenopus* oocytes, *J Physiol* 477 (Pt 3), 503–9, 1994.

95. Parker, I. and Ivorra, I., Inhibition by Ca^{2+} of inositol trisphosphate-mediated Ca^{2+} liberation: a possible mechanism for oscillatory release of Ca^{2+}, *Proc Natl Acad Sci USA* 87 (1), 260–4, 1990.

96. Taylor, C.W., Inositol trisphosphate receptors: Ca^{2+}-modulated intracellular Ca^{2+} channels, *Biochim Biophys Acta* 1436 (1–2), 19–33, 1998.

97. Sienaert, I., De Smedt, H., Parys, J.B., Missiaen, L., Vanlingen, S., Sipma, H., and Casteels, R., Characterization of a cytosolic and a luminal Ca^{2+} binding site in the type I inositol 1,4,5-trisphosphate receptor, *J Biol Chem* 271 (43), 27005–12, 1996.

98. Sienaert, I., Missiaen, L., De Smedt, H., Parys, J.B., Sipma, H., and Casteels, R., Molecular and functional evidence for multiple Ca^{2+}-binding domains in the type 1 inositol 1,4,5-trisphosphate receptor, *J Biol Chem* 272 (41), 25899–906, 1997.

99. Mignery, G.A., Johnston, P.A., and Sudhof, T.C., Mechanism of Ca^{2+} inhibition of inositol 1,4,5-trisphosphate ($InsP_3$) binding to the cerebellar $InsP_3$ receptor, *J Biol Chem* 267 (11), 7450–5, 1992.

100. Lin, C., Widjaja, J., and Joseph, S.K., The interaction of calmodulin with alternatively spliced isoforms of the type-I inositol trisphosphate receptor, *J Biol Chem* 275 (4), 2305–11, 2000.

101. Michikawa, T., Hirota, J., Kawano, S., Hiraoka, M., Yamada, M., Furuichi, T., and Mikoshiba, K., Calmodulin mediates calcium-dependent inactivation of the cerebellar type 1 inositol 1,4,5-trisphosphate receptor, *Neuron* 23 (4), 799–808, 1999.

102. Sienaert, I., Nadif Kasri, N., Vanlingen, S., Parys, J.B., Callewaert, G., Missiaen, L., and De Smedt, H., Localization and function of a calmodulin-apocalmodulin-binding domain in the N-terminal part of the type 1 inositol 1,4,5-trisphosphate receptor, *Biochem J* 365 (Pt 1), 269–277, 2002.

103. Yamada, M., Miyawaki, A., Saito, K., Nakajima, T., Yamamoto-Hino, M., Ryo, Y., Furuichi, T., and Mikoshiba, K., The calmodulin-binding domain in the mouse type 1 inositol 1,4,5-trisphosphate receptor, *Biochem J* 308 (Pt 1), 83–8, 1995.

104. Zhang, X. and Joseph, S.K., Effect of mutation of a calmodulin binding site on Ca^{2+} regulation of inositol trisphosphate receptors, *Biochem J* 360 (Pt 2), 395–400, 2001.

105. Adkins, C.E., Morris, S.A., De Smedt, H., Sienaert, I., Torok, K., and Taylor, C.W., Ca^{2+}-calmodulin inhibits Ca^{2+} release mediated by type-1, -2 and -3 inositol trisphosphate receptors, *Biochem J* 345 (Pt 2), 357–63, 2000.

106. Cardy, T.J. and Taylor, C.W., A novel role for calmodulin: Ca^{2+}-independent inhibition of type-1 inositol trisphosphate receptors, *Biochem J* 334 (Pt 2), 447–55, 1998.

107. Hirota, J., Furuichi, T., and Mikoshiba, K., Inositol 1,4,5-trisphosphate receptor type 1 is a substrate for caspase-3 and is cleaved during apoptosis in a caspase-3-dependent manner, *J Biol Chem* 274 (48), 34433–7, 1999.

108. Missiaen, L., DeSmedt, H., Bultynck, G., Vanlingen, S., Desmet, P., Callewaert, G., and Parys, J.B., Calmodulin increases the sensitivity of type 3 inositol-1,4,5-trisphosphate receptors to Ca^{2+} inhibition in human bronchial mucosal cells, *Mol Pharmacol* 57 (3), 564–7, 2000.

109. Missiaen, L., Parys, J.B., Weidema, A.F., Sipma, H., Vanlingen, S., De Smet, P., Callewaert, G., and De Smedt, H., The bell-shaped Ca^{2+} dependence of the inositol 1,4, 5-trisphosphate-induced Ca^{2+} release is modulated by Ca^{2+}/calmodulin, *J Biol Chem* 274 (20), 13748–51, 1999.

110. Sipma, H., De Smet, P., Sienaert, I., Vanlingen, S., Missiaen, L., Parys, J.B., and De Smedt, H., Modulation of inositol 1,4,5-trisphosphate binding to the recombinant ligand-binding site of the type-1 inositol 1,4,5-trisphosphate receptor by Ca^{2+} and calmodulin, *J Biol Chem* 274 (17), 12157–62, 1999.

111. Vanlingen, S., Sipma, H., De Smet, P., Callewaert, G., Missiaen, L., De Smedt, H., and Parys, J. B., Ca^{2+} and calmodulin differentially modulate myo-inositol 1,4, 5-trisphosphate (IP$_3$)-binding to the recombinant ligand-binding domains of the various IP$_3$ receptor isoforms, *Biochem J* 346 (Pt 2), 275–80, 2000.

112. Nosyreva, E., Miyakawa, T., Wang, Z.N., Glouchankova, L., Mizushima, A., Iino, M., and Bezprozvanny, I., The high-affinity calcium-calmodulin-binding site does not play a role in the modulation of type 1 inositol 1,4,5-trisphosphate receptor function by calcium and calmodulin, *Biochem J* 365, 659–667, 2002.

11 Ryanodine Receptors

Xander H.T. Wehrens, Stephan E. Lehnart, and Andrew R. Marks

CONTENTS

11.1 INTRODUCTION

Intracellular calcium release channels are present on sarcoplasmic and endoplasmic reticuli (SR, ER) of all cell types. There are two classes of these channels: ryanodine receptors (RyR) and inositol 1,4,5-trisphosphate receptors (IP$_3$R). RyRs are required for excitation–contraction (EC) coupling in striated (cardiac and skeletal) muscles. RyRs are made up of macromolecular signaling complexes that contain large cytoplasmic domains, which serve as scaffolds for proteins that regulate the function of the channel. These regulatory proteins include calstabin1

(FKBP12) and calstabin2 (FKBP12.6), 12 kDa subunits that stabilize the closed state of the channel and prevent aberrant calcium leak from the SR. Kinases and phosphatases are targeted to RyR2 channels and modulate RyR2 function in response to extracellular signals. Mutations in RyR genes have been associated with human diseases, including malignant hyperthermia, central core disease, and cardiac arrhythmia syndromes. In addition, defective regulation of RyRs may contribute to abnormal intracellular calcium release in heart failure. Recent studies in RyR and calstabin knockout mice have enhanced our understanding of the physiological roles of these proteins in cardiac development and EC coupling.

The sarco(endo)plasmic reticulum plays a critical role in intracellular Ca^{2+} handling, and enables triggered Ca^{2+} release deep inside the cell. Intracellular Ca^{2+} release is required for many cellular processes, including hormone secretion, fertilization, synaptic transmission, and excitation–contraction (EC) coupling in skeletal and cardiac muscle. In resting nonstimulated cells, cytosolic Ca^{2+} levels are typically low and rise approximately 10-fold upon stimulation following opening of CRC channels.

The Ca^{2+} release channels on the ER/SR organelles have evolved to be quite distinct from all other known ion channels. First, they are much larger than voltage-gated ion channels with a molecular weight of 260 kDa (inositol 1,4,5-trisphosphate receptor, IP_3R) to 560 kDa (ryanodine receptor, RyR). They are nonselective, high-conductance channels that function as Ca^{2+} release channels by virtue of their localization to organelles that contain very high concentrations of Ca^{2+} (millimolar concentrations). RyRs and IP_3Rs share structural features, including large N-terminal domains that form scaffolds for channel regulatory proteins in the cytoplasm, as well as Ca^{2+} channel transmembrane (TM) and pore-forming regions near the C-terminus. The precise number of TM segments per protein monomer has been controversial, as four-, six-, and ten-segment models have been proposed. Using site-specific RyR1 antibodies, Grunwald and Meissner demonstrated that RyR has an even number of TM segments, and that the carboxyl-terminus is located in the cytoplasm [1]. In more recent studies, MacLennan et al. defined the topology of RyR1 by fusing enhanced GFP (EGFP) in-frame to the carboxyl-terminus of recombinant RyR1, replacing a series of carboxyl-terminal deletions that started near the beginning or the end of predicted TM helices M1–M10 [2]. The constructs were expressed in HEK-293 (human embryonic kidney cell line 293) or mouse embryonic fibroblast (MEF) cells, and confocal microscopy of intact and saponin-permeabilized cells was used to determine the subcellular location of the truncated fusion proteins. Their results support a six-TM model, with two intervening membrane–associated loops [2]. For more detailed review, please refer to Du et al. [3].

The gene family of Ca^{2+} release channels comprises three isoforms of RyR and three IP_3Rs in vertebrates. In this chapter, we will focus on the RyR/Ca^{2+} release channels. We will review how a wide variety of biophysical, biochemical, and electrophysiological techniques have enhanced our understanding of the structure and regulation of ryanodine receptors. For a more comprehensive review of the current knowledge about RyRs, we refer the interested reader to several recent review articles [4–8].

11.2 RYANODINE RECEPTOR GENES AND ISOFORMS

Ryanodine receptors were first cloned from mammalian skeletal and cardiac muscle using peptide sequencing and cDNA library screening [9–11]. Analysis of the nucleotide sequences revealed that these two RyR subtypes are about 66% homologous [9, 11]. A third mammalian RyR isoform was cloned from rabbit brain and a mink lung epithelial cell line [12, 13]. Similar RyR isoforms have been cloned from nonmammalian vertebrate species (e.g., chicken, bullfrog, and fish) [14–16], although these species seem to express only two distinct isoforms. The nonmammalian vertebrate RyRα isoforms are homologous to the mammalian skeletal muscle isoform RyR1, while the RyRβ isoforms are more related to mammalian RyR3 [14,16]. Although RyRs were originally discovered in vertebrates, more recently they have been identified in invertebrates, including *Caenorhabditis elegans* and *Drosophila melanogaster* [17, 18].

In addition to gene duplication, additional means of generating diversity in RyR channels involves alternative splicing. Alternative splicing has been demonstrated for several vertebrate [19–21] and invertebrate RyR isoforms [22]. Considering that invertebrate species appear to have only 1 RyR gene, in contrast to the three mammalian RyR genes, it is likely that invertebrates and vertebrates used fundamentally different means of generating diversity in RyR function. Indeed, there is evidence for alternative splicing of the insect *H. virescens RyR* gene, resulting in functionally different channels [23]. Thus, the major means of generating diversity in invertebrate RyR channels may involve alternative splicing, whereas vertebrate RyR diversity may result primarily from the presence of multiple genes.

11.2.1 DIFFERENTIAL EXPRESSION OF RyR ISOFORMS

RNA blot hybridization analyses demonstrated that three RyR isoforms were differentially expressed in various rabbit tissues [13]. However, ribonuclease protection assays and *in situ* hybridization have shown that all three types of isoforms are widely expressed, and indicate that most tissues express more than one type of isoform [24]. Expression of RyR1 is relatively abundant in skeletal muscle, although it is also expressed at lower levels in cardiac and smooth muscle, cerebellum, testis, adrenal gland, and ovaries [9, 10, 14, 25, 26]. Whereas RyR1 is predominantly expressed in Purkinje cells in the brain, RyR2 is localized mainly in the somata of most neurons [25]. RyR2 is expressed robustly in the heart and brain, and at lower levels in the stomach, lung, thymus, adrenal gland, and ovaries [19, 26]. The RyR3 isoform is expressed in the brain, diaphragm, slow twitch skeletal muscle, as well as several abdominal organs [12–14, 26]. The nonmammalian RyRα isoform is expressed strongly in skeletal muscle and weakly in brain, whereas the RyRβ isoform is expressed in a variety of tissues, including skeletal and cardiac muscle, lung, stomach, and brain [16]. There is some evidence that alternative splicing of RyR genes may underlie the tissue-specific expression of certain isoforms [20].

11.2.2 Mutations in Ryanodine Receptor Genes

Allelic variants of certain RyR genes have been associated with phenotypic effects including genetic abnormalities. Mutations that specifically eliminate RyR1 channels in mice ($skrr^{ml}$) or RyRα in "crooked neck dwarf" chicken (cn) have recessive lethal phenotypes, resulting in perinatal death [27, 28]. In contrast, mice deficient of the *RyR3* gene are viable and exhibit functional EC coupling in skeletal muscle [29]. Missense mutations in the porcine and human *RyR1* genes have been linked to malignant hyperthermia and central core disease, inherited myopathies in which skeletal muscle contracture associated with hypermetabolism and elevation of body temperature are triggered by stress or anesthetics [30]. In humans, two distinct clinical syndromes causing exercise-induced sudden cardiac death, catecholaminergic polymorphic ventricular tachycardia (CPVT) and arrhythmogenic right ventricular cardiomyopathy type 2/dysplasia (ARVC/D-2), have been linked to mutations in the human *RyR2* gene [31–33]. Functional consequences of these arrhythmogenic mutations in *RyR2* will be discussed below (see Section 11.4.3).

11.3 RYANODINE RECEPTOR MACROMOLECULAR CHANNEL COMPLEX

RyRs are large homotetrameric ion channels comprised of four RyR monomers. Protein sequence analysis and computer modeling based upon hydropathy analyses have revealed that each RyR monomer has a large cytoplasmic aminoterminal domain, as well as TM and pore-forming regions near the carboxyl-terminus [2, 9, 11]. The amino-terminal region comprises ~90% of the RyR sequence and encodes a cytoplasmic domain that serves as a scaffold for proteins that modulate the channel function [34–41] (see Section 11.3.2).

11.3.1 Three-Dimensional Structure of Ryanodine Receptors

In retrospect, it is apparent that the darkly staining, rectangular-shaped so-called feet in conventional thin-section electron micrographs of skeletal muscle triad junctions, first visualized several decades ago, are in fact identical to the cytoplasmic domains of RyRs [42]. The biochemical identity of these "feet" was unknown until RyRs were purified and characterized in the late 1980s. In the past decade, cryo-electron microscopy (cryo-EM), in conjunction with computerized single particle image processing, has emerged as a powerful methodology for determining the 3D structure of RyR molecules [43, 44]. Cryo-EM has eliminated the artifacts associated with chemical fixation, dehydration, and contrast enhancement by heavy metals associated with conventional EM in the past. The first 3D reconstructions from micrographs of frozen-hydrated, detergent-solubilized RyRs were reported in the mid-1990s by two research groups [44, 45], and the best resolution of RyR reported to date is 22 Å [46].

The 3D reconstruction of the three RyR isoforms has revealed highly conserved structures, resembling the overall shape of a mushroom and consisting of two major components: a large, square prism-shaped cytoplasmic assembly (~290 × 290 × 130Å) composed of at least 10 distinct domains, and a differentiated small TM assembly of dimensions 120×120×70 Å (Figure 11.1). Sequence analysis [9, 21] and biochemical studies [1, 2] indicate that the amino-terminal ~4,000–4,500 amino acid residues form the large cytoplasmic assembly, while the remaining ~500–1,000 carboxyl-terminal residues comprise the TM regions.

Sequence-specific labels of sufficient size have been used to achieve a more detailed correlation of RyR's amino acid sequence with its 3D architecture. 3D cryo-EM has been utilized to localize the physical binding sites of some ligands that modulate RyR function. Figure 11.1 illustrates results that have been obtained for RyR–FKBP and RyR–calmodulin complexes [47–49]. For these experiments, RyR and ligand are mixed *in vitro* under conditions favoring complex formation, and then applied to specimen grids and quickly frozen for cryo-EM. By comparing reconstructions of RyR with and without ligand, a 3D-difference map can be generated [50].

Recently, the feasibility of expressing recombinant RyRs in mammalian cell lines for structural analysis by cryo-EM has been demonstrated [51]. By inserting autonomously folding peptides at specific sites in the amino acid sequence, the locations of surface-exposed amino-acid residues can be mapped onto the 3D architecture of the receptor [51–54]. Gene fusions of RyRs and glutathione S-transferase (GST) or green fluorescent protein (GFP) have been shown to be useful for 3D cryo-EM. Using this approach, every receptor image contains the label within each of the four RyR subunits, unlike the situation for noncovalently linked ligand/RyR complexes in which ligand dissociation often complicates the analysis.

Thus far, the amino-terminus has been localized on the 3D structure of RyR3 using a GST fusion protein, and amino acids Asp-4365, Thr-1366, and Thr-1874, which lie within the three divergent regions (DR1, DR2, and DR3, respectively)

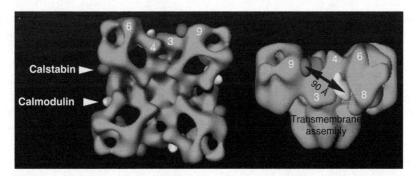

FIGURE 11.1 (See color insert following page 140.) Three-dimensional reconstruction of ryanodine receptor in complex with calmodulin and calstabin. Solid-body representations of 3D reconstruction of RyR together with the differences attributed to calmodulin (yellow) and calstabin (pink). Numerals indicate selected RyR domains. (Modified from Wagenknecht T, Radermacher M, Grassucci R, Berbowitz J, Xin HB, Fluscher S. *J Biol Chem* 1977, 272, 32463–71.)

have been located on the 3D structure of RyR2 using appropriate RyR2–GFP fusion proteins. The three divergent regions are much more variable relative to the average sequence identity among the three RyR isoforms, which is nearly 70% [55]. These divergent regions have been the focus of a number of structural and functional studies, as they are thought to be largely responsible for the differing properties of RyR isoforms. For example, sequence variations in the DR1 region, between RyR1 and RyR2, have been shown to account for these isoforms' differing sensitivities to Ca^{2+} inactivation [56].

11.3.2   The Calcium-Channel Stabilizing Subunit "Calstabin"

Ryanodine receptors are associated with isoform-specific channel-stabilizing sub-units known as calstabins (originally termed FK506-binding proteins, or FKBPs) [57–59]. Calstabin1 (also known as FKBP12) was originally identified as peptide KC7 during peptide sequencing of RyR1 [10]. Using co-immunoprecipita-tions, calstabin1 and calstabin2 (also known as FKBP12.6) have been shown to associate with RyR1 and RyR2, respectively, such that one calstabin protein is bound to each RyR monomer [57, 59–61]. Thus, there are four calstabin molecules bound to each RyR1 and RyR2 channel complex. RyR1 and RyR3 channels can bind both calstabin1 and calstabin2, although the affinity for calstabin1 is higher [59, 60, 62]. Therefore, *in vivo*, RyR1 and RyR3 have calstabin1 bound because of its higher abundance in the cytosol [59, 60]. On the other hand, RyR2 channels exhibit a relatively higher affinity for calstabin2, and bind calstabin2 *in vivo* [60, 63].

Cryo-EM studies of the RyR1 complex show that calstabin1 binds to RyR1 on the outer surface of the cytoplasmic domain [47]. Recent observations suggest that the binding site of calstabin2 on the 3D structure of RyR2 is similar to that of calstabin1 on RyR1 [64]. Site-directed mutagenesis studies have demonstrated that valine 2461 on RyR1 (corresponding to isoleucine 2427 on RyR2) is critical for calstabin1 binding [65]. The bond formed by V2461 and proline 2462 (or I2427-P2428 in RyR2) is likely constrained in a high energy, unstable, twisted-amide transition state intermediate of a peptidyl–prolyl bond (unable to isomerize to either *cis* or *trans* due to presumed steric hindrances in the RyR structure) [66]. It is believed that calstabin binds to this high energy, unstable twisted-amide with high affinity [65]. This model is supported by the finding that introducing flexibility around this peptidyl–prolyl bond with the mutation of V2461 to a glycine residue abolishes binding of both calstabin isoforms to RyR1 [65]. We have proposed that introduction of increased mobility around the peptidyl–prolyl bond by substituting a smaller amino acid, glycine, for either the valine or isoleucine allows for isomeri-zation to proceed by reducing steric hindrance at that site which reduces the binding affinity of calstabin to the channel [65]. In support of this model Bultynck et al. recently concluded, based on molecular modeling studies, that the proline in the calstabin-binding region on RyR induces a bend in the alpha helix, which imposes a twisted amide transition state on the peptidyl–proline bond and enables calstabin to bind to this domain [67]. In addition, it is likely that other regions on RyR are also involved in stabilizing the binding of calstabin to the channel since PKA phosphorylation of RyR2-Ser^{2809} which is 347 residues away from the peptidyl–prolyl bond at 2461–2462 reduces the binding affinity by adding a negative charge resulting in electrostatic repulsion of calstabin2 [68]. In agreement with this model, Masumiya et al. have proposed that other regions within the amino-terminal domain of RyR2 are required for the binding of GST-calstabin2 to RyR2, although they did not identify specific residues on RyR2 involved in calstabin2 binding [69].

11.3.3 OTHER REGULATORY PROTEINS ASSOCIATED WITH RYR

Co-immunoprecipitation experiments have revealed that protein kinases and phosphatases are targeted to the cytoplasmic scaffold domain of IP$_3$Rs and RyRs. Binding of these enzymes allows for rapid and localized modulation of the channel function in response to extracellular signals communicated via second messengers [68, 70, 71]. Specific association between RyR and the associated regulatory proteins was demonstrated using co-immunoprecipitation experiments in which RyRs were solubilized in 0.25–0.5% Triton-X100, and each of the components of the RyR macromolecular complex was shown to co-sediment with RyR on sucrose gradients using CHAPS solubilized membranes [68]. The association of kinases and phosphatases with RyR allows for local regulation of the channel activity [72, 73]. Highly conserved leucine/isoleucine zipper (LIZ) motifs in RyR mediate binding to cognate LIZs in the targeting proteins for kinases and phosphatases [70]. This feature of ion channel macromolecular complexes is shared by both intracellular Ca^{2+} release channels (RyRs and IP$_3$Rs) [70, 71], and by voltage-gated ion channels [74, 75]. The catalytic subunit of PKA (C) as well as its regulatory subunit (RII) are bound to the anchoring protein mAKAP (AKAP6), which in turn is bound to RyR1/RyR2 via LIZ motifs [68]. The protein phosphatase 1 (PP1) is bound to RyR1/RyR2 via spinophilin [70, 76], and protein phosphatase PP2A selectively associates with RyR2 through its adaptor protein PR130 [70]. Finally, we and others have recently shown that the Ca^{2+}/calmodulin-dependent protein kinase II (CaMKII) binds to RyR2, although the exact binding site has not yet been identified [40, 41].

Other proteins that bind to the cytoplasmic domain of RyR include calmodulin (CaM), a 16.7- kDa cytosolic protein that influences SR Ca^{2+} release [77–79], and sorcin, a ubiquitous 22-kDa Ca^{2+}-binding protein reported to associate with both RyR2 and the LTCC [80, 81]. Sorcin may reduce RyR2 open probability, but this effect can be relieved by PKA phosphorylation of sorcin [38]. A functional role for sorcin in the RyR channel complex is less well defined.

RyR also binds proteins at the luminal SR surface, including triadin, junctin, and calsequestrin (Figure 11.2). Junctin [82] and triadrin [83] are presumably involved in anchoring RyR to the SR membrane. Calsequestrin (CSQ) is a major Ca^{2+}-binding protein in the SR, and provides high-capacity low-affinity intraluminal Ca^{2+} binding [84, 85]. It has been suggested that Ca^{2+}-dependent conformational changes in CSQ may modulate RyR channel activity [86], although the exact nature of the CSQ-RyR modulation requires further investigations [87–89].

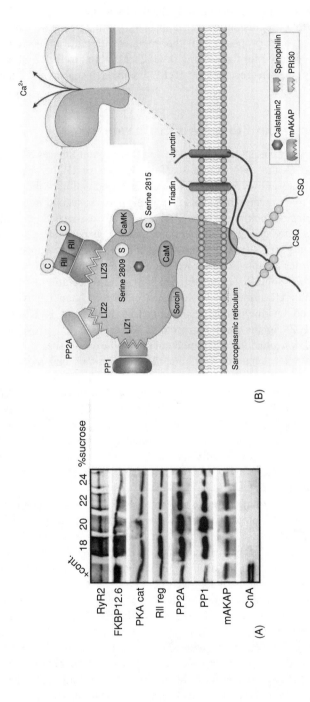

FIGURE 11.2 Ryanodine receptors are macromolecular signaling complexes. (A) Immunoblotting gradient fractions with specific antibodies showed that calstabin2 (FKBP12.6), PKA catalytic subunit, PKA regulatory subunit (RII), PP2A, PP1, and mAKAP, but not calcineurin (CaN) were detected in all fractions containing RyR2, suggesting that these molecules associate with RyR2 [68]. (B) Schematic representation of a ryanodine receptor monomer with associated regulatory subunits (each RyR channel tetramer consists of four identical RyR monomers). (Modified from Wehrens XH, Marks AR. *Nat rev Drug Discov* 2004, 3, 565–73.)

11.4 REGULATION OF RYANODINE RECEPTORS

RyRs are regulated by numerous natural and pharmacological ligands, and by covalent modifications such as phosphorylation, nitrosylation, and oxidation/reduction of cysteine sulfhydryl moieties. The functional effects of these modulations on RyR gating can be studied using measurements of ^{3}H-ryanodine binding or recordings of single channel properties in planar lipid bilayers. Two important physiological mechanisms of RyR regulation (Ca^{2+} and protein phosphorylation) will be discussed below; for more complete reviews of RyR modulation, see Meissner [90] and Fill and Copello. [5].

11.4.1 MODULATION OF RYANODINE RECEPTORS BY CALCIUM

In a seminal paper, Fabiato [91] demonstrated that SR Ca^{2+} release was a bell-shaped function of the trigger Ca^{2+} stimulus amplitude. The bimodal Ca^{2+} dependence suggested the presence of low-affinity (fast on rate, activation) and high-affinity (slow on rate, inactivation) Ca^{2+}-binding sites that are accessible from the cytoplasmic side. With this combination of two sites, large fast Ca^{2+} signals transiently activate SR Ca^{2+} release because Ca^{2+} would activate before inactivation catches up. Once the inactivation sites are occupied, the Ca^{2+} release process cannot respond to further Ca^{2+} stimulation [92].

Patch-clamp studies on isolated cardiac myocytes have not definitely established that Ca^{2+}-dependent inactivation actually occurs [93, 94]. For example, the effectiveness of a second trigger Ca^{2+} stimulus did not decrease when two stimuli were applied in rapid succession [95]. The identification, isolation, and functional characterization of RyR channels in planar lipid bilayers allowed for a more direct analysis of the cytosolic Ca^{2+} regulation of single RyR channels. Detailed studies of RyR channel opening, closing, and ligand regulation in planar lipid bilayers have led to several models for the termination of Ca^{2+} release [96–102]. Skeletal and cardiac RyR are activated by micromolar cytosolic $[Ca^{2+}]$, and inhibited by millimolar $[Ca^{2+}]$. This decrease in single RyR channel activity at high cytosolic Ca^{2+} levels is often interpreted as the molecular manifestation of Ca^{2+}-dependent inactivation. However, it is unlikely that cytosolic free Ca^{2+} levels of any mammalian cell exceed 1 mM, although local concentrations in restricted spaces may reach this level.

It is also possible that some of the Ca^{2+} that passes through the channel itself interacts with the activation site. This is called feed-through Ca^{2+} activation [103–107]. Consideration of feed-through Ca^{2+} activation is important because most RyR bilayer studies use huge Ca^{2+} driving forces. Whether or not feed-through Ca^{2+} activation has physiologically relevant manifestations in the cell remains to be determined. Laser flash photolysis of caged Ca^{2+} has been used to reveal how single RyR channels respond to more physiological Ca^{2+} stimuli [5, 96]. Both flash photolysis and mechanical solution changes have been applied to single RyR2 channels, and demonstrated that the Ca^{2+} deactivation kinetics of RyR channels allow them to turn off rapidly in response to a fast drop in local Ca^{2+} concentrations

[99, 100, 102]. Another mechanism through which Ca^{2+} can affect RyR function is through adaptation [92, 108, 109]. It appears that the RyR2 adaptation phenomenon is due to a transient shift in the Ca^{2+}-dependent modal gating, leading to a change in the single channel gating pattern.

11.4.2 PHOSPHORYLATION-DEPENDENT REGULATION OF RYANODINE RECEPTORS

The RyR macromolecular complex is regulated by protein phosphorylation that can evoke a repertoire of dynamic responses, including changes in open probability, the appearance of subconductance states, and reduced coordination among subunits within or among RyR tetramers [68, 110]. Based on phospho-peptide mapping, it was shown that protein kinase A (PKA) phosphorylates Serine 2809 on RyR2 [111, 112]. These results have subsequently been confirmed in several studies using GST-fusion proteins, mutations in the full-length recombinant RyR2 channel, and by a phospho-epitope specific antibody demonstrating that Serine 2809 is the unique PKA site on RyR2 [40, 68, 113–115].

There have been conflicting reports concerning the effects of exogenously applied PKA on the single channel behavior of RyR2 [68, 113–118] (Figure 11.3). Some studies have shown that phosphorylation by PKA increases the open probability (Po) of RyR2 by increasing the sensitivity of RyR2 to Ca^{2+}-dependent activation [40, 68, 113, 116], whereas other studies demonstrated first an increase in RyR2 open probability followed by a slight decrease in the steady-state Po of RyR2 channels; however, the PKA-mediated increase in RyR2 Po occurred in a time scale that is relevant to EC coupling whereas the decrease in RyR2 activity occurred after EC coupling would be completed [101]. When Ca^{2+} release events were measured in normal isolated cardiomyocytes, PKA-phosphorylation of RyR2 did not increase the Ca^{2+} spark frequency under conditions that simulate diastole in the heart, when RyR2 is expected to be tightly closed [119]. These data are consistent with the fact that healthy subjects or animals do not develop SR Ca^{2+} leak and arrhythmias during exercise [61, 113]. On the other hand, recent data show that PKA-phosphorylation of RyR2 enhances RyR2 activity and increases EC coupling gain during the early phase of EC coupling when only a small number of voltage-gated Ca^{2+} channels are open [120, 121], consistent with increased intracellular Ca^{2+} release following β-adrenergic stimulation that enhances cardiac contractility during exercise.

Based on phospho-peptide mapping, it was proposed that CaMKII phosphorylates the same residue on RyR2 as PKA (Ser 2809) [111, 112, 115]. Our recent studies, however, have demonstrated that CaMKII phosphorylates RyR2 Ser 2815 [40]. Mutating Ser2815 to alanine in full-length recombinant RyR2 channels abolishes CaMKII phosphorylation [40]. The identification of the unique CaMKII phosphorylation site on RyR2, which was also confirmed using a phospho-epitope specific antibody, has facilitated elucidation of the functional effects of CaMKII phosphorylation of RyR2 [40, 112, 116, 117, 122–124]. Single-channel studies of CaMKII phosphorylated wild-type RyR2 channels and RyR2-Ser2815asp mutant channels, which mimic constitutively CaMKII-phosphorylated RyR2, have shown

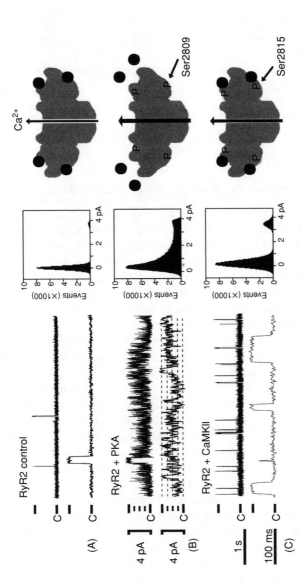

FIGURE 11.3 Effects of PKA or CaMKII phosphorylation on RyR2 channel function and composition of the macromolecular complex under diastolic conditions. (A) Nonphosphorylated RyR2 channel function under diastolic conditions. Histogram shows dominance of the closed state and cartoon represents binding of four calstabin2 subunits under nonphosphorylated conditions. (B) Exercise and stress activate the β-adrenergic receptor signaling pathway and protein kinase A (PKA). PKA phosphorylation of Ser-2809 on RyR2 increases the open probability and decreases the binding affinity for the channel-stabilizing subunit calstabin2, which is associated with partial channel openings. (C) Increasing the heart rate raises intracellular calcium (Ca^{2+}) levels, which activates Ca^{2+}/calmodulin-dependent protein kinase II (CaMKII). Autophosphorylation of CaMKII activates the enzyme, which phosphorylates Ser-2815 on RyR2 and increases the open probability without dissociating calstabin2 and without activating partial openings. (from Lehnart SE, Wehrens XH, Marks AR. *Biochem Biophys Res Commun* 2004, 322, 1267–79. With permission.)

that CaMKII phosphorylation of RyR2 at Ser2815 increases the open probability of the channel by augmenting the sensitivity to Ca^{2+}-dependent activation [40] (Figure 11.3). These results have been confirmed in single cardiomyocytes isolated from transgenic mice overexpressing CaMKIIδ, which results in enhanced CaMKII phosphorylation of RyR2 and an increased diastolic Ca^{2+} spark frequency [123, 124]. In contrast to phosphorylation by PKA, CaMKII-phosphorylation does not dissociate calstabin2 from the RyR2 channel [40].

The activity of RyR2 is also regulated by protein phosphatases that are targeted to the macromolecular channel complex [70, 117]. Terentyev et al. have reported that protein phosphatase 1 (PP1) may increase RyR2 activity, although this study is in contrast with several other studies that reported that PP1 decreases RyR2 activity [118, 125–129], and that PKA/CaMKII phosphorylation increases RyR2 activity [40, 68, 116, 123].

11.4.3 ABNORMAL RyR2 REGULATION IN CARDIAC ARRHYTHMIAS

Autosomal-dominant inherited missense mutations in the *RyR2* gene have been linked to exercise-induced arrhythmias, known as catecholaminergic polymorphic ventricular tachycardia (CPVT) [31, 32, 130]. Mutation carriers characteristically develop arrhythmias during exercise or emotional stress but not during rest. *RyR2* missense mutations have also been linked to "arrhythmogenic right ventricular dysplasia/cardiomyopathy type 2" (ARVD/C2) [33]. Importantly, arrhythmias in CPVT mutation carriers occur in the structurally normal heart, whereas ARDV/C2 is associated with abnormalities of the right ventricle.

Planar lipid bilayer studies of single RyR2 channels containing missense mutations found in CPVT patients demonstrated a gain-of-function defect following PKA phosphorylation, consistent with "leaky" RyR2 channels (diastolic SR Ca^{2+} leak) during stress/exercise [113, 131]. Consistent with these findings, intracellular Ca^{2+} leak was observed after β-adrenergic stimulation in atrial tumor cells expressing the same CPVT mutant RyR2 channels [132]. The observation that RyR2 missense mutations only result in leaky Ca^{2+} release channels under conditions that mimic activation of the β-adrenergic signaling cascade [113, 131] is in agreement with the finding in affected mutation carriers that arrhythmias only occur during stress or exercise (Figure 11.4) [130, 133].

An important finding in CPVT mutant RyR2 channels is a decreased binding affinity of the channel-stabilizing subunit calstabin2. For these experiments, wild-type or CPVT-mutant RyR2 were incubated with 20 to 1000 nM (final concentration) [^{35}S]-labeled calstabin2, followed by liquid scintillation counting of the precipitated RyR2 complex [113]. Enhanced dissociation of calstabin2 from the RyR2 channel complex during PKA phosphorylation may lead to arrhythmogenic diastolic SR Ca^{2+} leak in CPVT patients [68, 113, 134–136].

Other groups have proposed different mechanisms for RyR2-mediated CPVT arrhythmias [132, 137, 138]. However, these studies have all been done using heterologous expression of mutant channels in HEK293 cells that lack the critical components present in cardiomyocytes. Moreover, these authors have reported that

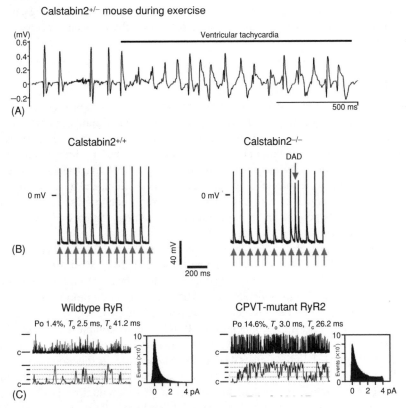

FIGURE 11.4 Reduced calstabin2 binding to RyR2 may cause cardiac arrhythmias due to intracellular Ca^{2+} leak. (A) Calstabin2 haploinsufficient mice (calstabin2$^{+/-}$) are vulnerable to sustained polymorphic ventricular tachycardia during exercise testing or injection with 0.5 mg/kg epinephrine. Reproduced from Wehrens et al. [61]. (B) Single myocytes isolated from wild-type or calstabin2-deficient (−/−) mice, subjected to rapid pacing at 12 Hz, are susceptible to delayed after-depolarizations. Action potentials were recorded in wild-type and calstabin2$^{-/-}$ cardiomyocytes. Upward arrows at the bottom of tracings indicate the timing of the induced action potentials; the downward arrow shows a DAD that produced an extra-systole in a calstabin2$^{-/-}$ cardiomyocyte. Modified from Wehrens et al. [113]. (C) RyR2 mutations associated with catecholaminergic polymorphic ventricular tachycardia (CPVT) in patients with exercise-induced arrhythmias and sudden cardiac death reduce the binding affinity of calstabin2 for RyR2. Following PKA phosphorylation, these RyR2 channels display a gain-of-function defect when studied in planar lipid bilayers. RyR2 activity was studied at cytoplasmic $[Ca^{2+}]$ of 150 nmol/l and $[Mg^{2+}]$ of 1.0 mmol/l representing diastolic conditions in the heart. Channel openings are upward deflections; the difference between horizontal bars indicates 4-pA current amplitude between open and closed states c, and 1 pA subconductance levels are indicated for phosphorylated channels. Examples of upper and lower current tracings represent 5000 ms or 200 ms. T_o, average open time; T_c, average closed time. (Modified from Lehnart et al. [131].)

the mutant channels are defective under unstimulated conditions, which is at odds with the clinical phenotype of the disease that is only manifested during stress. Further studies are warranted to further investigate the arrhythmogenic mechanisms in CPVT.

11.4.4   RYR1 MUTATIONS CAUSING MALIGNANT HYPERTHERMIA

Malignant hyperthermia (MH) is an autosomal-dominant muscle disorder in which genetically susceptible individuals develop skeletal muscle rigidity and high fever in response to inhalation anesthesia (e.g., halothane) and depolarizing muscle relaxants (e.g., succinylcholine) [139]. MH episodes may be life-threatening if not corrected immediately by suspension of administration of the triggering agent, treatment with the RyR channel modulator dantrolene sodium, and hyperventilation with 100% O_2. Central core disease (CCD) is a rare, nonprogressive myopathy characterized by hypotonia and proximal muscle weakness. Patients with CCD are also frequently found to be MH susceptible [140].

Genetic and biochemical data have demonstrated that mutations in RyR1 are the major cause of MH/CCD in swine and humans [30, 141, 142]. Most disease-linked mutations are distributed among three distinct regions of the RyR1 protein, known as MH/CCD region 1 (amino acids 35–614), MH/CCD region 2 (amino acids 2129–2458), and MH/CCD region 3 (C-terminal domain). MH/CCD regions 1 and 2 are located in the large cytosolic domain of the RyR1 channel complex, whereas MH/CCD region 3 contains all of the putative TM segments including the selectivity filter/pore-lining region [2]. Initial studies designed to characterize the functional effects of MH-linked mutations were conducted using muscle biopsies from pigs carrying the R615C RyR1 mutation (equivalent to the human R614C mutation) [143]. These studies found that MH-susceptible muscle exhibited higher ^{3}H-ryanodine binding, an increased sensitivity to activation by micromolar $[Ca^{2+}]$, and a higher resistance to Ca^{2+}-dependent inactivation. These abnormalities may be potentiated by inhalation anesthetics and depolarizing skeletal muscle relaxants, and, thus, result in supersensitive or overactive SR Ca^{2+} release channels [139, 144].

Some RyR1 mutations, particularly those in MH/CCD regions 1 and 2, result in the co-occurrence of both MH and CCD. Heterologous expression in HEK293 cells of RyR1 mutations associated with both MH and CCD leads to both a reduction in luminal endoplasmic reticulum (ER) Ca^{2+} content and an increase in resting cytoplasmic Ca^{2+} levels [145]. Similarly, homologous expression of the same mutant RyR1 proteins in RyR1-deficient (dyspedic) skeletal myotubes also results in SR Ca^{2+} depletion and elevations in resting cytosolic Ca^{2+} levels [146, 147]. These findings suggest that MH/CCD mutations in regions 1 and 2 result in the formation of "leaky" RyR1 channels. In contrast, SR Ca^{2+} depletion does not appear to occur following expression of RyR1 mutants that only result in MH and not CCD in either HEK293 cells [145] or dyspedic myotubes [148]. Thus, significant ER/SR Ca^{2+} leak only occurs for RyR1 disease mutants that result in coincidence of MH and CCD.

Functional measurements of the carboxyl-terminal I4897T mutation in rabbit RyR1 (corresponding to the human CCD-associated I4898T mutation) in HEK293 cells showed that this mutation also promotes Ca^{2+} leak and subsequent store depletion [149]. A similar conclusion was reached in experiments utilizing human B-lymphocytes (which express functional RyR1 Ca^{2+} release channels) obtained from patients with the I4898T CCD mutation [150]. The I4897T CCD mutation was also characterized following homologous expression in dyspedic skeletal myotubes [147, 151]. In contrast to the observations in nonmuscle cells, expression of RyR1-I4897T channels within a skeletal muscle environment did not promote SR Ca^{2+} leak and store depletion. The I4897T mutation was found to exert a dominant-negative reduction in voltage-gated SR Ca^{2+} release, consistent with the autosomal-dominant nature of this disorder. Based on these results, it was concluded that the I4897T mutation reduces Ca^{2+} release during skeletal muscle EC coupling via a mechanism distinct from that of SR Ca^{2+} leak. Rather, the mutation appears to produce a functional uncoupling of excitation from the efficient release of SR Ca^{2+} (termed EC uncoupling).

11.5 CELLULAR BASIS OF EXCITATION–CONTRACTION COUPLING

Electrical depolarization of the cell membrane leads to intracellular Ca^{2+} release and contraction of the myofilaments, a process known as EC coupling [152, 153]. During this process, electrical energy is converted into mechanical energy with Ca^{2+} serving as the second messenger. The initial trigger for EC coupling is generated by depolarization of the plasma membrane, which allows for Ca^{2+} entry through L-type Ca^{2+} channels (LTCCs) located on the transverse (T) tubules in cardiomyocytes or a conformational change of LTCC domains interacting with RyR1 in skeletal myocytes. Intracellular Ca^{2+} release from the SR via RyR channels, which elevates cytosolic Ca^{2+} concentrations from ~100 nM during diastole to about 1 μM during systole. This elevation in cytosolic $[Ca^{2+}]$ activates cardiac or skeletal muscle contraction, respectively.

11.5.1 EC Coupling in Cardiomyocytes

In cardiac myocytes, as opposed to skeletal muscle fibers, membrane depolarization induces Ca^{2+} entry through LTCCs. Ca^{2+} entry into cardiomyocytes was found absolutely necessary to trigger intracellular Ca^{2+} release and subsequent contraction using isolated heart or cardiomyocyte preparations [154]. The amplitude of the LTCC Ca^{2+} current (I_{Ca}) depends on the membrane voltage (E_m) resulting in a bell-shaped relationship [155, 156]. Importantly, the amplitude of the intracellular Ca^{2+} transient, which reflects SR Ca^{2+} release via RyR2, is graded by the LTCC Ca^{2+} influx [91, 153]. As a result, the bell-shaped dependence of I_{Ca} or the intracellular Ca^{2+} transient and E_m characterizes the relationship between the trigger Ca^{2+} and SR Ca^{2+} release via RyR2, which is better known as Ca^{2+}-induced Ca^{2+} release (CICR).

Electron microscopy and immunolabeling studies have identified intracellular junctions called dyads that represent close approximations of plasma membrane T-tubules and SR terminal cisternae containing LTCCs and RyR2s in heart muscle, respectively [157, 158]. Within the cardiac dyad, approximately 1 LTCC and 5–10 RyR2 channels are functionally coupled. No direct protein–protein interaction between LTCC and RyR2 was observed in the heart [158–160]. The structural arrangement of LTCCs and RyR2s in dyads allows for high-gain amplification of the LTCC trigger signal and for local control of graded SR Ca^{2+} release during CICR.

Using confocal microscopy and intracellular Ca^{2+} indicators in cardiomyocytes, cytosolic Ca^{2+} transients were observed to represent the summation of 10^4 up to 10^6 smaller intracellular Ca^{2+} release events called Ca^{2+} sparks [161–163]. Ca^{2+} sparks are optical images of elementary Ca^{2+} release events that are thought to occur from highly organized clusters of RyR2 Ca^{2+} release channels referred to as Ca^{2+} release units (CRUs) [160]. Ca^{2+} sparks are characterized by local refractoriness indicating that local SR Ca^{2+} release is independent of the duration of the LTCC-mediated Ca^{2+} influx [164, 165]. CICR results in Ca^{2+} sparks of similar or homogenous dimensions that are localized in the spatially restricted cleft between the membranes of the T-tubules and the junctional SR [160, 164, 166]. Ca^{2+} sparks are generated by multiples of RyR [167, 168]. Multi-RyR2 Ca^{2+} sparks enable a thermodynamically irreversible mode of channel gating and a supramolecular negative feedback mechanism that prevents regenerative forms of CICR [167]. A hundred or more RyR2 channels may become functionally coupled in a CRU in order to open and close simultaneously by the process of coupled gating which depends on calstabin2 [110].

11.5.2   EC Coupling in Skeletal Muscle Fibers

RyR1 is the major Ca^{2+} release channel isoform expressed in skeletal muscle [9], although low levels of RyR3 are also detected in slow-twitch skeletal muscle [169]. It has been proposed that RyR1 and RyR3 play distinct roles in skeletal muscle CRUs [170]. Skeletal muscle contraction is strongly E_m dependent as shown by voltage-clamp experiments in myofibers [171]. Even in the absence of extracellular Ca^{2+}, membrane depolarization is directly converted into intracellular Ca^{2+} release as evidenced by intracellular Ca^{2+} indicators [172]. Specific pharmacologic inhibition using Ca^{2+} channel blockers showed that activation of RyR1 is coupled to conformational changes in the pore-forming α_{1S} subunit of the skeletal muscle LTCC isoform. Fast voltage sensing via direct mechanical interactions between LTCC and RyR1, rather than a voltage-dependent Ca^{2+} transport protein and CICR, mediate this conformational coupling [172].

Indeed, overlay electron microscopy showed that every other RyR1 is physically linked with the four subunits to four adjacent LTCCs, allowing for tight conformational coupling [160, 173]. If the cell membrane containing LTCCs is removed by detergents or mechanical interventions, CICR can be elicited in skinned skeletal muscle fibers. However, voltage-clamp experiments have demonstrated that I_{Ca} carried by skeletal LTCC pore-forming α_{1S} subunit activates 10-fold slower ($\tau > 50\,ms$) than the cardiac α_{1C} subunit, and therefore Ca^{2+} influx is not expected to contribute to EC coupling under physiologic conditions in skeletal muscles [174,

175]. Consistent with voltage-dependent physical EC coupling, heterologous expression of the skeletal muscle LTCC (α_{1S} subunit in dysgenic myotubes which lack endogenous LTCCs) results in conformational RyR1 activation in the absence of extracellular Ca^{2+} influx [176, 177]. Moreover, elegant mutagenesis studies using a mutant skeletal muscle LTCC α_{1S} subunit, which does not permeate Ca^{2+}, have demonstrated voltage-coupling of skeletal muscle contraction despite functional knockout of I_{Ca} [178].

In skeletal muscle, local subcellular SR Ca^{2+} release events measured by confocal imaging are also called Ca^{2+} sparks. They initiate from the T tubule/SR junction interface sections called triads [179, 180]. The stochastic occurrence of spontaneous Ca^{2+} sparks under resting conditions is lower in skeletal myofibers than in cardiomyocytes indicating tight control of CRUs by E_m. Similar to cardiomyocytes, Ca^{2+} sparks in skeletal muscle were quantified to occur in multiples of RyR1 unitary conductance [181], and calstabin1 allows for the simultaneous opening and closing of RyR1 channels by the process of coupled gating [182]. Recently, heterologous expression studies in dysgenic myotubes combined with electrophysiologic I_{Ca} and intracellular Ca^{2+} imaging showed that the intracellular II–III LTCC loop contains a motif important for LTCC–RyR1 coupling [183, 184].

11.5.3 EC COUPLING ABNORMALITIES IN CARDIAC AND SKELETAL MUSCLE

Heart failure (HF) results in abnormal intracellular Ca^{2+} cycling [185, 186]. The intracellular Ca^{2+} transient becomes reduced in amplitude and the decay of the Ca^{2+} transient slowed as evidenced by fluorescence Ca^{2+} measurements in cardiomyocytes from patients with HF and animal models [187–189]. In isolated human heart muscle preparations contracting under isometric conditions, force production is depressed in heart failure muscle as a direct consequence of reduced intracellular Ca^{2+} transients [190, 191]. Intracellular Ca^{2+} cycling quantified using SR Ca^{2+} release experiments and intra-SR Ca^{2+} measurements showed depletion of the SR Ca^{2+} store [192–194]. Moreover, β-adrenergic augmentation of EC coupling gain was decreased as assessed by simultaneous patch-clamp and confocal intracellular Ca^{2+} measurements of $\Delta[Ca^{2+}]_i$ and I_{Ca} indicating defective CICR [188, 193, 195, 196]. Decreased shortening of failing cardiomyocytes as evidenced by video edge detection was consequently related to impaired coupling between LTCCs and RyR2s [188, 197], which is potentially due to an altered dyad architecture in failing hearts shown by microscopy [198, 199].

Despite reduced SR Ca^{2+} concentrations, a paradoxical increase of the diastolic Ca^{2+} spark frequency and duration was observed by confocal intracellular Ca^{2+} imaging in heart failure due to calmodulin kinase overexpression [123]. An increased diastolic Ca^{2+} spark frequency is consistent with increased RyR2 activity and SR Ca^{2+} leak and with elevated diastolic Ca^{2+} concentrations reported in failing myocardial muscle preparation earlier using conventional intracellular Ca^{2+} indicators [68, 200]. Heart failure is characterized by a hyperadrenergic state that causes PKA hyper-phosphorylation of RyR2 Ca^{2+} release channels as evidenced by back-phosphorylation and a phospho-epitope specific antibody, and hyperactive channel function that may contribute to SR Ca^{2+} store depletion and increased Ca^{2+} spark frequency [68, 136]. PKA hyperphosphorylation of RyR2 reduces the binding affinity of the channel for calstabin2, resulting in "leaky" channels that contribute to a diastolic SR Ca^{2+} leak that depletes SR Ca^{2+} and triggers delayed after-depolarizations that cause fatal cardiac arrhythmias [113, 131]. Co-immunoprecipitation experiments revealed a concomitant down-regulation of phosphatase (PP1 and PP2A) levels in the RyR2 complex in heart failure providing a rationale for chronically increased PKA phosphorylation levels [68, 201]. PKA hyper-phosphorylation in the junctional space was also reported for LTCC and NCX and is likely to contribute to defective EC coupling [75, 202].

It has been reported that RyR2 channels are not PKA phosphorylated in heart failure; however, this study used a back phosphorylation method in which the amount of RyR2 protein in the assay was not assessed [203]. Others have reported that PKA phosphorylation does not dissociate calstabin2 from RyR2 [114]; however, these studies have used molar excesses of calstabin2 in their experiments which overwhelms the PKA phosphorylation-induced reduction in binding affinity of calstabin2 to RyR2.

Interestingly, RyR1 channels from skeletal muscles of animals or patients with heart failure are also PKA hyper-phosphorylated, depleted of calstabin1, and are hyperactive in single-channel experiments [204]. Depletion of calstabin 1 is expected to decrease coupled gating between individual RyR1 channels and to increase SR Ca^{2+} leak in skeletal muscles [39, 182]. Consistent with RyR1 hyper-activity, SR Ca^{2+} release amplitude was found decreased and the frequency of spontaneous release events increased as evidenced by confocal imaging of Ca^{2+} sparks in skeletal myofibers in heart failure [204, 205]. Isolated skeletal muscles from animals with heart failure exhibited accelerated fatigue times consistent with defective intracellular Ca^{2+} cycling [204]. Since fatigue is a major symptom of heart failure patients and is of prognostic importance, future therapies may aim to improve fatigue to improve life quality and prognosis in heart failure patients.

Interestingly, studies have demonstrated an SR Ca^{2+} release defect in individuals susceptible to malignant hyperthermia (MH). The rate constants of $^{45}Ca^{2+}$ efflux from MH-susceptible SR vesicles were threefold larger than from normal SR, and single RyR1 channels were significantly more active at different cytosolic Ca^{2+} concentrations [206, 207]. Dantrolene inhibits RyR1 Ca^{2+} release in skeletal muscle, and remains the drug of choice to prevent and treat the acute MH crisis [208, 209]. However, MH-like symptoms can also be caused by exercise in the absence of drug triggers in susceptible individuals indicating additional mechanisms may activate aberrant Ca^{2+} release in myofibers [210].

11.6 STUDIES OF GENETICALLY ALTERED MICE

In recent years, the analysis of knockout mice lacking RyR subtypes has demonstrated their essential contribution to physiological functions in skeletal muscle, cardiac muscle, and neurons. Targeted deletion of individual RyR isoforms in mice has provided the opportunity to study their importance in the cellular and whole animal context. Calstabin knockout mice have added to our understanding of the function of these regulatory proteins on RyR function.

11.6.1 RYANODINE RECEPTOR KNOCKOUT MICE

Mice with a targeted disruption of the gene encoding *RyR1* survive during fetal developmental stages but die immediately after birth [28]. Death is probably due to respiratory failure because *RyR1* knockout neonates fail to breathe and do not move. In RyR1-deficient skeletal muscle, myofibril content was significantly reduced and electron-microscopic analysis frequently found abnormal triad junctions lacking "foot" structures [28, 211]. The contractile response to electrical stimulation in skeletal muscle from RyR1-knockout neonates is abolished, demonstrating that RyR1 is essential for both muscle maturation and EC coupling and cannot be substituted by other RyR isoforms [28].

In contrast to mature cardiac muscle, embryonic cardiomyocytes exhibit immature SR functions and RyR2-mediated Ca^{2+} release does not play a prominent role in EC coupling during fetal developmental stages. RyR2 knockout showed cardiac arrest and lethality at about embryonic day E10.5 with morphologic abnormalities of the heart tube [212]. Electronmicroscopic analysis revealed that the ER/SR tubule networks were partly swollen in E8.5 RyR2-deficient cardiac myocytes and were further vacuolated at E9.5 and E10.5 [213]. The authors concluded that without RyR2, cardiac myocytes become overloaded with cytoplasmic Ca^{2+}, similar to findings of the RyR1/RyR3 double knockout mice (see below).

Knockout mice lacking RyR3 demonstrated normal growth and reproduction [29]. Skeletal muscle from RyR3-knockout mice retained the skeletal–muscle type of EC coupling, but the Ca^{2+} signaling and force generation was slightly weakened in comparison with those in wild-type muscle [214–216]. RyR3 knockout mice, however, display a neurological phenotype, including increased locomotor activity [29], impaired learning and memory [217], and abnormal electrophysiological and biochemical properties of hippocampal neurons [218]. These observations indicate that RyR3-mediated Ca^{2+} release physiologically modulates several neuronal circuits. On the other hand, abnormalities caused by the loss of RyR3 in the brain might be compensated by residual RyR1 and RyR2.

11.6.2 CALSTABIN KNOCKOUT MICE

Mice with a targeted disruption of the RyR2 stabilizing subunit calstabin2 (FKBP12.6) display ventricular arrhythmias and sudden cardiac death when subjected to exercise and injection of epinephrine or isoproterenol [113] (Figure 11.4). Cardiomyocytes isolated from exercised calstabin2 knockout mice develop delayed after-depolarizations when paced at rates encountered in exercised mice (~750 beats per minute). Studies of an independently generated calstabin2 knockout mouse also demonstrated defects in cardiomyocyte Ca^{2+} signaling [219]. Single RyR2 channels isolated from exercised calstabin2-deficient mice exhibit significantly increased open probabilities compared with wild-type mice [113]. Together, these findings suggest that calstabin2 in the RyR2 macromolecular complex is required to keep PKA-phosphorylated RyR2 channels closed during diastole (e.g., during exercise) [61, 113, 220]. The absence of calstabin2 in the RyR2 complex predisposes mice to delayed after-depolarizations, ventricular arrhythmias, and sudden cardiac death during exercise and stimulation of the β-AR pathway.

Calstabin1 (FKBP12) modulates skeletal muscle RyR1 and has previously been shown to affect skeletal muscle EC coupling [221]. The majority of calstabin1 deficient mice die between embryonic day 14.5 and birth due to severe cardiac defects [222]. These mice have apparently normal skeletal muscle development throughout embryogenesis; however, they develop severe dilated cardiomyopathy (heart failure) and ventricular septal defects that are similar to a human congenital disorder, noncompaction of the left ventricular myocardium. Single-channel properties of RyR1 from calstabin1-deficient skeletal muscle demonstrated increased open probabilities and subconductance states [222], consistent with the effects of removal of calstabin1 from RyR1 [39]. Surprisingly, single-channel properties of RyR2 also demonstrated an increased open probability and subconductance states [222]; this is likely secondary to the heart failure that is present in these mice due to a large septal defect that occurs during development. Since most calstabin1 mice die before adulthood, the role of calstabin1 could not be assessed in these general calstabin1 knockout mice.

The generation of a skeletal muscle-restricted calstabin1-deficient mouse strain using Cre-lox mediated gene recombination has enabled the study of the specific effects of calstabin1 depletion on skeletal muscle function [223]. The loss of calstabin1 in myotubes alters EC coupling as evidenced by reduced SR Ca^{2+} release but increased L-type Ca^{2+} currents, suggesting a role for calstabin1 in retrograde EC coupling. Contractile properties were differentially affected according to muscle fiber type and chronic activity levels: in fast-type muscle with relatively low activity levels (EDL), decreased muscle contractility was observed, whereas slow-type muscle (soleus) and chronically active fast-type muscle (diaphragm) displayed adaptive responses to calstabin1 deficiency [223].

11.7 CONCLUDING REMARKS

A variety of sophisticated biochemical and complex imaging methods has allowed characterization of the functional importance of the RyR Ca^{2+} release during EC coupling in striated muscles. At the molecular level, RyR was shown to interact with multiple proteins by providing a cytoplasmic scaffold and that these interacting proteins control channel activity. In cardiomyocytes, PKA phosphorylation of RyR2 at Ser^{2809} increases channel activity and decreases the binding affinity of the calstabin2 subunit. In the heart, PKA phosphorylation of RyR2 at Ser^{2809} is thought to increase the gain of EC coupling providing augmented cardiac output during sympathetic activation. Whereas cell membrane Ca^{2+} influx activates RyR2 in the heart, in skeletal muscle physical coupling between LTCC and RyR1 translates membrane depolarization into SR Ca^{2+} release. A variety of cardiac and skeletal diseases affect the RyR channels and result in aberrant intracellular

Ca^{2+} release contributing to heart failure, sudden death, and MH. Future studies will address if these defects are accessible to pharmacologic treatment.

REFERENCES

1. Grunwald R, Meissner G. Lumenal sites and C terminus accessibility of the skeletal muscle calcium release channel (ryanodine receptor). *J Biol Chem* 1995;270: 11338–47.
2. Du GG, Sandhu B, Khanna VK, Guo XH, MacLennan DH. Topology of the Ca^{2+} release channel of skeletal muscle sarcoplasmic reticulum (RyR1). *Proc Natl Acad Sci USA* 2002;99:16725–30.
3. Du GG, MacLennan DH. Topology and transmembrane organization of ryanodine receptors. In: Wehrens XH, Marks AR, eds. *Ryanodine Receptors: Structure, Function and Dysfunction in Clinical Disease*. New York, USA: Springer, Inc.; 2005:9–23.
4. Wehrens XHT, Lehnart SE, Marks AR. Intracellular calcium release channels and cardiac disease. *Annu Rev Physiol* 2005;67:69–98.
5. Fill M, Copello JA. Ryanodine receptor calcium release channels. *Physiol Rev* 2002;82:893–922.
6. Bers DM. Macromolecular complexes regulating cardiac ryanodine receptor function. *J Mol Cell Cardiol* 2004;37:417–29.
7. Berridge MJ, Bootman MD, Roderick HL. Calcium signalling: dynamics, homeostasis and remodelling. *Nat Rev Mol Cell Biol* 2003;4:517–29.
8. Wehrens XH, Marks AR. *Ryanodine Receptors. Structure, Function and Dysfunction in Clinical Disease*. New York, NY: Springer; 2004.
9. Takeshima H, Nishimura S, Matsumoto T, Ishida H, Kangawa K, Minamino N, Matsuo H, Ueda M, Hanaoka M, Hirose T, Numa S. Primary structure and expression from complementary DNA of skeletal muscle ryanodine receptor. *Nature* 1989;339:439–45.
10. Marks AR, Tempst P, Hwang KS, Taubman MB, Inui M, Chadwick C, Fleischer S, Nadal-Ginard B. Molecular cloning and characterization of the ryanodine receptor/ junctional channel complex cDNA from skeletal muscle sarcoplasmic reticulum. *Proc Natl Acad Sci USA* 1989;86:8683–7.
11. Otsu K, Willard HF, Khanna VK, Zorzato F, Green NM, MacLennan DH. Molecular cloning of cDNA encoding the Ca^{2+} release channel (ryanodine receptor) of rabbit cardiac muscle sarcoplasmic reticulum. *J Biol Chem* 1990;265:13472–83.
12. Giannini G, Clementi E, Ceci R, Marziali G, Sorrentino V. Expression of a ryanodine receptor-Ca^{2+} channel that is regulated by TGF-β. *Science* 1992;257:91–3.
13. Hakamata Y, Nakai J, Takeshima H, Imoto K. Primary Structure and distribution of a novel ryanodine receptor/calcium release channel from rabbit brain. *FEBS*. 1992;312:229–35.
14. Ottini L, Marziali G, Conti A, Charlesworth A, Sorrentino V. Alpha and beta isoforms of ryanodine receptor from chicken skeletal muscle are the homologues of mammalian RyR1 and RyR3. *Biochem J* 1996;315:207–16.
15. Franck JPC, Morrissette J, Keen JE, Londraville RL, Beamsley M, Block BA. Cloning and characterization of fiber type-specific ryanodine receptor isoforms in skeletal muscles of fish. *Am J Physiol* 1998;275:C401–15.
16. Oyamada H, Murayama T, Takagi T, Iino M, Iwabe N, Miyata T, Ogawa Y, Endo M. Primary structure and distribution of ryanodine-binding protein isoforms of the bullfrog skeletal muscle. *J Biol Chem* 1994;269:17206–14.

17. Maryon EB, Coronado R, Anderson P. Unc-68 encodes a ryanodine receptor involved in regulating *C. elegans* body-wall muscle contraction. *J Cell Biol* 1996;134:885–93.

18. Takeshima H, Nishi M, Iwabe N, Miyata T, Hosoya T, Masai I, Hotta Y. Isolation and characterization of a gene for a ryanodine receptor/calcium release channel in *Drosophila melanogaster*. *FEBS Lett* 1994;337:81–7.

19. Nakai J, Imagawa T, Hakamata Y, Shigekawa M, Takeshima H, Numa S. Primary structure and functional expression from cDNA of the cardiac ryanodine receptor/calcium release channel. *FEBS* 1990;271:169–77.

20. Futatsugi A, Kuwajima G, Mikoshiba K. Tissue-specific and developmentally regulated alternative splicing in mouse skeletal muscle ryanodine receptor mRNA. *Biochem J* 1995;305:373–8.

21. Tunwell RE, Wickenden C, Bertrand BM, Shevchenko VI, Walsh MB, Allen PD, Lai FA. The human cardiac muscle ryanodine receptor–calcium release channel: identification, primary structure and topological analysis. *Biochem J* 1996;318:477–87.

22. Xu X, Bhat MB, Nishi M, Takeshima H, Ma J. Molecular cloning of cDNA encoding a *Drosophila* ryanodine receptor and functional studies of the carboxyl-terminal calcium release channel. *Biophys J* 2000;78:1270–81.

23. Puente E, Suner M, Evans AD, McCaffery AR, Windass JD. Identification of a polymorphic ryanodine receptor gene from heliothis virescens (*Lepidoptera noctuidae*). *Insect Biochem Mol Biol* 2000;30:335–47.

24. Giannini G, Sorrentino V. Molecular structure and tissue distribution of ryanodine receptors calcium channels. *Med Res Rev* 1995;15:313–23.

25. Kuwajima G, Futatsugi A, Niinobe M, Nakanishi S, Mikoshiba K. Two types of ryanodine receptors in mouse brain: skeletal muscle type exclusively in Purkinje cells and cardiac muscle type in various neurons. *Neuron* 1992;9:1133–42.

26. Giannini G, Conti A, Mammarella S, Scrobogna M, Sorrentino V. The ryanodine receptor/calcium channel genes are widely and differentially expressed in murine brain and peripheral tissues. *J Cell Biol* 1995;128:893–904.

27. Ivanenko A, McKemy DD, Kenyon JL, Airey JA, Sutko JL. Embryonic chicken skeletal muscle cells fail to develop normal excitation–contraction coupling in the absence of the alpha ryanodine receptor. Implications for a two-ryanodine receptor system. *J Biol Chem* 1995;270:4220–3.

28. Takeshima H, Iino M, Takekura H, Nishi M, Kuno J, Minowa O, Takano H, Noda T. Excitation–contraction uncoupling and muscular degeneration in mice lacking functional skeletal muscle ryanodine-receptor gene. *Nature* 1994;369:556–9.

29. Takeshima H, Ikemoto T, Nishi M, Nishiyama N, Shimuta M, Sugitani Y, Kuno J, Saito I, Saito H, Endo M, Iino M, Noda T. Generation and characterization of mutant mice lacking ryanodine receptor type 3. *J Biol Chem* 1996;271:19649–52.

30. Fujii J, Otsu K, Zorzato F, de Leon S, Khana V, Weiler J, O'Brien P, MacLennan D. Identification of a mutation in porcine ryanodine receptor is potentially causative of human malignant hyperthermia. *Science* 1991;253:448–51.

31. Priori SG, Napolitano C, Tiso N, Memmi M, Vignati G, Bloise R, Sorrentino VV, Danieli GA. Mutations in the cardiac ryanodine receptor gene (hRyR2) underlie catecholaminergic polymorphic ventricular tachycardia. *Circulation* 2001;103:196–200.

32. Laitinen PJ, Brown KM, Piippo K, Swan H, Devaney JM, Brahmbhatt B, Donarum EA, Marino M, Tiso N, Viitasalo M, Toivonen L, Stephan DA, Kontula K. Mutations of the cardiac ryanodine receptor (RyR2) gene in familial polymorphic ventricular tachycardia. *Circulation* 2001;103:485–90.

33. Tiso N, Stephan DA, Nava A, Bagattin A, Devaney JM, Stanchi F, Larderet G, Brahmbhatt B, Brown K, Bauce B, Muriago M, Basso C, Thiene G, Danieli GA, Rampazzo A. Identification of mutations in the cardiac ryanodine receptor gene in families affected with arrhythmogenic right ventricular cardiomyopathy type 2 (ARVD2). Hum Mol Genet 2001;10:189–94.

34. Coronado R, Morrissette J, Sukhareva M, Vaughan DM. Structure and function of ryanodine receptors. Am J Physiol 1994;266:C1485–504.

35. Diaz-Munoz M, Hamilton S, Kaetzel M, Hazarika P, Dedman J. Modulation of Ca^{2+} release channel activity from sarcoplasmic reticulum by annexin V1 (67-kDa Calcimedin). J Biol Chem 1990;265:15894–9.

36. Jones LR, Zhang L, Sanborn K, Jorgensen AO, Kelley J. Purification, primary structure, and immunological characterization of the 26-kDa calsequestrin binding protein (junctin) from cardiac junctional sarcoplasmic reticulum. J Biol Chem 1995;270:30787–96.

37. Knudson CM, Stang KK, Jorgenson AO, Campbell KP. Biochemical characterization and ultrastructural localization of a major junctional sarcoplasmic reticulum glycoprotein (triadin). J Biol Chem 1993;268:12646–5.

38. Lokuta AJ, Meyers MB, Sander PR, Fishman GI, Valdivia HH. Modulation of cardiac ryanodine receptors by sorcin. J Biol Chem 1997;272:25333–8.

39. Brillantes AB, Ondrias K, Scott A, Kobrinsky E, Ondriasova E, Moschella MC, Jayaraman T, Landers M, Ehrlich BE, Marks AR. Stabilization of calcium release channel (ryanodine receptor) function by FK506-binding protein. Cell 1994;77:513–23.

40. Wehrens XH, Lehnart SE, Reiken SR, Marks AR. Ca^{2+}/calmodulin-dependent protein kinase II phosphorylation regulates the cardiac ryanodine receptor. Circ Res 2004;94:e61–70.

41. Currie S, Loughrey CM, Craig MA, Smith GL. Calcium/calmodulin-dependent protein kinase IIdelta associates with the ryanodine receptor complex and regulates channel function in rabbit heart. Biochem J 2004;377:357–66.

42. Franzini-Armstrong C. Studies of the triad: I. Structure of the junction in frog twitch fibers. J Cell Biol 1970;47:488–99.

43. Chiu W, McGough A, Sherman MB, Schmid MF. High-resolution electron cryomicroscopy of macromolecular assemblies. Trends Cell Biol 1999;9:154–9.

44. Radermacher M, Rao V, Grassucci R, Frank J, Timerman AP, Fleischer S, Wagenknecht T. Cryo-electron microscopy and three-dimensional reconstruction of the calcium release channel/ryanodine receptor from skeletal muscle. J Cell Biol 1994; 127:411–23.

45. Serysheva, II, Orlova EV, Chiu W, Sherman MB, Hamilton SL, van Heel M. Electron cryomicroscopy and angular reconstitution used to visualize the skeletal muscle calcium release channel. Nature Struct Biol 1995;2:18–24.

46. Baker ML, Serysheva, II, Sencer S, Wu Y, Ludtke SJ, Jiang W, Hamilton SL, Chiu W. The skeletal muscle Ca^{2+} release channel has an oxidoreductase-like domain. Proc Natl Acad Sci USA 2002;99:12155–60.

47. Wagenknecht T, Radermacher M, Grassucci R, Berkowitz J, Xin HB, Fleischer S. Locations of calmodulin and FK506-binding protein on the three-dimensional architecture of the skeletal muscle ryanodine receptor. J Biol Chem 1997;272:32463–71.

48. Samso M, Trujillo R, Gurrola GB, Valdivia HH, Wagenknecht T. Three-dimensional location of the imperatoxin A binding site on the ryanodine receptor. J Cell Biol 1999;146:493–9.

49. Samso M, Wagenknecht T. Apocalmodulin and Ca^{2+}-calmodulin bind to neighboring locations on the ryanodine receptor. J Biol Chem 2002;277:1349–53.

50. Liu Z, Wagenknecht T. Three-dimensional reconstruction of ryanodine receptors. In: Wehrens XH, Marks AR, eds. *Ryanodine Receptors: Structure, Function and Dysfunction in Clinical Disease.* New York, Springer, Inc.; 2005:25–34.

51. Liu Z, Zhang J, Li P, Chen SR, Wagenknecht T. Three-dimensional reconstruction of the recombinant type 2 ryanodine receptor and localization of its divergent region 1. *J Biol Chem* 2002;277:46712–9.

52. Liu Z, Zhang J, Sharma MR, Li P, Chen SR, Wagenknecht T. Three-dimensional reconstruction of the recombinant type 3 ryanodine receptor and localization of its amino terminus. *Proc Natl Acad Sci USA* 2001;98:6104–9.

53. Zhang J, Liu Z, Masumiya H, Wang R, Jiang D, Li F, Wagenknecht T, Chen SR. Three-dimensional localization of divergent region 3 of the ryanodine receptor to the clamp-shaped structures adjacent to the FKBP binding sites. *J Biol Chem* 2003; 278:14211–8.

54. Liu Z, Zhang J, Wang R, Chen SR, Wagenknecht T. Location of divergent region 2 on the three-dimensional structure of cardiac muscle ryanodine receptor/calcium release channel. *J Mol Biol* 2004;338:533–45.

55. Sorrentino V, Volpe P. Ryanodine receptors: how many, where and why? *Trends Pharmacol Sci* 1993;14:98–103.

56. Du GG, Khanna VK, MacLennan DH. Mutation of divergent region 1 alters caffeine and Ca^{2+} sensitivity of the skeletal muscle Ca^{2+} release channel (ryanodine receptor). *J Biol Chem* 2000;275:11778–83.

57. Jayaraman T, Brillantes AM, Timerman AP, Erdjument-Bromage H, Fleischer S, Tempst P, Marks AR. FK506 binding protein associated with the calcium release channel (ryanodine receptor). *J Biol Chem* 1992;267:9474–77.

58. Lam E, Martin MM, Timerman AP, Sabers C, Fleischer S, Lukas T, Abraham RT, O'Keefe SJ, O'Neill EA, Wiederrecht GJ. A novel FK506 binding protein can mediate the immunosuppressive effects of FK506 and is associated with the cardiac ryanodine receptor. *J Biol Chem* 1995;270:26511–22.

59. Timerman AP, Ogunbumni E, Freund E, Wiederrecht G, Marks AR, Fleischer S. The calcium release channel of sarcoplasmic reticulum is modulated by FK-506-binding protein. Dissociation and reconstitution of FKBP-12 to the calcium release channel of skeletal muscle sarcoplasmic reticulum. *J Biol Chem* 1993;268:22992–9.

60. Timerman AP, Onoue H, Xin HB, Barg S, Copello J, Wiederrecht G, Fleischer S. Selective binding of FKBP12.6 by the cardiac ryanodine receptor. *J Biol Chem* 1996;271:20385–91.

61. Wehrens XH, Lehnart SE, Reiken SR, Deng SX, Vest JA, Cervantes D, Coromilas J, Landry DW, Marks AR. Protection from cardiac arrhythmia through ryanodine receptor-stabilizing protein calstabin2. *Science*.2004;304:292–6.

62. Van Acker K, Bultynck G, Rossi D, Sorrentino V, Boens N, Missiaen L, De Smedt H, Parys JB, Callewaert G. The 12 kDa FK506-binding protein, FKBP12, modulates the Ca^{2+}-flux properties of the type-3 ryanodine receptor. *J Cell Sci* 2004;117: 1129–37.

63. Jeyakumar LH, Ballester L, Cheng DS, McIntyre JO, Chang P, Olivey HE, Rollins-Smith L, Barnett JV, Murray K, Xin HB, Fleischer S. FKBP binding characteristics of cardiac microsomes from diverse vertebrates. *Biochem Biophys Res Commun* 2001;281:979–86.

64. Sharma MR, Jeyakuma LH, Fleischer S, Wagenknecht T. Three-dimensional visualisation of FKBP12.6 binding to cardiac ryanodine receptor (RyR2) in open buffer conditions. *Biophys J* 2002;82:644a.

65. Gaburjakova M, Gaburjakova J, Reiken S, Huang F, Marx SO, Rosemblit N, Marks AR. FKBP12 binding modulates ryanodine receptor channel gating. *J Biol Chem.* 2001;276:16931–5.

66. Marks AR. Cellular functions of immunophilins. *Physiol Rev* 1996;76:631–49.

67. Bultynck G, Rossi D, Callewaert G, Missiaen L, Sorrentino V, Parys JB, De Smedt H. The conserved sites for the FK506-binding proteins in ryanodine receptors and inositol 1,4,5-trisphosphate receptors are structurally and functionally different. *J Biol Chem* 2001;276:47715–24.

68. Marx SO, Reiken S, Hisamatsu Y, Jayaraman T, Burkhoff D, Rosemblit N, Marks AR. PKA phosphorylation dissociates FKBP12.6 from the calcium release channel (ryanodine receptor): defective regulation in failing hearts. *Cell* 2000;101:365–376.

69. Masumiya H, Wang R, Zhang J, Xiao B, Chen SR. Localization of the 12.6-kDa FK506-binding protein (FKBP12.6) binding site to the NH2-terminal domain of the cardiac Ca^{2+} release channel (ryanodine receptor). *J Biol Chem* 2003; 278:3786–92.

70. Marx SO, Reiken S, Hisamatsu Y, Gaburjakova M, Gaburjakova J, Yang YM, Rosemblit N, Marks AR. Phosphorylation-dependent regulation of ryanodine receptors. A novel role for leucine/isoleucine zippers. *J Cell Biol* 2001;153:699–708.

71. DeSouza N, Reiken S, Ondrias K, Yang YM, Matkovich S, Marks AR. Protein kinase A and two phosphatases are components of the inositol 1,4,5-trisphosphate receptor macromolecular signaling complex. *J Biol Chem* 2002;277:39397–39400.

72. Zaccolo M, Pozzan T. Discrete microdomains with high concentration of cAMP in stimulated rat neonatal cardiac myocytes. *Science* 2002;295:1711–15.

73. Berridge MJ. Microdomains and elemental events in calcium signalling. *Cell Calcium* 1996;20:95–6.

74. Marx SO, Kurokawa J, Reiken S, Motoike H, D'Armiento J, Marks AR, Kass RS. Requirement of a macromolecular signaling complex for beta adrenergic receptor modulation of the KCNQ1-KCNE1 potassium channel. *Science* 2002;295:496–9.

75. Hulme JT, Lin TW, Westenbroek RE, Scheuer T, Catterall WA. Beta-adrenergic regulation requires direct anchoring of PKA to cardiac CaV1.2 channels via a leucine zipper interaction with A kinase-anchoring protein 15. *Proc Natl Acad Sci USA* 2003;100:13093–8.

76. Allen P, Ouimet C, Greengard P. Spinophilin, a novel protein phosphatase 1 binding protein localized to dendritic spines. *Proc Natl Acad Sci USA* 1997;94:9956–61.

77. Porter Moore C, Zhang JZ, Hamilton SL. A role for cysteine 3635 of RYR1 in redox modulation and calmodulin binding. *J Biol Chem* 1999;274:36831–4.

78. Moore CP, Rodney G, Zhang JZ, Santacruz-Toloza L, Strasburg G, Hamilton SL. Apocalmodulin and Ca^{2+} calmodulin bind to the same region on the skeletal muscle Ca^{2+} release channel. *Biochemistry* 1999;38:8532–7.

79. Zhang H, Zhang JZ, Danila CI, Hamilton SL. A noncontiguous, intersubunit binding site for calmodulin on the skeletal muscle Ca^{2+} release channel. *J Biol Chem* 2003;278:8348–55.

80. Meyers MB, Pickel VM, Sheu SS, Sharma VK, Scotto KW, Fishman GI. Association of sorcin with the cardiac ryanodine receptor. *J Biol Chem* 1995;270:26411–8.

81. Meyers MB, Puri TS, Chien AJ, Gao T, Hsu PH, Hosey MM, Fishman GI. Sorcin associates with the pore-forming subunit of voltage-dependent L-type Ca^{2+} channels. *J Biol Chem* 1998;273:18930–5.

82. Zhang L, Kelley J, Schmeisser G, Kobayashi YM, Jones LR. Complex formation between junctin, triadin, calsequestrin, and the ryanodine receptor. Proteins of the

cardiac junctional sarcoplasmic reticulum membrane. *J Biol Chem* 1997;272: 23389–97.

83. Flucher BE, Andrews SB, Fleischer S, Marks AR, Caswell A, Powell JA. Triad formation: organization and function of the sarcoplasmic reticulum calcium release channel and triadin in normal and dysgenic muscle *in vitro*. *J Cell Biol* 1993;123: 1161–74.

84. Collins J, Tarcsafalvi A, Ikemoto N. Identification of a region of calsequestrin that binds to the junctional face membrane of sarcoplasmic reticulum. *Biochem Biophys Res Commun* 1990;167:189–93.

85. Viatchenko-Karpinski S, Terentyev D, Gyorke I, Terentyeva R, Volpe P, Priori SG, Napolitano C, Nori A, Williams SC, Gyorke S. Abnormal calcium signaling and sudden cardiac death associated with mutation of calsequestrin. *Circ Res* 2004;94:471–7.

86. Culligan K, Banville N, Dowling P, Ohlendieck K. Drastic reduction of calsequestrin-like proteins and impaired calcium binding in dystrophic mdx muscle. *J Appl Physiol* 2002;92:435–45.

87. Ohkura M, Furukawa K, Fujimori H, Kuruma A, Kawano S, Hiraoka M, Kuniyasu A, Nakayama H, Ohizumi Y. Dual regulation of the skeletal muscle ryanodine receptor by triadin and calsequestrin. *Biochemistry* 1998;37:12987–93.

88. Szegedi C, Sarkozi S, Herzog A, Jona I, Varsanyi M. Calsequestrin: more than 'only' a luminal Ca^{2+} buffer inside the sarcoplasmic reticulum. *Biochem J* 1999;337:19–22.

89. Beard NA, Laver DR, Dulhunty AF. Calsequestrin and the calcium release channel of skeletal and cardiac muscle. *Prog Biophys Mol Biol* 2004;85:33–69.

90. Meissner G. Regulation of mammalian ryanodine receptors. *Front Biosci* 2002; 7:d2072–80.

91. Fabiato A. Time and calcium dependence of activation and inactivation of calcium-induced release of calcium from the sarcoplasmic reticulum of a skinned canine cardiac purkinje cell. *J Gen Physiol* 1985;85:247–289.

92. Fill M, Zahradnikova A, Villalba-Galea CA, Zahradnik I, Escobar AL, Gyorke S. Ryanodine receptor adaptation. *J Gen Physiol* 2000;116:873–82.

93. Cleemann L, Morad M. Role of Ca^{2+} channel in cardiac excitation-contraction coupling in the rat: evidence from Ca^{2+} transients and contraction. *J Physiol* 1991;432: 283–312.

94. Yasui K, Palade P, Gyorke S. Negative control mechanism with features of adaptation controls Ca^{2+} release in cardiac myocytes. *Biophys J* 1994;67:457–60.

95. Nabauer M, Morad M. Ca^{2+}-induced Ca^{2+} release as examined by photolysis of caged Ca^{2+} in single ventricular myocytes. *Am J Physiol* 1990;258:C189–93.

96. Gyorke S, Fill M. Ryanodine receptor adaptation: control mechanism of Ca^{2+}-induced Ca^{2+} release in heart. *Science* 1993;260:807–9.

97. Gyorke S, Velez P, Suarez-Isla B, Fill M. Activation of single cardiac and skeletal ryanodine receptor channels by flash photolysis of caged Ca^{2+}. *Biophys J* 1994; 66:1879–86.

98. Laver DR, Curtis BA. Response of ryanodine receptor channels to Ca^{2+} steps produced by rapid solution exchange. *Biophys J* 1996;71:732–41.

99. Schiefer A, Meissner G, Isenberg G. Ca^{2+} activation and Ca^{2+} inactivation of canine reconstituted cardiac sarcoplasmic reticulum Ca^{2+}-release channels. *J Physiol* 1995;489:337–48.

100. Sitsapesan R, Montgomery RA, Williams AJ. New insights into the gating mechanisms of cardiac ryanodine receptors revealed by rapid changes in ligand concentration. *Circ Res* 1995;77:765–72.

101. Valdivia HH, Kaplan JH, Ellis-Davies GC, Lederer WJ. Rapid adaptation of cardiac ryanodine receptors: modulation by Mg^{2+} and phosphorylation. *Science* 1995;267: 1997–2000.

102. Velez P, Gyorke S, Escobar AL, Vergara J, Fill M. Adaptation of single cardiac ryanodine receptor channels. *Biophys J* 1997;72:691–7.

103. Blatter LA, Huser J, Ríos E. Sarcoplasmic reticulum Ca^{2+} release flux underlying Ca^{2+} sparks in cardiac muscle. *Proc Natl Acad Sci USA* 1997;94:4176–81.

104. Lipp P, Niggli E. Submicroscopic calcium signals as fundamental events of excitation — contraction coupling in guinea-pig cardiac myocytes. *J Physiol* 1996;492:31–8.

105. Parker I, Zang WJ, Wier WG. Ca^{2+} sparks involving multiple Ca^{2+} release sites along Z-lines in rat heart cells. *J Physiol* 1996;497:31–8.

106. Stern MD. Theory of excitation–contraction coupling in cardiac muscle. *Biophys J* 1992;63:497–517.

107. Wier WG, Egan TM, Lopez-Lopez JR, Balke CW. Local control of excitation–contraction coupling in rat heart cells. *J Physiol* 1994;474:463–71.

108. Lamb GD, Laver DR, Stephenson DG. Questions about adaptation in ryanodine receptors. *J Gen Physiol* 2000;116:883–90.

109. Gyorke S. Ca^{2+} spark termination: inactivation and adaptation may be manifestations of the same mechanism. *J Gen Physiol* 1999;114:163–6.

110. Marx SO, Gaburjakova J, Gaburjakova M, Henrikson C, Ondrias K, Marks AR. Coupled gating between cardiac calcium release channels (ryanodine receptors). *Circ Res* 2001;88:1151–58.

111. Witcher DR, Kovacs RJ, Schulman H, Cefali DC, Jones LR. Unique phosphorylation site on the cardiac ryanodine receptor regulates calcium channel activity. *J Biol Chem* 1991;266:11144–52.

112. Witcher DR, Strifler BA, Jones LR. Cardiac-specific phosphorylation site for multifunctional Ca^{2+}/calmodulin-dependent protein kinase is conserved in the brain ryanodine receptor. *J Biol Chem* 1992;267:4963–7.

113. Wehrens XH, Lehnart SE, Huang F, Vest JA, Reiken SR, Mohler PJ, Sun J, Guatimosim S, Song LS, Rosemblit N, D'Armiento JM, Napolitano C, Memmi M, Priori SG, Lederer WJ, Marks AR. FKBP12.6 deficiency and defective calcium release channel (ryanodine receptor) function linked to exercise-induced sudden cardiac death. *Cell* 2003;113:829–40.

114. Xiao B, Sutherland C, Walsh MP, Chen SR. Protein kinase A phosphorylation at serine-2808 of the cardiac Ca^{2+}-release channel (ryanodine receptor) does not dissociate 12.6-kDa FK506-binding protein (FKBP12.6) *Circ Res* 2004.

115. Rodriguez P, Bhogal MS, Colyer J. Stoichiometric phosphorylation of cardiac ryanodine receptor on serine 2809 by calmodulin-dependent kinase II and protein kinase A. *J Biol Chem* 2003;278:38593–600.

116. Hain J, Onoue H, Mayrleitner M, Fleischer S, Schindler H. Phosphorylation modulates the function of the calcium release channel of sarcoplasmic reticulum from cardiac muscle. *J Biol Chem* 1995;270:2074–81.

117. Lokuta AJ, Rogers TB, Lederer WJ, Valdivia HH. Modulation of cardiac ryanodine receptors of swine and rabbit by a phosphorylation–dephosphorylation mechanism. *J Phys* 1995;487:609–22.

118. Sonnleitner A, Fleischer S, Schindler H. Gating of the skeletal calcium release channel by ATP is inhibited by protein phosphatase 1 but not by Mg2+. *Cell Calcium* 1997;21:283–90.

119. Li Y, Kranias EG, Mignery GA, Bers DM. Protein kinase A phosphorylation of the ryanodine receptor does not affect calcium sparks in mouse ventricular myocytes. *Circ Res* 2002;90:309–16.

120. Ginsburg KS, Bers DM. Modulation of excitation–contraction coupling by isoproterenol in cardiomyocytes with controlled SR Ca^{2+} load and Ca^{2+} current trigger. *J Physiol* 2004;556:463–80.

121. Yoshida A, Takahashi M, Imagawa T, Shigekawa M, Takisawa H, Nakamura T. Phosphorylation of ryanodine receptors in rat myocytes during beta-adrenergic stimulation. *J Biochem (Tokyo)* 1992;111:186–90.

122. Dulhunty AF, Laver D, Curtis SM, Pace S, Haarmann C, Gallant EM. Characteristics of irreversible ATP activation suggest that native skeletal ryanodine receptors can be phosphorylated via an endogenous CaMKII. *Biophys J* 2001;81:3240–52.

123. Maier LS, Zhang T, Chen L, DeSantiago J, Brown JH, Bers DM. Transgenic CaMKII-deltaC overexpression uniquely alters cardiac myocyte Ca^{2+} handling: reduced SR Ca^{2+} load and activated SR Ca^{2+} release. *Circ Res* 2003;92:904–11.

124. Zhang T, Maier LS, Dalton ND, Miyamoto S, Ross J, Jr., Bers DM, Brown JH. The deltaC isoform of CaMKII is activated in cardiac hypertrophy and induces dilated cardiomyopathy and heart failure. *Circ Res* 2003;92:912–19.

125. Neumann J, Boknik P, Herzig S, Schmitz W, Scholz H, Gupta RC, Watanabe AM. Evidence for physiological functions of protein phosphatases in the heart: evaluation with okadaic acid. *Am J Physiol* 1993;265:H257–66.

126. duBell WH, Lederer WJ, Rogers TB. Dynamic modulation of excitation–contraction coupling by protein phosphatases in rat ventricular myocytes. *J Physiol* 1996;493:793–800.

127. duBell WH, Gigena MS, Guatimosim S, Long X, Lederer WJ, Rogers TB. Effects of PP1/PP2A inhibitor calyculin A on the E–C coupling cascade in murine ventricular myocytes. *Am J Physiol Heart Circ Physiol* 2002;282:H38–48.

128. Carr AN, Schmidt AG, Suzuki Y, del Monte F, Sato Y, Lanner C, Breeden K, Jing SL, Allen PB, Greengard P, Yatani A, Hoit BD, Grupp IL, Hajjar RJ, DePaoli-Roach AA, Kranias EG. Type 1 phosphatase, a negative regulator of cardiac function. *Mol Cell Biol* 2002;22:4124–35.

129. Santana LF, Chase EG, Votaw VS, Nelson MT, Greven R. Functional coupling of calcineurin and protein kinase A in mouse ventricular myocytes. *J Physiol* 2002;544:57–69.

130. Swan H, Piippo K, Viitasalo M, Heikkila P, Paavonen T, Kainulainen K, Kere J, Keto P, Kontula K, Toivonen L. Arrhythmic disorder mapped to chromosome 1q42–q43 causes malignant polymorphic ventricular tachycardia in structurally normal hearts. *J Am Coll Cardiol* 1999;34:2035–42.

131. Lehnart SE, Wehrens XHT, Laitinen PJ, Reiken SR, Deng SX, Chen Z, Landry DW, Kontula K, Swan H, Marks AR. Sudden death in familial polymorphic ventricular tachycardia associated with calcium release channel (ryanodine receptor) leak. *Circulation* 2004:r113–19.

132. George CH, Higgs GV, Lai FA. Ryanodine receptor mutations associated with stress-induced ventricular tachycardia mediate increased calcium release in stimulated cardiomyocytes. *Circ Res* 2003;93:531–40.

133. Priori SG, Napolitano C, Memmi M, Colombi B, Drago F, Gasparini M, DeSimone L, Coltorti F, Bloise R, Keegan R, Cruz Filho FE, Vignati G, Benatar A, DeLogu A. Clinical and molecular characterization of patients with catecholaminergic polymorphic ventricular tachycardia. *Circulation* 2002;106:69–74.

134. Ono K, Yano M, Ohkusa T, Kohno M, Hisaoka T, Tanigawa T, Kobayashi S, Matsuzaki M. Altered interaction of FKBP12.6 with ryanodine receptor as a cause of abnormal Ca^{2+} release in heart failure. *Cardiovasc Res* 2000;48:323–31.

135. Yano M, Kobayashi S, Kohno M, Doi M, Tokuhisa T, Okuda S, Suetsugu M, Hisaoka T, Obayashi M, Ohkusa T, Matsuzaki M. FKBP12.6-mediated stabilization of calcium-release channel (ryanodine receptor) as a novel therapeutic strategy against heart failure. *Circulation* 2003;107:477–84.

136. Yano M, Ono K, Ohkusa T, Suetsugu M, Kohno M, Hisaoka T, Kobayashi S, Hisamatsu Y, Yamamoto T, Noguchi N, Takasawa S, Okamoto H, Matsuzaki M. Altered stoichiometry of FKBP12.6 versus ryanodine receptor as a cause of abnormal Ca^{2+} leak through ryanodine receptor in heart failure. *Circulation* 2000;102:2131–6.

137. Jiang D, Xiao B, Yang D, Wang R, Choi P, Zhang L, Cheng H, Chen SR. RyR2 mutations linked to ventricular tachycardia and sudden death reduce the threshold for store-overload-induced Ca^{2+} release (SOICR). *Proc Natl Acad Sci USA* 2004;101:13062–7.

138. Jiang D, Xiao B, Zhang L, Chen SR. Enhanced basal activity of a cardiac Ca^{2+} release channel (ryanodine receptor) mutant associated with ventricular tachycardia and sudden death. *Circ Res* 2002;91:218–25.

139. Mickelson JR, Louis CF. Malignant hyperthermia: excitation–contraction coupling, Ca^{2+} release channel, and cell Ca^{2+} regulation defects. *Physiol Rev* 1996;76:537–92.

140. Shuaib A, Paasuke RT, Brownell KW. Central core disease. Clinical features in 13 patients. *Medicine (Baltimore)* 1987;66:389–96.

141. MacLennan DH, Duff C, Zorzato F, Fujii J, Phillips M, Korneluk RG, Frodis W, Britt BA, Worton RG. Ryanodine receptor gene is a candidate for predisposition to malignant hyperthermia. *Nature* 1990;343:559–61.

142. Zhang Y, Chen HS, Khanna VK, De Leon S, Phillips MS, Schappert K, Britt BA, Brownell KW, MacLennan DH. A mutation in the human ryanodine receptor gene associated with central core disease. *Nat Genet* 1993;5:46–9.

143. Shomer NH, Louis CF, Fill M, Litterer LA, Mickelson JR. Reconstitution of abnormalities in the malignant hyperthermia-susceptible pig ryanodine receptor. *Am J Physiol* 1993;264:C125–35.

144. Jurkat-Rott K, McCarthy T, Lehmann-Horn F. Genetics and pathogenesis of malignant hyperthermia. *Muscle Nerve* 2000;23:4–17.

145. Tong J, McCarthy TV, MacLennan DH. Measurement of resting cytosolic Ca^{2+} concentrations and Ca^{2+} store size in HEK-293 cells transfected with malignant hyperthermia or central core disease mutant Ca^{2+} release channels. *J Biol Chem* 1999;274:693–702.

146. Avila G, Dirksen RT. Functional effects of central core disease mutations in the cytoplasmic region of the skeletal muscle ryanodine receptor. *J Gen Physiol* 2001;118:277–90.

147. Avila G, O'Connell KM, Dirksen RT. The pore region of the skeletal muscle ryanodine receptor is a primary locus for excitation–contraction uncoupling in central core disease. *J Gen Physiol* 2003;121:277–86.

148. Yang T, Ta TA, Pessah IN, Allen PD. Functional defects in six ryanodine receptor isoform-1 (RyR1) mutations associated with malignant hyperthermia and their impact on skeletal excitation–contraction coupling. *J Biol Chem* 2003;278: 25722–30.

149. Lynch PJ, Tong J, Lehane M, Mallet A, Giblin L, Heffron JJ, Vaughan P, Zafra G, MacLennan DH, McCarthy TV. A mutation in the transmembrane/luminal domain of

the ryanodine receptor is associated with abnormal Ca^{2+} release channel function and severe central core disease. *Proc Natl Acad Sci USA* 1999;96:4164–9.

150. Tilgen N, Zorzato F, Halliger-Keller B, Muntoni F, Sewry C, Palmucci LM, Schneider C, Hauser E, Lehmann-Horn F, Muller CR, Treves S. Identification of four novel mutations in the C-terminal membrane spanning domain of the ryanodine receptor 1: association with central core disease and alteration of calcium homeostasis. *Hum Mol Genet* 2001;10:2879–87.

151. Avila G, O'Brien JJ, Dirksen RT. Excitation–contraction uncoupling by a human central core disease mutation in the ryanodine receptor. *Proc Natl Acad Sci USA* 2001;98:4215–20.

152. Endo M, Tanaka M, Ogawa Y. Calcium induced release of calcium from the sarcoplasmic reticulum of skinned skeletal muscle fibres. *Nature* 1970;228:34–6.

153. Fabiato A. Calcium-induced release of calcium from the cardiac sarcoplasmic reticulum. *Am J Physiol* 1983;245:C1–C14.

154. Rich TL, Langer GA, Klassen MG. Two components of coupling calcium in single ventricular cell of rabbits and rats. *Am J Physiol* 1988;254:H937–46.

155. McDonald TF, Cavalie A, Trautwein W, Pelzer D. Voltage-dependent properties of macroscopic and elementary calcium channel currents in guinea pig ventricular myocytes. *Pflugers Arch* 1986;406:437–48.

156. Cannell MB, Berlin JR, Lederer WJ. Effect of membrane potential changes on the calcium transient in single rat cardiac muscle cells. *Science* 1987;238:1419–23.

157. Cannell MB, Soeller C. Numerical analysis of ryanodine receptor activation by L-type channel activity in the cardiac muscle diad. *Biophys J* 1997;73:112–22.

158. Flucher BE, Franzini-Armstrong C. Formation of junctions involved in excitation–contraction coupling in skeletal and cardiac muscle. *Proc Natl Acad Sci USA* 1996;93:8101–6.

159. Carl SL, Felix K, Caswell AH, Brandt NR, Ball WJ, Jr., Vaghy PL, Meissner G, Ferguson DG. Immunolocalization of sarcolemmal dihydropyridine receptor and sarcoplasmic reticular triadin and ryanodine receptor in rabbit ventricle and atrium *J Cell Biol* 1995;129:672–82.

160. Franzini-Armstrong C, Protasi F, Ramesh V. Shape, size, and distribution of Ca^{2+} release units and couplons in skeletal and cardiac muscles. *Biophys J* 1999;77:1528–39.

161. Cannell MB, Cheng H, Lederer WJ. Spatial non-uniformities in $[Ca^{2+}]_i$ during excitation–contraction coupling in cardiac myocytes. *Biophys J* 1994;67:1942–56.

162. Cheng H, Lederer WJ, Cannell MB. Calcium sparks: elementary events underlying excitation–contraction coupling in heart muscle. *Science* 1993;262:740–44.

163. Wier WG. Confocal microscopy reveals local SR calcium release in voltage-clamped cardiac cells. *Adv Exp Med Biol* 1995;382:81–8.

164. Cannell MB, Cheng H, Lederer WJ. The control of calcium release in heart muscle. *Science* 1995;268:1045–49.

165. Sham JS, Song LS, Chen Y, Deng LH, Stern MD, Lakatta EG, Cheng H. Termination of Ca^{2+} release by a local inactivation of ryanodine receptors in cardiac myocytes. *Proc Natl Acad Sci USA* 1998;95:15096–101.

166. Lopez-Lopez JR, Shacklock PS, Balke CW, Wier WG. Local calcium transients triggered by single L-type calcium channel currents in cardiac cells *Science*. 1995;268:1042–5.

167. Wang SQ, Stern MD, Rios E, Cheng H. The quantal nature of Ca^{2+} sparks and *in situ* operation of the ryanodine receptor array in cardiac cells. *Proc Natl Acad Sci USA* 2004;101:3979–84.

168. Wang SQ, Wei C, Zhao G, Brochet DX, Shen J, Song LS, Wang W, Yang D, Cheng H. Imaging microdomain Ca^{2+} in muscle cells. *Circ Res* 2004;94:1011–22.

169. Conti A, Sorrentino V. Differential distribution of ryanodine receptor type 3 (RyR3) gene product in mammalian skeletal muscles. *Biophys J* 1996;70:A124.

170. Protasi F, Takekura H, Wang Y, Chen SR, Meissner G, Allen PD, Franzini-Armstrong C. RyR1 and RyR3 have different roles in the assembly of calcium release units of skeletal muscle. *Biophys J* 2000;79:2494–508.

171. Schneider MF, Chandler WK. Voltage dependent charge movement of skeletal muscle: a possible step in excitation–contraction coupling. *Nature* 1973;242:244–6.

172. Rios E, Brum G. Involvement of dihydropyridine receptors in excitation–contraction coupling in skeletal muscle. *Nature* 1987;325:717–20.

173. Franzini-Armstrong C. The sarcoplasmic reticulum and the control of muscle contraction. *FASEB J* 1999;13 Suppl 2:S266–70.

174. Dirksen RT, Beam KG. Single calcium channel behavior in native skeletal muscle. *J Gen Physiol* 1995;105:227–47.

175. Gonzalez-Serratos H, Valle-Aguilera R, Lathrop DA, Garcia MC. Slow inward calcium currents have no obvious role in muscle excitation–contraction coupling. *Nature* 1982;298:292–4.

176. Tanabe T, Beam KG, Adams BA, Niidome T, Numa S. Regions of the skeletal muscle dihydropyridine receptor critical for excitation–contraction coupling. *Nature* 1990;346:567–569.

177. Garcia J, Beam KG. Calcium transients associated with the T type calcium current in myotubes. *J Gen Physiol* 1994;104:1113–28.

178. Dirksen RT, Beam KG. Role of calcium permeation in dihydropyridine receptor function. Insights into channel gating and excitation–contraction coupling. *J Gen Physiol* 1999;114:393–403.

179. Tsugorka A, Ríos E, Blatter LA. Imaging elementary events of calcium release in skeletal muscle cells. *Science* 1995;269:1723–6.

180. Klein MG, Cheng H, Santana LF, Jiang YH, Lederer WJ, Schneider MF. Two mechanisms of quantized calcium release in skeletal muscle. *Nature* 1996;379:455–8.

181. Gonzalez A, Kirsch WG, Shirokova N, Pizarro G, Brum G, Pessah IN, Stern MD, Cheng H, Rios E. Involvement of multiple intracellular release channels in calcium sparks of skeletal muscle *Proc Natl Acad Sci USA* 2000;97:4380–5.

182. Marx SO, Ondrias K, Marks AR. Coupled gating between individual skeletal muscle Ca^{2+} release channels (ryanodine receptors). *Science* 1998;281:818–21.

183. Grabner M, Dirksen RT, Suda N, Beam KG. The II–III loop of the skeletal muscle dihydropyridine receptor is responsible for the bi-directional coupling with the ryanodine receptor. *J Biol Chem* 1999;274:21913–9.

184. Nakai J, Sekiguchi N, Rando TA, Allen PD, Beam KG. Two regions of the ryanodine receptor involved in coupling with L-type Ca^{2+} channels. *J Biol Chem* 1998;273:13403–13406.

185. Morgan JP, Erny RE, Allen PD, Grossman W, Gwathmey JK. Abnormal intracellular calcium handling, a major cause of systolic and diastolic dysfunction in ventricular myocardium from patients with heart failure. *Circulation* 1990;81:III21–32.

186. Arai M, Alpert NR, MacLennan DH, Barton P, Periasamy M. Alterations in sarcoplasmic reticulum gene expression in human heart failure. A possible mechanism for alterations in systolic and diastolic properties of the failing myocardium *Circ Res* 1993;72:463–9.

187. Beuckelmann DJ, Nabauer M, Erdmann E. Intracellular calcium handling in isolated ventricular myocytes from patients with terminal heart failure. *Circulation* 1992;85:1046–55.

188. Gomez AM, Valdivia HH, Cheng H, Lederer MR, Santana LF, Cannell MB, McCune SA, Altschuld RA, Lederer WJ. Defective excitation–contraction coupling in experimental cardiac hypertrophy and heart failure. *Science* 1997;276:800–6.

189. Gwathmey JK, Copelas L, MacKinnon R, Schoen FJ, Feldman MD, Grossman W, Morgan JP. Abnormal intracellular calcium handling in myocardium from patients with end-stage heart failure. *Circ Res* 1987;61:70–76.

190. Hasenfuss G, Mulieri LA, Leavitt BJ, Allen PD, Haeberle JR, Alpert NR. Alteration of contractile function and excitation–contraction coupling in dilated cardiomyopathy. *Circ Res* 1992;70:1225–32.

191. Pieske B, Maier LS, Bers DM, Hasenfuss G. Ca^{2+} handling and sarcoplasmic reticulum Ca^{2+} content in isolated failing and nonfailing human myocardium. *Circ Res* 1999;85:38–46.

192. Bers DM, Eisner DA, Valdivia HH. Sarcoplasmic reticulum Ca^{2+} and heart failure: roles of diastolic leak and Ca^{2+} transport. *Circ Res.* 2003;93:487–90.

193. Hobai IA, O'Rourke B. Decreased sarcoplasmic reticulum calcium content is responsible for defective excitation–contraction coupling in canine heart failure. *Circulation* 2001;103:1577–84.

194. Pieske B, Maier LS, Schmidt-Schweda S. Sarcoplasmic reticulum Ca^{2+} load in human heart failure. *Basic Res Cardiol* 2002;97:I63–71.

195. Esposito G, Santana LF, Dilly K, Cruz JD, Mao L, Lederer WJ, Rockman HA. Cellular and functional defects in a mouse model of heart failure. *Am J Physiol Heart Circ Physiol* 2000;279:H3101–12.

196. Wessely R, Klingel K, Santana LF, Dalton N, Hongo M, Jonathan Lederer W, Kandolf R, Knowlton KU. Transgenic expression of replication-restricted enteroviral genomes in heart muscle induces defective excitation–contraction coupling and dilated cardiomyopathy. *J Clin Invest* 1998;102:1444–53.

197. Gomez G, Rawson NE, Cowart B, Lowry LD, Pribitkin EA, Restrepo D. Modulation of odor-induced increases in [Ca(2+)](i) by inhibitors of protein kinases A and C in rat and human olfactory receptor neurons. *Neuroscience* 2000;98:181–9.

198. Gomez AM, Guatimosim S, Dilly KW, Vassort G, Lederer WJ. Heart failure after myocardial infarction: altered excitation–contraction coupling. *Circulation* 2001;104:688–93.

199. He J, Conklin MW, Foell JD, Wolff MR, Haworth RA, Coronado R, Kamp TJ. Reduction in density of transverse tubules and L-type Ca(2+) channels in canine tachycardia-induced heart failure. *Cardiovasc Res* 2001;49:298–307.

200. Beuckelmann DJ, Nabauer M, Erdmann E. Characteristics of calcium currents in isolated human ventricular myocytes from patients with terminal heart failure. *J Mol Cell Cardiol.* 1991;23:929–37.

201. Reiken S, Gaburjakova M, Guatimosim S, Gomez AM, D'Armiento J, Burkhoff D, Wang J, Vassort G, Lederer WJ, Marks AR. Protein kinase A phosphorylation of the cardiac calcium release channel (ryanodine receptor) in normal and failing hearts. Role of phosphatases and response to isoproterenol. *J Biol Chem* 2003;278:444–53.

202. Wei SK, Ruknudin A, Hanlon SU, McCurley JM, Schulze DH, Haigney MC. Protein kinase A hyperphosphorylation increases basal current but decreases beta-adrenergic responsiveness of the sarcolemmal Na+-Ca^{2+} exchanger in failing pig myocytes *Circ Res.* 2003;92:897–903.

203. Jiang MT, Lokuta AJ, Farrell EF, Wolff MR, Haworth RA, Valdivia HH. Abnormal Ca^{2+} release, but normal ryanodine receptors, in canine and human heart failure. *Circ Res* 2002;91:1015–22.

204. Reiken S, Lacampagne A, Zhou H, Kherani A, Lehnart SE, Ward C, Huang F, Gaburjakova M, Gaburjakova J, Rosemblit N, Warren MS, He KL, Yi GH, Wang J, Burkhoff D, Vassort G, Marks AR. PKA phosphorylation activates the calcium release channel (ryanodine receptor) in skeletal muscle: defective regulation in heart failure. *J Cell Biol* 2003;160:919–28.

205. Ward CW, Reiken S, Marks AR, Marty I, Vassort G, Lacampagne A. Defects in ryanodine receptor calcium release in skeletal muscle from post-myocardial infarct rats. *FASEB J* 2003;17:1517–9.

206. Fill M, Coronado R, Mickelson JR, Vilven J, Ma JJ, Jacobson BA, Louis CF. Abnormal ryanodine receptor channels in malignant hyperthermia. *Biophys J* 1990;57:471–75.

207. O'Sullivan GH, McIntosh JM, Heffron JJ. Abnormal uptake and release of Ca(2+) ions from human malignant hyperthermia-susceptible sarcoplasmic reticulum. *Biochem Pharmacol* 2001;61:1479–85.

208. Fruen BR, Mickelson JR, Louis CF. Dantrolene inhibition of sarcoplasmic reticulum Ca^{2+} release by direct and specific action at skeletal muscle ryanodine receptors. *J Biol Chem* 1997;272:26965–71.

209. Krause T, Gerbershagen MU, Fiege M, Weisshorn R, Wappler F. Dantrolene — a review of its pharmacology, therapeutic use and new developments. *Anaesthesia* 2004;59:364–73.

210. Louis CF, Balog EM, Fruen BR. Malignant hyperthermia: an inherited disorder of skeletal muscle Ca+ regulation. *Biosci Rep* 2001;21:155–68.

211. Takekura H, Nishi M, Noda T, Takeshima H, Franzini-Armstrong C. Abnormal junctions between surface membrane and sarcoplasmic reticulum in skeletal muscle with a mutation targeted to the ryanodine receptor. *Proc Natl Acad Sci USA* 1995;92:3381–5.

212. Takeshima H, Komazaki S, Hirose K, Nishi M, Noda T, Iino M. Embryonic lethality and abnormal cardiac myocytes in mice lacking ryanodine receptor type 2. *EMBO J* 1998;17:3309–16.

213. Uehara A, Yasukochi M, Imanaga I, Nishi M, Takeshima H. Store-operated Ca^{2+} entry uncoupled with ryanodine receptor and junctional membrane complex in heart muscle cells. *Cell Calcium* 2002;31:89–96.

214. Conklin MW, Ahern CA, Vallejo P, Sorrentino V, Takeshima H, Coronado R. Comparison of Ca^{2+} sparks produced independently by two ryanodine receptor isoforms (type 1 or type 3). *Biophys J* 2000;78:1777–85.

215. Balschun D, Wolfer DP, Bertocchini F, Barone V, Conti A, Zuschratter W, Missiaen L, Lipp HP, Frey JU, Sorrentino V. Deletion of the ryanodine receptor type 3 (RyR3) impairs forms of synaptic plasticity and spatial learning. *EMBO J* 1999;18:5264–73.

216. Clancy JS, Takeshima H, Hamilton SL, Reid MB. Contractile function is unaltered in diaphragm from mice lacking calcium release channel isoform 3. *Am J Physiol* 1999;277:R1205–9.

217. Kouzu Y, Moriya T, Takeshima H, Yoshioka T, Shibata S. Mutant mice lacking ryanodine receptor type 3 exhibit deficits of contextual fear conditioning and activation of calcium/calmodulin-dependent protein kinase II in the hippocampus. *Brain Res Mol Brain Res* 2000;76:142–50.

218. Shimuta M, Yoshikawa M, Fukaya M, Watanabe M, Takeshima H, Manabe T. Post-synaptic modulation of AMPA receptor-mediated synaptic responses and LTP by the type 3 ryanodine receptor. *Mol Cell Neurosci* 2001;17:921–30.

219. Xin HB, Senbonmatsu T, Cheng DS, Wang YX, Copello JA, Ji GJ, Collier ML, Deng KY, Jeyakumar LH, Magnuson MA, Inagami T, Kotlikoff MI, Fleischer S. Oestrogen protects FKBP12.6 null mice from cardiac hypertrophy. *Nature* 2002;416:334–8.
220. Wehrens XH, Marks AR. Altered function and regulation of cardiac ryanodine receptors in cardiac disease. *Trends Biochem Sci* 2003;28:671–8.
221. Lamb GD, Stephenson DG. Effects of FK506 and rapamycin on excitation–contraction coupling in skeletal muscle fibres of the rat. *J Phys*1996;494:569–76.
222. Shou W, Aghdasi B, Armstrong DL, Guo Q, Bao S, Charng MJ, Mathews LM, Schneider MD, Hamilton SL, Matzuk MM. Cardiac defects and altered ryanodine receptor function in mice lacking FKBP12. *Nature* 1998;391:489–92.
223. Tang W, Ingalls CP, Durham WJ, Snider J, Reid MB, Wu G, Matzuk MM, Hamilton SL. Altered excitation–contraction coupling with skeletal muscle specific FKBP12 deficiency. *FASEB J* 2004;18:1597–9.
224. Wehrens XH, Marks AR. Novel therapeutic approaches for heart failure by normalising calcium cycling. *Nat Rev Drug Discov* 2004;3:565–73.
225. Lehnart SE, Wehrens XH, Marks AR. Calstabin deficiency, ryanodine receptors, and sudden cardiac death. *Biochem Biophys Res Commun* 2004;322:1267–79.

12 Methods in Cyclic ADP-Ribose and NAADP Research

Anthony J. Morgan, Grant C. Churchill, Roser Masgrau, Margarida Ruas, Lianne C. Davis, Richard A. Billington, Sandip Patel, Michiko Yamasaki, Justyn M. Thomas, Armando A. Genazzani, and Antony Galione

CONTENTS

12.1   INTRODUCTION

In this chapter, we describe some of the methods used in our laboratory to study cADPR- and NAADP-mediated calcium signaling. We hope that the descriptions of the methods and consultation of the cited references will provide a detailed account of many of the techniques that will allow the reader to investigate the actions of these molecules in their chosen systems.

12.1.1   BACKGROUND TO cADPR AND NAADP

Although inositol 1,4,5-trisphosphate (IP_3) has long been held to operate as a ubiquitous intracellular messenger for Ca^{2+} mobilization [1], only recently has it gained wider acceptance that additional Ca^{2+} release mechanisms exist that are regulated by a family of pyridine nucleotide metabolites. The first indication of this concept came from the pioneering work of Lee and colleagues who showed that the "housekeeping" co-factors β-NAD^+ and β-$NADP^+$ could trigger Ca^{2+} release from sea urchin egg microsomal fractions by a mechanism apparently independent of IP_3 [2]. β-NAD^+ released Ca^{2+} after a lag period, reflecting the requirement for its metabolism to another species whereas β-$NADP^+$-induced Ca^{2+} release occurred without significant delay, which was attributable to the presence of a nucleotide contaminant. Subsequently, the structures of the active metabolites (Figure 12.1) were determined as cyclic adenosine diphosphate ribose (cADPR) and nicotinic acid adenine dinucleotide phosphate (NAADP), respectively, that have now been shown to mobilize Ca^{2+} in different cell types in a wide range of organisms including plants, invertebrates, and mammals [3].

Even from the earliest days, the properties of the β-NAD^+- and β-$NADP^+$-dependent Ca^{2+} release pathways were demonstrably distinct from that stimulated by IP_3, both in terms of their pharmacology as well as their intrinsic properties (see below). Most notably, the affinity of NAADP for its binding site is the highest among Ca^{2+}-mobilizing messengers (affinity 200–500 pM in sea urchin [4, 5]) with cADPR not far behind with an affinity of 2–17 nM in sea urchin [6, 7]; for comparison, the affinity of IP_3 for its receptor subtypes is of the order of 2–22 nM [8]. In addition to these different affinities, cADPR, NAADP, and IP_3 appear to activate different Ca^{2+} release mechanisms because desensitization of one pathway fails to affect release by

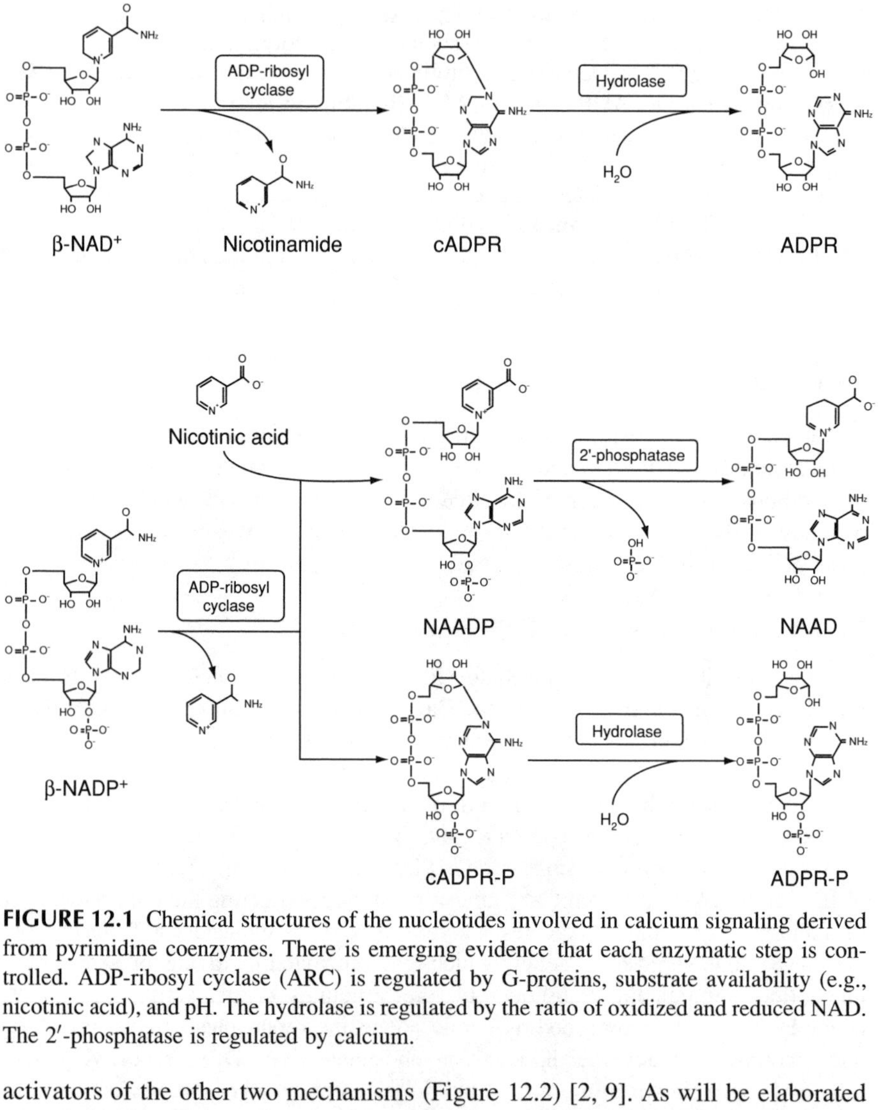

FIGURE 12.1 Chemical structures of the nucleotides involved in calcium signaling derived from pyrimidine coenzymes. There is emerging evidence that each enzymatic step is controlled. ADP-ribosyl cyclase (ARC) is regulated by G-proteins, substrate availability (e.g., nicotinic acid), and pH. The hydrolase is regulated by the ratio of oxidized and reduced NAD. The 2'-phosphatase is regulated by calcium.

activators of the other two mechanisms (Figure 12.2) [2, 9]. As will be elaborated upon below, such homologous desensitization points to the recruitment of very different intracellular Ca^{2+} release channels and stores by each messenger.

Our understanding of cADPR and NAADP metabolism and, crucially, its regulation is currently sketchy and sorely lags behind that of the IP$_3$/PLC system. However, since the first incarnation of this chapter in 1999, both cADPR and NAADP have subsequently been elevated to the status of *bona fide* second messengers, thanks in part to studies showing that external stimuli rapidly elevate their intracellular levels [10–12]; with such model systems in place, the exciting task ahead is to elucidate the underlying transduction mechanisms and physiologically relevant pathways of cADPR and NAADP synthesis.

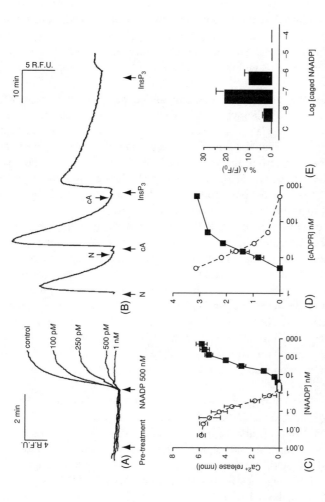

FIGURE 12.2 (A) Representative traces of Ca²⁺ release in response to 500 n*M* NAADP after pre-treatment with the indicated concentrations of sub-threshold NAADP in sea urchin homogenate. (B) Lack of cross-desensitization of the different messengers in sea urchin homogenate. NAADP and cADPR were used at 500 n*M*, IP₃ at 1 μ*M*. (C) and (D) Dose responses of NAADP and cADPR with respect to Ca²⁺ release (filled squares), and residual release in response to maximal, 500 n*M* (open symbols). Whereas cADPR (and IP₃ — not shown [4]) display the expected, reciprocal relationships between the initial and residual responses, NAADP exhibits its unique property where non-releasing concentrations inactivate. Data are taken from Ref. [4]. (E) Bell-shaped dose–response curve of NAADP-induced Ca²⁺ release in mammalian cells. Data show fluorescence responses to photolyzing caged NAADP in intact MIN6 pancreatic β cells (taken from Ref. [11]).

12.1.2   MECHANISM OF cADPR-MEDIATED Ca²⁺ RELEASE

12.1.2.1   Channels and Targets

The mechanism whereby cADPR stimulates Ca^{2+} release is of considerable interest and widespread applicability given that the intracellular channel mediator is almost certainly the ryanodine receptor (RyR) [13]. RyRs are large homotetrameric Ca^{2+} release channels (~560 kDa per subunit) [14, 15] that were first described as the major Ca^{2+} release pathways from the sarcoplasmic reticulum during excitation–contraction coupling in skeletal and cardiac muscle [15]. However, the expression of RyRs has proven more widespread than was originally thought with neurons and even nonexcitable cells such as endothelial cells and exocrine cells appended to a growing list [16].

It is true to say that the RyR's predominant *modus operandi* is that of Ca^{2+}-induced Ca^{2+} release (CICR) where the channel opens in response to modest, suprathreshold elevations of Ca^{2+}, and thereby serves as an *amplifier* of local, "trigger" Ca^{2+} release; in spatial terms alone, such amplification is paramount since it drives the transition from discrete, isolated Ca^{2+} events (Ca^{2+} sparks) to global, propagating Ca^{2+} waves [13]. In effect, cADPR activates the RyR by "promoting" its CICR mode or, put another way, by sensitizing the RyR to the ambient $[Ca^{2+}]$ [13] that results in an explosive Ca^{2+} release.

How precisely cADPR modulates RyRs has remained obscure because its binding site is, as yet, undefined, but would seem not to be on the channel protein itself (cADPR was shown to activate sheep skeletal RyRs in black lipid membranes, but since its "inactive" metabolite ADPR was also effective, interpretation is difficult [17]). Adjunct candidate cADPR-binding proteins include photoaffinity-labelled 100- and 140-kDa soluble sea urchin egg proteins [18], as well as members of the family of FKBPs (FK-506 Binding Protein). Targets for the immunosuppressant, FK-506, these proteins normally bind to and stabilize (suppress) RyR [19] (and IP₃R [20]) channel activity, for which reason they have recently been renamed *calstabins* [21] (with a given isoform associating with its particular, cognate RyR subtype [19]).

This family was first tangibly implicated in cADPR action when FKBP12.6 (calstabin 2) was shown to bind cADPR and dissociate from pancreatic β cell RyRs, thereby relieving them from calstabin suppression and promoting Ca^{2+} release [22]. However appealing a mechanism, subsequent attempts to confirm this model have met with mixed results: cADPR fails to affect the association of recombinant FKBP isoforms with RyR3 [23]; whereas in airway smooth muscle, cADPR-stimulated Ca^{2+} release is absent in both FKBP12.6-null mice [24] as well as in arterial preparations functionally or physically stripped of FKBP12.6 [25]. However, as some authors concede, such data (or inhibition by FK-506 [25, 26]) do not prove that FKBP is the long-sought cADPR receptor rather than an essential co-factor, and therefore we are still no closer to identifying the cADPR receptor. Other proteins such as calmodulin (CaM), CaM-Kinase II, and PKA clearly modulate cADPR-induced Ca^{2+} release [21], but they are unlikely to bind cADPR (though this has not been empirically determined, to the best of our knowledge) and merely serve to alter the gain of the system, even perhaps by altering the association of calstabin to phosphorylated RyR [27].

As well as the mechanism of cADPR activation, that of desensitization of the cADPR response to successive stimuli has been recently probed in sea urchin, and found to involve CaM [28]. Basically, this essential cADPR co-factor dissociates from the ER (presumably the RyR complex to which it binds [14]) upon maximal cADPR challenge, thereby rendering the channel refractory to further cADPR addition. Only upon the removal (hydrolysis) of cADPR will CaM gradually reassociate with the RyR and sensitivity be restored.

12.1.2.2 cADPR and RyR Subtypes

In the wake of the RyR hypothesis, the obvious question was whether all RyR were responsive to cADPR or only a particular isoform. In mammals, there are three genetically distinct isoforms that share >60% sequence homology, RyR1 (predominantly skeletal muscle), RyR2 (mostly cardiac and in the brain), and RyR3, expressed ubiquitously but at low concentrations [15, 19]. In sea urchin, only one highly homologous isoform exists that has been cloned and that is probably more similar to mammalian type 2 and 3 than to type 1 [29].

As is the case for IP_3 receptors [20], there is a spectrum of expression whereby some cell types express only one isoform while in others, multiple isoforms are co-expressed [16, 21]. Originally, the mechanistic differences between RyRs of skeletal and cardiac muscle prompted investigators to compare cADPR responsiveness and concluded that only the cardiac isoform (RyR2) could be activated and that the skeletal (RyR1 and 3) could not [30]. As is often the case, the intervening years have proven to be more confusing than enlightening with subsequent reports of cADPR stimulating [17, 31] or having no effect [32] upon skeletal muscle RyR (which ADPR might stimulate [17, 33]). Similarly, RyR3 has now been shown to respond to cADPR in muscle preparations [34] and T cells [35]. Less ambiguously, cADPR stimulation of RyR2 shows a more consistent pattern across different cell types, e.g., Refs. [24, 36–38].

The situation with skeletal muscle is even more complex than first imagined since excitation–contraction coupling potentially involves both RyR1 and RyR3 subtypes (formerly α- and β-RyR) that may not act equivalently [15] (if both expressed in adult [31]). Functionally speaking, RyRs amplify the exciting sarcolemmal action potential by two modes of action, a depolarization-induced Ca^{2+} release (DICR) and the CICR. The former is attributable to a physical interaction between a RyR closely apposed with the sarcolemmal voltage sensor, the L-type Ca^{2+} channel; upon depolarization, the L channel conformational change is transmitted to the RyR bound to it and DICR activated. By contrast, physical intimacy is not obligate for CICR, by definition, since RyR are conventionally activated by the Ca^{2+} that enters the cytosol via adjacent L channels.

In molecular terms, type 1 receptors appear to be the primary mediators of DICR since (a) they are the only isoform to bind the L channel, and (b) their CICR capability is efficiently stabilized (suppressed) by their cognate FKBP isoform (calstabin 1) [39]. By contrast, RyR3 are more suited to the CICR mode in skeletal muscle, not least because they exhibit a higher CICR gain than type 1 [39]. If true, then it would be expected that cADPR might act preferentially at type 3 over type 1, and indeed this has been shown to be the case in lipid bilayers [34]. However, in another study, cADPR remains a potent stimulator of Ca^{2+} release in skeletal muscle of RyR3 knockout mice [31], implying that RyR1 are indeed responsive and/or that isoform switching has occurred to compensate for the lack of type 3. Again, these confusions reinforce the fact that a fuller appreciation of the nature of the cADPR-binding sites is much needed to tease out differential RyR regulation.

12.1.2.3   Non-RyR Sites of Action?

In spite of an overwhelming amount of data in favor of the RyR as the primary target of cADPR (>500 publications excluding reviews), it has recently been challenged by some who suggest that cADPR may act at multiple sites [40], and may even stimulate Ca^{2+}-ATPases of the sarcoplasmic reticulum (SERCA) [41] or the plasma membrane (PMCA) [42]. However, Ca^{2+} removal was only enhanced by very high concentrations of cADPR (300 μM) in the latter study (interestingly, a smooth muscle preparation in which cADPR failed to promote Ca^{2+} release anyway). These proposals require further molecular analysis or confirmation.

12.1.3   Mechanism of NAADP-Induced Ca^{2+} Release

NAADP is at least as potent as cADPR but its mechanism of Ca^{2+} release is generally thought to be independent of both IP_3Rs and RyRs and represents a new class of Ca^{2+} release channel. Such a conclusion is based upon multiple lines of evidence: (a) the antagonist profile is inconsistent with the involvement of these well-understood intracellular channels (Section 12.6.1); (b) in sharp contrast to both IP_3 and RyRs, NAADP-gated Ca^{2+} channels do not appear to be regulated by cytosolic Ca^{2+} in the dual manner that is a hallmark of IP_3 and ryanodine receptors (significantly, they do not undergo CICR, nor are affected by the luminal Ca^{2+} concentration [43, 44]); (c) NAADP-induced Ca^{2+} release (NICR) undergoes a unique form of self-inactivation not manifest by the alternative systems (see below); (d) homologous desensitization of the Ca^{2+} release to IP_3 and cADPR does not affect NICR (or *vice versa*); (e) NICR is selectively blocked by potassium and L-type Ca^{2+} channel antagonists [45]; (f) biochemically, their native size (400–470 kDa) is substantially smaller than IP_3 and ryanodine receptors (1000–2000 kDa) [46]; (g) in contrast to Ca^{2+} stores expressing IP_3Rs and RyRs, those sensitive to NAADP are thapsigargin insensitive [47, 48].

The reason for the thapsigargin insensitivity of an NAADP-dependent store remained enigmatic until the revelation that it was not the ER but acidic organelles (e.g., lysosome-related organelles [10, 48–50], and secretory vesicles [49, 51, 52]) that housed the NAADP-regulated channel (but not SERCA). The properties of these stores are inherently different from the ER, in terms of their location, and Ca^{2+} handling, and thereby afford the cell the ability to selectively tune NAADP-induced Ca^{2+} release without affecting the other systems, or to deliver Ca^{2+} to a different subcellular locale. Indeed, we recently demonstrated that different agonists can mobilize different Ca^{2+} stores ("organelle selection") within mammalian cell by judicious recruitment of the appropriate second messenger complement [10, 49]; in other words, agonists coupling to IP_3 and/or cADPR mobilize the ER/SR whereas those coupled to NAADP mobilize acidic vesicles. Furthermore, the distribution of these acidic Ca^{2+} stores can be different from cell type to cell type and is reflected in the spatial Ca^{2+} response (polarized versus global; Figure 12.5C) [49] and/or regions of the ER/SR with which it interacts [10].

Perhaps the most unusual feature of NAADP-gated Ca^{2+} release mechanism is its inactivation properties. Unlike the mechanisms regulated by IP_3 and cADPR which, after induction of Ca^{2+} release, appear to become refractory to subsequent activation (Figure 12.2D), very low concentrations of NAADP (IC_{50} 200 pM) are able to inactivate NAADP-induced Ca^{2+} release fully at concentrations well below those required to activate Ca^{2+} release (EC_{50} 30 nM), at least, in sea urchin egg and plants [4, 53, 54] (Figure 12.2A and C). The mechanism and physiological significance of this most unusual phenomenon are still unclear, but this inactivation — together with essentially irreversible binding of NAADP to its receptor [5] — means that a spatio-temporal "imprint" of prior NAADP exposure lingers long after NAADP has gone, thereby engendering a primitive form of "memory" [55]. Whether this occurs in the same form in mammalian cells, however, is open to conjecture, especially in view of the reversible kinetics of NAADP binding [56] as well as differences in the details of its self-inactivation (i.e., only occurring by supramaximal rather than low [NAADP]), manifesting itself as a bell-shaped Ca^{2+} release concentration-dependence curve (Figure 12.2E) [11, 51, 57, 58].

The very fact that NAADP-induced Ca^{2+} release cannot amplify itself through CICR raised the pertinent question as to how NAADP-induced Ca^{2+} waves spread across the cell [59]. The most straightforward explanation is that NAADP-induced Ca^{2+} release is probably an early event in the Ca^{2+} signal, equivalent to the "trigger release" mentioned above (Section 12.1.2.1), which is, in turn, locally amplified by CICR at the IP_3 and/or ryanodine receptors [59]. Furthermore, the diffusion coefficient of NAADP is likely greater than that for Ca^{2+} itself and therefore can diffuse greater distances from a point source (concentration permitting) and hence leave a Ca^{2+} wave in its wake [59]. Not only sea urchin eggs, but also pancreatic acinar cells [60] and smooth muscle [10] show this functional and dynamic interaction between Ca^{2+} messengers and stores that has been termed "channel chatter" [61] and whose intensity determines whether or not a Ca^{2+} signal "goes global" [62].

Despite a wealth of data, both biochemical and functional, that NAADP activates a hitherto unknown channel, a few recent reports apparently undermine this paradigm by suggesting that NAADP directly activates the RyR itself [63–65]. The fact that NAADP may stimulate release from the nuclear envelope and/or contiguous ER [66] also prompts the authors to cast doubt upon the existence of a separate, acidic NAADP-sensitive store. We cannot currently account for such discrepancies, particularly when we have found no corroborative evidence in intact preparations of the same cell type [49], but proffer "channel chatter" as one issue worthy of consideration. Further work is essential to clarify this most important of questions.

12.2 PURIFICATION AND SYNTHESIS OF cADPR AND ANALOGS

cADPR is synthesized from its precursor β-NAD$^+$ catalyzed by ADP-ribosyl cyclase enzymes (see Section 12.7.1).

By purifying these enzymes and by incubating them with substrates, cADPR can be synthesized and then must be separated from its metabolites by HPLC. First, we describe how to prepare the HPLC columns before discussing the specifics of the cADPR synthetic reaction.

12.2.1 HPLC

Anion exchange HPLC has been used routinely to separate and purify synthesized compounds and to clean up commercially purchased compounds, particularly, caged compounds. We use a variation of the method first reported by Axelson et al. [67] in which compounds are eluted with volatile agents that are removed upon drying the collected fractions, thus eliminating a desalting step. Either trifluoroacetic acid (TFA) or triethylammonium bicarbonate (TEAB) is used. TFA is generally used because it can be made quickly and is much cheaper than commercial TEAB (on the other hand, TEAB can be made by bubbling CO_2 into triethylammonia, which is stirred on ice, until the pH reaches 7.5–8.8). Owing to the extreme volatility of TFA, we favor its dilution based on mass rather than volume: assuming FW = 114.02 and $D^{20} = 1.5351$, addition of 8.55 g (~5–6 ml) of TFA directly into a beaker of 500 ml of water placed upon a tared balance gives a 150 mM solution. To avoid inhalation of fuming, neat TFA, dilution should be carried out in a fume cupboard. A macroporous strong anion exchange resin (AG MP-1, BioRad) is used as the anion exchange resin because it packs well and exhibits high flow rates at low pressures. Importantly, this allows easy packing of your own columns and makes a guard column unnecessary, as a fouled column can be repacked quickly and cheaply. If better resolution is required or compounds that are sensitive to acidic conditions are to be separated, a commercial anion-exchange column with a smaller particle size (10 μm) is used. We use a MonoQ 5/5 (Pharmacia) and elute the nucleotides with a TEAB gradient [68] in which solvent A is water and B is TEAB (1 M, pH 8.8) and progresses as follows (time in min and %B): 0, 0%, 12, 0%, 16, 20%, 36, 30%, 37, 100%, 40, 100%, 40.1, 0%, 45, 0%.

Fines (small particles), which would inhibit flow rates and increase pressures, were removed from the AG MP-1 by repeatedly suspending it in a large volume of water (10 g into 100 ml) and pouring off the material remaining in suspension after about 90% of the resin had settled. To exchange its counterion, the AG MP-1 was rinsed with 1 M NaOH (20 ml per g resin) to replace Cl$^-$ with OH$^-$, rinsed with double deionized H_2O until the pH was <9 to remove excess NaOH, and then rinsed with 150 mM TFA (20 ml per g resin) to replace OH$^-$ with trifluoroacetate. AG MP-1 that was prepared this way was stored at 4°C. Columns were packed by joining two columns (borosilicate glass, 3×50 mm, Omnifit) with a sleeve (Omnifit), securing them vertically to a ring stand, attaching a funnel to the top column with a short piece of snug-fitting tubing, fitting an endcap with a frit (Omnifit) to the bottom column, and pouring a slurry of AG MP-1 suspended in 150 mM TFA into the funnel. After the resin had settled by gravity (20–40 min), the funnel was removed and an endcap without a frit was screwed onto the column and an adapter (10–32 internal to 1/4″-28 external, Upchurch) was used to couple the column to peek tubing from the HPLC outport, and 150 mM TFA was pumped through the column at 5 ml/min for 15 min to pack the resin. The sleeve was then removed and the top of the lower packed column was fitted with an endcap with a frit and a column-to-peek adapter. Retention times were slightly different between columns so each freshly packed column was calibrated with standards.

12.2.2 cADPR Synthesis

ADP-ribosyl cyclase from *Aplysia* ovotestis is the usual choice. The reason for this is that it behaves as a simple cyclase enzyme with a high synthetic catalytic activity. In contrast, several of the mammalian cyclases are bifunctional enzymes, also expressing cADPR hydrolase activities that would degrade cADPR to ADP-ribose (ADPR). *Aplysia* cells can be obtained from various marine organism suppliers (Marinus Inc., Long Beach, CA) as they are often used for neurobiological investigations. Knowledge of the mollusc's gross anatomy is required, and a crude extract of ovotestis can be aliquoted and stored prior to use [69]. Alternatively, the enzyme can be purchased separately from biochemical suppliers. *Aplysia* ADP-ribosyl cyclase (Sigma) was found to have cADPR hydrolase contamination that is dealt with in one of two ways: either by (a) dilution to ameliorate the hydrolase or (b) purification by cation exchange chromatography on a carboxymethyl resin (CM Sepharose Fast-Flow, Pharmacia) packed into a 3 × 150 mm column as described above for AG MP-1 resin. ADP-ribosyl cyclase is suspended in 20 mM Tris-HCl, pH 7.4 (buffer A), loaded onto a column equilibrated in the same buffer, and eluted at a flow of 1 ml/min with a linear gradient of NaCl. Buffer B is 1 M NaCl in 20 mM Tris-HCl, pH 7.4. Gradients are 0 min, 0%B; 5 min, 0%B; 60 min 50%B; 60.1–70 min, 0%B. Protein peaks are detected by monitoring absorbance at 280 nm. An initial large peak (fall through) is discarded and a broad peak at 13–17 min (about 75–110 mM NaCl) is collected as 1-ml fractions that have hydrolase-free cyclase activity. These fractions were used directly as stock solutions, as the NaCl was sufficiently diluted so as to not interfere with the cyclase reaction (used at 1:100 dilution).

To synthesize cADPR, the following reaction is set up and allowed to proceed for 2 h at room temperature: 2 mM β-NAD$^+$ and 100 ng/ml ADP-ribosyl cyclase in 10 mM Tris-HCl, pH 7.2. The compounds are separated by anion exchange HPLC using a nonlinear TFA gradient at a flow rate of 1 ml/min as described previously [70] where solvent A is water; solvent B is 150 mM TFA. Gradients are 0 min, 0%B; 6 min, 2%B; 11 min, 4%B; 16 min, 8%B; 21 min, 16%B; 26 min, 32%B; 26.1–30 min, 100%B; 30.1–40 min, 0%B. Protein is not removed from the sample before injection, but small particles are removed by either filtration through 0.22-μm syringe filters or by centrifugation at 13,000g for 10 min. The approximate elution times are: nicotinamide (2 min), β-NAD$^+$ (9 min), cADPR (14 min), and ADPR (21 min) (see Figure 12.3). The peak-contained cADPR is collected and concentrated to dryness in a vacuum concentrator and stored at $-80°C$. Conversions are typically around 50% (note: cADPR stocks are stored long term as lyophilized powder, or only short term as frozen solutions because the N^1 glycosidic bond is labile and nonenzymatically hydrolyzed in neutral aqueous solution to yield inactive ADP-ribose; $t_{1/2}$ ~12 h at room temperature). For preparative work a column can be overloaded with about 10 μmol nucleotides per 1 ml AG MP-1 resin and still provide adequate separation of reactants and products. Cyclic ADP-ribose concentration is determined by absorption at 254 nm based on a molar absorption coefficient of 14,300 $M^{-1}cm^{-1}$, and the identity of cADPR is confirmed by cross-desensitization in sea urchin homogenate bioassay (see Figure 12.2 and Sections 12.7.2 and 12.8.3). The purity of the purified cADPR is typically greater than 98% based on HPLC peak areas.

12.2.3 Synthesis of [^{32}P]cADPR

Synthesis of [^{32}P]cADPR is desirable for the identification of cADPR-binding sites (Section 12.8.4) and may also be used in a radioreceptor assay for determination of endogenous cADPR levels (Section 12.8.4). This compound has a much higher specific activity (~1000 Ci/mmol) than its tritiated commercially available counterpart (30–70 Ci/mmol; Amersham).

Note that for enzymatic syntheses of radiolabelled cADPR or NAADP, we often check that thawed aliquots of frozen enzymes are still functional by carrying out cold, unlabelled reactions first. This is both safer and cheaper than wasting expensive [^{32}P]β-NAD$^+$.

To prepare [^{32}P]cADPR, the cyclization reaction is carried out for 2 h at room temperature in the vial in which the radiolabelled compound is shipped. The reaction solution contains 250 μCi [^{32}P]β-NAD$^+$ (Amersham), 100 ng/ml of re-purified ADP-ribosyl cyclase, and 5 mM Tris-HCl (pH 7.4) in a total volume of 250 μl.

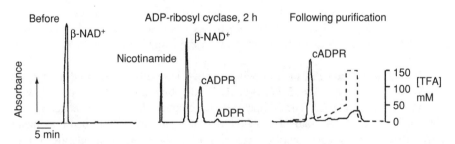

FIGURE 12.3 Use of HPLC to monitor and purify cADPR synthesized by ADP-ribosyl cyclase. cADPR is synthesized by incubating β-NAD$^+$ with purified *Aplysia* ADP-ribosyl cyclase for 2 h. Details of the reaction conditions and HPLC separation are given in the text. Solid lines are the absorbance at 254 nm and the dashed line is the TFA gradient.

Cyclic ADP-ribose was separated from its substrate using an HPLC method, described previously [31], which limits the radioactive contamination to the column and some tubing that is dedicated to radioactive work and prevents contamination of the injection port and the absorbance detector. This method also enables all the procedures with radioactive compounds to be performed in the same space a few feet away from the HPLC equipment behind Perspex (plexiglass) shielding. All work with ^{32}P is carried out behind Perspex shielding with gloves and goggles.

The reaction mixture is loaded onto a column with a peristaltic pump (flow rate of about 100 μl/min) by inserting the end of peek tubing (attached via a peristaltic tubing adapter, Upchurch) into the reaction vial. The reaction vial is washed five times with 300 μl of water until the radioactivity, remaining in the reaction vial (as detected with a Geiger counter), does not decrease any further and the majority of the radioactivity is at the top of the column (note: a column that is inadequately washed with water prior to use may contain residual TFA that causes the radioactive nucleotides to elute into or through the column during loading, resulting in loss of precious [^{32}P]cADPR or incomplete resolution of products). The connection between the peek tubing attached to the column and the peristaltic tubing is then broken, and reattached via a female–female peek union (Upchurch) to a length of peek tubing coming from the outport of the HPLC pump. Five ml of water is pumped through the column to ensure there are no leaks and then the TFA gradient is started. The bottom of the column has a short piece of peek tubing attached and once the effluent of the first 5–6 min is discarded as waste, 24 × 1-ml fractions are collected in microfuge tubes (with the lids cut off) placed in a rack within an acrylic, radiation-shielded storage box (Sigma). Each of the tubes already contains a pre-titrated volume (15–60 μl) of 1 M Tris-HCl (pH 8.8) that neutralizes the TFA at which cADPR elutes (10–30% solution B). Peaks are determined by scintillation counting 1 μl of each fraction in 10 ml water using Cerenkov radiation (Figure 12.4). β-NAD$^+$ elutes at 11 min, cADPR at 19 min, and ADPR at 23 min. Fractions are then stored frozen at −20°C. The use of [^{32}P]cADPR is described in Section 12.8.4.

12.2.4 CAGED cADPR

Caged compounds are inactive precursors of the molecule of interest, be it a second messenger or a Ca^{2+} chelator [71] that are photolyzed by UV light, breaking down on a microsecond timescale to liberate the active species. It is crucial that these precursors are not only inactive, but do not antagonize other processes, e.g., caged cADPR is not an antagonist of the cADPR binding [72]. If measuring Ca^{2+} simultaneously, then indicator dyes such as fluo-3 or Oregon-green-BAPTA-1 should be used, since these are excited in the visible spectrum and not the UV range that is required for fura-2, which may lead to the unintended uncaging of the molecule.

Among their useful properties is the fact that UV light can be delivered to a nonopaque specimen at precise times and at precise intensities. The former permits repetitive (or cumulative) production of the active species (provided that the excess caged compound is not exhausted; Figure 12.5B); the fact that the degree of photolysis is proportional to the light intensity (and quantum efficiency of the caged

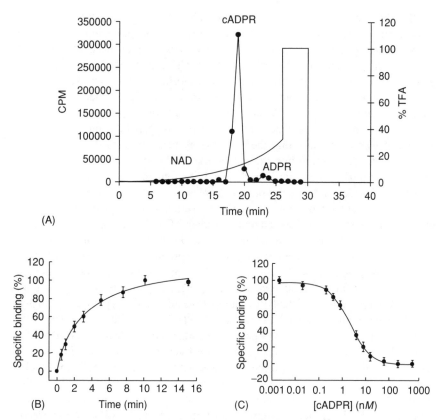

(A)

(B)   Time (min)

(C)   [cADPR] (nM)

FIGURE 12.4 (A) HPLC separation and purification of synthetic [^{32}P]cADPR. After incubation of [^{32}P]β-NAD$^+$ with ARC, products are loaded onto and HPLC column. One-ml fractions were eluted with the TFA gradient and conditions detailed in the text, and 1 μl taken for Cerenkov determination of radioactivity. (B) Association kinetics of [^{32}P]cADPR binding to the sea urchin receptor. (C) Homologous displacement of [^{32}P]cADPR by cold cADPR in the same system; affinity ~2 nM, B_{max} = 12 fmol/mg protein, and n^H ~0.97 (data of B and C from Thomas, J. M., Masgrau, R., Churchill, G. C., and Galione. *Biochem J*, 359, 451–7, 2001).

species) allows precise titration of the active compound's concentration. With computer-controlled modern UV laser systems (e.g., 351 and 364 nm lines, Enterprise, Coherent on a Zeiss 510LSM confocal microscope), one can even choose to photolyze in subcellular domains (rather than globally) to investigate regional effects of the agent in larger cells such as eggs and myocytes (Figure 12.5A). The use of a multiphoton laser to uncage will restrict the uncaging volume further still [73].

A major disadvantage of most caged compounds is that they are cell impermeant and therefore require access to the cell interior (Section 12.10.1); exceptions to this include AM ester forms of the caged Ca^{2+}, NP-EGTA, and the Ca^{2+} chelator, diazo-2 (Molecular Probes). This usually restricts use to single cells, but this inconvenience is more than offset by the facility of temporally separating

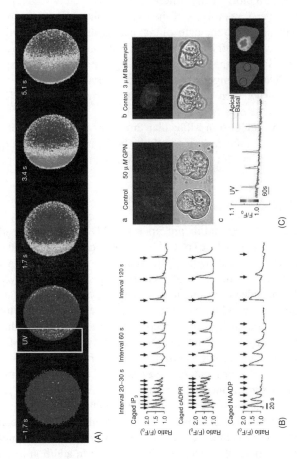

FIGURE 12.5 (A) Photolysis of caged NAADP in an intact sea urchin egg. An egg was previously microinjected with the Ca^{2+} dye 10 μM Oregon Green-BAPTA-dextran and 50 μM caged NAADP and visualized using confocal microscopy. Local uncaging was effected in the region indicated using a UV laser limited by a shutter and images recorded at the times indicated. Note the cortical flash and subsequent Ca^{2+} wave [12]. (B) Traces from eggs microinjected with the caged messengers indicated and global repetitive photolysis initiated by UV laser at each arrow. Only NAADP shows a profound inactivation independent of the frequency of uncaging. (C) Effect of GPN and bafilomycin A1 upon apical Lysotracker Red staining of pancreatic acinar cells (fluorescent and corresponding bright field images — upper panels). Local, apical Ca^{2+} oscillations evoked by global photolysis of caged NAADP in the same cell type (data from Yamasaki, M., et al., J Biol Chem 279, 7234–40, 2004).

injection and delivery of the active species (Figure 12.5) that obviates diffusional barriers and injection artifacts.

Use of caged compounds comes with its own peculiar pitfalls. Controls should be performed to check that the UV irradiation of cells does not produce the observed effects of photolysis of the caged compound. In addition, to check that other products of photolysis such as the freed uncaging group are innocuous, the same caged derivative of a biologically "inert" substance such as caged phosphate can be used [57, 71]. Indeed, photolysis of the caging 2-nitrophenyl group releases 1-(2-nitrosophenyl)ethanone into the cytosol that can oxidize cellular thiols; modification of sulfhydryls can be offset by including reducing agents extracellularly such as dithiothreitol. Photodamage by the UV light itself can be reduced by application of free radical scavengers, but again appropriate controls are required. If these steps are followed, then the effects of uncaging caged cADPR can be ascribed more clearly to the effects of cADPR itself.

Caged cADPR is available from Molecular Probes as a cell impermeant NPE-ester and should be stored in the dark. It should be borne in mind that the quantum yields of these compounds are four- to sixfold lower than those for other adenine nucleotides [74].

12.2.5 cADPR ANALOGS

A number of analogs of cADPR and its precursor β-NAD$^+$ have been synthesized and deployed in the biological analyses of Ca^{2+} release or cADPR metabolism, both in cell-free systems and in intact cells [75]. Many are derived from a chemoenzymatic synthesis that exploits the broad substrate specificity of *Aplysia* ARC such that chemically (or indeed enzymatically) synthesized β-NAD$^+$ analogs can be cyclized into their corresponding cADPR congenors. More recently, total synthetic routes for cADPR and analogs have been deduced that overcome the limitations imposed by ARC enzymology and that broaden the possibilities. Although a plethora of analogs exist that allows the dissection of structure–activity relationships at either the Ca^{2+} release or metabolic sites, we shall restrict our comments to those of more practical significance to the physiologist, and allow the reader to delve further into more detailed reviews should they so choose [75].

Anyone studying the function of a given receptor, whether of the intracellular or plasmalemmal variety, is accustomed to having a variety of pharmacological tools at their disposal, and the field of cADPR is no exception: in addition to the usual agonists and antagonists (some with additional useful properties, see below), fluorescent cADPR analogs have also been developed that permit real time *in vitro* determinations of ARC and cADPR hydrolase activities.

12.2.5.1 Agonists and Antagonists

Analogs substituted on the position 8 of the adenine ring usually (but not always) act as competitive antagonists as assessed from Ca^{2+} release assays using the sea urchin homogenate [74] or mammalian permeabilized cells [38, 41, 76–78]. Dealing with antagonists first, a number of 8-substituted analogs have been synthesized with a rank order of antagonist potency that decreases with increasing size of the molecular volume of the substituent and which is exemplified by the relative IC_{50}s of 8-NH_2-, 8-N_3-, and 8-Br-cADPR (Table 12.1). Although the amino analog is the most potent, the brominated antagonist has the distinct advantage of being cell permeant that has been used to good effect in sea urchin eggs [79], various smooth muscle beds [80–86], cardiac muscle [87, 88], and neutrophils [89]. As might be expected, the azido analog (or rather, a [^{32}P]-labelled version) has been used successfully to photoaffinity label polypeptides of MW 100 and 140 kDa in sea urchin egg homogenate that are much smaller than the RyR itself [18].

Modification at the N^7 position by replacing the nitrogen with a carbon results in the 7-deaza series of compounds. These are more resistant to hydrolysis by cADPR hydrolases [90] and by combining modifications at the 7 and 8 positions has led to the synthesis of 7-deaza-8-bromo-cADPR [91], which as well as being a nonhydrolyzable, competitive antagonist, is also membrane permeant [79]; with intact cells, however, it has to be used at concentrations ~100 times the IC_{50} determined in homogenate to effect a complete block of the Ca^{2+} signal [79]. Unfortunately, this useful compound is not yet commercially available and therefore entails a stepwise, chemo-enzymatic synthesis [79].

In the case of agonists, it has been useful to render cADPR hydrolysis-resistant, particularly in preparations of considerable hydrolase capacity. Just as with the N^7 modification, the 3-deaza series are also weak hydrolase substrates, and the full

TABLE 12.1
Summary of Activities and Potencies of cADPR Analogs

Analog	Agonist	Antagonist	Hydrolysis Resistant	Cell Permeant	EC_{50}	Commercial Source	Notes
cADPR	●	○	○	○	8–48 nM	Molecular Probes, Sigma, Calbiochem	
8-Br-N^1-cIDPR	●	○	●	●	>500 μM		Intact cell data
3-deaza-cADPR	●	○	●	○	1 nM	Sigma	Synthesis [92]
8-NH_2-cADPR	○	●	○	○	9 nM	Molecular Probes	
8-N_3-cADPR	○	●	○	○	0.45 μM	Sigma	Photoaffinity label
8-Br-cADPR	○	●	○	●	1 μM	Sigma	
7-deaza-8-Br-cADPR	○	●	●	●	0.7 μM		Broken cell system

Closed circles represent activity, whereas open circles represent inactivity.

agonist 3-deaza-cADPR is even more potent at releasing Ca^{2+} than is the naturally occurring cADPR [92]. This may in part reflect an increased effective concentration of the analog relative to cADPR at the Ca^{2+} release site since the former is not enzymatically removed, an explanation that also might account for the greater potency of the agonist $2''$-NH_2-cADPR [75]. Recently, ARC-catalyzed cyclization of a brominated, hypoxanthine analog of β-NAD^+ (8-Br-NHD^+, Biolog, Bremen, Germany) resulted in the production of 8-Br-N^1-cIDPR that was surprising in two regards: not only was it an 8-substituted *agonist*, but it was also cell permeant [93]. Coupled with its resistance to hydrolysis, it is hoped that the preliminary demonstration of Ca^{2+} release in intact T cells will soon be followed by similar and more detailed studies in this and other systems, and perhaps the development of more potent congenors.

It is worthwhile to point out at this stage, that although we may think of cADPR receptors as created equal, some may be more equal than others. That is to say, that differences in binding requirements likely exist between phyla or even species: we do not yet fully understand why, but the relative efficacy or potency of agonist analogs can be different in echinoderms and mammals, e.g., cADPR-$2'$-phosphate is a more efficient stimulator of Ca^{2+} release in T cells than in sea urchin [94], whereas the converse applied to cADPcR, a carbocyclic analog. Similarly, the partial agonist 7-deaza-cADPR is more potent than cADPR in the sea urchin *Strongylocentrotus purpuratus*, but is less potent than cADPR in the related *Lytechinus pictus*. Clearly, this serves to illustrate that caution should be exercised in assuming that all receptors will be the same, and that interpretation of analog work should be effected with care.

12.2.5.2 Fluorescent cADPR Analogs

By incubating cell extracts with a non-fluorescent β-NAD^+ analog, NGD (its guanine equivalent), ARC activity can be continuously monitored by following increases in fluorescence that accompany the accumulation of the fluorescent product cGDPR, the cADPR analog [95] (see Section 12.7.3.1). In cGDPR, a glycosidic linkage to the N^7 rather than N^1 results in a compound that is inactive at causing Ca^{2+} release and, importantly, is resistant to hydrolysis, making it ideal for monitoring ARC activity [96].

To indirectly assess cADPR hydrolase activity, one can monitor the rate of disappearance of the fluorescent cADPR analog, cIDPR (with an inosine base) that is itself inactive with respect to Ca^{2+} release, and that can be synthesized from NHD^+ or obtained from commercial sources (see Section 12.7.3.2).

12.3 PURIFICATION AND SYNTHESIS OF NAADP AND ANALOGS

12.3.1 SYNTHESIS OF NAADP

Unlike cADPR, which can be only about 90% pure from commercial sources, due primarily to hydrolysis to ADPR, NAADP is commonly 95% or more pure.

Nevertheless, if one is to use large amounts for subsequent chemistry and biology, or wishes to test very pure compound, it is easily made in large quantities to a purity of $\geq 99\%$. It should be re-emphasized that purity of NAADP is more critical than it is for cADPR, because of its potent self-inactivation of Ca^{2+} release, particularly in those systems such as sea urchin where sub-threshold concentrations completely inactivate.

To synthesize NAADP, the following are combined: commercial NADP (typically, 1 g; 13 mM final concentration), 100 mM nicotinic acid, pH 4.5, and ADP-ribosyl cyclase (1 µg/mL) in a final volume of 100 ml. The reaction is periodically monitored with HPLC over an analytical (3×150 mm) AG MP-1 column and terminated when the conversion is maximal, taking into account losses to ADPR-P. If the reaction seems to stop, more ARC is added. If there is a worrying accumulation of ADPR-P, use a lower (e.g., 0.1 µg/ml) concentration of ARC in the next synthesis (see Section 12.2.2 on dilution). We commonly get 85–95% conversions. The reaction is then separated on a semipreparative column packed with AG MP-1. We use two semipreparative columns, a 10×250 mm and a 3×25 cm, which can be overloaded to separate 80 and 600 mg of nucleotides, respectively. The peak containing NAADP is collected in 50-ml centrifuge tubes. At the end of all the HPLC runs, the tubes are pooled and evaporated to dryness in a rotary evaporator. The sample is then washed three times with high-purity methanol to remove residual TFA by evaporating to dryness three times, then taken up in about 20 ml water. A serial dilution is made and the absorbance at 254 nm is taken and converted to a concentration using an absorption coefficient of 18,000 $M^{-1}cm^{-1}$. The solution is then aliquoted into either 10 or 1 mg samples and dried in a speed vac (Savant) and stored at $-80°C$. If other than the acid form is desired, the solution is either neutralized with Na^+ or K^+ carbonate, or passed over Dowex 50 of the K^+ or Na^+ form (10 mg sample to 20 g Dowex). Overall recovered yield is normally about 80% based on weight.

12.3.2 SYNTHESIS OF [^{32}P]NAADP

Radiolabelled NAADP was synthesized by a variation of a two-step method reported previously (Figure 12.6) [5, 53] and using the same apparatus we use for preparing [^{32}P]cADPR (Sections 12.2.1 and 12.2.3). In the first step, [^{32}P]NADP was synthesized by incubating [^{32}P]NAD (1000 Ci/mmol, Amersham Pharmacia) at a minimal dilution (8 µM in final reaction) for 2 h at 37°C with human NAD kinase (kindly provided by Matthias Ziegler, University of Bergen, Norway) [97] 50 mU/ml (where 1 U equals 1 µmol/min using 2 mM NAD, 10 mM MgATP at 37°C), MgATP 5 mM, $MnCl_2$ 0.5 mM, Hepes 63 mM. For the second step, the reaction was diluted with 200 mM nicotinic acid (pH 4.4) to a final concentration of 100 mM, and ADP-ribosyl cyclase 1 µg/ml (Sigma) and incubated for 1 h at 23°C.

The reaction was terminated by pumping it onto an HPLC column, and reactants were separated on an anion-exchange resin (AG MP-1, Biorad) using a concave upward gradient of trifluoroacetic acid (Figure 12.4, Section 12.2.2). One-ml fractions were collected in 1.5-ml centrifuge tubes (with the lids cut off) containing

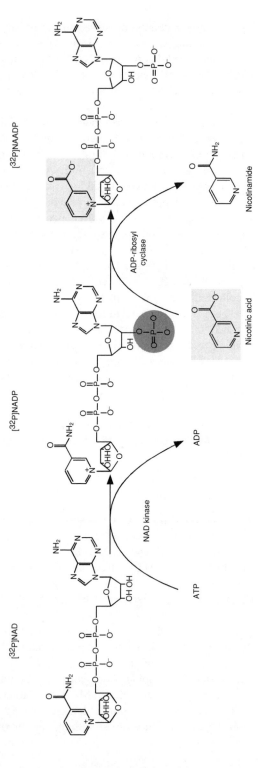

FIGURE 12.6 Scheme of stepwise enzymatic synthesis of [^{32}P]NAADP highlighting 2′-phosphate and base exchange reaction using human β-NAD$^+$ kinase and ARC as detailed in the text. [^{32}P]β-NAD$^+$ is radiolabelled at the pyrophosphate bridge. Chemical modifications are highlighted in dark and light shades.

20 ml of 1 M Hepes-NaOH, pH 7. The tubes were placed in a Perspex stand that fits into a Perspex box to contain the beta radiation. One μl of each fraction was scintillation counted in 10 ml water using Cerenkov radiation to identify radioactive peaks (Figure 12.7). The [^{32}P]NAADP-containing fractions were pooled, stored at 4°C and used for the binding reactions.

12.3.3   CAGED NAADP

12.3.3.1   Synthesis

While caged cADPR is still commercially available, caged NAADP now has to be synthesized in house, where we use a modification of those methods employed by Lee et al. [68, 98], who modified those originally developed to cage ATP [99]. Basically, a stable precursor of the caging reagent is made and then converted to an unstable reactive reagent immediately before addition to the compound to be caged (Figure 12.8).

For the first reaction (synthesis of 2-nitroacetophenone hydrazone), the conditions are identical to those reported by Walker et al. [99]. Into a 250-ml round-bottom flask we place 100 ml of 95% ethanol, 10 g of 2-nitroacetophenone, 3.2 ml of glacial acetic acid, 5.6 g of hydrazine monohydrate. Heat the mixture under reflux for 3 h and then cool, remove the ethanol with rotary evaporation, add 100 ml water and 100 ml chloroform. Then transfer to a separatory funnel (250 ml), shake, remove the upper (aqueous) layer and discard, repeating the process twice more. Place the chloroform phase into a 250-ml Erlenmeyer flask and dry by adding about 3 g of anhydrous MgSO$_4$. Swirl for a few minutes and then filter through Whatman No. 1 paper into a 250-ml round-bottom flask. Remove chloroform with rotary

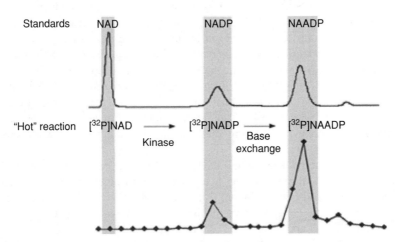

FIGURE 12.7 HPLC resolution of intermediates in synthesis of NAADP. Upper trace represents the A$_{240}$ of unlabelled standards and the lower trace, the equivalent [^{32}P]-labelled metabolites.

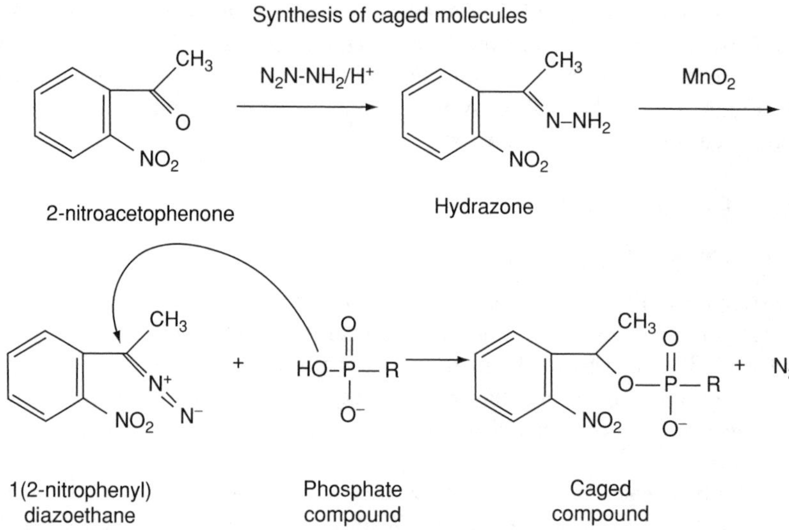

Synthesis of caged molecules

FIGURE 12.8 Chemical structures of the reactants in caged NAADP synthesis as detailed in the text.

evaporation. Transfer product (a yellow oil) to foil-wrapped container. Store at −20°C until needed. We have stored this product for 2 years without any problem.

In preparation for the actual caging reaction, NAADP (10 mg) is taken up in 1 ml water and the pH adjusted to 1.5 using a little concentrated HCl (if a salt form of NAADP is used, about 7 μl is required) and either pH paper or a microprobe. The solution is then taken up in 2 ml with water.

The second reaction (Figure 12.8), synthesis of 1(2-nitrophenyl)diazoethane, is performed immediately before the caging reaction and in a fume cupboard. Note that the product of this reaction, as are all diazoethane derivatives, is thermally unstable and light sensitive, and it can explode. This danger is minimal if it is used in low amounts (less than 4 mmol), the temperature is always maintained below 35°C, and it is kept in solution. A face shield is required as a minimal amount of personal protection. All work should be behind the door of a fume cupboard. Into a 5-ml round-bottom flask, place (in this order): a magnetic stir bar, 20 μl 2-nitroacetophenone hydrazone, 2 ml diethylether, and 60 mg MnO$_2$ (while stirring vigorously). Stir for 20 min.

The diazo is filtered directly into the acidified NAADP solution to remove the MnO$_2$ and Mn(OH)$_2$ through a filter made from a 5-ml pipette tip that is plugged with cotton wool and contains about 2 g of diatomaceous earth. The solution is forced through with pressure from the pipette. Wash with 1 ml of ether. The solution should be dark red-orange and free of black particles. If some of the MnO$_2$ does contaminate the NAADP solution, start over, as the nucleotides will be oxidized and destroyed. In our experience, a gummy product that adheres to the stirbar appears to be indicative of oxidation and, thus, bad filtration.

The efficiency of stirring and maintenance of acidic pH are crucial for the reaction to proceed and to achieve good yields. If the pH increases, it can be brought back to pH 1.6 with addition of HCl. All other methods state to do these reactions in the dark. We have not found this to be essential, but do shield the reaction with a bit of aluminum foil. When the caging reagent turns yellowish, the ether layer is drawn off and replaced with a fresh layer. Initially, this happens quite rapidly (30 min) and slows as the reaction proceeds. The progress of the reaction is monitored with HPLC by withdrawing 2 µl of the reaction, mixing it with 90 µl water and running it over either an AG MP-1 column eluted with TFA or a MonoQ column eluted with TEAB (Figure 12.9A). Typically, we get about 80% conversion after 3–6 additions of diazo, with the differences apparently due to the efficiency of stirring, as the reaction is thought to occur at the interface of the aqueous and organic phases.

At completion, the reaction mixture is extracted three times with 2 ml diethyl ether and the aqueous layer is run over a MonoQ column and eluted with TEAB. The first two peaks are collected. We are not sure if this represents isomers of singly

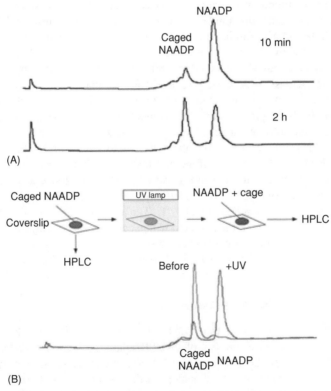

FIGURE 12.9 (A) Monitoring the progress of the final reaction synthesizing caged NAADP using HPLC coupled with the A_{240}. Samples were withdrawn at the times indicated before the reaction has gone to completion. (B) *In vitro* assessment of the effect of photolysis upon caged NAADP demonstrating its breakdown into NAADP as assessed by HPLC.

caged NAADP or singly and doubly caged NAADP. What we know is that it works. That is, it is regenerated to NAADP upon exposure to UV light. To confirm that the NAADP has been caged, 2 μl of the reaction is mixed with 98 μl of water and placed in a weigh boat and exposed to UV light (365 nm) from a hand-held source for 20 min (Figure 12.9B). Exposure longer than this causes photodamage to the NAADP, while a shorter one gives incomplete hydrolysis.

12.3.3.2 Purification

For invertebrates and plants, the NAADP response shows unusual desensitization whereby low concentrations completely prevent subsequent activation. This is a potential problem when using caged NAADP as inevitably there will be some carry-through of free NAADP in the caged NAADP due to either carry-through during HPLC purification or hydrolysis during dry down. Lee et al. [68] reported that purification required two runs over an AG MP-1 column with TFA and then one final run over a MonoQ column with TEAB. In contrast, we can obtain caged NAADP that is free enough of NAADP for work with even sea urchin eggs, which are sensitive to inhibition at nanomolar concentrations, with a single run over a MonoQ column. Contamination levels are usually assessed empirically by injecting caged NAADP and determining if there was a subsequent response. This test should detect anything above 0.001% contamination, that is, 1 nM in a 100 μM solution.

In aliquots that have been stored at $-80°C$ for several months, the level of free NAADP can increase to desensitizing levels. Rather than having to re-purify with HPLC and drying down the caged NAADP, we use the trick of adding alkaline phosphatase [68]. To prevent subsequent injection of this enzyme into a cell, we used alkaline phosphatase attached to agarose beads (Sigma). It is provided as a slurry in 1 M Na_2SO_4 and so we put about 40 μl into a centrifuge tube and add 460 μl of TEAB (1 M, pH 8.8), centrifuge at 10,000g for 10 s, decant and discard the supernatant and repeat this five times. About 5 μl of the loose agarose pellet is transferred (cut the fine point off a pipette tip) to an aliquot of caged NAADP and this is then incubated for at least 1 h at 37°C. The agarose beads are allowed to settle by gravity and then the solution is mixed with a calcium dye and injected. From this point onward, the solution is not frozen but stored at 4°C with the beads. The caged NAADP has been used for up to 1 month without any trace of contaminating free NAADP.

12.3.4 Synthetic NAADP Analogs

In much the same way that the broad substrate specificity of ARC has been exploited to synthesize a number of cADPR analogs (Section 12.2.5), this trait has also been used for the synthesis of NAADP analogs. In addition to the enzyme being able to use various structural analogs of the nucleotide (NADP), it also shows promiscuity in its use of structural analogs of the pyridine (nicotinic acid) in the base–exchange reaction. Unfortunately, while such analogs have been able to give interesting information concerning the interaction between the NAADP receptor protein and its ligand, this enzymatic approach has been unable to produce analogs with antagonistic properties. In fact, all of the active analogs have been agonists with a lower potency than NAADP. Studies into the structure–activity relationship have found that even small changes to any of three groups on the NAADP molecule result in a significant loss of affinity/efficacy [9, 46, 100]. Most strikingly, there appears to be an absolute requirement for the $2'$ phosphate group and even moving it to the $3'$ position results in a loss of both affinity and efficacy. Secondly, the high-affinity binding of NAADP is dependent on the electronic configuration of the group at the $3'$ position of the pyridine ring.

Two fluorescent analogs of NAADP have been synthesized using ARC and fluorescent NADP (nicotinamide 1, N^6-ethenoadenine dinucleotide phosphate) [101]. The first of these is the direct result of the base–exchange reaction and is named etheno-NAADP (nicotinic acid 1, N^6-ethenoadenine dinucleotide phosphate). However, the excitation maximum of this compound overlaps with the absorption maximum for proteins making it unsuitable for use in cell imaging studies. Conversion of this compound to nicotinic acid 1, N^6-etheno-2-aza-adenine dinucleotide phosphate (etheno-aza-NAADP) overcomes this problem as the excitation peak of this compound lies at around 360 nm.

Nicotinamide 1, N^6-ethenoadenine dinucleotide phosphate is no longer commercially available but it or etheno-NAADP can be made from N(A)ADP [102]. Briefly, 5 mg N(A)ADP in 100 μl water is mixed with 200 μl of citrate phosphate buffer (pH 4.0, containing 62 parts 0.1 M citric acid and 38 parts 0.2 M Na_2HPO_4) in a glass tube. Fifty μl of 2-chloroacetaldehyde is then added in the fume hood and the mixture is heated at 80°C for 40 min.

The most interesting compounds in terms of their use so far are structurally unrelated to NAADP. The triazine dyes are used industrially as textile dyes but their ability to function as nucleotide mimetics has also led to their extensive use for the study and purification of nucleotide binding proteins. Several of these dyes display IC_{50}s in the low micromolar range in [^{32}P]NAADP-binding assays and act as antagonists of the receptor. Furthermore, immobilized Reactive Green 5 and Reactive Green 19 have been used in a partial purification of the receptor [103].

12.4 STUDYING cADPR- AND NAADP-INDUCED Ca²⁺ RELEASE

12.4.1 THE SEA URCHIN EGG

The sea urchin egg has been used as a model system in cell biology for well over a century. The benefits of Ca^{2+} signaling studies are its large size (100 μm) facilitating microinjection studies, and the production by each animal of several millilitres of packed cells, all of one type, allowing complementary biochemical studies. As described below in Sections 12.4.2.2 and 12.4.3.3, homogenates and microsomes contain robust Ca^{2+} release channels for each second messenger system, making them the most reliable system for directly studying Ca^{2+} release from intracellular stores.

Most of the work from our laboratory has focused upon two species of sea urchins that live off the coasts of the United States, *L. pictus* (a.k.a. the Californian Blond) and *S. purpuratus*. Both species survive the long flights to the United

Kingdom well, but we have observed a tendency for *S. purpuratus* eggs to show a refractoriness to NAADP in some — but not all — shipments. We are still unsure why this occurs, but perhaps this reflects some form of stress in transit because homogenates prepared Stateside have not shown this insensitivity. NAADP responses are also present in *Paracentrotus lividus*, a common sea urchin in the Mediterranean Sea, and in *Psammechinus miliaris*, a sea urchin that lives off the coast of Britain. Loading of cell impermeant molecules (e.g., Ca^{2+} dyes) as AM esters is generally not possible with these species as they have a very low esterase activity, with the exception of the species *Arbacia punctulata*. Therefore, any of these species, and probably other sea urchins as well, is suited to studies to investigate cADPR- and NAADP-mediated signaling.

12.4.1.1 Collection of Sea Urchin Eggs

Unfertilized *L. pictus* sea urchin eggs can be collected during the gravid season between the months of May and September. Urchins survive shipping if correctly packaged in oxygenated sealed bags of sea water interspersed with bags of ice. A useful source of *L. pictus* is Marinus Inc., Long Beach, CA and urchins shipped as described above, even survive intercontinental journeys. These urchins survive well in marine aquaria at their natural temperature of 17°C, and can be kept for many months. Eggs can be obtained by stimulating ovulation of female *L. pictus* urchins by five intracoelomic injections of 0.5 *M* KC1 solution (up to 1 ml total volume) into each chamber (Figure 12.10). From the aboral surface, yellow-orange eggs (or white sperm from males) will begin to appear, and the urchin is inverted and placed upon a 20-ml universal tube (Figure 12.10B) full to the brim with artificial sea water (ASW: 435 m*M* NaCl; 40 m*M* $MgCl_2$–$6H_2O$; 15 m*M* $MgSO_4$–$7H_2O$; 11 m*M* $CaCl_2$; 10 m*M* KC1; 2.5 m*M* $NaHCO_3$, 20 m*M* Tris and 1 m*M* EDTA, pH 8.0). Eggs tend to shed over a period varying from 5 to15 min with volumes of 1–5 ml that are allowed to settle (note: healthy eggs are a vibrant yellow-orange; a dull-colored, low yield invariably indicates poor eggs).

Sperm are normally collected "dry" (since ASW activates them) by placing inverted male urchins on a plastic petri dish and later transferring the ejaculate in a microfuge tube stored at 4°C. Sperm can be kept overnight in the refrigerator (the next day, we check sperm motility and ability to fertilize fresh eggs), but they do not survive longer times.

Most of the supernatant is discarded, the eggs are gently resuspended and then dejellied by filtering through a 85-μm Nitex mesh (Plastok Associates Ltd., Merseyside, U.K.) or 100-μm nylon net filters (Millipore). To achieve this, we make a crude filtration unit by taking a 15-ml conical plastic centrifuge tube (Alpha Labs) that has its base sawn off to make it open, and a large central hole drilled in its cap; across the threaded end of the of the tube a small piece of the nylon mesh is loosely placed and the annular cap screwed on so that the mesh becomes taut across the cap aperture (Figure 12.10C). This unit is placed inside a larger, 50-ml tube, the nylon mesh pre-wetted with a little ASW, and the egg suspension gently decanted into the top of the filter unit. Dejellied eggs are transferred into a fresh 15-ml conical

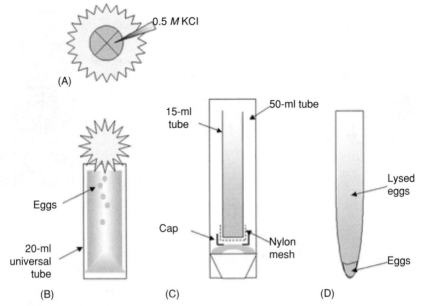

FIGURE 12.10 Schematic of harvesting and de-jellying of sea urchin eggs. (A) Sea urchin eggs are injected on their oral side with 1 ml of 0.5 M KCl. (B) Female urchins are inverted and placed on a 20-ml universal containing ASW, and the shed eggs allowed to settle. (C) Apparatus for de-jellying. The egg suspension is passed through a 100 μm nylon mesh that "nicks" and removes the jelly coat. (D) De-jellied eggs are centrifuged such that the supernatant contains egg-jelly and any lysed eggs (an infrequent occurrence).

centrifuge tube and washed by centrifugation at 800g at 10°C in approximately ten times their volume. Any lysed or damaged eggs remain in the supernatant and should be discarded (Figure 12.10D). The eggs are then gently resuspended in a similar volume of ASW, kept cool, and used on the same day (urchin fecal bacteria contaminate and eventually spoil eggs).

12.4.2   SEA URCHIN EGG HOMOGENATES

12.4.2.1   Preparation of Sea Urchin Egg Homogenates

Homogenates (50% v/v) of unfertilized *L. pictus* eggs are prepared in a similar manner to that described by Clapper et al. [2] (Figure 12.11). After de-jellying (as above, but on a larger scale), eggs are washed twice in EGTA/Ca^{2+}-free ASW, and twice in Ca^{2+}-free ASW (each comprising 470 mM NaCl; 27 mM MgCl$_2$–6H$_2$O; 28 mM MgSO$_4$–7H$_2$O; 10 mM KCl and 2.5 mM NaHCO (pH 8.0), plus 1 mM EGTA for the first two washes). Finally, eggs are washed with an "intracellular medium" (Glu-IM) consisting of: 250 mM potassium gluconate; 250 mM N-methylglucamine; 20 mM HEPES (pH 7.2) and 1 mM MgCl$_2$. Eggs are then homogenized in Glu-IM plus: 2 mM ATP; 20 U/ml creatine phosphokinase; 20 mM phosphocreatine; Complete^{TM}, EDTA-free protease inhibitor cocktail tablets (Roche), using a Dounce glass tissue

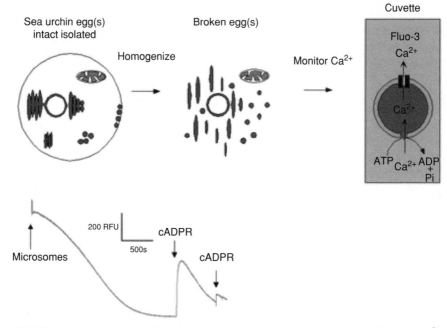

FIGURE 12.11 Scheme depicting the preparation of egg homogenates (Upper left). Ca^{2+} concentration is depicted by the gray scale intensity with darker being higher. Homogenates are placed in a fluorimeter cuvette containing a Ca^{2+} reporter such as fluo-3. Organelles sequester Ca^{2+} in an ATP-dependent fashion (a Ca^{2+} pump is shown for simplicity's sake) and release it back to the extravesicular space in response to messengers (upper right). Lower left shows a representative trace where microsomes sequester Ca^{2+} and lower the bulk cuvette $[Ca^{2+}]$ until it reaches a steady baseline. Addition of 10 μM cADPR evokes a considerable Ca^{2+} release that has desensitized to itself. Similar principles hold for whole homogenate and subcellular fractions.

homogenizer, size "A" pestle. Cortical granules were removed and discarded by pelletting at 13,000g, 4°C for 10 s. Homogenates are aliquoted (0.5 or 1 ml) and stored at -70°C, until used for Ca^{2+}-release or binding studies.

12.4.2.2 Fluorimetry of Ca^{2+} Release from Egg Homogenates

Ca^{2+} loading of intracellular stores was achieved by incubating homogenates for 3 h at 17°C in IM containing the additions for ATP-regenerating system, mitochondrial inhibitors (1 μg/ml oligomycin and antimycin, 1 mM sodium azide) and protease inhibitors as outlined above. Use of the regenerating system allows the resynthesis of ATP and keeps ATP levels reasonably constant. During the first 45–60 min, homogenate is diluted 1:1, and thereafter, was sequentially diluted 1:1 every 30 min, an overall dilution from 50% v/v to the final 2.5% v/v. In the final 30 min, 3 μM fluo-3 was added as the Ca^{2+} fluorescent indicator. This method for preparation has been found empirically to achieve better loading of vesicles and highly reproducible lower steady-state resting Ca^{2+} levels than by a one-step dilution. Free

Ca^{2+} was measured at 17°C using 500 μl of homogenate in a stirring Perkin-Elmer LS-50B/PTI fluorimeter (Figure 12.11) at excitation and emission wavelengths of 490 and 535 nm, respectively. Additions were made to the homogenate in 5 μl volumes, and changes in relative fluorescence units (RFU) could be calibrated by determining F_{min} and F_{max} by sequential addition of 0.5 mM EGTA and 5 mM Ca^{2+} and using the equation $[Ca^{2+}] = K_d*(F-F_{min})/(F_{max}-F)$ (where K_d = dissociation constant of the fluorescent dye, and F the fluorescence at any time).

12.4.3 SEA URCHIN EGG FRACTIONATION

12.4.3.1 Preparation of Microsomes

A Percoll density gradient was used as an established means of obtaining a band of Ca^{2+}-sequestering and releasing microsomes that have been shown to be predominantly endoplasmic reticulum derived [104]. In addition, it was used to fractionate the sea urchin egg homogenate according to density (Figure 12.12A). A 25% Percoll buffer was prepared for the fractionation, containing an intracellular medium with potassium acetate substituted in place of potassium gluconate, as this provided a more suitable density base for separation of subcellular components. The final fractionation buffer therefore contained 250 mM N-methyl glucamine; 250 mM potassium acetate; 1 mM $MgCl_2$; 20 mM HEPES; 100 μM EGTA; plus an ATP-regenerating system comprising 500 μM ATP; 1 U/ml creatine phosphokinase and 1 mM phosphocreatine. We usually prepare this acetate-based buffer as a 2 or 4 × concentrated stock to allow us to compensate for dilution by large Percoll volumes (especially when 50% is used in a second round of separation — see below).

 $L.$ $pictus$ egg homogenate (1 ml of 50% v/v) was layered onto the Percoll fractionation medium (9 ml) and incubated at room temperature for 30 min (which improves subsequent separation) before centrifugation at 27,000g, for 30 min at 10°C in a 50Ti Beckman rotor. The resulting fractions were either collected by gently removing successive 1-ml layers from the top, or (as is more usual) the microsomal band (Figure 12.12B) was removed by puncturing the vessel wall using a syringe and needle. This yields approximately 1 ml of microsomes. The so-called yolk fraction appears as a broad, denser band (Figure 12.12B) and can also be collected in a similar manner (volume 1.5–2 ml) for further fractionation (see below). Fractions were stored at −70°C until required.

12.4.3.2 Preparation of NAADP-Responsive Acidic Vesicles

As part of the investigation into the nature of a selective NAADP-sensitive Ca^{2+} store, we focused attention upon the yolk fraction collected using the 25% gradient above, and further fractionated this on a second, Percoll gradient (this time, 50% v/v). Freshly prepared yolk is preferred to a thawed sample. All of the yolk collected above is layered onto 9 ml of acetate IM plus 50% Percoll, and also centrifuged 27,000g, for 30 min at 10°C. This procedure yields three narrow bands (Figure 12.12C) that we termed non-particulate, particulate, and dense and which are collected with a syringe. It is the dense fraction that is enriched in lysosomal markers as well as

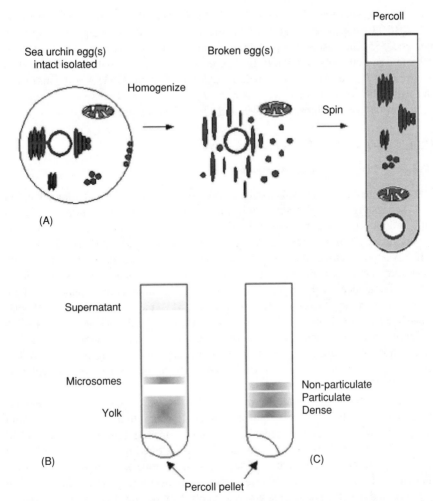

FIGURE 12.12 (A) Schematic representation of subcellular fractionation of eggs upon a Percoll density gradient. Where the denser organelles sediment toward the bottom of the tube. (B) Pattern of distribution following separation of whole homogenate on a 25% Percoll gradient. (C) Pattern of distribution following separation of the yolk fraction on a 50% Percoll gradient.

NAADP-sensitive Ca^{2+} release and binding sites (with little contamination from the stores that respond to the other messengers, cADPR and IP_3) [48]. In this way, a selective enrichment of NAADP-dependent stores is effected.

12.4.3.3 Testing Fractions for Ca^{2+} Releasing Properties

Percoll fractions (approximately 50% v/v) are defrosted and immediately used for fluorimetry. Ca^{2+} sequestration was initiated by mixing 100 μl of the fraction with 400 μl of the IM supplemented with MgATP and the regenerating system as detailed in Section 12.4.2.1. Mitochondrial inhibitors and protease inhibitors (Section 12.4.2.2) can be included to prevent mitochondrial Ca^{2+} release and protein degradation, respectively. After addition of the Percoll fraction, the fall in fluorescence was monitored in the fluorimeter until it attains a stable, steady-state basal value (Figure 12.11). The increased density of microsomal membranes in these fractions compared to the whole homogenate means that long preincubations to allow Ca^{2+} sequestration are not necessary, and the microsomes will sequester Ca^{2+} within 5–15 min (Figure 12.11). Ca^{2+} sequestration and release to various agents was detected fluorimetrically, using 3 μM fluo-3 as a Ca^{2+} indicator, according to the method described in Section 12.4.2.2.

12.5 PHARMACOLOGY OF cADPR SIGNALING

The pharmacology of cADPR-induced Ca^{2+} release has been most extensively investigated in the sea urchin egg system [105] but has been confirmed in other mammalian systems such as the pancreatic acinar cell [60] and T cell [94]. As has already been alluded to, cADPR was found to activate and modulate CICR via a ryanodine-sensitive Ca^{2+} release channel [105], independently of the activation of inositol trisphosphate receptors and NAADP receptors. The very fact that each receptor homologously desensitizes without affecting the other two systems is compelling evidence that each channel is a different protein (Figure 12.2). Pharmacological analyses confirmed that cADPR activated RyR and not IP_3 receptors (IP_3Rs). First, the competitive IP_3R antagonist heparin blocked IP_3-induced Ca^{2+} release but not that triggered by cADPR [57, 106]. Second, the pharmacology of cADPR-induced Ca^{2+} release mirrored that of the RyR itself. Blockers of RyRs such as Mg^{2+}[107], ruthenium red [108], dantrolene [108], and the partial antagonist, procaine (1–5 mM) [109–111] selectively inhibited cADPR-induced Ca^{2+} release. Moreover, cross-desensitization was observed between maximal release induced by cADPR and known activators of the RyR: caffeine and ryanodine induced Ca^{2+} release in both intact sea urchin eggs [48] and homogenates that abrogated Ca^{2+} release to cADPR but not to IP_3; conversely, sea urchin egg microsomes that had discharged their Ca^{2+} contents in response to cADPR were rendered insensitive to caffeine and ryanodine but not to IP_3 [105]. At more modest, submaximal concentrations, potentiators of CICR at RyR such as Ca^{2+}, Sr^{2+}, and caffeine acted synergistically with cADPR to enhance Ca^{2+} release [112, 113].

Binding studies have also proven informative because cADPR can enhance [^3H]ryanodine binding, a parameter that is invariably associated with an increase in the open conductance state of the RyR, a CICR-activated conformation that favors ryanodine binding [13]. On the other side of the coin, [^{32}P]cADPR was not affected by any CICR activator or inhibitor [7] (with the exception of a small inhibition by high Ca^{2+} [6, 7]) that indicates that their effects upon cADPR-induced Ca^{2+} release is further downstream of ligand–protein interaction, and presumably operates at the level of channel open/closed probability.

As mentioned in Section 12.2.5, the development and commercial availability of highly selective cADPR analogs, including caged compounds, competitive antagonists, metabolically resistant agonists and antagonists, and membrane-permeable analogs, have led to important advances in cADPR research. These compounds have therefore been of great benefit in probing the fundamental mechanisms governing cADPR-mediated signaling mechanisms and pathways.

While the pharmacology of cADPR-mediated Ca^{2+} release has been most extensively studied in sea urchin egg homogenate and microsomal systems [74], the techniques used can and have been applied to mammalian microsomes [22, 31, 114–116] and permeabilized cells [38, 41, 76–78, 117]. The fluorimetric technique (see Sections 12.4.2.2 and 12.4.3.3) for measurement of Ca^{2+} has been widely used, but in mammalian systems ^{45}Ca efflux studies have also been employed [30, 118]. The choice of intracellular medium for such studies may be important. Ca^{2+} release in the sea urchin homogenate system, for example, is more sensitive to releasing agents in the N-methylglucamine/potassium gluconate medium than in a KCl medium, while replacement of gluconate with mannitol, improves the sensitivity of the IP_3 response [119]. In trying to optimize Ca^{2+} release in broken cell systems, it is clear that different media should be tested.

12.5.1 TROUBLESHOOTING AND PRACTICAL CONSIDERATIONS

It is sound pharmacological principle that confirmation or dismissal of the involvement of cADPR in a physiological response should never be concluded from the action of one compound alone because of the obvious issue of drug selectivity. The frequently promiscuous nature of agents commonly used in Ca^{2+} signaling has been formally documented [120] but in the context of this discussion, the more noteworthy examples are that N-desulfated heparin should be used as a control for heparin side effects (e.g., plasmalemmal receptor uncoupling [121]) as it is not an IP_3 receptor blocker [122, 123] (note: other cell permeant, so-called selective blockers of the IP_3 receptor have had their selectivity seriously questioned and should be used cautiously i.e., Xestospongin C [124–127] and 2-aminoethoxydiphenylborate (2-APB) [128–132]). More directly relevant to cADPR, some RyR agents exhibit confounding side effects: ruthenium red is also a potent blocker of the mitochondrial Ca^{2+} uniporter [133] (whereas its cell permeant analog Ru360 is more selective for the uniporter over the RyR [133]). Interpretation of results with ryanodine are not always straightforward either, but not because of multiple binding sites (with the exception of a proposed mitochondrial site of action at the uniporter [134]), but rather because of its binding characteristics at the RyR itself [13, 21]: first, it exhibits a bell-shaped concentration–response curve where low concentrations (nanomolar to micromolar) activate the RyR while high concentrations (>10 μM) block, but the precise ranges may vary from preparation to preparation and so will have to be empirically determined; second, ryanodine only binds to the open conformation of the RyR and its effects are therefore highly use-dependent. Therefore, under basal conditions, for example, the open conductance state is fleeting, and binding may be very slow. These potential problems are another persuasive reason for using highly selective cADPR agonists and antagonists.

In addition to questionable pharmacology, there are some practical problems associated with experimental approaches. Using fluorimetry to measure Ca^{2+} release, there are several potential pitfalls and artifacts that should be avoided, and extensive controls should be performed to avoid the drawing of invalid conclusions. Care should be taken that drugs do not interfere with Ca^{2+} indicator fluorescence (e.g., by absorbing the exciting light and/or being fluorescent itself), as well as tested for any physical interactions with the dyes and, if necessary, corrections for any interference made. Trying a different Ca^{2+} indicator may circumvent such issues.

Several checks can be performed in a fluorimeter to confirm that an apparent cADPR response is authentic: one trivial explanation for an increase in fluorescence is Ca^{2+} contamination of cADPR itself, and care should be taken in dissolving samples in the purest water available. In addition, low (μM) concentrations of EGTA can be added to the samples, but should not be so high as to reduce or abolish the detection of the Ca^{2+} release response. A few beads of Chelex can be added to aliquots of cADPR-containing solutions, which can be then spun down and the supernatant decanted for use. However, since Chelex can also sequester other molecules, steps such as verifying the continued presence of cADPR or other molecules in the treated solutions by HPLC could be carried out.

Alternatively, check for desensitization. A second application of a maximal cADPR concentration does not induce further Ca^{2+} release due to CaM dissociation in sea urchin (Section 12.1.2.1; Figure 12.2). Inactivation has also been observed in other but not all cell preparations but provides a useful check for Ca^{2+} contamination as a second fluorescent peak may be indicative of such contamination.

Another useful check is to heat-inactivate solutions containing cADPR. Heating the solution to 90°C for several hours nonenzymatically converts cADPR to ADPR that does not mobilize Ca^{2+} from intracellular stores. However, ADPR may have other cellular effects, for example, the potent gating of plasma membrane cation channels [135–137] and activation of the skeletal muscle RyR [17, 33].

Finally, verification of an authentic Ca^{2+} release by cADPR can also come from the analogs mentioned above (see Section 12.2.5). 8-NH_2-cADPR, while blocking cADPR-induced Ca^{2+} release, does not impair the ability of caffeine or ryanodine to elicit Ca^{2+} release from sea urchin egg homogenates. These data may point to the complex nature of the regulation of RyRs that are subject to allosteric regulation by small molecules as well as by the many different proteins with which they have been shown to interact [21]. For example, the calmodulin requirement of cADPR-induced Ca^{2+} release in purified sea urchin egg microsomes [28, 138] may provide one explanation for the failure of some investigators to observe cADPR-mediated Ca^{2+} release in some broken cell systems.

12.6 PHARMACOLOGY OF NAADP SIGNALING

To verify that a response is indeed due to NAADP, pharmacological analysis is the only choice in the absence of molecular biological information regarding its receptor. However, the paucity of specific tools reduces the investigator to three complementary approaches: (a) eliminating the cADPR and IP$_3$ systems to see if the response is affected, (b) exploiting the self-inactivation by NAADP, and (c) debilitation of the NAADP-dependent acidic Ca^{2+} stores.

12.6.1 TARGETING RECEPTORS

From Sections 12.2.5 and 12.5, it is clear that a number of agents are available to block cADPR and IP$_3$, albeit with cautious interpretation. In a "clean" system like sea urchin homogenate, NAADP responses have been shown to be unaffected by pretreatment with concentrations of heparin and 8-NH$_2$-cADPR that fully block IP$_3$ and ryanodine receptors. However, in systems displaying considerable interaction and the "channel chatter" referred to earlier (e.g., pancreatic acinar cells, smooth muscle), NAADP responses are amplified by the CICR systems, making it difficult to determine whether NAADP acts directly at RyR or IP$_3$ receptors or at a unique receptor.

Conversely, introduction of NAADP into the preparation at inactivating concentrations is currently the best diagnostic tool that we have. Delivery can be as a bolus injection, cytosolic dialysis with a patch pipette, or even photolysis of a caged precursor to liberate inactivating concentrations [11, 12]. In sea urchin eggs and cauliflower, concentrations of NAADP (nM) that are subthreshold with respect to Ca^{2+} release potently inactivate intact eggs or broken cells [139] to a subsequent maximal NAADP challenge, without affecting cADPR or IP$_3$. This inactivation is persistent in maximally stimulated homogenate, but recovers in the intact egg within ~20 min [55]. As alluded to already, mammalian cells display a bell-shaped concentration-dependency and require high doses of NAADP (10–100 μM) to inactivate release in systems as diverse as pancreatic acinar and β cells, and T cells [11, 51, 139]. However, some suggest that self-inactivation may not be as universal as first thought (see Ref. [139] for review).

In terms of other analogs and compounds, thio-NAADP was reported as the first NAADP antagonist, but this was later withdrawn when contaminating, inactivating levels of NAADP itself were discovered [140]. More promisingly, triazine dyes (Section 12.3.4) are agonists at the NAADP receptor [103], which are the first chemically unrelated structures to have activity at the NAADP-binding site.

NAADP-induced Ca^{2+} release is sensitive to other agents whose site of action is indeterminate: L-type Ca^{2+} channel-specific agents and K$^+$ channel blockers appear to act as NAADP receptors over intracellular receptors [45]. This is also true for T-type Ca^{2+} channel blockers, such as amiloride (100 μM), flunarizine (100 μM), and pimozide (25 μM) (Genazzani and Galione, unpublished data). Unfortunately, the concentrations of these drugs needed to block the NAADP response appear to be at least tenfold higher than those required to block their main targets, making these compounds unsuited for use in intact cells. Calmodulin antagonists, such W7 and J8, also abolish NAADP-induced Ca^{2+} release but at similar concentrations to those needed to block cADPR- and IP3-induced Ca^{2+} release. TMB-8 has also been reported to block NAADP-induced Ca^{2+} release in sea urchin eggs [43] although also in this case the concentrations required (500 μM) are unlikely to distinguish between the different mechanisms.

Important differences between NAADP-induced Ca^{2+} release and the other two mechanisms is depicted by the different sensitivities to pH [141], Ca^{2+} [43, 44] and monovalent cations.

12.6.2 Targeting Stores

If the actions of NAADP do indeed center upon acidic rather than ER stores, this provides another avenue for the investigator. Depletion of Ca^{2+} stores with select-ive agents has been a much-used and informative protocol. Often performed in Ca^{2+}-free media, emptying the ER with SERCA inhibitors (thapsigargin, cyclopia-zonic acid (CPA), or t-butylhydroquinone (tBHQ)) has implicated this organelle in a raft of physiological responses (note: while thapsigargin is the most potent (EC_{50} ~10 nM), its irreversibility, avidity for sticking to and leaching from perfusion tubing, as well as side effects at higher concentrations [142] often makes CPA and tBHQ the drugs of choice). Ca^{2+} ionophores such as ionomycin (1–5 μM) can also be used to deplete intracellular stores, albeit more discriminately, as they can act predominantly at internal membranes [143]. Clearly, direct measurements of ER $[Ca^{2+}]$ (Section 12.10.1.3) will verify organelle involvement. NAADP would not be expected to deplete ER or be sensitive to ER depletion.

On the other hand, the nature of acidic Ca^{2+} stores is sufficiently different to permit selective targeting over the ER. First, these stores can be fluorescently labelled with a panoply of differently colored vital stains including acridine orange, and members of the Lysosensor or Lysotracker families (Molecular Probes). Con-sequently, organellar distribution and/or luminal pH can be monitored in live cells [10, 48, 49] (Figure 12.5C). Second, these vesicles do *not* accumulate Ca^{2+} via the action of a P-type ATPase such as SERCA, but rather through an ill-defined mechanism dependent upon the H^+ gradient (e.g., Ca^{2+}/H^+ exchange). While SERCA inhibitors will have no effect therefore, blockers of the V-type H^+-ATPases that generate luminal acidity have profound consequences for Ca^{2+} handling: application of cell permeant pump blockers such as bafilomycin A1 or concanamy-cin A (1–10 μM) increase the vesicular pH and concomitantly inhibit NAADP-dependent Ca^{2+} release [10, 48–50].

In addition to these specific inhibitors, vesicular pH can be collapsed in intact cells by protonophores such as FCCP/CCCP, monensin (Na^+/H^+), and nigericin (K^+/H^+) (which also mandates for caution in assuming a mitochondrial site of action for FCCP and CCCP). Alternatively, extracellular application of 1–10 mM NH_4Cl (and the passage of membrane-permeant NH_3) directly and rapidly alkali-nizes the lumen. One thing to bear in mind is that after the inhibition of H^+ pumps, the collapse of ionic gradients may not be rapid: slow H^+ and/or Ca^{2+} leakage rates hinder an immediate effect upon NAADP-induced Ca^{2+} release in sea urchin [48]. Even when H^+ gradients are rapidly collapsed with protonophores or NH_4Cl, the leakiness of stores with respect to Ca^{2+} may still be a crucial factor. It might be thought that a more direct and rapid maneuver to deplete acidic Ca^{2+} stores would be the use of ionomycin. However, ionomycin acts as a Ca^{2+}/H^+ exchanger and is therefore ineffective at depleting acidic Ca^{2+} pools (a fact worth remembering when interpreting effects of ionomycin in intact cells or homogenates); only when the pH gradient is collapsed will ionomycin be successful in mobilizing Ca^{2+} i.e., in combination with other ionophores or V-type pump inhibitors [144].

Acidic Ca^{2+} stores (lysosome-related organelles) have another unique property that we have exploited to target their function, namely, they express proteases and exopeptidases. The membrane-permeant drug, L-glycyl-L-phenylalanine 2-naphthylamide (GPN), is a substrate of the lysosomal exopeptidase cathepsin C. GPN is cleaved into two hydrophilic products that are trapped and accumulate in the vesicle lumen until it osmotically lyses [48]. Upon bursting, these vesicles release their stored Ca^{2+} as brief, asynchronous events [48] or a larger Ca^{2+} release [145]. Although lysis is a conceptually cruder method of interfering with NAADP-dependent Ca^{2+} stores, essentially identical results have been observed with GPN compared with other techniques, and it does not interfere with [^{32}P]NAADP binding [48] nor affect mitochondrial membrane potential in intact sea urchin eggs as judged by tetramethylrhodamine ethyl ester fluorescence (Morgan and Galione, unpublished data).

It should always be verified that agents designed to abrogate acidic store function are having their desired effect, and we have found that organelles loaded with the Lysotracker or Lysosensor dyes alkalinize or release their fluorescence in response to bafilomycin A1 or GPN, respectively [48, 49], with an appropriate time course and concentration range (Figure 12.5).

12.7 ENZYMOLOGY OF cADPR AND NAADP METABOLISM

12.7.1 ENZYMATIC SYNTHESIS AND DEGRADATION OF cADPR AND NAADP

Our knowledge surrounding cADPR metabolism is superior to that for NAADP as there is little doubt that the cyclization of β-NAD^+ is a physiologically relevant route of cADPR synthesis, which is counteracted by hydrolase activity that converts it to ADPR (Figure 12.1). On the other hand, the NAADP story is not so certain; NADP, NAAD[1], and NAADPH [146] are all potential precursors, converted through base exchange, phosphorylation, and oxidation reactions, respectively. We shall briefly present an overview of our current understanding of their molecular identities.

12.7.1.1 ADP-Ribosyl Cyclases

To date, three different types of ARC proteins have been cloned, ARC from the ovotestis of the mollusc *Aplysia californica* [147] and the mammalian orthologs, CD38 [148] and CD157/BST-1 [149]. Historically, CD38 was cloned first, but its enzymatic function only became apparent after cloning of the *Aplysia* enzyme whose role was to synthesize cADPR. The amino acid sequence alignment of

---

[1] However, human β-NAD^+ kinase cannot use NAAD as a substrate [97] which mandates for a specific protein if such a route exists.

these three ARC members clearly indicate significant areas of homology with an overall identity between 25 and 30% and predicted molecular weights of around 35 kDa. Conservation of 10 out of 11 to 16 cysteine residues suggested that the tertiary structures of these proteins were similar and that they shared functional properties. This was indeed proved when the crystal structures of *Aplysia* ARC [150] and CD157 [151] were determined. These enzymes crystallize as dimers formed by two identical monomers associating in a head-to-head manner and enveloping a central cavity [150, 151].

Several lines of evidence implicate CD38 in the *in vivo* synthesis of cADPR, not least the parallel induction of CD38 and cADPR synthesis or levels in HL60 cells, myometrial cells, and astrocytes [152–155]. In uterine muscle, steroid hormone regimens aimed to mimic pregnancy affect ARC activity in general [156], and CD38 activity in particular, whose expression is the outcome of a fine balance between several fluctuating hormones throughout gestation [157–159]. In the same way, in smooth muscle, thyroid hormones, vitamin D_3, and retinoic acid upregulate ARC expression [160], and this is not restricted to the animal kingdom: transcriptional control is also manifest in plants in response to phytohormones such as abscisic acid [161].

Moreover, stimulated production of cADPR observed in certain cell types upon treatment with specific agonists is no longer seen in the equivalent tissues derived from CD38 knockout mice or in cells where CD38 expression has been disrupted by RNA interference [89, 157, 162–164].

However, CD38 and CD157 are unlikely to be the whole story as the literature is scattered with indications of other isoforms for which we still have no molecular handle: e.g., in contrast to CD38, a skeletal muscle particulate ARC is selectively inhibited by Cu^{2+} and Zn^{2+} [165]; soluble ARC activities preclude the involvement of typical, particulate CD38-type proteins and which have been found in sea urchin eggs [166, 167], brain homogenates [168], human T cells [169], and pancreatic acinar cells [170].

12.7.1.2 cADPR Hydrolases

In contrast to the *Aplysia* ARC, which converts NAD^+ almost exclusively into cADPR [69, 171], CD38 and CD157 are bifunctional enzymes that not only catalyze the production of cADPR, but can also hydrolyze it to ADPR [172–174]. In fact, the yield of cADPR production is low (<2% of the reaction products [175, 176]), with 98% reflecting the net conversion of NAD to ADPR, at least *in vitro*.

Fascinatingly, these functionally antagonistic enzymatic activities are manifestly independent of one another [177, 178], raising the possibility that these enzymatic activities are differentially regulated in the cell to provide a physiological means of regulating synthesis/hydrolysis of cADPR. Precedent for this assertion indeed exists both *in vitro* and *in vivo*: hydrolase activity is selectively inhibited by ATP [173], NADH [81], and ADPR [179]. In sea urchin, ARC activity is stimulated by cGMP, but hydrolase activity is not [166]. The whole story is likely even more complex because, *in vivo*, hormonal-induced expression of CD38 [157,

159, 180] and posttranslational modification induced by retinoic acid in HL60 cells [181] selectively increases ARC activity but not hydrolase activity. How its intrinsic hydrolase activity is suppressed when there is clearly more CD38 expressed is currently unknown.

12.7.1.3 NAADP Metabolism

Rather surprisingly, ARC was found to be a promiscuous enzyme able to use the phosphorylated version of β-NAD$^+$, β-NADP$^+$ as a substrate, although not in a cyclization reaction, but rather serving to catalyze the exchange of its nicotinamide group with free nicotinic acid (base–exchange reaction) to produce NAADP [182] (Figure 12.1). However, the base–exchange reaction requires a substantial concentration of nicotinic acid far beyond physiological values (half maximal rate at 5–10 mM [166, 167, 182]) and has thus far only been demonstrated *in vitro*. That aside, the ARC family of enzymes is a potential source of two pharmacologically distinct Ca^{2+}-mobilizing metabolites, both cADPR and NAADP.

For this potential to be more realistic, these relative activities would need to be separately regulated (perhaps reciprocally), by analogy with ARC and hydrolase activities (see above). One critical factor in regulating which catalytic pathway ARC takes is pH: above neutrality, cyclization, and cADPR production is favored, whereas NAADP synthesis via base exchange is optimal at about pH 4.5 [166, 167, 182]. In a more recent study, a skeletal muscle particulate form of ARC (not CD38) differed in the absolute pH values for these optima, but the trends were similar [165].

While we currently know little about the physiological mechanisms underpinning NAADP production, we know even less about potential removal processes (which are clearly just as crucial for second messenger dynamics). In a sea urchin membrane fraction NAADP removal was demonstrated [167] although the route was unknown. To date, only one study in brain [183] — again a particulate fraction — has documented an enzymatic route in any detail at all, a Ca^{2+}-dependent 2′-phosphatase activity that converts NAADP to the receptor-inactive NAAD [184] (Figure 12.1). If physiologically relevant, this represents a form of Ca^{2+}-dependent negative feedback.

12.7.1.4 Subcellular Location

The above underscores the importance of known ARCs in the production of cADPR *in vivo* but this has to be somehow reconciled with its cellular localization: in contrast to the soluble *Aplysia* ARC, both CD38 and CD157 are membrane-bound ectoenzymes that differ in their mode of anchoring; while CD157 is glycosylphosphatidylinositol-anchored [149], CD38 is a type II membrane protein with a single transmembrane domain [148]. The apparent topological paradox of ectocellular synthesis and intracellular activity of cADPR has been discussed at length by others and might be explained by two observations: (1) influx of cADPR across the plasma membrane to reach its target stores, as suggested by experiments and (2) ligand-induced internalization of CD38, following membrane oligomerization of CD38 with consequent import of cADPR to an intracellular compartment [185–189].

Apart from the localization of CD38 in the plasma membrane, there are increasing numbers of reports on its localization in the membranes of internal organelles, such as endoplasmic reticulum, mitochondria, and the nuclear envelope, in a variety of cell types [170, 190–195]. We have also highlighted above that there are hitherto unknown soluble forms.

12.7.2 DETECTION OF METABOLISM BY DISCONTINUOUS BIOASSAY

The ability of a cell homogenate system to synthesize or degrade cADPR can be readily determined by incubating the homogenate with β-NAD$^+$ (usually 2.5 mM), or cADPR (5 μM), respectively. The synthesis and degradation of NAADP were determined by incubating with β-NADP$^+$ plus nicotinic acid, or NAADP itself, respectively. Incubations were maintained at 17°C in IM buffer for sea urchin egg homogenates or at 37°C in sucrose-HEPES buffer (250 mM sucrose, 20 mM HEPES, pH 7.2) for mammalian cell homogenates.

Synthesis or degradation of the respective Ca^{2+} releasing agents was monitored over time, where 5 μl of the reaction mixture was withdrawn and assayed immediately, using the sea urchin egg homogenate as a bioassay (see Section 12.8.3, and Figure 12.13), for activity that stimulated the RyR or the NAADP-sensitive Ca^{2+} release mechanism, respectively. The specificity of release was confirmed by desensitization to standard chemicals, or by blockade using specific antagonists (see Sections 12.2.5, 12.5, and 12.6). For substrate incubations, an increase in the appropriate Ca^{2+} releasing agent with time indicated the presence of synthetic enzymes in the homogenate tested. The presence of degradative enzymes, e.g., cADPR hydrolases was predicted from a decrease in the Ca^{2+} release from discontinuous monitoring over time, for incubations with either cADPR or NAADP. We have found that inclusion of the ADP-ribosyl cyclase inhibitor nicotinamide (10 mM) [196] in the test homogenate prevents any possible conversion of unreacted β-NAD$^+$ to cADPR by egg homogenate enzymes.

12.7.3 FLUORESCENT ANALOGS

12.7.3.1 NGD Assay: Continuous Assay for ADP-Ribosyl Cyclase Activity

The ADP-ribosyl cyclase activity of a given cell homogenate was detected by a continuous assay, using NGD instead of β-NAD$^+$ as substrate (Section 12.2.5.2) [95]; fortunately, NADase, which may conceivably contaminate crude tissue extracts does not interfere with ARC rate analysis *per se* because it does not produce fluorescent cGDPR but the non-fluorescent GDPR [95]. However, the degradation of ARC substrate by these enzymes is overcome somewhat by ensuring NGD is in excess.

The production of the fluorescent product, cyclic GDP-ribose, can be measured using a Perkin Elmer LS50-B fluorimeter, with excitation and emission wavelengths of 300 and 410 nm, respectively. Reaction mixtures contained 50–100 μl of the homogenate tested, made up to a final volume of 500 μl using sucrose–HEPES or IM buffer, pH 7.2. When a basal fluorescence was recorded, the reaction was initiated with the addition of 50–250 μM NGD (K_m of CD38 and Aplysia

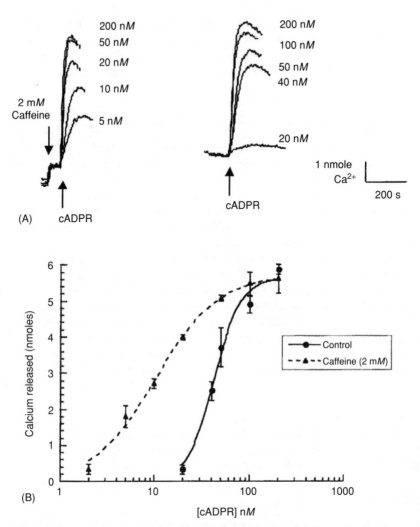

FIGURE 12.13 cADPR bioassay. (A) Fluorimetric traces obtained from 2.5% egg homogenate following addition of cADPR standards ± 2 mM caffeine. (B) Effect of 2 mM caffeine on cADPR concentration response curve. In the presence of 2 mM caffeine, EC_{30} value for cADPR was shifted from 42.5 ± 3.2 nM to $10.5 + 0.9$ nM. The pre-addition of caffeine was made 60 s prior to cADPR application.

ARC are of the order of 1–3 μM [95]). The assay is not always appropriate (or as sensitive as the bioassay described in Sections 12.7.2 and 12.8.3) since, for example, in sea urchin egg homogenates, ARC activity was not able to be detected using this assay, either because the production was too low or the homogenate too turbid to measure changes in fluorescence. The NGD assay implicitly assumes that the ARC isoform is capable of cyclizing this non-physiological substrate, which may not always be the case [197].

12.7.3.2   cIDPR Assay: Continuous Assay for cADPR Hydrolase

The fluorescent cADPR analog, cIDPR, is used as a substrate which, when hydrolyzed is converted into the non-fluorescent IDPR [40]; therefore, the hydrolase activity is proportional to the rate of disappearance of fluorescence. Note that cIDPR does not release Ca^{2+} nor antagonize cADPR action [96]. Cyclic IDPR can be purchased from Sigma and is made up as a 1-mM stock in distilled water, and used at a final concentration of 10 μM. The fluorimeter excitation and emission wavelengths are set as 275 and 404 nm, respectively [96]. Of a total volume of 500 μl, 10–200 μl of a broken cell preparation are added, depending upon the activity of the hydrolase, but mindful of the turbidity issue raised above which interferes with fluorescence. Once cIDPR is added, the fluorescence in the stirred cuvette is allowed to settle prior to hydrolase addition, and only then is the rapid, initial rate of change of fluorescence measured , thereby avoiding product inhibition or substrate depletion. Experiments are conducted at a temperature optimal for the species examined. Usually, frozen fractions are thawed and used immediately since hydrolase activity can deteriorate over time.

12.7.4   HPLC

Reaction products formed as a result of metabolism can also be verified by HPLC protocols (see Sections 12.2.1 and 12.2.2). In addition, when ascertaining synthesis of either cADPR or NAADP, it is important to check that their respective precursors NAD and NADP are not contaminated with these molecules. If so, they can be purified by the techniques described in Section 12.2. It is worth remembering that the Ca^{2+}-mobilizing activity of NAADP was discovered as a contaminant of commercially available β-NADP$^+$.

12.8   MEASUREMENT OF cADPR LEVELS IN CELLS AND TISSUES

12.8.1   Background

A number of different techniques have been used to assess levels of cADPR in cells and tissues. These include microsomal bioassays using sea urchin egg homogenates [198] or mammalian microsomes [199], thin layer chromatography [200, 201], radioimmunoassay [152, 202], radioreceptor assay [203], HPLC [204] and, more recently, a novel fluorescence cycling assay [205].

Thanks to these techniques, changes in cADPR levels have been shown to couple to a wide range of stimuli and cell types including acetylcholine in pancreatic acinar cells [163] and NG-108–15 cells [201], glucose in pancreatic β cells [37], hypoxia in smooth muscle [81], fertilization of sea urchin eggs [206, 207], T cell receptor activation [169], ischemia in cardiac muscle [208], endothelin in smooth muscle [38], and possibly intermediary metabolism [193]. Physiologically, changes in cADPR are important for brain development [195], osteoclastic bone resorption [209], the diabetic phenotype in β cells from ob/ob mice [37], humoral immunity [164], and neutrophil chemotaxis [210] to name but a few.

12.8.1.1 Kinetics

If cADPR is an acute regulator of $[Ca^{2+}]_i$ signals (rather than a chronic regulator of cell excitability [13]), then its cellular levels will be dynamic and rapidly increase in response to extracellular stimuli. Gratifyingly, this has now been confirmed in a number of systems, but unfortunately, few studies have examined the accumulation kinetics in any depth compared with the type of information available for IP_3 [211, 212]. In intact cells, a sustained cADPR level is certainly evident within 10 min of T cell receptor stimulation [169], and within 5 min of β cell glucose stimulation [37] but the lack of shorter time points prevented conclusions about the onset kinetics. In killer cells, IL-8 enhances both $[Ca^{2+}]_i$ and cellular motility, partially via stimulation of PLCβ but also by elevating cGMP and cADPR [89]; the kinetic profile of these messengers closely mirror each other, significantly raised by the earliest (30 s) time point, and peaking at 1 min [89].

That said, researchers have examined the promotion of ARC activity with a superior temporal resolution that should extrapolate to rapid increases in cADPR: within 30–60 s, cyclase activity was significantly enhanced by G protein-dependent muscarinic receptors in neural-derived [201] or adrenal chromaffin cells [213], the latter replicated by Ca^{2+} influx in the same study, implying some form of rapid Ca^{2+} regulation of ARC. Similarly, sympathetic innervation in heart is not only chronotropic but also ionotropic and accordingly, β-adrenoceptor stimulation enhances ARC activity within 30 s [214], whereas the effect of Angiotensin II takes about 1 min [215].

12.8.1.2 Signaling Cascades Modulating ARC

The various and complex signaling cascades culminating in elevated IP_3 levels have been intricately dissected and are well understood, thanks in part to the pioneering molecular investigations of Rhee, Simon, Schlessinger and colleagues [212]. Phospholipase C isoforms, G proteins, tyrosine kinases, and SH2 domains are now the stuff of undergraduate textbooks, but, in sharp contrast, the waters of pyridine nucleotide regulation are very muddy indeed.

The first clue as to how ADP-ribosyl cyclase activities can be regulated came from pharmacological studies in the sea urchin egg that examined the nitric oxide (NO)- or cGMP-induced Ca^{2+} release that occurred after an appreciable delay [196, 200, 216]. The pharmacology of Ca^{2+} release was indicative of a cADPR-mediated phenomenon, and, ultimately, the delay proved to correspond to the time for cADPR to be synthesized from β-NAD^+ [196, 200]. Subsequently, it was shown that just after fertilization, the initial Ca^{2+} signal induces a sluggish NO response; this in turn prolongs the otherwise brief $[Ca^{2+}]_i$ spike via cADPR production [207] (though the precise details may be species-dependent [206]). Although likely to involve protein kinase G (PKG), the precise molecular mechanism underlying enhanced cADPR accumulation is still unknown as we have no handle upon sea urchin ARC or its regulation i.e., is it ARC or an upstream, regulatory protein that is the PKG substrate?

12.8.1.3 Mammalian Cells: NO and cADPR

This is all very well, but does this have any bearing upon the pathway for agonist-stimulated increases in cADPR seen in mammalian cells? Certainly the reports linking NO/cGMP and cADPR-induced Ca^{2+} release have been manifold, e.g., in secretory cells [217], neurons [203, 218, 219], and neurosecretory cells [220], but these reports only implicate and do not demonstrate stimulated cADPR levels. Therefore, it is encouraging that there has been some more concrete substantiation of this contention: Sternfeld et al. [170] suggested that in rat pancreatic acinar cells, both acetylcholine and cholesystokinin rapidly stimulated an, as yet, undefined cytosolic ARC isoform (and not the particulate CD38), via cGMP. Interestingly, this precisely mirrors the selectivity of cGMP for stimulating soluble over particulate ARC in sea urchin [166] and raises the potential of multiple, yet conserved, activation mechanisms.

However, the effects of NO upon Ca^{2+} signals are pleiotropic (e.g., Refs [221–223]) and do not always enhance cADPR-mediated Ca^{2+} release: cADPR synthesis and Ca^{2+} release are actually *inhibited* in smooth muscle from either coronary arteries [76] or airways [224], but this probably involves the cGMP-*independent*, direct S-nitrosylation of CD38 thiol groups that inhibits its ARC but not hydrolase activity [224].

12.8.1.4 Other Pathways

In Jurkat T cells, it is also a soluble form of ARC that is regulated by the T cell receptor, CD3, that initiates a cascade heavily reliant upon tyrosine kinases [169]. Although this has not been empirically tested, it is likely that cADPR production is similarly dependent (note: certainly the particulate ARC form, CD38, colocalizes with the tyrosine kinase, Lck, in Jurkat lipid rafts [225]). In excitable adrenal chromaffin cells [213], and longitudinal smooth muscle [108], Ca^{2+} influx may also support ARC activity (perhaps in concert with cAMP [213]).

In view of the above, it is tempting to speculate that soluble forms of ARC may be regulated by protein phosphorylation (PKG, tyrosine kinases) whereas the particulate form is driven by different signals. Support for this fanciful notion comes from investigations of particulate ARC in several neural-derived cells, including NG-108–15. Members of the seven-transmembrane spanning receptors couple to ARC in a manner highly reminiscent of the adenylyl cyclase system (another particulate enzyme that synthesizes a cyclic nucleotide); while some subtypes stimulate ARC activity via a cholera toxin-sensitive G protein, others are inhibitory via a pertussis toxin-sensitive G protein. To date, both muscarinic acetylcholine [201] and metabotropic glutamate receptors [226] have revealed this dual regulation of ARC. Given that there is no absolute correspondence between the effects upon ARC and upon adenylyl cyclase activity, it is unlikely that changes in cyclic AMP levels mediate the regulation of ARC leaving one to wonder whether some ARC isoforms directly couple to G proteins, as does the prototypic adenylyl cyclase.

12.8.1.5   Regulation of cADPR Removal

To those familiar with the cAMP/cGMP cascade, the consequences of inhibiting the hydrolysis by phosphodiesterases are well documented, and have attained clinical importance with the uplifting effects of rolipram and sildenafil (Viagra). If the turnover of the second messenger is sufficiently high, then inhibition of its removal will result in a rapid elevation of the messenger. Might this paradigm ever be successfully exploited to enhance cADPR? One model invokes product inhibition of the hydrolase by ADP-ribose serving to prolong the cADPR signal [179], but little has been done to verify its role. According to Okamoto and Takasawa, the answer is an emphatic "yes" in pancreatic β cells [227]; insulin is secreted from these cells in response to elevated blood glucose and to the atypical signaling cascade that it recruits. Insulin granules are exocytosed in response to slow, oscillating increases in $[Ca^{2+}]_i$ that are the culmination of a complex dialogue between Ca^{2+} influx and release from internal stores. cADPR was implicated in the latter, but rather than stimulating an ARC activity, cADPR hydrolase is competitively inhibited by the increased cellular ATP (resulting from glucose uptake and metabolism) and cADPR increases within 5 min [227].

Such is the potential importance, that the diabetic aetiology may partly owe its characteristics to defects in this pathway [227], and indeed, studies have shown that diabetic patients possessed higher levels of CD38 autoantibodies compared to controls [228], although only a small fraction of patients was found to exhibit this change. The role of cADPR in insulin secretion remains controversial but may reflect the origin or state of pancreatic material that may account for the failure of other groups to observe glucose-induced cADPR changes by radioimmunoassay for cADPR [229].

In addition to the acute regulation of cADPR levels by the modulation of ADP-ribosyl cyclase activities, modulation of the expression of ADP-ribosyl cyclase enzymes may influence cADPR concentrations in cells (Section 12.7).

12.8.2   PREPARATION AND EXTRACTION OF CADPR FROM CELLS AND TISSUES

Cells and tissues can be treated with agonists and then plunged into ice-cold perchloric acid solution (0.5 M) to stop the reaction and precipitate proteins. [^{3}H]cADPR (10,000 cpm) can be added to the samples to assess recovery. The samples are then quickly vortexed or sonicated and centrifuged at 15,000g for 10 min at 4°C. Supernatants are then neutralized with $KHCO_3$ (2 M), and then centrifuged again at 15,000g for 10 min at 4°C. The supernatant can then be frozen by snap freezing in liquid nitrogen and stored at −70°C until analysis. The pellet of precipitated potassium perchlorate is discarded. Alternatively, if samples are destined for determination of both NAADP and cADPR, then the extraction procedure for NAADP (Section 12.9.2.1) is also appropriate for cADPR levels.

12.8.3   BIOASSAY FOR CADPR

As described in Section 12.7.2, a bioassay using the sea urchin egg homogenate has been very useful in examining cADPR metabolism. The advantage of the bioassay is in its specificity. Cross-desensitization between a sample and authentic cADPR in terms of Ca^{2+} release, or inhibition by 8-NH_2-cADPR, provides good evidence that the Ca^{2+}-mobilizing effects can be ascribed as due to cADPR. Standard curves for Ca^{2+} release by known amounts of authentic cADPR can be used to determine amounts of cADPR in tissue samples. A detailed description of this method has been described previously [198]. The drawback of the technique is in its sensitivity and its susceptibility to perturbation by other factors in the tissue sample such as the level of divalent cations. As mentioned above (Section 12.5), Ca^{2+} ions potentiate cADPR-induced Ca^{2+} release while magnesium ions inhibit it. Chelation of Ca^{2+} in the samples by adding Chelex beads or magnesium by adding EDTA, offers a solution to these difficulties. The low sensitivity to cADPR means that tissue samples need to be concentrated and in so doing molecules or ions are concentrated that could interfere with the assay. We have found that a tissue substance that is concentrated along with cADPR, interferes with the fluorescence of a number of Ca^{2+} indicators that we routinely use for fluorimetry. One potential way around this is to add an HPLC step to resolve cADPR from other substances, but unfortunately using the protocols described in Section 12.2, it co-elutes with cADPR. The sensitivity of the assay can be improved by up to tenfold by adding 1–2 mM caffeine (see Figure 12.13) or 50 μM thimerosal to test homogenates. These are threshold concentrations of these agents that potentiate cADPR-induced Ca^{2+} release.

12.8.4 [^{32}P]cADPR BINDING AND RADIORECEPTOR ASSAY

12.8.4.1 cADPR-Binding Sites

As an example of cADPR binding we present data for binding to whole-egg homogenates from the sea urchin *L. pictus,* where a constant amount of hot (labelled) cADPR is displaced by increasing concentration of cold (unlabelled) cADPR (homologous displacement) (Figure 12.4C). We also can see from the association kinetics, that binding is rapid (Figure 12.4B). Binding is carried out in a 250 μl total volume in 5-ml plastic tubes behind suitable shielding. In 200 μl of Glu-IM (for composition, see Section 12.4.2.1), ~50,000 cpm (200 pM) of [^{32}P]cADPR, added together with a series of concentrations of cold cADPR, are incubated on ice (or room temperature if hydrolase activity is not a problem with the preparation under test). A 10–30 min incubation is suitable for equilibrium binding in sea urchin [7]. Binding is initiated by the addition of 50 μl of homogenate diluted to typically give 2.5% (v/v) final (~3 mg protein/assay). Bound and free radioligand are separated by filtration with a cell harvester (Brandel) onto glass fiber filter paper (Whatman B), that is washed immediately before filtration with ice-cold Glu-IM, and washed twice with 2–4 ml of ice-cold Glu-IM immediately after filtration (note: binding can be increased by soaking filters with 0.3% (v/v) phenylethylimine in water for 1–2 h prior to filtration). The Brandel system is adequate for longer time points (minutes), but for faster kinetic analyses (seconds), binding is started and stopped (filtered) on an individual tube basis using a sampling vacuum manifold that takes 25-mm disk filters (Millipore). The resulting filters are placed into scintillation tubes with 10 ml water and counted in a scintillation counter using Cerenkov radiation. Specific binding was calculated as the difference between total binding and non-specific (defined with 10 μM cold cADPR).

It is noteworthy that, in spite of the profound effects of cADPR in mammalian systems, reports of cADPR binding in the same tissues are scarce indeed. With the exception of a few reports in longitudinal smooth muscle [108], and synaptosomes [230], we and others have been baffled that cADPR binding is so hard to demonstrate in mammalian cells. Undoubtedly, the commercial lack of [^{32}P]cADPR has hampered the field to an extent, but that is not the whole story; low protein expression, considerable hydrolase activity and loss of essential factors in a broken cell preparation might all be compounding factors.

12.8.4.2   cADPR Radioreceptor Assay

Because of the great selectivity of the sea urchin binding site for cADPR over other contaminating nucleotides [6, 7], this binding assay is exploited to determine cADPR mass in extracts from tissues treated with various hormones and drugs. As mentioned above (Section 12.2.3), we favor [^{32}P]cADPR synthesized in-house over commercial [^{3}H]cADPR by virtue of its superior specific radioactivity. By comparison with known cADPR standards, the levels contained in tissue extracts can be easily determined.

Although cADPR is alkaline-sensitive, it is stable enough in acid to allow tissue extraction in perchloric acid ($HClO_4$; PCA) in a protocol essentially identical to that below for NAADP levels (Section 12.9.2). Tissue samples are incubated in physiological salts and the reaction terminated by addition of PCA and sonication with a probe sonicator. Neutralization is effected by addition of $KHCO_3$ (*not* KOH which is harsher and degrades cADPR, presumably through the formation of microdomains of locally high pH during neutralization). After centrifugation the extracted nucleotides are used to displace [^{32}P]cADPR in the sea urchin binding assay described above, and compared with a standard curve of cold cADPR displacement (cf. NAADP).

12.9   MEASUREMENT OF NAADP IN CELLS AND TISSUES

12.9.1   Background

If cADPR can be considered the poor cousin of IP_3 in terms of understanding the regulation of its concentration, NAADP is positively the dark horse of the family. That NAADP levels will be tightly controlled is self-evident, particularly considering the self-inactivation phenomena detailed above.

But the levels of NAADP do change, and change rapidly in response to stimuli as recent studies reveal, and in so doing, belatedly elevate NAADP to the status of *bona fide* second messenger. Some doubt surrounded the presence of endogenous NAADP in cellular extracts until the work of Billington et al. showed supramaximal, micromolar levels in (resting) sea urchin sperm [231]. Subsequently, it was confirmed that this concentrated source of NAADP was transferred into the egg as a bolus of messenger when the sperm fused with the egg at fertilization [12]. Not only does the sperm supply the initial anterograde signal to evoke a Ca^{2+} wave at egg activation, but also a retrograde signal from the egg to the sperm — and this is the crux of the argument — stimulates a second wave of NAADP *production* in the sperm [12], the first demonstration that NAADP is synthesized in response to an extracellular stimulus (in this case, egg jelly). This was soon followed in mammalian cells by reports of glucose-stimulated NAADP levels in a pancreatic β cell line [11] and in smooth muscle in response to endothelin [10]. Even low levels of NAADP have now been found in erythrocytes, hepatocytes, and *E. coli* thanks to improvements in assay sensitivity [232].

It is too early in the day to wonder about signaling cascades, but in sea urchin, NAADP production was enhanced by cAMP in the particulate fraction that contrasted with the cGMP-dependent enhancement of cADPR production in the soluble fraction [167]. This alone indicates that these two messengers do not share synthetic routes.

12.9.2 RADIORECEPTOR ASSAY FOR NAADP

12.9.2.1 Acid Extraction

A scheme of the entire radioreceptor assay procedure is presented in Figure 12.14. To extract nucleotides from a suspension of cells, an equal volume of ice-cold $HClO_4$ (1.5 M) was added and the cells disrupted with a probe sonicator (Jencons, VibraCell at amplitude 60) for three bursts of 3 s. If the level of NAADP was too low to be detected, then the suspension of cells was first centrifuged ($9000g$ for 7 s) to pellet them and then an approximately equal (to the pellet) volume of ice-cold $HClO_4$ (1.5 M) was added, followed by sonication. The disrupted cells were incubated on ice for 1 h and then centrifuged to pellet the denatured protein ($9000g$ for 10 min). The protein pellet was stored at $-80°C$ for determination by protein assay. Typically, the pellet was taken up in 0.1–1 ml 100 mM NaOH. The acidic supernatant was transferred to a new tube and neutralized with an equal volume of $KHCO_3$ (2 M). The samples were incubated on ice for 1–2 h then centrifuged to remove the $KClO_4$ precipitate. The resulting neutralized samples were transferred to new tubes and stored at $-80°C$ until NAADP determination with the radioreceptor assay.

For cells adhering to culture dishes, the buffer was removed and replaced with ice-cold 1.5 M $HClO_4$ which terminated the reactions. The cells are then scraped, disrupted by sonication, and left on ice for 30–60 min. At this point, we took an aliquot to determine the protein content, and the rest of the sample was centrifuged and processed as described above.

Recovery of NAADP with this acid extraction procedure was 65% as determined by extracting a sample with known amount of NAADP (cf. Ref. [233]) and determining the radioactivity recovered. A similar recovery was obtained using cold NAADP and a Ca^{2+}-release bioassay that uses the sea urchin homogenate [234].

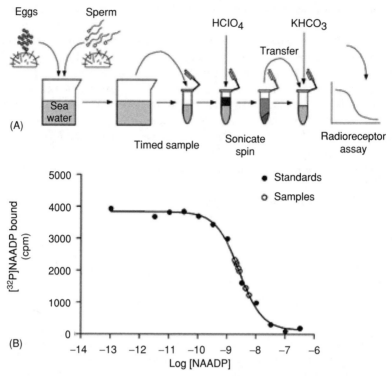

FIGURE 12.14 The NAADP radioreceptor assay. (A) Schematic of the entire process for measuring NAADP levels as a result of egg insemination. (B) Typical standard curve for NAADP homologous displacement of [^{32}P]NAADP from the sea urchin binding site (filled symbols). Appropriately diluted unknown samples from tissue extracts are shown in open symbols and lie on the most sensitive, "linear" portion of the curve.

12.9.2.2 NAADP Radioreceptor Assay

The NAADP-binding protein from sea urchin (*L. pictus*), which is highly specific for NAADP [5, 53, 184] was used. Homogenates of sea urchin eggs were prepared as described (Section 12.4.2). Binding reactions typically contained 0.5% (v/v) homogenate, 0.15–0.2 nM [^{32}P]NAADP (synthesized, Section 12.3.2), 2.5–50 μl of standard or sample (tissue extract) and a volume of Glu-IM (Section 12.4.2.1) sufficient to bring the total volume to 250 μl. Because of the irreversibility of NAADP binding, it is crucial that the binding protein is added *last* to tubes already containing [^{32}P]NAADP with and without unlabelled NAADP. Binding reactions were incubated at room temperature for 15–20 min. Reactions were terminated by rapid filtration (Brandel cell harvester) through GF/B filters (Whatmann) with two 2–4-ml washes with ice-cold Glu-IM. Specific binding was taken by subtracting non-specific from the total bound (non-specific was taken as that bound in the presence of 1 μM NAADP). The levels of NAADP are easily determined by comparison with the known NAADP standards. Note that the samples need to be appropriately concentrated or diluted to fall on the linear part of the displacement curve. Figure 12.14B shows a typical standard curve, with sample values plotted for comparison.

12.9.3 NAADP RECEPTOR BINDING

12.9.3.1 Membrane-Associated Receptors

To characterize NAADP-binding sites in membrane fractions from various sources, we essentially adopt the protocol detailed in the radioreceptor section (Section 12.9.2.2). However, it should be recognized that in mammalian tissues, where binding is reversible [56], slightly different conditions may be required. For example, in a brain membrane fraction [56], or MIN6 crude membranes [11] 1 nM [^{32}P]NAADP was incubated with different concentrations of cold NAADP and 400–500 μg protein per tube in 20 mM HEPES, 1 mM EDTA (pH 7.2) at 37°C.[2] Binding takes ≥30 min to reach equilibrium and bound and free [^{32}P]NAADP separated by rapid filtration on GF/B filters with washes effected by the same HEPES/EDTA buffer.

12.9.3.2 Detergent-Solubilized Receptors

Binding to soluble (or solubilized) proteins necessitates different methods to separate bound and free ligand because soluble proteins may only be poorly retained on glass filters. To solubilize NAADP receptors from sea urchin egg homogenate, we use the protocol as detailed previously [46]: homogenate (50% v/v) is washed twice by centrifugation at 20,000g for 10 min at 4°C in Glu-IM. Washed homogenates (17% v/v) are solubilized in 1% (v/v) Triton X-100 or 1% (w/v) CHAPS on ice for 60 min and then centrifuged at 100,000g for 60 min at 4°C. In 1.5 ml microfuge tubes, the supernatant containing detergent-solubilized receptors are added to [^{32}P]NAADP (0.2 nM) with various concentrations of cold NAADP and binding allowed to proceed at room temperature for 1–3 h in a total of 250 μl of Glu-IM (final detergent concentration 0.2% w/v).

Bound and free ligand are separated using polyethylene glycol (PEG) precipitation of proteins. Essentially, 400 μg of carrier protein (γ-globulin; 20 μl of a 20 mg/ml stock in water, stored −20°C) are added to each assay tube, and the binding–reaction stopped by addition of 250 μl of ice-cold PEG (30% (w/v) stock in Glu-IM, ~8000 MW; PEG is prepared by stirring at room temperature for >20 min until entirely dissolved, and stored 4°C). After vortexing, samples are left for >20 min to precipitate protein. To remove free (unbound) [^{32}P]NAADP, the tubes are centrifuged at 12,000g for 5 min and as much of the supernatant removed as possible by aspirating under vacuum. Carry over of residual free [^{32}P]NAADP is removed by washing the surface of the compact protein pellet with 500 μl of 15%

---

[2] This particular buffer composition and temperature may not be optimal. In our experience, intracellular buffers with high K$^+$ increases [^{32}P]NAADP binding in sea urchin egg [235]. Lower temperatures [231] may also improve binding by reducing NAADP degradation.

PEG that is likewise aspirated. The pellet containing [^{32}P]NAADP bound to its receptor is dissolved by adding 1 ml of water and left for 30 min at room temperature to allow the pellet to soften. The pellet can then be resuspended by gentle trituration with a 1-ml pipette and then transferred to a scintillation vial containing 10 ml of water for Cerenkov counting.

12.10   INVESTIGATING cADPR AND NAADP SIGNALING IN SINGLE CELLS

Broadly speaking, cADPR and NAADP (and their pharmacological tools) are sufficiently similar in their physicochemical properties that the techniques employed to probe the function of these molecules are also essentially identical. To avoid repetition, we shall outline experimental details that can be applied to either molecule, but will highlight essential differences where appropriate.

Two main approaches have been employed to investigate the effects and roles of cADPR or NAADP in Ca^{2+} signaling in intact cells (a) to examine the effects of cADPR/NAADP and agonist analogs on Ca^{2+} signals and/or cellular responses and (b) to monitor the effects of cADPR antagonists or NAADP-induced desensitization. The major technical difficulty is that, in general, pyridine nucleotides and their analogs are membrane impermeant, and have to be directed to their intracellular targets by microinjection [38, 58, 169, 236, 237], patch pipettes [36, 57, 238, 239], liposomes [50], permeabilization [38, 52, 66, 77, 78, 240–242], or use of homogenates and subcellular fractions [54, 243]. These procedures introduce new problems that have to be accounted for or circumvented when drawing particular conclusions about the actions or roles of cADPR and NAADP in cellular Ca^{2+} signaling.

12.10.1   TECHNIQUES FOR INTRODUCING CELL IMPERMEANT COMPOUNDS

12.10.1.1   Microinjection

Both cADPR and NAADP are highly soluble in water and so can be present in microinjection solutions at the high concentrations ($\leq$10 mM) desirable when injecting only small percentages (1–2%) of total cell volumes (i.e., its concentration will have to be 50–100 fold greater in the pipette than required finally in the cell). Pressure injection into cells has been a popular way of introducing the molecule of interest, usually performed in concert with measurements of $[Ca^{2+}]_i$ with fluorescent dyes in single cells, using epifluorescent or confocal imaging.

Micropipettes are pulled on the day of use from capillary glass with an internal filament (e.g., 1 mm (OD) $\times$ 0.78 mm (ID); Harvard Apparatus Ltd.) in a one-stage action on a Flaming–Brown-style puller (P-87, Sutter Instruments) and stored in a dust-free container (e.g., on Bluetak within a petri dish). The filament assists back-filling (of as little as 0.2 μl) with a Pipetman (P2, Gilson) equipped with an ultra fine plastic pipette tip (Alpha Laboratories). An ideal tip diameter of ~0.2 μm cannot be accurately determined optically, but rather by electrical resistance or using so-called "bubble" techniques [244, 245], which measure the threshold pressure for bubble formation from a tip placed in ethanol/methanol. The latter are particularly useful for the rapid (and cheap!) routine checks that ameliorate the tedious day-to-day variations that can plague the experimenter (preparation of poor micropipettes is probably the most common cause of a bad day's experimentation).

It is usual for back-filled pipette solutions to be buffered in terms of pH and, in some cases, Ca^{2+}: we find that injection wounding of eggs is reduced by buffers prepared at appropriate intracellular pH and salt composition, and occasionally, by including low concentrations of EGTA (10–50 μM). Practically, we seldom resort to the latter since Ca^{2+}-binding fluorescent dyes are often included in the pipette at concentrations around 1–2 mM. Their presence is doubly useful because we can fluorescently monitor injection success (Figure 12.15). Sea urchin eggs lack sufficient esterase to enable loading via the ester form of a calcium-sensitive dye; therefore, the dyes must be microinjected and Dextran-conjugated dyes are preferred because they do not compartmentalize, a particular problem with sea urchin eggs [246].

Owing to the small tip diameter, pipette blockage can be a common frustration, which not even the full pressure "Clean" mode of an injector pressure unit can shift; nor can drag (scrap) the electrode along cell-free regions of the bottom of the coverslip, which otherwise frees membrane debris. Consequently, we remove particulate matter *prior* to back-filling with microfuge filtration units based on size (0.02–0.22 μm; Millipore) or molecular weight (MWCO 100 kDa; Microcon filtration units, Millipore). We have found this invaluable with some of the less soluble Ca^{2+} dyes such as Rhod Dextran. However, the significant loss of material to the retention ("dead") volumes of these units may make it prohibitively expensive with precious material such as caged compounds or analogs; under these

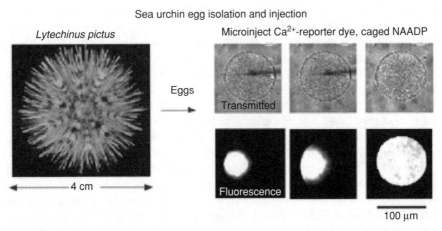

FIGURE 12.15 Photograph and size of the sea urchin species *L. pictus*, and the size of its eggs (transmitted image on a confocal microscope, upper panels). The microinjection pipette is seen as an out-of-focus shadow emanating from a few microns above the egg. The injection motion and entry into the cytoplasm brings this into sharper focus and corresponds to the point at which fluorescence is injected into and spreads rapidly the cell.

circumstances, we might filter the bulk of the concentrated pipette solution and then, to a small aliquot, add a volume of precious analog. The combined solution is then centrifuged in a microfuge at maximum RCF (13,000–17,000g) for 5 min to pellet residual particulate matter, and we carefully remove the supernatant for use.

Microelectrodes are controlled using a hydraulic (Narishige) or electronic (Eppendorf) micromanipulator attached to an inverted microscope on a vibration-resistant air table (Newport). Upon attaching the electrode holder to the manipulator, the microelectrode tip is placed approximately at the center of the objective field of view. Accurately locating the electrode close to the cells is difficult when using high power objectives that cover so little of the field. This process is greatly facilitated by using a Bertrand lens (the lens otherwise used to center the microscope's phase-contrast rings): this lens effectively "de-magnifies" the field of view greatly in all planes allowing the user to find the electrode with ease (note: the Bertrand image is inverted so that the manipulator controls appear to be reversed). As the electrode is lowered, the Bertrand lens is constantly refocused until the electrode is very close to the objective, when the user is forced to revert to the magnified, normal image through the binocular.

For large sea urchin eggs, we inject axially, perpendicular to the plasma membrane, using pressures of 200–500 hPa and injection times of 0.5–2 s provide successful injections (note: resilience of the plasma membrane to penetration by the injection tip can be overcome by a gentle tap of the microscope during the injection motion). The volume injected (ideally 1% of the total cell volume) is varied by altering the time or pressure. With smaller mammalian cells, shorter injection times are preferable, and electrode movement not advisable.

To examine the effect of cADPR, for example, a double injection protocol is required where the cell is pre-injected with the salt form of a Ca^{2+} dye, allowed to recover, and then a different pipette is used to inject cADPR while monitoring the fluorescence response of the Ca^{2+} dye. In the case of the second pipette, patency can optionally be monitored by including a fluorescent yet inert agent that is spectrally distinct from the pre-injected Ca^{2+} dye (e.g., Texas Red when fluo-3 is used). Low EGTA in the pipette might be appropriate under these conditions (see above).

Injections can be prone to artifacts associated with Ca^{2+} influx after puncturing the plasma membrane, but this may be reduced by lowering extracellular Ca^{2+} in the bathing media. However, in many cells extracellular Ca^{2+} facilitates the healing of the microinjection wound and thus reduces further Ca^{2+} influx or loss of cellular components. Thus, it is important to distinguish between the action of cADPR and the effect of microinjection *per se*. Careful controls are required, such as injections of microinjection solutions without cADPR, or solutions containing heat-inactivated cADPR (see Section 12.5). If imaging is performed, then the dynamics of the spatial spread of the Ca^{2+} signal is an important indicator of whether cADPR is affecting Ca^{2+} release from intracellular stores. This is most easily examined in large cells such as sea urchin eggs where the injection of Ca^{2+} (or puncturing the plasma membrane) contrasts with the actions of injecting cADPR because they do not elicit a propagating Ca^{2+} wave. Interpretation of results is much more difficult if cells display only a single Ca^{2+} transient after cADPR injections. However, the generation of regular Ca^{2+} spiking is not often associated with injection artifacts, and is more indicative of a real effect. The use of caged cADPR (see Section 12.2.4) can circumvent many of the objections raised about the validity of results obtained from microinjection experiments.

It should be borne in mind that injection delivers a finite bolus of the compound of interest. If its chemical instability, metabolism, or cellular extrusion are issues, then sufficient concentrations must be injected to compensate for these processes (clearly cADPR and some of its analogs (Section 12.2.5) are hydrolase substrates).

Following pressure injection, the intracellular concentration of a compound is commonly estimated from the volume injected, which is assumed to be the same as the volume ejected from a micropipette either in air or under a drop of oil. Unfortunately, partial blockage of the micropipette tip and hydrostatic pressure inside the cell can result in inaccurate estimates of concentration. Alternatively, we use the fluorescence of the calcium indicator dye to estimate the intracellular dye concentration and, thus, the intracellular concentrations of any other compounds in the micropipette. We feel this approach yields more exact and reproducible estimates of intracellular concentrations.[3] To determine the intracellular concentration of the calcium-sensitive dye, an egg is injected and the fluorescence monitored and compared to a calibration curve relating fluorescence to dye concentration. To prepare calibration curves, a calcium-sensitive dye is serially diluted with a Ca-EGTA buffer (100 nM free calcium) prepared by the pH-metric method described [247]. These solutions are loaded into pulled micropipettes and imaged after positioning the pipette to ensure that its contents fill the confocal slice. The settings for laser power and photomultiplier tube amplification are varied throughout the range used in the actual experiments to generate a family of curves relating fluorescence intensity to dye concentration.

12.10.1.2 Patch Pipette

The whole cell patch configuration is also a useful method for applying cADPR or NAADP into cells by adding to the patch pipette solution at the desired concentration. This can be combined with Ca^{2+} measurements or electrophysiological recordings (see Chapter 6). After patching on to the cell's plasma membrane, suction is needed to break the membrane inside the pipette allowing continuity of cytoplasm with pipette solution. Although this allows cADPR to diffuse to intracellular sites of action, it should also be borne in mind that any cytosolic components that might be required to confer full sensitivity of cADPR on its release mechanism, may be dramatically diluted over time. To minimize these possible effects, substances such as ATP or calmodulin can be included in the patch solution (for cardiac myocytes, we have used a pipette solution containing (mM): KC1, 140; NaCl, 5; $MgCl_2$, 2;

---

[3] Note: this approach works for caged compounds (caged cADPR, caged NAADP) and inhibitors (e.g., heparin and 8-amino-cADPR), but obviously is obscured by the effects of Ca^{2+}-releasing compounds (i.e., free NAADP).

K$_2$ATP, 1; HEPES, 5 (pH 7.2); fluo-3 (100 μM) and caged cADPR (1 mM)). If cADPR is to have effects, these should be seen after "breakthrough." In the NG-108–15 neuroblastoma cell line, cytosolic application of cADPR through a patch pipette enhances and globalizes the Ca^{2+} signals elicited by Ca^{2+} influx through plasma membrane voltage-sensitive Ca^{2+} channels [248]. Unlike the finite bolus delivered by pressure injection, cell dialysis via patch provides an effectively "infinite" reservoir of compound to circumvent the removal issues raised in the previous section. Again, as mentioned for microinjection studies rigorous controls should be conducted. For the reasons stated above, it is sometimes advantageous to include caged cADPR in the patch pipette, to temporally separate "breakthrough" from cADPR-evoked Ca^{2+} release.

12.10.1.3 Permeabilized Cells

The measurement of Ca^{2+} release from cADPR- or NAADP-sensitive stores in permeabilized cells is achieved by radioactivity (^{45}Ca^{2+}) [44, 118, 249, 250] or by fluorescent dyes. Only the latter can be used with single cells whereas both techniques are amenable to monitoring fluxes in populations of cells. These techniques are dealt with in great detail in other chapters of this book (Chapters 1, 2, and 14) and so we will confine ourselves to a few remarks regarding single cell work.

Permeabilization of the plasma membrane should be effected as gently as possible so as to disrupt the internal cellular architecture as little as possible. Carried out in cells washed with an intracellular medium with low free [Ca^{2+}], its progress is monitored by the loss of a previously loaded cytosolic fluorescent dye (e.g., AM esters of Ca^{2+} dyes) or by the uptake of cell impermeant dyes (Trypan Blue, Sytox Green). The presence of millimolar ATP during permeabilization is not obligate, and should be avoided in cells that express purinoceptors, particularly those coupled to Ca^{2+} mobilization for obvious reasons. However, it is clearly to be added soon after in order to replete intracellular Ca^{2+} stores. If using adherent cells, it is wiser to permeabilize them already attached to the coverslip on which they will be imaged rather than in suspension because the former protects them against disruption of store contiguity and internal architecture that result in artefacts [251].

A number of permeabilization agents are available that exploit different mechanisms, some of which may be more appropriate than others to preserve function: cholesterol-binding agents should be used at the lowest, empirically determined concentration (10–100 μg/ml digitonin or saponin) because their selectivity over internal membranes is pretty narrow (only threefold in some systems). Ideally, they should be washed out as soon as permeabilization is complete or they will continue to disrupt internal stores as the experiment progresses. β-Escin is commonly used in "skinned" muscle preparations that allow the entry of larger molecules ($<$30,000 Da) [252] while still retaining coupling to cell surface receptors [78]. Smaller holes are produced by toxins such as Streptolysin O [242] or α toxin [252] allowing passage of ~1000 Da.

The permeabilization procedure, by definition, results in loss of small MW molecules from the cytosol into the extracellular medium, including cytosolic Ca^{2+}

dyes that might have been loaded as AM esters. Monitoring the appearance of Ca^{2+} released into this "exposed" cytosol is therefore not technically straightforward, but it is possible. With a confocal laser scanning microscope, add a low concentration of Ca^{2+}-reporting dye to the entire chamber and monitor fluorescence from a cellular region. The rapid diffusion of Ca^{2+} chelators such as fluorescent dyes necessitates the use of high MW (500 kDa) dextran conjugates or membrane-binding analogs (e.g., fura-C_{18}) whose mobility is greatly reduced, thereby briefly "trapping" the Ca^{2+} close to its release sites [253] (note: be aware that with some techniques, the permeabilization pore size may exclude large MW dyes). Another issue is that the intracellular medium cannot be strongly buffered with EGTA, or it will out-compete the Ca^{2+} indicator. Overnight treatment of nominally Ca^{2+}-free buffers with a cation exchange resin such as Chelex 100 (200–400 mesh, BioRad) will lower their free Ca^{2+} concentration from contaminating (μM) levels down to even 100 nM with careful use of plastic rather than glassware (which otherwise leaches Ca^{2+}). Note: checking the $[Ca^{2+}]$ before and after treatment is advised as we have experienced batch-to-batch variation of the resin; we measure $[Ca^{2+}]$ fluorimetrically using as low a concentration of fura-2 as possible (e.g., 0.5 μM). By using the measurement of cytosolic $[Ca^{2+}]$, cADPR has been shown to increase Ca^{2+} spark frequency in single permeabilized myocytes [240] and to release Ca^{2+} in skinned arterial smooth muscle [76].

In principle, a less technically demanding approach is the measurement of Ca^{2+} within the lumen of Ca^{2+} stores themselves, so that Ca^{2+} release in response to cADPR or NAADP will be manifest as a decrease in the Ca^{2+} concentration therein. Briefly, such measurements rely on the indicator being serendipitously trapped (in the case of small MW fluorescent dyes) or molecularly targeted (in the case of photoproteins such as ER-cameleons or aequorins) within the lumen of the organelle. Clearly, accurate reporting of the $[Ca^{2+}]$ is best achieved by reporters of low Ca^{2+} affinity, concordant with the Ca^{2+} dynamic range evoked. Compared to the technique above, the advantage of a reporter immobilized within the lumen is that (a) the free $[Ca^{2+}]$ of the bathing media can be clamped at any desired level using pre-determined Ca^{2+}/chelator combinations[4] without the need for excessive precautions against Ca^{2+} contamination (b) rapid Ca^{2+} diffusion and dilution into the medium is no longer an issue, and which potentially increases the signal to noise. The use of photoproteins or dyes has seen some success with pyridine nucleotides, with cADPR in dye-loaded retinal [77] and pancreatic acinar cells [66] that also respond to NAADP [63, 66].

Not all is rosy, however, with either the protein or dye approaches: first, the former might necessitate specialist equipment (e.g., a luminometer) as well as established molecular biological techniques for efficient transfection (note: unfortunately, these techniques are difficult to apply to the transcriptionally "inert" sea urchin egg). Second, loading of Ca^{2+} stores with fluorescent dyes by incubating with moderate-high concentrations of AM esters is rather a hit and miss affair: depending upon the

---

[4] Calculated with programs such as Maxchelator (Dr. Chris Patton, Stanford; http://www.stanford.edu/~cpatton/maxc.html).

relative esterase/extrusion activities, some cell types readily compartmentalize dyes into the ER (e.g., hepatocytes [254]), whereas others are frustratingly resistant, or load the dye into an undesired cellular compartment. It is often easier to change cell type than find conditions that favor store loading in the chosen preparation.

As already noted, a major difference between cADPR- and NAADP-stimulated Ca^{2+} release is that only cADPR is affected by the cytosolic $[Ca^{2+}]$ (Sections 12.5 and 12.6). The free $[Ca^{2+}]$ of the bathing intracellular medium will therefore probably be more critical for optimization of cADPR responses than for those to NAADP (a condition more controllable with compartmentalized dyes). In addition, the efficiency of Ca^{2+} reporters can be severely compromised in environments of low pH [255], a pertinent matter for NAADP, whose vesicular stores are demonstrably acidic (Sections 12.1.3 and 12.6.2). As ever, caution should be exercised in dismissing the role of either cADPR or NAADP in broken cell systems (*vis à vis* intact cells) where their responses are not as robust as IP_3 — at least in mammalian systems — which presumably reflect the loss of some crucial factor (or association) upon permeabilization (cf. calmodulin in sea urchin homogenate).

12.11   CONCLUSION

In this chapter, we have discussed the wide range of techniques that we use to study calcium signaling by cADPR and NAADP. We hope that many of these techniques described will prove applicable and useful to the reader's chosen system of study, and may aid in enhancing our understanding of cADPR and NAADP signaling and calcium signaling in general. New pharmacological tools that continue to emerge will be invaluable in dissecting the role of these messengers in cellular control processes. In particular, the development of potent and selective membrane permeant agonists and antagonists will be particularly welcomed. New methods are now available to measure changes in tissue and cellular levels of NAADP and cADPR that are beginning to be linked to activation of specific cell surface receptors, confirming their roles as calcium-mobilizing intracellular messengers. Over the next few years we expect the molecular components and targets of these new signaling pathways to be clarified and defined. In particular, a better understanding of agonist regulation of the synthesis of these messengers, and the precise identification of their molecular targets is anticipated. We hope that the approaches discussed here will provide some of the tools to achieve these aims.

REFERENCES

1. Berridge, M. J., Bootman, M. D., and Roderick, H. L. Calcium signalling: dynamics, homeostasis and remodelling. *Nat Rev Mol Cell Biol* 4, 517–29 (2003).
2. Clapper, D. L., Walseth, T. F., Dargie, P. J., and Lee, H. C. Pyridine nucleotide metabolites stimulate calcium release from sea urchin egg microsomes desensitized to inositol trisphosphate. *J Biol Chem* 262, 9561–8 (1987).
3. Guse, A. H. Regulation of calcium signaling by the second messenger cyclic adenosine diphosphoribose (cADPR). *Curr Mol Med* 4, 239–48 (2004).

4. Genazzani, A. A., Empson, R. M., and Galione, A. Unique inactivation properties of NAADP-sensitive Ca^{2+} release. *J Biol Chem* 271, 11599–602, 1996.

5. Patel, S., Churchill, G. C., and Galione, A. Unique kinetics of nicotinic acid-adenine dinucleotide phosphate (NAADP) binding enhance the sensitivity of NAADP receptors for their ligand [In Process Citation]. *Biochem J* 352 Pt 3, 725–9, 2000.

6. Lee, H. C. Specific binding of cyclic ADP-ribose to calcium-storing microsomes from sea-urchin eggs. *J Biol Chem* 266, 2276–81, 1991.

7. Thomas, J. M., Masgrau, R., Churchill, G. C., and Galione, A. Pharmacological characterization of the putative cADP-ribose receptor. *Biochem J* 359, 451–7, 2001.

8. Wilcox, R. A., Primrose, W. U., Nahorski, S. R., and Challiss, R. A. New developments in the molecular pharmacology of the myo-inositol 1,4,5-trisphosphate receptor. *Trends Pharmacol Sci* 19, 467–75, 1998.

9. Lee, H. C. and Aarhus, R. Structural determinants of nicotinic acid adenine dinucleotide phosphate important for its calcium-mobilizing activity. *J Biol Chem* 272, 20378–83, 1997.

10. Kinnear, N. P., Boittin, F. X., Thomas, J. M., Galione, A., and Evans, A. M. Lysosome-sarcoplasmic reticulum junctions: a trigger zone for calcium signalling by NAADP and endothelin-1. *J Biol Chem* 279, 54319–26, (2004).

11. Masgrau, R., Churchill, G. C., Morgan, A. J., Ashcroft, S. J., and Galione, A. NAADP. A new second messenger for glucose-induced Ca^{2+} responses in clonal pancreatic beta cells. *Curr Biol* 13, 247–51, 2003.

12. Churchill, G. C. et al. Sperm deliver a new second messenger. NAADP. *Curr Biol* 13, 125–8, 2003.

13. Morgan, A. J. and Galione, A. Cyclic ADP-ribose and NAADP in *Structures, Metabolism and Functions* (ed. Lee, H. C.) 167–197 (Kluwer, Dordrecht, 2002).

14. Balshaw, D. M., Yamaguchi, N., and Meissner, G. Modulation of intracellular calcium-release channels by calmodulin. *J Membr Biol* 185, 1–8, 2002.

15. Murayama, T. and Ogawa, Y. Roles of two ryanodine receptor isoforms coexisting in skeletal muscle. *Trends Cardiovasc Med* 12, 305–11, 2002.

16. Sutko, J. L. and Airey, J. A. Ryanodine receptor Ca^{2+} release channels: does diversity in form equal diversity in function? *Physiol Rev* 76, 1027–71, 1996.

17. Sitsapesan, R. and Williams, A. J. Cyclic ADP-ribose and related compounds activate sheep skeletal sarcoplasmic reticulum Ca^{2+} release channel. *Am J Physiol* 268, C1235–40, 1995.

18. Walseth, T. F., Aarhus, R., Kerr, J. A., and Lee, H. C. Identification of cyclic ADP-ribose-binding proteins by photoaffinity labeling. *J Biol Chem* 268, 26686–91, 1993.

19. Chelu, M. G., Danila, C. I., Gilman, C. P., and Hamilton, S. L. Regulation of ryanodine receptors by FK506 binding proteins. *Trends Cardiovasc Med* 14, 227–34, 2004.

20. Patterson, R. L., Boehning, D., and Snyder, S. H. Inositol 1,4,5-trisphosphate receptors as signal integrators. *Annu Rev Biochem* 73, 437–65, 2004.

21. Wehrens, X. H., Lehnart, S. E., and Marks, A. R. Intracellular calcium release channels and cardiac disease. *Annu Rev Physiol* 67, 69–98, 2005.

22. Noguchi, N. et al. Cyclic ADP-ribose binds to FK506-binding protein 12.6 to release Ca^{2+} from islet microsomes. *J Biol Chem* 272, 3133–6, 1997.

23. Bultynck, G. et al. The conserved sites for the FK506-binding proteins in ryanodine receptors and inositol 1,4,5-trisphosphate receptors are structurally and functionally different. *J Biol Chem* 276, 47715–24, 2001.

24. Wang, Y. X. et al. FKBP12.6 and cADPR regulation of Ca^{2+} release in smooth muscle cells. *Am J Physiol Cell Physiol* 286, C538–46, 2004.

25. Tang, W. X., Chen, Y. F., Zou, A. P., Campbell, W. B., and Li, P. L. Role of FKBP12.6 in cADPR-induced activation of reconstituted ryanodine receptors from arterial smooth muscle. *Am J Physiol Heart Circ Physiol* 282, H1304–10, 2002.

26. Hashii, M., Minabe, Y., and Higashida, H. cADP-ribose potentiates cytosolic Ca^{2+} elevation and Ca^{2+} entry via L-type voltage-activated Ca^{2+} channels in NG108–15 neuronal cells. *Biochem J* 345 Pt 2, 207–15, 2000.

27. Marx, S. O. et al. PKA phosphorylation dissociates FKBP12.6 from the calcium release channel (ryanodine receptor): defective regulation in failing hearts. *Cell* 101, 365–76, 2000.

28. Thomas, J. M., Summerhill, R. J., Fruen, B. R., Churchill, G. C., and Galione, A. Calmodulin dissociation mediates desensitization of the cADPR-induced Ca^{2+} release mechanism. *Curr Biol* 12, 2018–22, 2002.

29. Shiwa, M., Murayama, T., and Ogawa, Y. Molecular cloning and characterization of ryanodine receptor from unfertilized sea urchin eggs. *Am J Physiol Regul Integr Comp Physiol* 282, R727–37, 2002.

30. Meszaros, L. G., Bak, J., and Chu, A. Cyclic ADP-ribose as an endogenous regulator of the non-skeletal type ryanodine receptor Ca^{2+} channel. *Nature* 364, 76–9, 1993.

31. Fulceri, R. et al. Ca^{2+} release induced by cyclic ADP ribose in mice lacking type 3 ryanodine receptor. *Biochem Biophys Res Commun* 288, 697–702, 2001.

32. Copello, J. A., Qi, Y., Jeyakumar, L. H., Ogunbunmi, E., and Fleischer, S. Lack of effect of cADP-ribose and NAADP on the activity of skeletal muscle and heart ryanodine receptors. *Cell Calcium* 30, 269–84, 2001.

33. Bastide, B., Snoeckx, K., and Mounier, Y. ADP-ribose stimulates the calcium release channel RyR1 in skeletal muscle of rat. *Biochem Biophys Res Commun* 296, 1267, 2002.

34. Sonnleitner, A., Conti, A., Bertocchini, F., Schindler, H., and Sorrentino, V. Functional properties of the ryanodine receptor type 3 (RyR3) Ca^{2+} release channel. *Embo J* 17, 2790–8, 1998.

35. Kunerth, S. et al. Amplification and propagation of pacemaker Ca^{2+} signals by cyclic ADP-ribose and the type 3 ryanodine receptor in T cells. *J Cell Sci* Pt 117, 2141–49, 2004.

36. Rakovic, S., Galione, A., Ashamu, G. A., Potter, B. V. L., and Terrar, D. A. A specific cyclic ADP-ribose antagonist inhibits cardiac excitation–contraction coupling. *Curr Biol* 6, 989–96, 1996.

37. Takasawa, S. et al. Cyclic ADP-ribose and inositol 1,4,5-trisphosphate as alternate second messengers for intracellular Ca^{2+} mobilization in normal and diabetic beta-cells. *J Biol Chem* 273, 2497–500, 1998.

38. Barone, F. et al. A pivotal role for cADPR-mediated Ca^{2+} signaling: regulation of endothelin-induced contraction in peritubular smooth muscle cells. *Faseb J* 16, 697–705, 2002.

39. Murayama, T. and Ogawa, Y. RyR1 exhibits lower gain of CICR activity than RyR3 in the SR: evidence for selective stabilization of RyR1 channel. *Am J Physiol Cell Physiol* 287, C36–45, 2004.

40. Prakash, Y. S., Kannan, M. S., Walseth, T. F., and Sieck, G. C. cADP ribose and $[Ca^{2+}](i)$ regulation in rat cardiac myocytes. *Am J Physiol Heart Circ Physiol* 279, H1482–9, 2000.

41. Lukyanenko, V., Gyorke, I., Wiesner, T. F., and Gyorke, S. Potentiation of Ca^{2+} release by cADP-ribose in the heart is mediated by enhanced SR Ca^{2+} uptake into the sarcoplasmic reticulum. *Circ Res* 89, 614–22, 2001.

42. Bradley, K. N., Currie, S., MacMillan, D., Muir, T. C., and McCarron, J. G. Cyclic ADP-ribose increases Ca^{2+} removal in smooth muscle. *J Cell Sci* Pt 278, 11057–64, 2003.

43. Chini, E. N. and Dousa, T. P. Nicotinate-adenine dinucleotide phosphate-induced Ca^{2+}-release does not behave as a Ca^{2+}-induced Ca^{2+}-release system. *Biochem J* 316, 709–11, 1996.

44. Bak, J., Billington, R. A., and Genazzani, A. A. Effect of luminal and extravesicular Ca^{2+} on NAADP binding and release properties. *Biochem Biophys Res Commun* 295, 806–11, 2002.

45. Genazzani, A. A. et al. Pharmacological properties of the Ca^{2+}-release mechanism sensitive to NAADP in the sea urchin egg. *Br J Pharmacol* 121, 1489–95, 1997.

46. Berridge, G., Dickinson, G., Parrington, J., Galione, A., and Patel, S. Solubilization of receptors for the novel Ca^{2+}-mobilizing messenger, nicotinic acid adenine dinucleotide phosphate. *J Biol Chem* 277, 43717–23, 2002.

47. Genazzani, A. A. and Galione, A. Nicotinic acid-adenine dinucleotide phosphate mobilizes Ca^{2+} from a thapsigargin-insensitive pool. *Biochem J* 315, 721–5, 1996.

48. Churchill, G. C. et al. NAADP mobilizes Ca^{2+} from reserve granules, a lysosome-related organelle, in sea urchin eggs. *Cell* 111, 703–8, 2002.

49. Yamasaki, M. et al. Organelle selection determines agonist-specific Ca^{2+} signals in pancreatic acinar and beta cells. *J Biol Chem* 279, 7234–40, 2004.

50. Brailoiu, E. et al. NAADP potentiates neurite outgrowth. *J Biol Chem* 280, 5646–50, 2004.

51. Johnson, J. D. and Misler, S. Nicotinic acid-adenine dinucleotide phosphate-sensitive calcium stores initiate insulin signaling in human beta cells. *Proc Natl Acad Sci USA* 99, 14566–71, 2002.

52. Mitchell, K. J., Lai, F. A., and Rutter, G. A. Ryanodine receptor type I and nicotinic acid adenine dinucleotide phosphate (NAADP) receptors mediate Ca^{2+} release from insulin-containing vesicles in living pancreatic beta-cells (MIN6). *J Biol Chem* 278, 11057–64, 2003.

53. Aarhus, R. et al. Activation and inactivation of Ca^{2+} release by $NAADP^{+}$. *J Biol Chem* 271, 8513–6, 1996.

54. Navazio, L. et al. Calcium release from the endoplasmic reticulum of higher plants elicited by the NADP metabolite nicotinic acid adenine dinucleotide phosphate. *Proc Natl Acad Sci USA* 97, 8693–8, 2000.

55. Churchill, G. C. and Galione, A. Prolonged inactivation of NAADP-induced Ca^{2+} release mediates a spatiotemporal Ca^{2+} memory. *J Biol Chem* 276, 11223–5, 2001.

56. Patel, S., Churchill, G. C., Sharp, T., and Galione, A. Widespread distribution of binding sites for the novel Ca^{2+}-mobilizing messenger, nicotinic acid adenine dinucleotide phosphate, in the brain [In Process Citation]. *J Biol Chem* 275, 36495–7, 2000.

57. Cancela, J. M., Churchill, G. C., and Galione, A. Coordination of agonist-induced Ca^{2+}-signalling patterns by NAADP in pancreatic acinar cells. *Nature* 398, 74–6, 1999.

58. Berg, I., Potter, B. V., Mayr, G. W., and Guse, A. H. Nicotinic acid adenine dinucleotide phosphate ($NAADP^{+}$) is an essential regulator of T-lymphocyte Ca^{2+}-signaling. *J Cell Biol* 150, 581–8, 2000.

59. Churchill, G. C. and Galione, A. Spatial control of Ca^{2+} signaling by nicotinic acid adenine dinucleotide phosphate diffusion and gradients [in process citation]. *J Biol Chem* 275, 38687–92, 2000.

60. Cancela, J. M. Specific Ca^{2+} signaling evoked by cholecystokinin and acetylcholine: the roles of NAADP, cADPR, and IP3. *Annu Rev Physiol* 63, 99–117, 2001.

61. Patel, S., Churchill, G. C., and Galione, A. Coordination of Ca^{2+} signalling by NAADP. *Trends Biochem Sci* 26, 482–9, 2001.

62. Cancela, J. M., Van Coppenolle, F., Galione, A., Tepikin, A. V., and Petersen, O. H. Transformation of local Ca^{2+} spikes to global Ca^{2+} transients: the combinatorial roles of multiple Ca^{2+} releasing messengers. *Embo J* 21, 909–19, 2002.

63. Gerasimenko, J. V. et al. NAADP mobilizes Ca^{2+} from a thapsigargin-sensitive store in the nuclear envelope by activating ryanodine receptors. *J Cell Biol* 163, 271–82, 2003.

64. Hohenegger, M., Suko, J., Gscheidlinger, R., Drobny, H., and Zidar, A. Nicotinic acid adenine dinucleotide phosphate, NAADP, activates the skeletal muscle ryanodine receptor. *Biochem J* 367, 423–31, 2002.

65. Mojzisova, A., Krizanova, O., Zacikova, L., Kominkova, V., and Ondrias, K. Effect of nicotinic acid adenine dinucleotide phosphate on ryanodine calcium release channel in heart. *Pflugers Arch* 441, 674–7, 2001.

66. Krause, E., Gobel, A., and Schulz, I. Cell side specific sensitivities of intracellular Ca^{2+} stores for IP3, cyclic ADP-ribose and NAADP in permeabilized pancreatic acinar cells from mouse. *J Biol Chem* 23, 23, 2002.

67. Axelson, J. T., Bodley, J. W., and Walseth, T. F. A volatile liquid chromatography system for nucleotides. *Anal Biochem* 116, 357–60, 1981.

68. Lee, H. C., Aarhus, R., Gee, K. R., and Kestner, T. Caged nicotinic acid adenine dinucleotide phosphate. Synthesis and use. *J Biol Chem* 272, 4172–8, 1997.

69. Hellmich, M. R. and Strumwasser, F. Purification and characterization of a molluscan egg-specific NADase, a second-messenger enzyme. *Cell Regul* 2, 193–202, 1991.

70. Lee, H. C., Walseth, T. F., Bratt, G. T., Hayes, R. N., and Clapper, D. L. Structural determination of a cyclic metabolite of NAD with intracellular calcium-mobilizing activity. *J Biol Chem* 264, 1608–15, 1989.

71. Adams, S. R. and Tsien, R. Y. Controlling cell chemistry with caged compounds. *Annu Rev Physiol* 55, 755–84, 1993.

72. Aarhus, R., Gee, K., and Lee, H. C. Caged cyclic ADP-ribose — synthesis and use. *J Biol Chem* 270, 7745–9, 1995.

73. Brown, E. B., Shear, J. B., Adams, S. R., Tsien, R. Y., and Webb, W. W. Photolysis of caged calcium in femtoliter volumes using two-photon excitation. *Biophys J* 76, 489–99, 1999.

74. Walseth, T. F. et al. Preparation of cyclic ADP-ribose antagonists and caged cyclic ADP-ribose. *Methods Enzymol* 280, 294–305, 1997.

75. Potter, B. V. and Walseth, T. F. Medicinal chemistry and pharmacology of cyclic ADP-ribose. *Curr Mol Med* 4, 303–11, 2004.

76. Yu, J. Z., Zhang, D. X., Zou, A. P., Campbell, W. B., and Li, P. L. Nitric oxide inhibits $Ca(2+)$ mobilization through cADP-ribose signaling in coronary arterial smooth muscle cells. *Am J Physiol Heart Circ Physiol* 279, H873–81, 2000.

77. Churchill, G. C. and Louis, C. F. Imaging of intracellular calcium stores in single permeabilized lens cells. *Am J Physiol* 276, C426–34, 1999.

78. Prakash, Y. S., Kannan, M. S., Walseth, T. F., and Sieck, G. C. Role of cyclic ADP-ribose in the regulation of $[Ca^{2+}]i$ in porcine tracheal smooth muscle. *Am J Physiol* 274, C1653–60, 1998.

79. Sethi, J. K., Empson, R. M., Bailey, V. C., Potter, B. V., and Galione, A. 7-Deaza-8-bromo-cyclic ADP-ribose, the first membrane-permeant, hydrolysis-resistant cyclic ADP-ribose antagonist. *J Biol Chem* 272, 16358–63, 1997.

80. Dipp, M. and Evans, A. M. Cyclic ADP-ribose is the primary trigger for hypoxic pulmonary vasoconstriction in the rat lung *in situ*. *Circ Res* 89, 77–83, 2001.

81. Wilson, H. L. et al. ADP-ribosyl cyclase and cyclic ADP-ribose hydrolase act as a redox sensor: a primary role for cADPR in hypoxic pulmonary vasoconstriction. *J Biol Chem* 276, 11180–8, 2001.

82. Sanz, E. et al. Mechanisms of relaxation by urocortin in renal arteries from male and female rats. *Br J Pharmacol* 140, 1003–7, 2003.

83. White, T. A., Kannan, M. S., and Walseth, T. F. Intracellular calcium signaling through the cADPR pathway is agonist specific in porcine airway smooth muscle. *Faseb J* 17, 482–84, 2003.

84. Zhang, A. Y., Yi, F., Teggatz, E. G., Zou, A. P., and Li, P. L. Enhanced production and action of cyclic ADP-ribose during oxidative stress in small bovine coronary arterial smooth muscle. *Microvasc Res* 67, 159–67, 2004.

85. Geiger, J., Zou, A. P., Campbell, W. B., and Li, P. L. Inhibition of cADP-ribose formation produces vasodilation in bovine coronary arteries. *Hypertension* 35, 397–402, 2000.

86. Ge, Z. D. et al. Cyclic ADP-ribose contributes to contraction and Ca^{2+} release by M(1) muscarinic receptor activation in coronary arterial smooth muscle. *J Vasc Res* 40, 28–36, 2003.

87. Xie, G. H. et al. Increase of intracellular Ca^{2+} during ischemia/reperfusion injury of heart is mediated by cyclic ADP-ribose. *Biochem Biophys Res Commun* 307, 713–18, 2003.

88. Iino, S., Cui, Y., Galione, A., and Terrar, D. A. Actions of cADP-ribose and its antagonists on contraction in guinea pig isolated ventricular myocytes. Influence of temperature. *Circ Res* 81, 879–84, 1997.

89. Rah, S. Y., Park, K. H., Han, M. K., and Im, M. J. Activation of CD38 by IL8 signaling regulates intracellular Ca^{2+} level and motility of lymphokine-activated killer cells. *J Biol Chem* 280, 2888–95, 2004.

90. Bailey, V. C., Sethi, J. K., Fortt, S. M., Galione, A., and Potter, B. V. L. 7-Deaza cyclic adenosine 5'-diphosphate ribose: first example of a Ca^{2+}-mobilizing partial agonist related to cyclic adenosine 5'-diphosphate ribose. *Chem Biol* 4, 51–61, 1997.

91. Bailey, V. C., Sethi, J. K., Galione, A., and Potter, B. V. L. Synthesis of 7-Deaza cyclic adenosine 5'-diphosphate ribose: the first hydrolysis resistant antagonist at the cADPR receptor. *Chem Comm* 4, 695–6, 1997.

92. Wong, L., Aarhus, R., Lee, H. C., and Walseth, T. F. Cyclic 3-deaza-adenosine diphosphoribose: a potent and stable analog of cyclic ADP-ribose. *Biochim Biophys Acta* 1472, 555–64, 1999.

93. Wagner, G. K., Black, S., Guse, A. H., and Potter, B. V. First enzymatic synthesis of an N^1-cyclised cADPR (cyclic-ADP ribose) analogue with a hypoxanthine partial structure: discovery of a membrane permeant cADPR agonist. *Chem Commun (Camb)*, 1944–5, 2003.

94. Guse, A. H. Biochemistry, biology, and pharmacology of cyclic adenosine diphosphoribose (cADPR). *Curr Med Chem* 11, 847–55, 2004.

95. Graeff, R. M., Walseth, T. F., Fryxell, K., Branton, W. D., and Lee, H. C. Enzymatic synthesis and characterizations of cyclic GDP-ribose: a procedure for distinguishing enzymes with ADP-ribosyl cyclase activity. *J Biol Chem* 269, 30260–7, 1994.

96. Graeff, R. M., Walseth, T. F., Hill, H. K., and Lee, H. C. Fluorescent analogs of cyclic ADP-ribose: synthesis, spectral characterization, and use. *Biochemistry* 35, 379–86, 1996.

97. Lerner, F., Niere, M., Ludwig, A., and Ziegler, M. Structural and functional characterization of human nad kinase. *Biochem Biophys Res Commun* 288, 69–74, 2001.

98. Gee, K. R. and Lee, H. C. Characterization and application of photogeneration of calcium mobilizers CADP-release and nicotinic acid adenine dinucleotide phosphate from caged analogs in *Methods in Enzymology* (Marriott, G. ed.) 403–15 Academic Press, New York, 1998.

99. Walker, J. W., Reid, G. P., and Trentham, D. R. Synthesis and properties of caged

nucleotides in *Biomembranes* (Fleischer, S. ed.) 208–21, Academic Press, New York, 1989.

100. Billington, R. A., Tron, G. C., Reichenbach, S., Sorba, G., and Genazzani, A. A. Role of the nicotinic acid group in NAADP receptor selectivity. *Cell Calcium* 37, 81–6, 2005.

101. Lee, H. C. and Aarhus, R. Fluorescent analogs of NAADP with calcium mobilizing activity. *Biochim Biophys Acta* 1425, 263–71, 1998.

102. Bobalova, J., Bobal, P., and Mutafova-Yambolieva, V. N. High-performance liquid chromatographic technique for detection of a fluorescent analogue of ADP-ribose in isolated blood vessel preparations. *Anal Biochem* 305, 269–76, 2002.

103. Billington, R. A., Bak, J., Martinez-Coscolla, A., Debidda, M., and Genazzani, A. A. Triazine dyes are agonists of the NAADP receptor. *Br J Pharmacol* 142, 1241–6, 2004.

104. Clapper, D. L. and Lee, H. C. Inositol trisphosphate induces calcium release from nonmitochondrial stores in sea urchin egg homogenates. *J Biol Chem* 260, 13947–54, 1985.

105. Galione, A., Lee, H. C., and Busa, W. B. Ca^{2+}-induced Ca^{2+} release in sea urchin egg homogenates: modulation by cyclic ADP-ribose. *Science* 253, 1143–6, 1991.

106. Lee, H. C., Aarhus, R., and Walseth, T. F. Calcium mobilization by dual receptors during fertilization of sea urchin eggs. *Science* 261, 352–5, 1993.

107. Graeff, R. M., Podein, R. J., Aarhus, R., and Lee, H. C. Magnesium ions but not ATP inhibit cyclic ADP-ribose-induced calcium release. *Biochem Biophys Res Commun* 206, 786–91, 1995.

108. Kuemmerle, J. F. and Makhlouf, G. M. Agonist-stimulated cyclic ADP ribose. Endogenous modulator of Ca^{2+}-induced Ca^{2+} release in intestinal longitudinal muscle. *J Biol Chem* 270, 25488–94, 1995.

109. Muir, S. R. and Sanders, D. Pharmacology of Ca^{2+} release from red beet microsomes suggests the presence of ryanodine receptor homologs in higher plants. *FEBS Lett* 395, 39–42, 1996.

110. Vu, C. Q., Lu, P. J., Chen, C. S., and Jacobson, M. K. 2'-phospho-cyclic ADP-ribose, a calcium-mobilizing agent derived from NADP. *J Biol Chem* 271, 4747–54, 1996.

111. Petr, J., Urbankova, D., Tomanek, M., Rozinek, J., and Jilek, F. Activation of *in vitro* matured pig oocytes using activators of inositol triphosphate or ryanodine receptors. *Anim Reprod Sci* 70, 235–49, 2002.

112. Lee, H. C. Potentiation of calcium- and caffeine-induced calcium release by cyclic ADP-ribose. *J Biol Chem* 268, 293–9, 1993.

113. Guo, X. and Becker, P. L. Cyclic ADP-ribose-gated Ca^{2+} release in sea urchin eggs requires an elevated. *J Biol Chem* 272, 16984–9, 1997.

114. Ozawa, T. Ryanodine-sensitive Ca^{2+} release mechanism of rat pancreatic acinar cells is modulated by calmodulin. *Biochim Biophys Acta* 1452, 254–62, 1999.

115. Sasamori, K. et al. Cyclic ADP-ribose, a putative Ca^{2+}-mobilizing second messenger, operates in submucosal gland acinar cells. *Am J Physiol Lung Cell Mol Physiol* 287, 69–78, 2004.

116. White, A. M., Watson, S. P., and Galione, A. Cyclic ADP-ribose-induced Ca^{2+} release from rat brain microsomes. *FEBS Lett* 318, 259–63, 1993.

117. Gromada, J., Jorgensen, T. D., and Dissing, S. Cyclic ADP-ribose and inositol 1,4, 5-triphosphate mobilizes Ca^{2+} from distinct intracellular pools in permeabilized lacrimal acinar cells. *FEBS Lett* 360, 303–6, 1995.

118. Ozawa, T. and Nishiyama, A. Characterization of ryanodine-sensitive Ca^{2+} release from microsomal vesicles of rat parotid acinar cells: regulation by cyclic ADP-ribose. *J Membr Biol* 156, 231–9, 1997.

119. Jones, K. T., Cruttwell, C., Parrington, J., and Swann, K. A mammalian sperm cytosolic phospholipase C activity generates inositol trisphosphate and causes Ca^{2+} release in sea urchin egg homogenates. *FEBS Lett* 437, 297–300, 1998.

120. Taylor, C. W. and Broad, L. M. Pharmacological analysis of intracellular Ca^{2+} signalling: problems and pitfalls. *Trends Pharmacol Sci* 19, 370–5 (1998).

121. Dasso, L. L. and Taylor, C. W. Heparin and other polyanions uncouple alpha 1-adrenoceptors from G-proteins. *Biochem J* 280 Pt 3, 791–5, 1991.

122. Tones, M. A. et al. The effect of heparin on the inositol 1,4,5-trisphosphate receptor in rat liver microsomes. Dependence on sulphate content and chain length. *FEBS Lett* 252, 105–8, 1989.

123. Thorn, P. and Petersen, O. H. Calcium oscillations in pancreatic acinar cells, evoked by the cholecystokinin analogue JMV-180, depend on functional inositol 1,4,5-trisphosphate receptors. *J Biol Chem* 268, 23219–21, 1993.

124. Solovyova, N., Fernyhough, P., Glazner, G., and Verkhratsky, A. Xestospongin C empties the ER calcium store but does not inhibit InsP3-induced Ca^{2+} release in cultured dorsal root ganglia neurones. *Cell Calcium* 32, 49–52, 2002.

125. Ozaki, H. et al. Inhibitory mechanism of xestospongin-C on contraction and ion channels in the intestinal smooth muscle. *Br J Pharmacol* 137, 1207–12, 2002.

126. De Smet, P. et al. Xestospongin C is an equally potent inhibitor of the inositol 1,4,5-trisphosphate receptor and the endoplasmic-reticulum Ca^{2+} pumps. *Cell Calcium* 26, 9–13, 1999.

127. Castonguay, A. and Robitaille, R. Xestospongin C is a potent inhibitor of SERCA at a vertebrate synapse. *Cell Calcium* 32, 39–47, 2002.

128. Bootman, M. D. et al. 2-aminoethoxydiphenyl borate (2-APB) is a reliable blocker of store-operated Ca^{2+} entry but an inconsistent inhibitor of $InsP_3$-induced Ca^{2+} release. *Faseb J* 16, 1145–50, 2002.

129. Lemonnier, L., Prevarskaya, N., Mazurier, J., Shuba, Y., and Skryma, R. 2-APB inhibits volume-regulated anion channels independently from intracellular calcium signaling modulation. *FEBS Lett* 556, 121–6, 2004.

130. Peppiatt, C. M. et al. 2-Aminoethoxydiphenyl borate (2-APB) antagonises inositol 1,4,5-trisphosphate-induced calcium release, inhibits calcium pumps and has a use-dependent and slowly reversible action on store-operated calcium entry channels. *Cell Calcium* 34, 97–108, 2003.

131. Harks, E. G. et al. Besides affecting intracellular calcium signaling, 2-APB reversibly blocks gap junctional coupling in confluent monolayers, thereby allowing measurement of single-cell membrane currents in undissociated cells. *Faseb J* 17, 941–3, 2003.

132. Chung, M. K., Lee, H., Mizuno, A., Suzuki, M., and Caterina, M. J. 2-aminoethoxy-diphenyl borate activates and sensitizes the heat-gated ion channel TRPV3. *J Neurosci* 24, 5177–82, 2004.

133. Matlib, M. A. et al. Oxygen-bridged dinuclear ruthenium amine complex specifically inhibits Ca^{2+} uptake into mitochondria *in vitro* and *in situ* in single cardiac myocytes. *J Biol Chem* 273, 10223–31, 1998.

134. Beutner, G., Sharma, V. K., Giovannucci, D. R., Yule, D. I., and Sheu, S. S. Identification of a ryanodine receptor in rat heart mitochondria. *J Biol Chem* 276, 21482–8, 2001.

135. Wilding, M., Russo, G. L., Galione, A., Marino, M., and Dale, B. ADP-ribose gates the fertilization channel in ascidian oocytes. *Am J Physiol* 275, C1277–83, 1998.

136. Sano, Y. et al. Immunocyte Ca^{2+} influx system mediated by LTRPC2. *Science* 293, 1327–30, 2001.

137. Perraud, A. L. et al. ADP-ribose gating of the calcium-permeable LTRPC2 channel revealed by Nudix motif homology. *Nature* 411, 595–9, 2001.

138. Lee, H. C., Aarhus, R., Graeff, R., Gurnack, M. E., and Walseth, T. F. Cyclic ADP ribose activation of the ryanodine receptor is mediated by calmodulin. *Nature* 370, 307–9, 1994.

139. Cancela, J. M., Charpentier, G., and Petersen, O. H. Co-ordination of Ca^{2+} signalling in mammalian cells by the new Ca^{2+}-releasing messenger NAADP. *Pflug Arch* 446, 322–7, 2003.

140. Dickey, D. M., Aarhus, R., Walseth, T. F., and Lee, H. C. Thio-NADP is not an antagonist of NAADP. *Cell Biochem Biophys* 28, 63–73, 1998.

141. Chini, E. N., Liang, M., and Dousa, T. P. Differential effect of pH upon cyclic-ADP-ribose and nicotinate-adenine dinucleotide phosphate-induced Ca^{2+} release systems. *Biochem J* 335, 499–504, 1998.

142. Nelson, E. J. et al. Inhibition of L-type calcium-channel activity by thapsigargin and 2,5-t-butylhydroquinone, but not by cyclopiazonic acid. *Biochem J* 302 Pt 1, 147–54, 1994.

143. Morgan, A. J. and Jacob, R. Ionomycin enhances Ca^{2+} influx by stimulating store-regulated cation entry and not by a direct action at the plasma membrane. *Biochem J* 300 Pt 3, 665–72, 1994.

144. Marchesini, N., Luo, S., Rodrigues, C. O., Moreno, S. N., and Docampo, R. Acidocalcisomes and a vacuolar H^+-pyrophosphatase in malaria parasites. *Biochem J* 347 Pt 1, 243–53, 2000.

145. Srinivas, S. P., Ong, A., Goon, L., and Bonanno, J. A. Lysosomal Ca^{2+} stores in bovine corneal endothelium. *Invest Ophthalmol Vis Sci* 43, 2341–50, 2002.

146. Billington, R. A. et al. Production and characterization of reduced NAADP. *Biochem J* 378, 275–80, 2004.

147. Glick, D. L. et al. Primary structure of a molluscan egg-specific NADase, a second-messenger enzyme. *Cell Regul* 2, 211–18, 1991.

148. Jackson, D. G. and Bell, J. I. Isolation of a cDNA encoding the human CD38 (T10) molecule, a cell surface glycoprotein with an unusual discontinuous pattern of expression during lymphocyte differentiation. *J Immunol* 144, 2811–15, 1990.

149. Kaisho, T. et al. BST-1, a surface molecule of bone marrow stromal cell lines that facilitates pre-B-cell growth. *Proc Natl Acad Sci USA* 91, 5325–9, 1994.

150. Prasad, G. S. et al. Crystal structure of Aplysia ADP ribosyl cyclase, a homologue of the bifunctional ectozyme CD38. *Nat Struct Biol* 3, 957–64, 1996.

151. Yamamoto-Katayama, S. et al. Crystallographic studies on human BST-1/CD157 with ADP-ribosyl cyclase and NAD glycohydrolase activities. *J Mol Biol* 316, 711–23, 2002.

152. Takahashi, K. et al. Accumulation of cyclic ADP-ribose measured by a specific radio-immunoassay in differentiated human leukemic HL-60 cells with all trans-retinoic acid. *FEBS Lett* 371, 204–8, 1995.

153. Munshi, C., Graeff, R., and Lee, H. C. Evidence for a causal role of CD38 expression in granulocytic differentiation of human HL-60 cells. *J Biol Chem* (2002).

154. Barata, H. et al. The role of cyclic-ADP-ribose-signaling pathway in oxytocin-induced Ca^{2+} transients in human myometrium cells. *Endocrinology* 145, 881–89, 2004.

155. Bruzzone, S. et al. Glutamate-mediated overexpression of CD38 in astrocytes cultured with neurones. *J Neurochem* 89, 264–72, 2004.

156. Chini, E. N., de Toledo, F. G., Thompson, M. A., and Dousa, T. P. Effect of estrogen upon cyclic ADP ribose metabolism: beta-estradiol stimulates ADP ribosyl cyclase in rat uterus. *Proc Natl Acad Sci USA* 94, 5872–6, 1997.

157. Thompson, M. et al. Role of CD38 in myometrial Ca^{2+} transients: modulation by progesterone. *Am J Physiol Endocrinol Metab* 287, E1142–E1148, 2004.
158. Tliba, O., Panettieri Jr, R. A., Tliba, S., Walseth, T. F., and Amrani, Y. TNF-α differentially regulates the expression of pro-inflammatory genes in human airway smooth muscle cells by activation of IFN-β-dependent CD38 pathway. *Mol Pharmacol* 66, 322–29, 2004.
159. Dogan, S., Deshpande, D. A., Kannan, M. S., and Walseth, T. F. Changes in CD38 expression and adenosine-diphosphate-ribosyl cyclase activity in rat myometrium during pregnancy: influence of sex steroid hormones. *Biol Reprod* 71, 97–103, 2004.
160. de Toledo, F. G., Cheng, J., Liang, M., Chini, E. N., and Dousa, T. P. ADP-ribosyl cyclase in rat vascular smooth muscle cells: properties and regulation. *Circ Res* 86, 1153–9, 2000.
161. Sanchez, J. P., Duque, P., and Chua, N. H. ABA activates ADPR cyclase and cADPR induces a subset of ABA-responsive genes in Arabidopsis. *Plant J* 38, 381–95, 2004.
162. Kato, I. et al. CD38 disruption impairs glucose-induced increases in cyclic ADP-ribose, $[Ca^{2+}]i$, and insulin secretion. *J Biol Chem* 274, 1869–72, 1999.
163. Fukushi, Y. et al. Identification of cyclic ADP-ribose-dependent mechanisms in pancreatic muscarinic Ca^{2+} signaling using CD38 knockout mice. *J Biol Chem* 276, 649–55, 2001.
164. Partida-Sanchez, S. et al. Regulation of dendritic cell trafficking by the ADP-ribosyl cyclase CD38; Impact on the development of humoral immunity. *Immunity* 20, 279–91, 2004.
165. Bacher, I., Zidar, A., Kratzel, M., and Hohenegger, M. Channeling of substrate promiscuity of the skeletal muscle ADP-ribosyl cyclase isoform. *Biochem J* Pt 381, 147–54, 2004.
166. Graeff, R. M., Franco, L., De Flora, A., and Lee, H. C. Cyclic GMP-dependent and -independent effects on the synthesis of the calcium messengers cyclic ADP-ribose and nicotinic acid adenine dinucleotide phosphate. *J Biol Chem* 273, 118–25, 1998.
167. Wilson, H. L. and Galione, A. Differential regulation of nicotinic acid-adenine dinucleotide phosphate and cADP-ribose production by cAMP and cGMP. *Biochem J* 331, 837–43, 1998.
168. Matsumura, N. and Tanuma, S. Involvement of cytosolic NAD^+ glycohydrolase in cyclic ADP-ribose metabolism. *Biochem Biophys Res Commun* 253, 246–52, 1998.
169. Guse, A. H. et al. Regulation of calcium signalling in T lymphocytes by the second messenger cyclic ADP-ribose. *Nature* 398, 70–3, 1999.
170. Sternfeld, L., Krause, E., Guse, A. H., and Schulz, I. Hormonal control of ADP-ribosyl cyclase activity in pancreatic acinar cells from rats. *J Biol Chem* 278, 33629–36, 2003.
171. Lee, H. C. Specific binding of cyclic ADP-ribose to calcium-storing microsomes from sea urchin eggs. *J Biol Chem* 266, 2276–81, 1991.
172. Howard, M. et al. Formation and hydrolysis of cyclic ADP ribose catalyzed by lymphocyte antigen-CD38. *Science* 262, 1056–9, 1993.
173. Takasawa, S. et al. Synthesis and hydrolysis of cyclic ADP-ribose by human leukocyte antigen CD38 and inhibition of the hydrolysis by ATP. *J Biol Chem* 268, 26052–4, 1993.
174. Hirata, Y. et al. ADP ribosyl cyclase activity of a novel bone marrow stromal cell surface molecule, BST-1. *FEBS Lett* 356, 244–8, 1994.
175. Schuber, F. and Lund, F. E. Structure and enzymology of ADP-ribosyl cyclases: conserved enzymes that produce multiple calcium mobilizing metabolites. *Curr Mol Med* 4, 249–61, 2004.

176. Ziegler, M. New functions of a long-known molecule. Emerging roles of NAD in cellular signaling. *Eur J Biochem* 267, 1550–64, 2000.

177. Tohgo, A. et al. Essential cysteine residues for cyclic ADP-ribose synthesis and hydrolysis by CD38. *J Biol Chem* 269, 28555–7, 1994.

178. Graeff, R., Munshi, C., Aarhus, R., Johns, M., and Lee, H. C. A single residue at the active site of CD38 determines its NAD cyclizing and hydrolyzing activities. *J Biol Chem* 22, 22, 2001.

179. Genazzani, A. A., Bak, J., and Galione, A. Inhibition of cADPR-hydrolase by ADP-ribose potentiates cADPR synthesis from β-NAD$^+$. *Biochem Biophys Res Commun* 223, 502–7, 1996.

180. Dogan, S. et al. Estrogen increases CD38 gene expression and leads to differential regulation of Adenosine Diphosphate (ADP)-ribosyl cyclase and cyclic ADP-ribose hydrolase activities in rat myometrium. *Biol Reprod* 66, 596–602, 2002.

181. Umar, S., Malavasi, F., and Mehta, K. Post-translational modification of CD38 protein into a high molecular weight form alters its catalytic properties. *J Biol Chem* 271, 15922–7, 1996.

182. Aarhus, R., Graeff, R. M., Dickey, D. M., Walseth, T. F., and Lee, H. C. ADP-ribosyl cyclase and CD38 catalyze the synthesis of a calcium-mobilizing metabolite from NADP. *J Biol Chem* 270, 30327–33, 1995.

183. Berridge, G., Cramer, R., Galione, A., and Patel, S. Metabolism of the novel Ca^{2+}-mobilizing messenger nicotinic acid-adenine dinucleotide phosphate via a 2′-specific Ca^{2+}-dependent phosphatase. *Biochem J* 365, 295–301, 2002.

184. Billington, R. A. and Genazzani, A. A. Characterization of NAADP$^+$ binding in sea urchin eggs. *Biochem Biophys Res Commun* 276, 112–16, 2000.

185. Zocchi, E. et al. NAD$^+$-dependent internalization of the transmembrane glycoprotein CD38 in human Namalwa B cells. *FEBS Lett* 396, 327–32, 1996.

186. Franco, L. et al. The transmembrane glycoprotein CD38 is a catalytically active transporter responsible for generation and influx of the second messenger cyclic ADP-ribose across membranes. *Faseb J* 12, 1507–20, 1998.

187. Zocchi, E. et al. Ligand-induced internalization of CD38 results in intracellular Ca^{2+} mobilization: role oi NAD$^+$ transport across cell membranes. *Faseb J* 13, 273–83, 1999.

188. Chidambaram, N. and Chang, C. F. NADP$^+$-dependent internalization of recombinant CD38 in CHO cells. *Arch Biochem Biophys* 363, 267–72, 1999.

189. Han, M. K. et al. Antidiabetic effect of a prodrug of cysteine, L-2-oxothiazolidine-4-carboxylic acid, through CD38 dimerization and internalization. *J Biol Chem* 277, 5315–21, 2002.

190. Adebanjo, O. A. et al. A new function for CD38/ADP-ribosyl cyclase in nuclear Ca^{2+} homeostasis. *Nat Cell Biol* 1, 409–14, 1999.

191. Khoo, K. M. and Chang, C. F. Characterization and localization of CD38 in the vertebrate eye. *Brain Res* 821, 17–25, 1999.

192. Khoo, K. M. et al. Localization of the cyclic ADP-ribose-dependent calcium signaling pathway in hepatocyte nucleus. *J Biol Chem* 275, 24807–17, 2000.

193. Sun, L. et al. A novel mechanism for coupling cellular intermediary metabolism to cytosolic Ca^{2+} signaling via CD38/ADP-ribosyl cyclase, a putative intracellular NAD$^+$ sensor. *Faseb J* 16, 302–14, 2002.

194. Meng Khoo, K. and Fong Chang, C. Identification and characterization of nuclear CD38 in the rat spleen. *Int J Biochem Cell Biol* 34, 43–54, 2002.

195. Ceni, C. et al. CD38-dependent ADP-ribosyl cyclase activity in developing and adult mouse brain. *Biochem J* 370, 175–83, 2003.

196. Sethi, J. K., Empson, R. M., and Galione, A. Nicotinamide inhibits cyclic ADP-ribose-mediated calcium signalling in sea urchin eggs. *Biochem J* 319, 613–17, 1996.

197. Ceni, C. et al. Evidence for an intracellular ADP-ribosyl cyclase/NAD+-glycohydrolase in brain from CD38 deficient mice. *J Biol Chem* 278, 40670–78, 2003.

198. Walseth, T. F., Wong, L., Graeff, R. M., and Lee, H. C. Bioassay for determining endogenous levels of cyclic ADP-ribose. *Methods Enzymol* 280, 287–94, 1997.

199. Takasawa, S., Nata, K., Yonekura, H., and Okamoto, H. Cyclic ADP-ribose in insulin secretion from pancreatic beta cells. *Science* 259, 370–3, 1993.

200. Galione, A. et al. cGMP mobilizes intracellular Ca^{2+} in sea urchin eggs by stimulating cyclic ADP-ribose synthesis [see comments]. *Nature* 365, 456–9, 1993.

201. Higashida, H. et al. Muscarinic receptor-mediated dual regulation of ADP-ribosyl cyclase in NG108–15 neuronal cell membranes. *J Biol Chem* 272, 31272–7, 1997.

202. Graeff, R. M., Walseth, T. F., and Lee, H. C. Radioimmunoassay for measuring endogenous levels of cyclic ADP-ribose in tissues. *Methods Enzymol* 280, 230–41, 1997.

203. Reyes-Harde, M., Empson, R., Potter, B. V., Galione, A., and Stanton, P. K. Evidence of a role for cyclic ADP-ribose in long-term synaptic depression in hippocampus. *Proc Natl Acad Sci USA* 96, 4061–6, 1999.

204. da Silva, C. P., Potter, B. V., Mayr, G. W., and Guse, A. H. Quantification of intracellular levels of cyclic ADP-ribose by high-performance liquid chromatography. *J Chromatogr B Biomed Sci Appl* 707, 43–50, 1998.

205. Graeff, R. and Lee, H. C. A novel cycling assay for cellular cADP-ribose with nanomolar sensitivity. *Biochem J* 361, 379–84, 2002.

206. Kuroda, R. et al. Increase of cGMP, cADP-ribose and inositol 1,4,5-trisphosphate preceding Ca^{2+} transients in fertilization of sea urchin eggs. *Development* 128, 4405–14, 2001.

207. Leckie, C. et al. The NO pathway acts late during the fertilization response in sea urchin eggs. *J Biol Chem* 278, 12247–54, 2003.

208. Ge, Z. D., Li, P. L., Chen, Y. F., Gross, G. J., and Zou, A. P. Myocardial ischemia and reperfusion reduce the levels of cyclic ADP-ribose in rat myocardium. *Basic Res Cardiol* 97, 312–19, 2002.

209. Sun, L. et al. CD38/ADP-ribosyl cyclase: a new role in the regulation of osteoclastic bone resorption [published erratum appears in *J Cell Biol* 146(6) (1999 Sep 20) following 1399]. *J Cell Biol* 146, 1161–72, 1999.

210. Partida-Sanchez, S. et al. Chemotaxis and calcium responses of phagocytes to formyl peptide receptor ligands is differentially regulated by cyclic ADP ribose. *J Immunol* 172, 1896–1906, 2004.

211. Nahorski, S. R., Young, K. W., John Challiss, R. A., and Nash, M. S. Visualizing phosphoinositide signalling in single neurons gets a green light. *Trends Neurosci* 26, 444–52, 2003.

212. Rhee, S. G. Regulation of phosphoinositide-specific phospholipase C. *Annu Rev Biochem* 70, 281–312, 2001.

213. Morita, K., Kitayama, S., and Dohi, T. Stimulation of cyclic ADP-ribose synthesis by acetylcholine and its role in catecholamine release in bovine adrenal chromaffin cells. *J Biol Chem* 272, 21002–9, 1997.

214. Higashida, H. et al. Sympathetic potentiation of cyclic ADP-ribose formation in rat cardiac myocytes. *J Biol Chem* 274, 33348–54, 1999.

215. Higashida, H. et al. Angiotensin II stimulates cyclic ADP-ribose formation in neonatal rat cardiac myocytes. *Biochem J* 352 Pt 1, 197–202, 2000.

216. Willmott, N. et al. Nitric oxide induced mobilization of intracellular calcium via the cyclic ADP-ribose signalling pathway. *J Biol Chem* 271, 3699–705, 1996.

217. Looms, D., Tritsaris, K., Nauntofte, B., and Dissing, S. Nitric oxide and cGMP activate Ca^{2+}-release processes in rat parotid acinar cells. *Biochem J* 355, 87–95, 2001.

218. Mothet, J. P. et al. Cyclic ADP-ribose and calcium-induced calcium release regulate neurotransmitter release at a cholinergic synapse of Aplysia. *J Physiol (Lond)* 507, 405–14, 1998.

219. Pollock, J., Crawford, J. H., Wootton, J. F., Seabrook, G. R., and Scott, R. H. Metabotropic glutamate receptor activation and intracellular cyclic ADP-ribose release Ca^{2+} from the same store in cultured DRG neurones. *Cell Calcium* 26, 139–48, 1999.

220. Clementi, E., Riccio, M., Sciorati, C., Nistico, G., and Meldolesi, J. The type 2 ryanodine receptor of neurosecretory PC12 cells is activated by cyclic ADP-ribose. Role of the nitric oxide/cGMP pathway. *J Biol Chem* 271, 17739–45, 1996.

221. Murthy, K. S. and Makhlouf, G. M. cGMP-mediated Ca^{2+} release from IP3-insensitive Ca^{2+} stores in smooth muscle [see comments]. *Am J Physiol* 274, C1199–205, 1998.

222. Yao, X. and Huang, Y. From nitric oxide to endothelial cytosolic Ca^{2+}: a negative feedback control. *Trends Pharmacol Sci* 24, 263–6, 2003.

223. Li, N., Sul, J. Y., and Haydon, P. G. A calcium-induced calcium influx factor, nitric oxide, modulates the refilling of calcium stores in astrocytes. *J Neurosci* 23, 10302–10, 2003.

224. White, T. A., Walseth, T. F., and Kannan, M. S. Nitric oxide inhibits ADP-ribosyl cyclase through a cGMP-independent pathway in airway smooth muscle. *Am J Physiol Lung Cell Mol Physiol* 283, L1065–71, 2002.

225. Munoz, P. et al. CD38 signaling in T cells is initiated within a subset of membrane rafts containing Lck and the CD3-zeta subunit of the T cell antigen receptor. *J Biol Chem* 278, 50791–802, 2003.

226. Higashida, H. et al. Subtype-specific coupling with ADP-ribosyl cyclase of metabotropic glutamate receptors in retina, cervical superior ganglion and NG108–15 cells. *J Neurochem* 85, 1148–58, 2003.

227. Okamoto, H. and Takasawa, S. Recent advances in the Okamoto model: the CD38-cyclic ADP-ribose signal system and the regenerating gene protein (Reg)-Reg receptor system in beta-cells. *Diabetes* 51 Suppl 3, S462–73, 2002.

228. Ikehata, F. et al. Autoantibodies against CD38 (ADP-ribosyl cyclase/cyclic ADP-ribose hydrolase) that impair glucose-induced insulin secretion in noninsulin-dependent diabetes patients. *J Clin Invest* 102, 395–401, 1998.

229. Malaisse, W. J. et al. Cyclic ADP-ribose measurements in rat pancreatic islets. *Biochem Biophys Res Commun* 231, 546–8, 1997.

230. Singh, A. K. Early developmental changes in intracellular Ca^{2+} stores in rat brain. *Comp Biochem Physiol A Mol Integr Physiol* 123, 163–72, 1999.

231. Billington, R. A., Ho, A., and Genazzani, A. A. Nicotinic acid adenine dinucleotide phosphate (NAADP) is present at micromolar concentrations in sea urchin spermatozoa. *J Physiol* 544, 107–12, 2002.

232. Churamani, D., Carrey, E. A., Dickinson, G. D., and Patel, S. Determination of cellular nicotinic acid adenine dinucleotide phosphate (NAADP) levels. *Biochem J* Pt 380, 449–54, 2004.

233. Walseth, T. F., Aarhus, R., Zeleznikar, R. J., Jr., and Lee, H. C. Determination of endogenous levels of cyclic ADP-ribose in rat tissues. *Biochim Biophys Acta* 1094, 113–20, 1991.

234. Burdakov, D. and Galione, A. Two neuropeptides recruit different messenger pathways to evoke Ca^{2+} signals in the same cell. *Curr Biol* 10, 993–6, 2000.

235. Dickinson, G. D. and Patel, S. Modulation of NAADP receptors by K^+ ions: evidence for multiple NAADP receptor conformations. *Biochem J* 375, 805–12, 2003.

236. Lee, H. C. and Aarhus, R. A derivative of NADP mobilizes calcium stores insensitive to inositol trisphosphate and cyclic ADP-ribose. *J Biol Chem* 270, 2152–7, 1995.

237. Guse, A. H., Berg, I., da Silva, C. P., Potter, B. V., and Mayr, G. W. Ca^{2+} entry induced by cyclic ADP-ribose in intact T-lymphocytes. *J Biol Chem* 272, 8546–50, 1997.

238. Albrieux, M., Lee, H. C., and Villaz, M. Calcium signaling by cyclic ADP-ribose, NAADP, and inositol trisphosphate are involved in distinct functions in ascidian oocytes. *J Biol Chem* 273, 14566–74, 1998.

239. Cancela, J. M. and Petersen, O. H. The cyclic ADP ribose antagonist 8-NH2-cADP-ribose blocks cholecystokinin-evoked cytosolic Ca^{2+} spiking in pancreatic acinar cells. *Pflug Arch* 435, 746–8, 1998.

240. Lukyanenko, V. and Gyorke, S. Ca^{2+} sparks and Ca^{2+} waves in saponin-permeabilized rat ventricular myocytes. *J Physiol (Lond)* 521, 575–85, 1999.

241. Li, N., Teggatz, E. G., Li, P. L., Allaire, R., and Zou, A. P. Formation and actions of cyclic ADP-ribose in renal microvessels. *Microvasc Res* 60, 149–59, 2000.

242. Inngjerdingen, M., Al-Aoukaty, A., Damaj, B., and Maghazachi, A. A. Differential utilization of cyclic ADP-ribose pathway by chemokines to induce the mobilization of intracellular calcium in NK cells. *Biochem Biophys Res Commun* 262, 467–72, 1999.

243. Lee, H. C., Aarhus, R., and Graeff, R. M. Sensitization of calcium-induced calcium release by cyclic ADP-ribose and calmodulin. *J Biol Chem* 270, 9060–6, 1995.

244. Hola, M., Revest, P. A., Knight, D. E., and Brooks, R. F. A simple technique for estimating the tip diameter of microinjection pipettes. *Lab Prac* 40, 51–2, 1991.

245. Schnorf, M., Potrykus, I., and Neuhaus, G. Microinjection technique: routine system for characterization of microcapillaries by bubble pressure measurement. *Exp Cell Res* 210, 260–7, 1994.

246. Gillot, I. and Whitaker, M. Calcium signals in and around the nucleus in sea urchin eggs. *Cell Calcium* 16, 269–78, 1994.

247. Tsien, R. Y. and Pozzan, T. in *Methods Enzymol* 230–44, 1989.

248. Empson, R. M. and Galione, A. Cyclic ADP-ribose enhances coupling between voltage-gated Ca^{2+} entry and intracellular Ca^{2+} release. *J Biol Chem* 272, 20967–70, 1997.

249. Yusufi, A. N., Cheng, J., Thompson, M. A., Burnett, J. C., and Grande, J. P. Differential mechanisms of Ca^{2+} release from vascular smooth muscle cell microsomes. *Exp Biol Med (Maywood)* 227, 36–44, 2002.

250. Wells, J., Zhang, G. H., and Martinez, J. R. Comparison of calcium mobilization in response to noradrenaline and acetylcholine in submandibular cells of newborn and adult rats. *Arch Oral Biol* 42, 633–40, 1997.

251. Renard-Rooney, D. C., Hajnoczky, G., Seitz, M. B., Schneider, T. G., and Thomas, A. P. Imaging of inositol 1,4,5-trisphosphate-induced Ca^{2+} fluxes in single permeabilized hepatocytes. Demonstration of both quantal and nonquantal patterns of Ca^{2+} release. *J Biol Chem* 268, 23601–10, 1993.

252. Burdyga, T. V., Taggart, M. J., Crichton, C., Smith, G. L., and Wray, S. The mechanism of Ca^{2+} release from the SR of permeabilised guinea-pig and rat ureteric smooth muscle. *Biochim Biophys Acta* 1402, 109–14, 1998.

253. Belan, P. V. et al. A new technique for assessing the microscopic distribution of cellular calcium exit sites. *Pflug Arch* 433, 200–8, 1996.

254. Hajnoczky, G. and Thomas, A. P. Minimal requirements for calcium oscillations driven by the IP$_3$ receptor. *Embo J* 16, 3533–43, 1997.
255. Christensen, K. A., Myers, J. T., and Swanson, J. A. pH-dependent regulation of lysosomal calcium in macrophages. *J Cell Sci* 115, 599–607, 2002.

13 Methods for Studying Calcium Pumps

Leonard Dode, Luc Raeymaekers, Ludwig Missiaen, Bente Vilsen, Jens P. Anderson, and Frank Wuytack

CONTENTS

13.1   INTRODUCTION

All mechanisms of intracellular Ca^{2+} signaling are based on the well-known asymmetric distribution of Ca^{2+} concentration ($[Ca^{2+}]$) across membranes in resting cells: from as low as submicromolar in cytosol to millimolar range in the extracellular milieu and intracellular stores. Temporal increases in cytosolic free Ca^{2+} concentration ($[Ca^{2+}]_{cyt}$) known as Ca^{2+} transients represent highly versatile intracellular Ca^{2+} signals able to initiate, propagate, and regulate a large range of cellular functions of outmost importance for living cells [1]. Ca^{2+} entry from outside the cell or Ca^{2+} release from internal Ca^{2+} stores represent the "on" reactions responsible for introduction of Ca^{2+} into the cytoplasm upon cell stimulation. Ca^{2+} signal is removed by the "off" reactions, i.e., $[Ca^{2+}]_{cyt}$ is restored to the resting level ($\sim 0.1\ \mu M$) via concerted action of specialized molecules such as Ca^{2+} exchangers and Ca^{2+} pumps (or Ca^{2+}-ATPases) [2]. Failure to conveniently remove the activator Ca^{2+} pulse could in turn trigger cytotoxic processes and, ultimately, cell death [3].

Sarco(endo)plasmic reticulum Ca^{2+}-ATPases (SERCAs) are single-subunit membrane-spanning P-type ATPases that mediate the ATP-driven transport of cytoplasmic Ca^{2+} against a concentration gradient into the lumen of intracellular stores such as sarcoplasmic (SR) or endoplasmic reticulum (ER) with a stoichiometry of two Ca^{2+} per ATP hydrolyzed and in exchange for protons [4–8]. In addition to SERCAs, secretory-pathway Ca^{2+}-ATPases (SPCAs) accumulate Ca^{2+} in the Golgi apparatus, whereas plasma membrane Ca^{2+}-ATPases (PMCAs) extrude Ca^{2+}

across plasma membrane, and in contrast to SERCAs, in both cases the free energy released by ATP hydrolysis allows only one Ca^{2+} to be transported per cycle. Although an active transporter of Ca^{2+}, SPCA also is the only known eukaryotic pump able to efficiently scavenge cytosolic Mn^{2+} at the expense of ATP with a net Mn^{2+}:ATP coupling ratio of 1:1 [9–14]. All P-type ATPases, including among others the calcium pumps (SERCA, PMCA, and SPCA), the sodium pump (Na^+, K^+-ATPase), and the gastric pump (H^+, K^+-ATPase), are defined by the compulsory formation of a phosphorylated intermediate during the reaction cycle (see Section 13.3.1), hence the "P-type" terminology [15–18]. P-type ATPase fingerprinting relies on the identification of a specific phosphorylation motif, <u>D</u>KTGT[L/I/V/M][T/I/S], with the underlined aspartate residue (e.g., Asp^{351} in vertebrate SERCAs) invariably phosphorylated [17].

This review deals with analytical methods used for *in vitro* and *in vivo* characterization of the aforementioned Ca^{2+} pumps. Many available techniques can be classified either as techniques mainly useful for biochemical and biophysical surveys monitoring conformational changes during ion transport across membranes (e.g., chemical derivatization [19], intrinsic tryptophan fluorescence measurements [19, 20–24], and infrared spectroscopy [19, 25–27]), or as methods particularly applicable to the functional studies of Ca^{2+} pumps as part of the Ca^{2+} signaling machinery and within the molecular context of the cell biology [5–9, 13, 14, 28–35]. Both approaches have produced a huge amount of data. In line with the subject of this book, this chapter mainly focuses on the latter category of methods including detection and study of the functional properties of different isoforms. It is beyond the scope of this review to quote all papers reporting results obtained with one of the mentioned techniques, but references to comprehensive reviews and excellent reports have been included.

13.2 IDENTIFICATION OF ISOFORMS, MEASUREMENT OF EXPRESSION LEVELS, AND SUBCELLULAR DISTRIBUTION OF CALCIUM PUMPS AND THEIR REGULATORS

P-type ATPases are grouped into five types, each type further divided into subtypes depending on the cations (e.g., K^+, Cu^{2+}, Cu^+, Ca^{2+} and H^+, Na^+ and K^+, H^+ and K^+, Mn^{2+} and H^+, Mg^{2+} etc.) or other substrate (aminophospholipids) specificities. According to this classification, SERCAs and SPCAs belong to the same P_{2A} group, whereas PMCAs belong to the P_{2B} group [17, 18]. For more information with regard to the other types and subtypes belonging to the P-type ATPase superfamily, the interested reader can consult the P-type ATPase database available online at http://www.patbase.kvl.dk [36].

In the last two decades, several breakthroughs in DNA recombinant methodology such as cDNA cloning and polymerase chain reaction (PCR) have helped the advance of our scientific knowledge [37]. These methods and others (see below) provided evidence for the expression of at least 36 human calcium pumps encoded by 9 distinct genes: 3 SERCA genes (*ATP2A1–3*), 4 PMCA genes (*ATP2B1–4*), and

2 SPCA genes (*ATP2C1–2*). The diversity of Ca^{2+}-ATPase isoforms is brought about through the alternative processing of the primary gene transcript in a cell type- or tissue-specific manner or following a developmentally modulated pathway. Recent reviews and articles have dealt with the topic of isoform diversity for SERCA [9, 13], PMCA [13, 38], and SPCA [9, 12, 14] pumps. It appears that the alternative splicing of all genes mentioned above affects the region encoding their C-termini with the notable exception of PMCA 2, 3, and 4 genes, which make use of an additional splice site located in the region encoding part of the N-termini [9, 11, 38]. In the case of the SERCA family, the isoforms generated for each gene are identical up to amino acid (aa) 993. For instance, SERCA1 is expressed only as one of its developmentally spliced variants in the fast-twitch skeletal muscle, the adult SERCA1a (994 aa) and the neonatal SERCA1b (1001 aa); the last amino acid (Gly^{994}) in SERCA1a is replaced by a highly charged octapeptide DPEDERRK [39]. As a result of alternative splicing, three SERCA2 isoforms are expressed as follows: SERCA2a (997 aa) is the main Ca^{2+} pump in cardiac, smooth, and slow-twitch skeletal muscles, SERCA2b (1042 aa) represents the ubiquitously expressed SERCA isoform, and, finally, SERCA2c (999 aa) found in primary human monocytes and a few cell lines [9, 40]. For human SERCA3 gene, a total of six isoforms can be generated by means of alternative splicing and are found mainly in non-muscle cells and tissues: SERCA3a (999 aa), SERCA3b (1043 aa), SERCA3c (1029 aa), SERCA3d (1044 aa), SERCA3e (1052 aa), and SERCA3f (1033 aa) [9, 41–44].

13.2.1 STUDIES AT THE mRNA LEVEL

13.2.1.1 Detecting and Quantifying RNA

A first impression of the relative amounts of Ca^{2+} pumps and their regulators in different cell types or experimental conditions can be obtained by assessing the levels of their corresponding mRNAs. However, mRNA levels do not necessarily reflect the actual protein levels, since protein expression can be regulated at stages downstream of the gene transcription. Indeed, considerable post-transcriptional regulation takes place during cardiac SERCA2 expression [45, 46]. Likewise, several conditions are known where levels in phospholamban protein (PLN), an *in vivo* SERCA2 regulator, changed without changes in its corresponding mRNA levels [47, 48].

The mRNA levels corresponding to the different Ca^{2+} pumps and their regulators can be determined and quantified via classical techniques including: Northern hybridization, ribonuclease protection assays (RPA), *in situ* hybridization, and reverse transcriptase-PCR.

Northern hybridization, which now belongs to the standard repertoire of molecular biology, allows the rapid simultaneous analysis of a large number of RNA preparations. It gives at the same time information on the size of the RNAs transcribed from the genes. It suffices to refer to the excellent laboratory manual written by Sambrook and Russell (2001) for further details on the background and methodology [37]. Profiling the expression of a particular gene over a broad range of tissues can conveniently be done by dot blot hybridization using the multiple tissue expression (MTE) array from Clontech (BD Biosciences). This array is composed of poly(A)$^+$ RNA from tens of selected tissues. The amount of each mRNA applied onto the blot is normalized to a panel of eight housekeeping genes, thus minimizing the tissue-specific variations often related to the expression of any single housekeeping gene. The dot blot is then hybridized with cDNA probes specific for the gene under investigation. Dot blot hybridization analyses have shown that the tissue distribution pattern of human SERCA3a mRNA is mainly nonmuscle with high levels in thymus, trachea, bone marrow, spleen, pancreas, peripheral leukocytes, salivary glands, and colon, and intermediate and low levels in other tissues [42]. Moreover, the expression of SERCA3b and SERCA3c appears to be restricted more to human kidney, thymus, colon, salivary glands, and trachea, but the expression levels are lower than those of the corresponding SERCA3a [42]. Similarly, the human SPCA1 seems to be ubiquitously expressed, thus having a housekeeping function, whereas human SPCA2 mRNA appears to be confined to secretory tissues, keratinocytes, and epithelial cells of the gastrointestinal tract [49].

RPA allow measuring the abundance of specific mRNAs by hybridization of test RNAs to complementary radiolabeled RNA probes. The nonhybridized sequences are then digested with single-strand-specific ribonucleases and the digestion-resistant fragments separated by electrophoresis. By selecting different lengths of radiolabeled probes the mRNA levels of several different genes can be visualized and quantified on a single gel. Recently, the RPA methodology was employed to measure the levels of SERCA2 proteins from a number of tissues isolated from knock-out mice in which the muscle-specific SERCA2a isoform was replaced by the nonmuscle SERCA2b splice variant, thus causing mild concentric hypertrophy and impairment of contraction–relaxation in the heart [47].

The fastest, but maybe the least quantitative way to measure mRNA levels is by means of reverse transcriptase PCR. The method is so extremely sensitive that it can sometimes be used for single-cell PCR [50]. Recent developments in the field of real-time PCR have greatly improved the quantitative aspect of the assay; such an approach has been used to assess the RNA copy number for the hSPCA1d mutants transiently expressed in COS–1 cells [51].

Ratio RT-PCR is the method of choice when the relative expression of two related genes is investigated at the mRNA level (e.g., SERCA1/SERCA2, SERCA2/SERCA3, and SPCA1/SPCA2 ratios). In this method, a common set of primers is used to co-amplify homologous fragments of the related sequences [29]. The relative amount of both messengers is then deduced by restriction analysis of the ^{32}P-radiolabeled PCR products.

The levels of mRNA obtained by one of the above-described techniques are of course determined by the rate of mRNA production (gene transcription) and RNA decay. The next logical step is therefore to assess both the rate of gene transcription and the rate of RNA decay.

To investigate whether a particular gene is transcriptionally regulated *nuclear run-on assays* are used. Nuclear run-on assays allow determining the number of RNA polymerases traversing the gene at a given time [37]. These assays have been effective in assessing the transcription of Ca^{2+} pumps and phospholamban [52, 53].

mRNA stability is another important determinant of the messenger level. The rate of decay of a messenger is determined by factors like the length of the $poly(A)^+$ tail and the presence, often in the $3'$-untranslated end of the mRNA of stabilizing/destabilizing sequences. In an elegant series of experiments, Dr. Grover's team demonstrated, firstly, that SERCA2a mRNA is more stable than that of SERCA2b and, secondly, that the difference in stability is related to the differential processing of the $3'$-end of the primary SERCA2 transcript [54].

13.2.1.2 Alternative Splicing

Interestingly, with the notable exception of SPCA2 so far, the $3'$-end of the primary transcripts of all SERCA, PMCA, and SPCA1 genes are subject to alternative RNA processing [9–14]. For SERCA2 the structural and functional difference between the muscle-specific SERCA2a and housekeeping SERCA2b variants, which are the result of such an alternative splicing event is best characterized. Yet, the mechanisms controlling this tissue-specific alternative processing are still poorly understood. The regulation of alternative splicing can be most conveniently studied by means of an artificial minigene construct comprising only the relevant downstream part of the pig SERCA2 gene [55, 56]. When introduced in myoblast-like cells the RNA transcribed from such a minigene appears to be spliced similarly as the parent full-length gene and, most importantly, the muscle-specific alternative splicing is induced along with muscle differentiation. The minigene is easily amenable to directed mutagenesis and, therefore, allows easy exploration of specific DNA motifs that can potentially control the alternative splicing of the pre-mRNA derived from the minigene. Once the relevant regulatory motifs (known as *cis*-active DNA elements) have been pinpointed, they can be altered by targeted mutagenesis in such a way as to prevent the alternative processing. In this way the *Atp2a2* gene of mice was altered creating a new allele for which the alternative splicing that underlies the generation of the muscle-specific SERCA2a isoform was prevented. By default these mice (termed $SERCA2^{b/b}$) could only express the housekeeping SERCA2b variant, even in cardiac muscle that normally expresses 98% SERCA2a and 2% SERCA2b [47]. Such an approach allows investigating the *in vivo* consequences of replacing the native cardiac SERCA2a isoform with the SERCA2b.

13.2.2 STUDIES AT THE PROTEIN LEVEL

13.2.2.1 Non-Discriminating and Isoform-Specific Antibodies

Antibodies allow the study of tissue- and cell-type specific expression patterns of total Ca^{2+}-ATPase levels or of specific isoforms, the quantification of protein content, or may be useful as tools to probe protein domains for structure–function relationships. In the past, a large number of antibodies have been generated against SERCA and PMCA pumps, but antibody availability is more limited for the more recently discovered SPCA pumps. Several criteria should be considered in the selection of existing antibodies or in the design of new ones. Isoform-specific antibodies usually are either monoclonals or polyclonals generated against short peptide sequences occurring only in the isoform of interest. Selecting specific sequences conserved among any of the concerned Ca^{2+}-ATPase family, but avoiding motifs also found in other ion-motive P-type pumps (Na^+, K^+-ATPases or H^+, K^+-ATPases) is a prerequisite for generating non-discriminating antibodies. As a general rule, immunogenic peptides should be hydrophilic. N-terminal or C-terminal peptides usually fulfill this criterion. Antibodies obtained from such peptides are very useful to discriminate between splice variants especially when, for instance, the C-terminus is often subject to alternative splicing, but they should be avoided for the detection of all splice products of a specific gene. The selection of internal peptides suitable for immunization may be less straightforward. It may be worthwhile to make use of the known 3D structures of SERCA1a [7, 57–61] for the identification of surface-localized regions either by alignment of the polypeptide sequence of interest with that of SERCA1a or by homology model building [62] (e.g., http://swissmodel.expasy.org/). The availability of antibodies raised in different animal species is very useful for dual-labeling experiments in immunocytochemical localizations. For a list of antibodies useful for isoform identification of SERCAs and PMCAs the reader is referred to a recently published review [13]. Because space limitations do not permit an extensive catalogue, the next paragraphs highlight the most common antibodies used in some important applications, with an emphasis on commercially available antibodies.

13.2.2.1.1 SERCA

Apart from pan-specific antibodies, antibodies specific to each of the protein products of the three SERCA genes are commercially available, e.g., from Abcam (http://www.abcam.com), Affinity Bioreagents (http://www.bioreagents.com), BIOMOL (http://www.biomol.com), Chemicon International (http:/www.chemicon.com), and Santa Cruz Biotechnology (http://www.scbt.com).

Many noncommercial antibodies have been described in the literature including antibodies specific to splice variants of each of the SERCA genes. Suitable monoclonal antibodies of high titer and good specificity (designated A52 and IIH11) against the adult fast-twitch skeletal muscle isoform, SERCA1a, have earlier been described [63, 64]. A52 antibody has predominantly been employed for determining the amount of wild-type and mutant SERCA1a proteins transiently expressed in COS–1 cells by means of a sandwich enzyme-linked immunosorbent assay (ELISA) quantitative procedure [65, 66]. The monoclonal antibody A52 recognizes a specific region of SERCA1a encompassing amino acids 657–672, which lies close to Cys^{670} and Cys^{674} residues [66]. These residues are preferentially labeled by an iodoacetamide directed fluorescent label without loss of catalytic activity [19, 67]. Recently, a polyclonal antiserum specifically reacting with the neonatal SERCA1b form has been obtained [68].

Specific antibodies against the muscle-specific form SERCA2a and against the housekeeping variant SERCA2b have initially been prepared in our laboratory [69, 70]. A SERCA2a-specific antibody is also available from BIOMOL. A monoclonal antibody (IID8) able to recognize both isoforms was also documented [64].

Six variants of the human SERCA3 gene have been shown (see above in Section 13.2) to be generated by alternative splicing near the C-terminus [41–44]. Antibodies specific for human isoforms SERCA3a, SERCA3b, SERCA3c, and SERCA3f have been documented [34, 41, 44, 71]. In addition, a human SERCA3-specific monoclonal antibody (PL/IM430) was described (see Section 13.4.3 and the references therein). Rats do not express variants similar to human SERCA3b and SERCA3c, but instead express an isoform labeled SERCA3b/c, which has been demonstrated with a specific antibody [43]. Interestingly, a polyclonal antiserum (N89) made against the N-terminal region of rat SERCA3 [29] is able to cross-react with SERCA3 from cat, pig, and human [41].

The model plant *Arabidopsis thaliana* possesses more Ca^{2+}-ATPase genes than animals (but no SPCA-related genes) [72]. Many of the gene products have not yet been characterized. By using antibodies it has been possible to study the distribution of some of these pumps in different organs of the plant, at the subcellular level in protoplasts, and in subcellular fractions [73, 74].

13.2.2.1.2   SPCA
Antibodies have been described against the first characterized member of this family, the Pmr1 pump of *Saccharomyces cerevisiae* [75], against the SPCA pump of *Caenorhabditis elegans*, [76] and against the human homologues SPCA1 51 and SPCA2 [49, 77].

13.2.2.1.3   PMCA
Sequence analysis of the four PMCA genes has shown that a large number of alternatively spliced transcript can in theory be produced. Over 20 different messengers have been detected [38]. The catalogue of isoform diversity is far less complete at the protein level. More detailed studies using new specific antibodies are needed, in addition to the extensive studies that have already been published. For example, the tissue distribution of each of the four different PMCA gene products has been studied [78], and patterns of alternative splicing have been analyzed at the protein level [79]. Pan-specific antibodies and antibodies specific to each of the four PMCA gene products are commercially available from Abcam and from Affinity Bioreagents.

13.2.2.2   Discrimination of Calcium Pumps by Non-Immunological Techniques

13.2.2.2.1   Electrophoretic Size Fractionation of SERCA3 and SERCA2 Isoforms
The difference in molecular mass between SERCA2a and SERCA2b is sufficient to cause a difference in migration distance in SDS polyacrylamide gels that can be used to discriminate and quantitate the relative concentrations of these isoforms [80]. For an unknown reason, SERCA2 and SERCA3 isoforms also migrate at different distances in SDS gels. This behavior, in combination with a pan-specific antibody, has been exploited by Mountian et al. to measure the relative concentrations of these pumps in endothelial cells and in platelets [81]. Another way to

recognize SERCA isoforms is by their proteolytic products. Such a difference in the size of partial tryptic fragments has been documented for SERCA1 and SERCA2 [70].

13.2.2.2.2 Phosphoprotein Formation

Demonstration of the phosphoprotein intermediate is another sensitive way to detect P-type Ca^{2+}-transport ATPases. The enzyme is maximally labeled by a short incubation on ice in the presence of Ca^{2+} and $[\gamma-^{32}P]ATP$ of high specific activity. The reaction is quenched with trichloroacetic acid (TCA) solution containing cold ATP and/or inorganic phosphate, followed by several washes of the precipitated protein by centrifugation. The aspartyl phosphate that is formed breaks down in alkaline conditions, although the SERCA phosphoprotein, but not that of PMCA, is partially preserved after electrophoresis in conventional SDS Laemmli buffer. For better preservation of the aspartyl phosphate, gel-electrophoresis protocols in acid conditions have been widely applied [82, 83]. As a control for the identity of the observed radioactive product a parallel reaction in EGTA-containing medium without Ca^{2+} is carried out. The phosphoprotein can be discriminated from hydroxyl phosphates generated by protein-kinase activity by its sensitivity to 0.2 M hydroxylamine [84].

13.2.2.3 Phospholamban and Sarcolipin

In the unphosphorylated state, phospholamban (PLB) inhibits the SERCA2 Ca^{2+} pump. Phosphorylation at Ser^{16} by cAMP-dependent protein kinase or at Thr^{17} by Ca^{2+}-calmodulin-dependent protein kinase relieves the inhibition. It is therefore of great functional importance to determine the phosphorylation status of PLB. Antibodies are commercially available that recognize all forms of PLB, or either of the Ser^{16}- or the Thr^{17}-phosphorylated forms (e.g., Abcam, Santa Cruz).

Antibodies to sarcolipin have been made by our group and used for sarcolipin quantitation in different muscles from mouse, rat, rabbit, and pig [85].

13.3 STUDIES OF CALCIUM PUMP ACTIVITY AND ITS CATALYTIC MECHANISM

The adult rabbit fast-twitch skeletal muscle represents a rich source for the isolation of sarcoplasmic reticulum membranes and, subsequently, for the purification of its Ca^{2+}-ATPase (SERCA1a isoform) [86–89]. Such an enzymatic preparation has been successfully used for biochemical and biophysical characterization including enzymatic studies, chemical derivatization, spectroscopic studies, electron microscopy, and x-ray crystallography [7, 8, 19, 57–61, 90]. The mechanism and energetics governing the active Ca^{2+} translocation across membranes have been extensively studied for SERCA1a (for a short description of the reaction scheme cycle, see Section 13.3.1) [4, 91, 92]. To further dissect the structure–function relationships, mutant proteins are generally produced by means of oligonucleotide-directed mutagenesis [37] followed by comprehensive functional characterization using a panel of preparative and analytical methods outlined in Figure 13.1 and briefly presented below (see Sections 13.3.2–13.3.5). Although, most of this methodology has been described in relation to SERCAs, the principles and even entire procedures can be applicable for the study of SPCAs, PMCAs, and other P-type ATPases with or without modifications.

13.3.1 CATALYTIC CYCLE

Figure 13.2 illustrates a simplified Ca^{2+}-ATPase reaction scheme for a SERCA prototype (i.e., two Ca^{2+} ions transported/ATP hydrolyzed) based on the classical E_1/E_2 model with the enzyme reversibly cycling between two distinct conformational states: the E_1 states either displaying high-affinity Ca^{2+}-binding sites cytoplasmically oriented or containing occluded Ca^{2+}, and the E_2 states with lumenally facing low-affinity Ca^{2+}-binding sites [4] and possibility to occlude protons in its dephosphorylated form [7]. However, one should be aware that, first, the depicted scheme cannot take into account the multitude of conformational changes brought about by interactions with Ca^{2+}, H^+, Mg^{2+}, ATP, ADP, or P_i, and, second, the existence of an ADP-insensitive E_2-P·$(Ca^{2+})_2$ intermediate has yet to be confirmed [92, 93]. Nonetheless, over the years this scheme has been used in relation to a plethora of extensive biochemical and mutational studies mainly performed on the skeletal muscle SERCA1a isoform [5–8, 28, 32, 33]. Recently, methods have been developed for transient kinetic analysis of expressed Ca^{2+}-ATPase wild type and mutants [94, 95], and we have used both steady-state and transient kinetic analyses to dissect the functional differences between SERCA1 and the other members of the SERCA family and extended the validity and usefulness of the scheme to SERCA2 and SERCA3 isoforms [34, 35, 44] and to Darier disease (SERCA2b) mutants [35].

During the catalytic cycle, several major conformational states associated with ATP hydrolysis and Ca^{2+} translocation can now be experimentally distinguished such as E_1·$(Ca^{2+})_2$, E_1~P·$(Ca^{2+})_2$, E_2-P, and E_2 [4, 91]. The cooperative binding of two calcium ions [96–98] in exchange for protons describing the E_2 to E_1·$(Ca^{2+})_2$ transition (reactions 1 and 2 in Figure 13.2) is accompanied by conformational changes. These rearrangements can be experimentally monitored by changes in the infrared spectra [25–27], circular dichroism [99], and intrinsic tryptophan fluorescence [20–24] of the pump as well as by changes to extrinsic spectroscopic probes covalently attached to the pump [96, 100, 101]. In the presence of ATP as energy source, these conformational changes initiate the transfer of γ-phosphoryl group from ATP to the aspartate residue (Asp^{351}) of the phosphorylation domain (reaction 3) leading to E_1~P·$(Ca^{2+})_2$, a high-energy phosphorylated intermediate [102, 103]. The slow conversion of this intermediate to a low-energy E_2-P phosphoenzyme intermediate (reactions 4 and 5) constitutes a crucial rate-limiting step in the Ca^{2+} transport activity. Prior to the hydrolysis of E_2-P to E_2 (reaction 6), calcium ions are delivered into the ER/SR lumen [103]. It is believed that the events involving lumenal proton binding and proton ejection from lumen into the cytosol take place along reactions 5, 6, and 1 [104–111]. Naturally occurring proton leakage from native SR membranes raises questions as to the physiological importance of

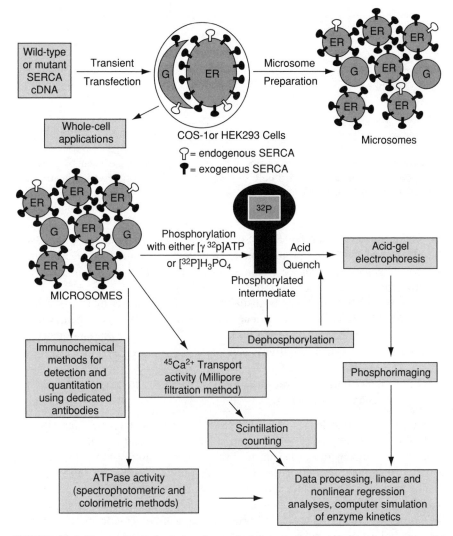

FIGURE 13.1 Flow scheme depicting the methodology typically employed to analyze the structure–function relationships in SERCA, the prototypic Ca^{2+}-ATPase. Forty-eight to seventy-two hours following transfection of COS-1 or HEK293 cells with the desired cDNAs, the cells transiently expressing wild-type or mutant SERCA proteins in the ER are used either for whole-cell applications (immunocytochemistry and Ca^{2+} fluxes) or for preparation of microsomes (see Section 13.3.2). SERCA proteins overexpressed in microsomes can be quantified by immunochemical or phosphorylation methods (see Section 13.3.3). A number of functional assays make use of microsomes to study the overall catalytic reactions (rate of $^{45}Ca^{2+}$ translocation and the turnover rate of ATPase activity) as well as the ability to phosphorylate with either $[\gamma\text{-}^{32}P]ATP$ or inorganic phosphate $[^{32}P]H_3PO_4$ and to dephosphorylate from their corresponding phosphoenzymes under different buffer conditions. Following size-fractionation by acid-gel electrophoresis, the acid-stable phosphorylated intermediates can be quantified by phosphorimaging (see Sections 13.3.4 and 13.3.5).

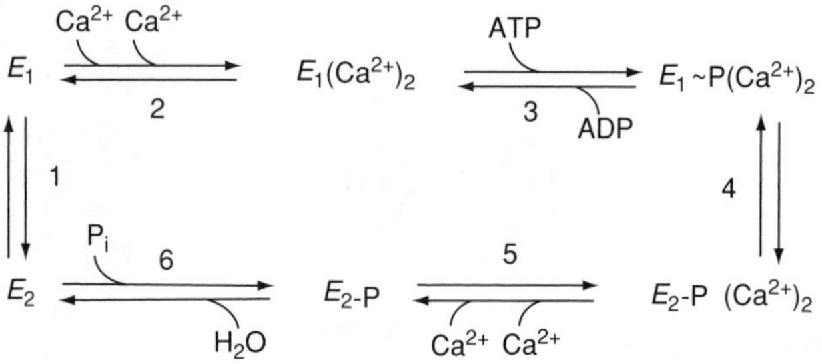

FIGURE 13.2 Minimal reaction scheme for the catalytic cycle mediated by a SERCA prototype. Ca^{2+} ions are occluded only in the E_1~P·$(Ca^{2+})_2$ phosphorylated intermediate. For details, see Section 13.3.1.

the proton counter transport by SERCA [112]. Recently, it has been proposed that in the absence of Ca^{2+}, the proton counter transport is required for neutralizing a number of negatively charged residues clustered around the Ca^{2+}-binding sites, thereby promoting the stability of the ATPase [7].

E_1~P·$(Ca^{2+})_2$ and E_2-P can be distinguished experimentally by their sensitivity toward ADP [113, 114]. In the presence of ADP, E_1~P·$(Ca^{2+})_2$ can generate ATP by transferring its phosphate group back to ADP. In contrast, the phosphorylated intermediate E_2-P remains ADP-insensitive.

13.3.2 HETEROLOGOUS EXPRESSION SYSTEMS IN YEAST, INSECT, AND MAMMALIAN CELLS

Overexpression of SERCA, PMCA, or SPCA pumps in mammalian COS–1 or HEK293 cells usually does not pose particular problems and a plethora of expression vectors ranging from plasmid to viruses are available. Both cell lines carry a stably integrated portion of the SV40 viral genome coding for the viral large T antigen protein. This protein boosts the replication of any transfected recombinant plasmid (i.e., plasmid containing the nucleotide sequence of the desired Ca^{2+}-ATPase cDNA) carrying the SV40 origin of replication. This leads to a massive plasmid replication and subsequently to high levels of transiently expressed protein in COS–1 and HEK293 cells [37]. pMT2 [115] and pSV57 [116, 117] vectors controlled by late SV40 promoter and pcDNA3 series vectors (commercially available from Invitrogen) under the control of cytomegalovirus (CMV) promoter are among the widely used mammalian expression vectors for heterologous expression of various SERCA and SPCA isoforms [28, 31, 33–35, 73, 74, 118–120]. Transfection of recombinant plasmid DNA by calcium phosphate coprecipitation is one of the easiest and most widely used methods of DNA delivery into mammalian cells [37, 121]. Other recent transfection methods include electroporation and liposome-mediated gene transfer [37, 122, 123]. Commercially available reagents for the latter method are usually expensive, but a home-made polyethylenimine-based reagent [124] can be a much cheaper alternative (e.g., the authors have successfully used the latter reagent in transfecting HEK293 cells with various human wild-type and mutant SERCA cDNAs).

Recently, recombinant adenovirus technology has also been employed as an effective method for the delivery of exogenous wild-type SERCA1a and SERCA2a cDNAs in COS–1 cells and in cultured neonatal cardiomyocytes [125, 126] and mutant SERCA1a cDNAs in COS–1 cells [125]. For instance, in neonatal cardiac myocytes the efficiency of infection was maximal (100%) when compared with only 5% to 10% efficiency obtained by other transfection methods [126].

The refinement of the heterologous protein expression methods in S. cerevisiae (versatile promoter toolkit, choice of host strains, availability of selectable markers, and designer strains) combined with several biological and genetic advantages (shorter cell cycle, inexpensive and simple growth conditions, functional complementation, and high-throughput screens of mutant phenotypes or drug targets) make yeast a good choice as a model organism for the study of Ca^{2+} transporters [14]. The yeast system has been successfully used for the biochemical characterization of the yeast SPCA (PMR1) [127–130], human SPCA1a [131], rabbit SERCA1a [132], and Ca^{2+} pumps from protozoa, plants, and other yeast species [14]. (See also Chapter 9.)

Strikingly, attempts to construct a stable expression plasmid for PMCA1 (i.e., the most widespread PMCA isoform in mammals) have been frustrated by unexpected DNA stability problems, which are not encountered with the other PMCA isoforms [133]. Full-length clones of PMCA1 cDNA appear to be unstable in bacteria and attempts to reconstruct the full open reading frame from fragments result in the various deletion-containing or rearranged products. Guerini et al. succeeded in transferring the human PMCA1 cDNA into a baculovirus expression vector thus allowing its expression in insect $Sf9$ cells [133]. After the transfer into the virus, the problems with the DNA stability were no longer encountered. However, pulse-chase studies showed that the PMCA1 protein, again in contrast to other PMCA protein isoforms, was more prone to calpain-dependent degradation in these cells. Thus, the expression of PMCA1 appears to be particularly plagued by technical hurdles at several stages.

Depending on the intended purpose, the transfected cells may be either harvested and processed for downstream functional analyses concerning the overexpressed protein or used as such for whole-cell applications. Processing involves total cell extract preparation or isolation of different membrane fractions. For the preparation and purification of various membrane fractions we refer to a chapter in a previous volume of this series [134]. Following mild lysis and differential centrifugation, the endoplasmic reticulum and Golgi membranes containing overexpressed Ca^{2+}-ATPases are recovered into the so-called "microsomal" fraction (Figure 13.1). This fraction is characterized by the presence of "right side out" vesicles formed spontaneously due to rearrangements of ER and Golgi membranes during preparatory steps [135]. These vesicular structures are able to maintain the in vivo permeability barriers.

The whole-cell approach is currently used for analyses such as immunocytochemistry (see Section 13.3.3) and measurements of Ca^{2+} fluxes in permeabilized cells (see Section 13.4.2).

13.3.3 DETECTION AND QUANTITATION OF OVEREXPRESSED CALCIUM PUMPS

To determine the subcellular localization and the expression level of any given Ca^{2+}-ATPase transiently overexpressed in COS–1 and HEK293 cell lines, immunodetection analyses such as immunocytochemistry on intact cells and Western blotting on different cellular fractions (e.g., microsomal fraction for SERCAs and SPCAs) are usually performed according to established procedures [37].

A strategy to determine the extent of transfection efficiency in mammalian cell lines (COS–1, HEK293, etc.) could rely on the detection of a fluorescent marker (e.g., enhanced green fluorescence protein, EGFP, available from Clontech), whose encoding cDNA is co-transfected with that of the calcium pump under investigation, independently or as part of the same plasmid (bicistronic vector). Positively transfected fluorescent cells are visualized by means of a fluorescent microscope. Interestingly, when a recombinant EGFP adenovirus vector is used to infect COS–1 cells, a 97–100% efficiency of infection is obtained at 100 plaque-forming units (pfu) *per* cell [125]. Cells infected with recombinant SERCA1a adenovirus vector (100 pfu/cell) express higher levels of SERCA1a when compared with cells transfected by the DEAE-dextran method [125]. Nevertheless, for the most part of the methods depicted in Figure 13.1, transfection of COS–1 and HEK293 cell lines using either the calcium phosphate precipitation method or the polyethylenimine method usually yields sufficiently high levels of transiently expressed SERCA and SPCA pumps [28, 34, 35, 51].

Immunocytochemistry on whole cells involves a short fixation (10 min) usually in 4% paraformaldehyde followed by a mild permeabilization in 0.2% TritonX–100 to allow the diffusion of the primary SERCA- or SPCA-specific antibodies. Secondary antibodies are usually carriers of specific labeled groups allowing the visualization of the detected epitope e.g., by wide-field or confocal fluorescence microscopy.

Availability of pure SERCA1a protein required for accurate standardization and of large amounts of monoclonal SERCA1-specific antibody (see Section 13.2.2.1.1) have contributed to the development of the sandwich ELISA for quantifying overexpressed wild-type and mutant SERCA1a proteins [65, 66]. A variation of this method could imply the use of SDS polyacrylamide gel electrophoresis to separate the SERCA1a standards and the various samples to be quantified followed by Western blotting and quantification by means of chemiluminiscence imaging.

In the absence of pure Ca^{2+}-ATPase standards, the amount of expressed Ca^{2+}-ATPase can be accurately determined by measurement of its capacity for phosphorylation by [γ-^{32}P]ATP or inorganic phosphate ([^{32}P]H$_3$PO$_4$ or ^{32}P$_i$), the total phosphorylation level indicative of its active site concentration (pmol Ca^{2+}-ATPase/mg microsomal protein) [136]. Using this procedure, the expression levels of wild-type and mutant SERCA proteins in COS–1 and HEK293 cells range between 150 and 400 pmol active enzyme/mg microsomal protein, which is several hundred-fold higher than the level of endogenous SERCA2b pump [34, 35, 136]. The contribution of the endogenous enzyme to the enzymatic measurements described below is thus negligible.

Another quantitative method takes into account the ability of thapsigargin (TG), a sesquiterpene lactone isolated from the roots of *Thapsia garganica,* to specifically inhibit SERCA pumps (see Section 13.4.1) at subnanomolar concentrations [137]. TG sensitivity can be determined by measuring the level of phosphoenzyme (*EP*) formed as a function of total TG concentration using the equation,

$$EP = 100\% \cdot (1 - F_b) \tag{13.1}$$

where F_b represents the fraction of inactive enzyme that is bound to TG

$$F_b = 0.5 \cdot [R + R_0 + 1 - (R^2 + R_0^2 + 2\,R \cdot R_0 + 2 \cdot R_0 - 2 \cdot R + 1)^{0.5}] \tag{13.2}$$

A simple one-site binding model is considered in deriving Equation 13.2, where R is the molar ratio between the total [TG] and the enzyme concentration, and R_0 is the ratio between the dissociation constant (K_d) for the enzyme–TG complex and the enzyme concentration [94]. Figure 13.3 demonstrates the usefulness of this method by showing an example of sensitivity of SERCA1a phosphorylation (in the presence of Ca^{2+} and [γ^{32}-P]ATP) to inhibition by TG. The line shows the best fit of the Equation 13.1 to the data. This fitting allows determination of both the K_d for TG binding and the enzyme concentration of the microsomal preparation used.

13.3.4 OVERALL REACTIONS: ION TRANSPORT AND ATPASE ACTIVITY

The overall catalytic performance of any SERCA is described by the following reaction:

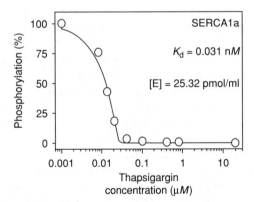

FIGURE 13.3 A typical experiment showing the thapsigargin sensitivity of SERCA1a at 25°C upon phosphorylation with [γ-^{32}P]ATP for 5 s in the presence of 1.5 m*M* free [Ca^{2+}] (theoretical background shown in Section 13.3.3; more experimental details in Ref. [94]). The phosphorylation level is shown as a percentage of that obtained in the absence of thapsigargin. The line shows the best fit of Equation 13.1 to the data. The respective K_d for thapsigargin binding (in n*M*) and the enzyme concentration (in pmol/ml) characterizing the microsomal preparation used are 0.031 and 25.32.

$$2(Ca^{2+})_{cytosol} + n(H^+)_{lumen} + (ATP)_{cytosol} \leftrightarrow 2(Ca^{2+})_{lumen} + n(H^+)_{cytosol}$$
$$+ (ADP)_{cytosol} + (P_i)_{cytosol}$$

where the hydrolysis of ATP energizes the translocation of two Ca^{2+} from cytosol into the SR or ER lumen in exchange for protons. The number of protons ($n = 1$ or 3) greatly depends on the reaction conditions known to affect the positive cooperativity of Ca^{2+}-binding reaction such as pH, temperature, and presence of Mg^{2+}. The enzyme is fully protonated (i.e., $n = 3$) at pH 6.0, but Ca^{2+} binding becomes rate-limiting due to the slow deprotonation process, whereas at pH 9.0, Ca^{2+} binding is fast because the enzyme is totally deprotonated. At physiological pH values, the proportion between the protonated and deprotonated enzyme species determines the rate of the Ca^{2+} binding transition [97, 99, 138]. Faster rates are always obtained in the presence of Mg^{2+} ions, which stimulate enzyme deprotonation. Furthermore, Mg^{2+} binds to the catalytic site and represents the co-factor required for the phosphorylation reaction in the presence of ATP or P_i [139–141].

The overall catalytic activity of Ca^{2+}-transporting ATPases can be monitored in the steady state by specifically measuring the rate of ATP-dependent Ca^{2+} transport (see Sections 13.3.4.2.) and/or by assaying the rate of Ca^{2+}-dependent ATP hydrolysis (see Section 13.3.4.3). Early studies of Ca^{2+} uptake by SR vesicles demonstrated that the optimal molar ratio of 2:1 between calcium translocation and ATP utilization only occurs in the initial catalytic cycles characterizing the pre-steady-state (or transient state), whereas a lower coupling ratio is reached in the steady state [86, 96]. Such limitations on the steady-state Ca^{2+} filling of SR vesicles arise due to excessive ATP-dependent accumulation of lumenal Ca^{2+}. High $[Ca^{2+}]$ in the lumen can inhibit the calcium pump in the forward direction by binding to lumenally facing low-affinity Ca^{2+} binding sites ("back inhibition"), and, in the presence of a high [ADP]/[ATP] ratio, can lead to Ca^{2+} efflux coupled to ATP synthesis (reversal of the pump) [142]. This process is accompanied by heat absorption from the medium [143]. Passive Ca^{2+} efflux through the pump can take place without concomitant ATP synthesis. "Slippage" of the pump ("true" uncoupling of ATPase and transport activities) depends on the uncoupled cleavage of the phosphoenzyme intermediates accumulated at steady state in the presence of high lumenal Ca^{2+} and high [ADP]/[ATP] ratio in the medium [142–144]. In skeletal muscle, slippage or uncoupled ATPase activity appears to be an important pathway through which the pump is able to modulate the production of heat in the presence of a calcium gradient [143, 145].

In view of the above-mentioned considerations, any improvements on the relieving of back inhibition will increase the coupling ratio as well as Ca^{2+} translocation efficiency. Efficient measures include the addition of oxalate, phosphate, or calcium ionophore, or the presence of an ATP-regenerating enzyme-coupled system in the reaction mixture. Such intervening measures will have a dramatic effect on the reduction of slippage and/or efflux of Ca^{2+}. Therefore, the measurements of Ca^{2+} transport and ATPase activities at steady state would benefit from such enhancements, especially when leaky ER vesicles (with passive Ca^{2+} leak) isolated from transfected COS–1 or HEK293 are used.

Topics related to the updated methodology described in these sections (13.3.4 and 13.3.5) have been treated previously and the reader is further referred to volume 157 of *Methods in Enzymology* edited by Fleischer and Fleischer [146].

13.3.4.1   Assays for Ca^{2+} Transport in Isolated Membrane Vesicles

Membrane fragments formed by cell homogenization have a strong tendency to form closed vesicular structures that maintain the *in vivo* permeability barriers. This property has been fully exploited to measure the Ca^{2+}-transporting activities of the Ca^{2+} pumps embedded in sarco(endo)plasmic-, Golgi-, and plasma membrane-derived vesicles. Due to the small size of the vesicles, the lumenal Ca^{2+} concentration reaches inhibiting levels soon after the initiation of Ca^{2+} transport, resulting in a rapidly decreasing rate of transport until no further net accumulation of Ca^{2+} occurs. The phase of linear time-dependent Ca^{2+} uptake can be much prolonged by the inclusion of oxalate or phosphate in the Ca^{2+}-uptake medium. Both anions promote the formation of lumenal insoluble Ca^{2+} complexes in the presence of Ca^{2+} at concentrations exceeding the solubility product, thereby increasing the capacity for active Ca^{2+} accumulation inside the vesicles. Oxalate in the uptake solution is often used at a concentration of 5 mM. Phosphate could be used at higher concentrations (50–100 mM), although low concentrations (1 mM) of physiological significance have proved to allow calcium filling of the vesicles through the formation of soluble complexes with lumenal calcium [142]. Permeation of oxalate and phosphate through endoplasmic reticulum or Golgi-derived membranes is very rapid. However, Ca^{2+} uptake in plasma membrane vesicles is insensitive to oxalate, but it is nonetheless stimulated by phosphate [147].

Ca^{2+} transport can be determined either by measuring the change in the lumenal Ca^{2+} content, or by measuring the change of Ca^{2+} concentration in the medium. The latter method requires additional information on the Ca^{2+}-buffering capacity of the medium, which is defined as the ratio between the change in the total [Ca^{2+}] and the change in the free [Ca^{2+}]. The free Ca^{2+} concentration can be precisely controlled by means of a specific Ca^{2+}-chelating agent such as ethylene glycol-bis(β-aminoethylether)-N,N,N',N'-tetraacetic acid (EGTA) in the presence of various metals and ligands in the medium. The relevant computer calculations and stability constants for Ca^{2+} and Mg^{2+} complexes with ligands such as ATP, EGTA, and oxalate have been extensively treated [148, 149]. The various free CaCl$_2$ concentrations in the presence of oxalate can be calculated with the CaBuf computer program (developed by G. Droogmans and available from the Laboratory of Physiology, Leuven at ftp://ftp.cc.kuleuven.ac.be/pub/droogmans/cabuf.zip) and in the absence of oxalate with MAXC software using the dissociation constants therein [150]. Any change (normally a decrease) in the free [Ca^{2+}] can be correlated to a change in the levels of light absorption at two selective wavelengths undergone by metallochromic indicators (e.g., murexide, antipyrilazo III, and arsenazo III) [146, 151, 152]. Although they have been successfully used in measurements of the initial rates of Ca^{2+} uptake in SR vesicles [151], the general use of metallochromic absorbance dyes was short-lived since they were not sensitive enough to the submicromolar cytosolic [Ca^{2+}] found in cells. In this respect, further investigation of calcium homeostasis and dynamics became possible after the introduction of a new generation of Ca^{2+} indicators with greatly improved fluorescence properties derived from BAPTA (an EGTA-like chemical compound where the aliphatic amino groups of EGTA critical for conferring its pH-sensitivity are switched to aromatic amino groups in BAPTA), which is a largely pH-insensitive Ca^{2+} chelator [153]. All BAPTA-based indicators bind to Ca^{2+} with a 1:1 stoichiometry. The Ca^{2+}-free and Ca^{2+}-bound forms of the indicators will exhibit either a shift in excitation or emission maxima used for [Ca^{2+}] quantification and imaging purposes (ratiometric indicators: fura–2, indo–1, fura red, mag-fura–2, mag-fura–5) or changes in their fluorescence intensity (non-ratiometric indicators: fluo–3, rhod–2) [153, 154]. (See also Chapter 1.) The high-affinity fura–2 indicator was used to measure Ca^{2+} uptake into the SR of saponin permeabilized cardiac myocytes (expressing SERCA2a) or into microsomes isolated from HEK293 cells (overexpressing SERCA2a and SERCA2b) at Ca^{2+} concentrations of up to 1 μM [155, 156]. The low- to moderate-affinity indicators (e.g., mag-fura–2, mag-fura–5, and mag-indo–1 available from Molecular Probes both as cell-impermeant potassium salts or as cell-permeant acetoxymethyl esters) have been employed for measuring intracellular Ca^{2+} levels in the range between 1 and 100 μM due to their enhanced linearity for Ca^{2+} [157]. It can be envisioned that the latter group of fluorescent indicators in their salt form would be useful to measure the rate of Ca^{2+} uptake in microsomes at saturating Ca^{2+} concentrations of up to 100 μM.

The most widely used method to monitor the rate of ATP-driven Ca^{2+} uptake in ER-derived membrane vesicles (microsomes) is by measurement of the total luminal Ca^{2+} content using a radioactive Ca^{2+} tracer [$^{45}Ca^{2+}$]$CaCl_2$ in combination with Millipore filtration (HAWP filters of 0.45 μm pore size) to separate the vesicles (~10 μg microsomal protein per each time point or per each [Ca^{2+}] at which the rate of Ca^{2+} uptake is determined) from the reaction medium (175 μl) containing 20 mM MOPS pH 6.8, 100 mM KCl, 5.5 mM $MgCl_2$, 5 mM ATP, 0.5 mM EGTA, 5 mM potassium oxalate, 0.45 mM $CaCl_2$ (free [Ca^{2+}] of 3.6 μM at 37°C), and 1 μCi [$^{45}Ca^{2+}$]$CaCl_2$/ml after incubation at 27°C or 37°C for serial time intervals (e.g., 1, 2, 5, 10, 15, and 20 min) [158, 159]. The reaction is initiated either by addition of microsomes to the complete reaction mixture or by addition of the energy-providing component (5 mM ATP) to the rest of the reaction solution already containing the microsomes. Normally, prior to filtration, each reaction (150 μl) is terminated by addition of 5 ml ice-cold stop solution containing 1 mM $LaCl_3$ (see Section 13.4.2.4) and 150 mM KCl. After filtration, the filters are washed twice with 5 ml ice-cold stop solution and scintillation-counted using either the Multi-Purpose Scintillation Counter LS6500 (Beckman, USA) or the Liquid Scintillation Analyzer Tri-Carb 2900TR (Packard, USA). Nevertheless, we have found that the reaction aliquot (150 μl) can be directly filtered and immediately washed without any significant loss of the Ca^{2+} uptake rate and, optionally, the presence of microsomes on the Millipore filters can be color tracked by the use of minute amounts of ruthenium red (ryanodine receptor inhibitor) in the reaction solution [160]. Depending on the Ca^{2+}-ATPase and the incubation temperature, the linear phase of the time-course of oxalate-stimulated Ca^{2+} transport at saturating free Ca^{2+} concentrations (3–10 μM) lasts up to 15–20 min. Ca^{2+}-dependent Ca^{2+} uptake can only be studied within this time interval for which a linear relationship exists between time and transport rates. The various Ca^{2+} concentrations needed to set the free Ca^{2+} concentrations can be calculated using the CaBuf program mentioned above. In the presence of 5 mM oxalate, accurate measurements of oxalate-stimulated Ca^{2+} transport rates can be carried out at free Ca^{2+} concentrations of up to 30 μM for all wild-type Ca^{2+}-ATPases and most of their derived mutants. Above this level, precipitation of Ca^{2+} oxalate crystals occurs both outside and inside the vesicles, thus generating a high background on the Millipore filters. For some SERCA1a mutants (e.g., $Phe^{760} \longrightarrow Gly$, $Tyr^{763} \longrightarrow Gly$), the rate of Ca^{2+} transport had to be studied at higher Ca^{2+} concentrations [161]. To adjust free Ca^{2+} concentrations above 10 μM and up to 200 μM, a Ca^{2+}-selective electrode can be used [161]. Additionally, the high filter background can be avoided provided some precautions are taken such as: all supersaturated Ca^{2+}-oxalate solutions have to be freshly prepared and kept at 37°C before use and the Ca^{2+} uptake should be measured at 37°C instead of 27°C [159].

As an example, Figure 13.4 shows comparatively the time-course (panel A) and Ca^{2+}-dependence (panel B) of the oxalate-stimulated ATP-driven $^{45}Ca^{2+}$ transport rates determined for SERCA1a and SERCA3 isoforms [160]. The specific rates of Ca^{2+} uptake corresponding to the initial linear phase were calculated after normalization of the data to the active site concentration of each expressed SERCA determined by phosphorylation. The transport rates of the SERCA3 isoforms do not differ significantly from one another or from that of SERCA1 (panel A), but the apparent affinities for Ca^{2+} displayed by SERCA3 isoforms are several-fold lower than that of SERCA1a (panel B). This is in good agreement with the apparent Ca^{2+} affinity data measured at steady state using other assays monitoring for instance the Ca^{2+}-dependent activities of ATPase hydrolysis and phosphorylation capacity [34]. The Ca^{2+} titration data can be fitted by nonlinear regression analysis using either the SigmaPlot (SPSS Inc.) or the Origin (MicroCal Software Inc.) computer software. The analysis of substrate concentration [S] dependence is based on the Hill function described below:

$$v = \frac{V \times [S]^n}{K_{0.5}^n + [S]^n} \tag{13.3}$$

where $K_{0.5}$ represents the substrate (e.g., Ca^{2+}, ATP, and P_i) concentration at which the rate v is half of the limiting (or maximal) rate V; n is the Hill coefficient indicative of enzyme cooperativity ($n = 1$ for noncooperative enzymes; $n > 1$ for positively cooperative enzymes).

Measurement of ATP-dependent Ca^{2+} uptake by radioactive tracer and Millipore filtration has been successfully used in the past to characterize the ability of SERCA1a to pump Ca^{2+} into SR vesicles [146]. The method also has the advantage of being applicable to the comparative study of Ca^{2+} transport activity mediated by wild-type SERCA pumps and their corresponding mutants overexpressed in COS–1

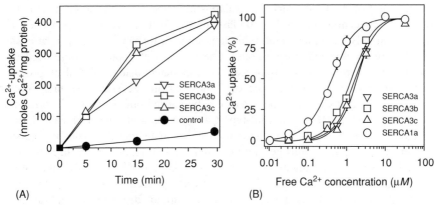

(A)   (B)

FIGURE 13.4 Ca^{2+}-transport activity. Panel (A) shows the time-course of $^{45}Ca^{2+}$ uptake into microsomes isolated from cells transfected with the pMT2 vector without insert (control) and with DNA inserts encoding the indicated human SERCA3 isoforms, measured at 27°C in a medium containing 20 mM MOPS, pH 6.8, 100 mM KCl, 5.5 mM MgCl$_2$, 5 mM ATP, 0.5 mM EGTA, 5 mM potassium oxalate (to trap Ca^{2+} inside the vesicles), and 0.497 mM $^{45}CaCl_2$ giving a free Ca^{2+} concentration of 10 μM. The specific Ca^{2+}-uptake activity (nmoles Ca^{2+} transported/mg microsomal protein) of a typical experiment is shown following subtraction of background levels determined in the absence of ATP. Panel (B) shows the Ca^{2+}-dependence of the Ca^{2+}-transport activity catalyzed by human SERCA3 and rabbit SERCA1 isoforms at 27°C after 10 min incubation in the same medium as in panel (A) except that the total concentration of $^{45}CaCl_2$ was varied to produce the indicated free Ca^{2+}-concentrations. The specific Ca^{2+}-uptake activity (nmoles Ca^{2+}/mg protein) is shown relative to the maximum value obtained for each of the indicated SERCA protein. The lines show the best fits of the Hill equation (Equation 13.3) to the data. The respective $K_{0.5}$ and Hill coefficients ($n \pm$ SE) extracted by the regression analysis are as follows: human SERCA3a (1.69 $\pm$ 0.14 μM; 1.63 $\pm$ 0.16), SERCA3b (1.34 $\pm$ 0.11 μM; 1.59 $\pm$ 0.15), SERCA3c (1.82 $\pm$ 0.18 μM; 1.70 $\pm$ 0.19), and rabbit SERCA1a (0.40 $\pm$ 0.02 μM; 1.37 $\pm$ 0.10) [160].

and HEK293 cells as has been previously documented [6, 32–35, 42, 159]. Furthermore, the inhibition of pump's activity (see Section 13.4) by various concentrations of compounds such as TG and vanadate in steady-state conditions can also be studied using this method [162].

13.3.4.2   Assays for Ca^{2+} and Mn^{2+} Transport in Permeabilized Cells

This part of the chapter focuses not only on measuring the Ca^{2+} uptake in permeabilized cells, but also on the Ca^{2+}-release mechanisms, since they are relevant to characterize the compartment in which a Ca^{2+} pump is expressed. For example, the SERCA pumps of the Golgi apparatus are present in an inositol trisphosphate (IP$_3$)-sensitive compartment, while the SPCA pumps are not [163].

While most experiments have been done on cultured cells, Ca^{2+} uptake has also successfully been measured on small tissue samples and on tissue sections. Cultured cells can be grown in monolayers or in suspension. Experiments on monolayers have the major advantage that bathing solutions can be easily replaced. Agonists like IP_3 can therefore be added and subsequently removed, and the loading of the Ca^{2+} stores and the subsequent efflux can be done using media with different composition. This is especially useful to study effects of e.g., cytosolic Ca^{2+} or ATP on intracellular Ca^{2+}-release channels, after an initial loading of the stores at an optimal $[Ca^{2+}]$ and [ATP] for pumping. The major disadvantage is that the technique can only be applied to adherent cells. Coating of the surface with e.g., poly-L-lysine or gelatin often helps to increase the adhesion. Experiments on cell suspensions, in contrast, can also be done on non-adherent cells. The major disadvantage is that changing of the medium requires centrifugation and resuspension of the cells.

The plasma membrane must first be permeabilized for gaining direct access to the intracellular Ca^{2+} stores. Digitonin and other saponins are most frequently used in the Ca^{2+}-signaling community. These detergents interact with cholesterol in the plasma membrane to form pores, which make the membrane freely permeable to small molecules and ions. The cholesterol content of the Golgi apparatus gradually decreases from the *trans-* to the *cis-*Golgi [164], implying that the distal parts of the Golgi apparatus also may become permeabilized. Membranes containing negligible amounts of cholesterol such as those of the endoplasmic reticulum remain functionally intact [165, 166]. Other techniques of permeabilization include the use of α-toxin, a transmembrane pore-making protein produced by *Staphylococcus aureus*, streptolysin O from β-hemolytic streptococci and electropermeabilization. Whereas α-toxin and electropermeabilization create small pores of approximately 2 nm, digitonin, saponin, and streptolysin O form bigger holes and therefore also allow the introduction of large molecules, such as enzymes. Digitonin, saponin, and streptolysin O also result in the loss of cytosolic constituents that might be necessary for signal-transduction pathways in the cell. Even vesicular fractions may leave the cell into the medium, since low-speed centrifugation of saponin-treated hepatocytes resulted in the appearance of IP_3-sensitive Ca^{2+}-pools in the cell-free supernatant fraction [167].

Ca^{2+} uptake or release from the stores can be measured by following the decrease or increase, respectively, of the free $[Ca^{2+}]$ in the external medium. These experiments require a relatively high cell density, since enough Ca^{2+} stores have to be present to remove the micromolar concentrations of contaminating Ca^{2+} from the medium. Such high cell densities can only be obtained with cell suspensions. Since Ca^{2+} is the variable that is measured, it is difficult to study the effect of addition of Ca^{2+} itself, e.g., in studies on the effect of Ca^{2+} on Ca^{2+}-release channels. But since the external $[Ca^{2+}]$ is not clamped, Ca^{2+} can feed back on its own release and therefore more closely mimic the situation in an intact cell. The experiments are now mostly done with fluorescent Ca^{2+} indicators in a fluorimeter, one after the other. The quality of the cells often decreases as a function of time, making these measurements more qualitative than quantitative. This approach is more suited to assess Ca^{2+}-release mechanisms than the Ca^{2+} pumps.

Ca^{2+} uptake or release from the stores can also be measured by monitoring the free $[Ca^{2+}]$ in the lumen of the stores with low-affinity Ca^{2+} indicators. Hofer and Machen (1993) introduced the acetoxymethyl-ester derivative of the fluorescent dye mag-fura–2 in the subcellular compartments in intact cells and then permeabilized the cells to release the cytoplasmic dye, while keeping the mag-fura–2 in the store [168]. This technique also allows measuring the Ca^{2+} handling in subcellular microenvironments in permeabilized cells [169]. The major disadvantage of the method is that all internal compartments are loaded and not just the endoplasmic reticulum.

In permeabilized cells, the Ca^{2+} uptake or release from the stores can also be followed by measuring the Ca^{2+} content of the stores using radioactive tracer ($^{45}Ca^{2+}$). It is a suitable technique to study the transport properties of an over-expressed SPCA pump. In our personal experience, this approach is more reproducible for SPCA pumps than $^{45}Ca^{2+}$-uptake measurements using SPCA-containing microsomes. We routinely do the fluxes in 12-well clusters, thereby allowing 12 simultaneous measurements. The technique can be easily modified to study $^{54}Mn^{2+}$ uptake, since some pumps like the SPCAs also transport manganese ions [76]. The latter measurements have so far not been possible with lumenally trapped fluorescent or luminescent indicators. The technique however does not allow to specifically focus on the endoplasmic reticulum or the Golgi apparatus. Neither does it have the fast time resolution of fluorescence of luminescence measurements. However, a maximal time resolution of 9 ms has been achieved using specifically built equipment [170].

It is possible to permeabilize the plasma membrane of cells without prior isolation if the tissue samples are sufficiently small to allow fast diffusion of the permeabilizing agent. This method allows studying the $^{45}Ca^{2+}$ uptake in conditions much closer to the *in vivo* situation. For instance, $^{45}Ca^{2+}$ uptake has been measured in small bundles of smooth-muscle cells [171, 172]. The anatomical distribution of Ca^{2+} pools and their sensitivity to SERCA inhibitors and Ca^{2+}-releasing agonists can be determined on tissue sections by incubation in a $^{45}Ca^{2+}$-containing uptake solution followed by autoradiography. This method has been applied to brain and whole-body sections of rat [173, 174].

The reader is referred for further details on the measurement of Ca^{2+} fluxes in permeabilized cells to an excellent review by Taylor (1999) [175].

Aequorin Ca^{2+} probes can be genetically targeted to subcellular organelles like the endoplasmic reticulum or the Golgi apparatus (Chapter 5), but they are luminescent and therefore not bright enough for single-cell imaging without specialized equipment [176–178]. Aequorin must first be reconstituted with the coelenterazine cofactor, and to do so the stores must initially be completely depleted of Ca^{2+}, which may affect their organization. Ca^{2+} indicators based on the *Aequorea victoria* green fluorescent protein also allow a specific targeting to subcellular organelles and are suitable for imaging (Chapter 4). The recent dyes have a sufficiently large dynamic range to adequately monitor the $[Ca^{2+}]$ in the lumen of the endoplasmic reticulum [179].

13.3.4.3 Assays of Ca^{2+}-ATPase Activity in Isolated Membrane Vesicles

During the course of Ca^{2+}-activated ATPase activity, both ADP and P_i are produced. When the resulting ADP is further used as a substrate to regenerate ATP and produce lactate and oxidized nicotinamide adenine dinucleotide (NAD^+) from phosphoenolpyruvate and reduced nicotinamide adenine dinucleotide (NADH) in an enzyme-coupled assay (see Figure 13.5) then the Ca^{2+}-dependent ATPase activity can be spectrophotometrically determined at 340 nm from recordings of NADH oxidation (i.e., calculated from the slopes of the linear phase showing the decrease in NADH absorbance at 340 nm as a function of time).

Typically, a standard reaction mixture (2.5 ml) consists of 10–40 μg microsomal protein, 0.15 mM NADH, 1 mM phosphoenolpyruvate, 10 IU of pyruvate kinase/ml, and 107 IU of lactate dehydrogenase/ml, to which additional components are added: 25 mM TES/Tris pH 7.5, 100 mM KCl (K^+ in this concentration range acts as an enhancer of the turnover rate), 6 mM $MgCl_2$ (Mg^{2+} required as a cofactor for binding to the catalytic site as MgATP; free $[Mg^{2+}]$ is 1 mM here), 5 mM ATP, 1 mM EGTA, and $CaCl_2$ to set the free $[Ca^{2+}]$ to 100 μM. After the assay mixture is preincubated for 5 min at 37°C, the reaction is initiated by the addition of either ATP or Ca^{2+} or microsomes and the decrease in optical density at 340 nm can be recorded by means of a spectrophotometer with a thermostated multicuvette holder. In addition, calcium ionophore A23187 (1–2 μM) is used to stimulate the efflux of Ca^{2+} from the vesicles, thus equilibrating the Ca^{2+} concentration on both sides of the vesicular membranes and relieving the pump's "back inhibition" exerted by Ca^{2+} accumulated in the vesicles through active pumping. The millimolar absorptivity (ε) of NADH at 340-nm wavelength is 6.22 mM^{-1}cm^{-1} and the concentration (c expressed in mM) of NADH can be calculated by applying Beer's Law (Equation 13.4):

$$A = \epsilon \cdot l \cdot c \tag{13.4}$$

where A represents the absorbance measured at 340 nm and l is the path length (in cm). The Ca^{2+}-dependent ATPase activity is calculated following subtraction of the background Mg^{2+}-ATPase activity measured in the absence of Ca^{2+}, but in the presence of 2–4 mM EGTA. Variations in the medium composition, choice of buffers (MOPS or TES), solute concentrations, and reaction conditions (temperature and pH) have been described [136, 159, 180, 181].

Production of inorganic phosphate during ATPase activity can also be determined. One spectrophotometric assay example involves the employment of the

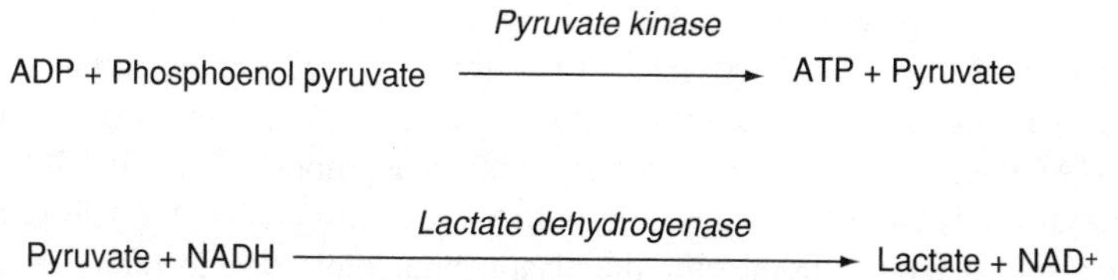

Pyruvate kinase

ADP + Phosphoenol pyruvate ⟶ ATP + Pyruvate

Lactate dehydrogenase

Pyruvate + NADH ⟶ Lactate + NAD+

FIGURE 13.5 The standard enzyme-coupled assay measuring the reduction in NADH absorbance at 340 nm associated to the corresponding increase in the ATPase activity as a function of time.

purine–nucleoside phosphorylase (PNP), which enables the transformation of inorganic phosphate and 2-amino–6-mercapto–7-methylpurine riboside (MESG) into ribose–1-phosphate and 2-amino–6-mercapto–7-methylpurine (AMM). The production of AMM is proportional to the P_i released and it is monitored by measuring the increase in the absorbance of AMM at 360 nm [156]. This method was recently used to test the sensitivity to replacement of K^+ with Cs^+ of the Ca^{2+}-ATPase activity mediated by SERCA2 isoforms (transiently expressed in HEK293 cells) in the presence of 3.5 μM calcium ionophore and ~1 μM free calcium concentration [156].

Determination of P_i produced during the ATPase reaction can more directly be assessed by either one- or two-step methods depending on whether acid quenching and color development are performed in one step or sequentially, respectively. For example, a one-step method makes use of an acidic molybdo-vanadate reagent (Lin-Morales reagent) required to both quench the ATPase reaction mixture and form with the P_i released a complex, whose absorbance is read at 350 nm wavelength [146, 182–184]. Variations concerning the buffer composition used to monitor the ATPase activity mediated by PMCA (expressed in red blood cell membranes) and SERCA1a from light SR membranes have been described [146, 184].

The two-step procedure currently used by authors for microdetermination of inorganic phosphate in the presence of labile organophosphates relies on the Baginski method [185]. Briefly, following incubation at 37°C, the hydrolytic activity of microsomal Ca^{2+}-ATPases (~0.5 pmol/reaction medium) is quenched by addition of 1.0 ml of ice-cold freshly prepared molybdate solution (0.5 M HCl, 170 mM ascorbic acid, and 4 mM ammonium heptamolybdate) to 0.5 ml reaction medium containing 50 mM TES/Tris, pH 7.0, 100 mM KCl, 7 mM MgCl$_2$, 5 mM ATP, 1 mM EGTA, various concentrations of CaCl$_2$ to obtain free Ca^{2+} concentrations in the range in between 0.01 and 1000 μM, and 1 μM calcium ionophore A23187 (if ionophore sensitivity of Ca^{2+}-ATPase activity is studied, then the reaction should be performed both in the presence and in the absence of ionophore). If more samples are to be terminated at serial time intervals, it is advisable to keep the reaction tubes on ice until all samples have been quenched. In the next step, aliquots of 1.5 ml of a second solution (0.35 mM acetic acid, 70 mM sodium citrate, and 150 mM sodium-m-arsenite) are added. Reaction tubes are further incubated for 5 min at 37°C to complete the reaction (arsenite–citrate reagent binds excess molybdate and prevents it from reacting with any additional phosphate formed from hydrolysis) and allow color development prior to absorbance measurements at 850 nm against a reagent blank (no microsomes added in the blank) [136]. The rates of ATP hydrolysis are calculated by relating to standards of known P_i concentration that are similarly treated. The reactions are usually initiated by the addition of microsomal membranes to the reaction mixture. Importantly, the contaminant ADP and P_i present in minute amounts in any nonradioactive ATP batches commercially available could interfere with the determination of ATPase activity and in chase procedures monitoring the disappearance of ^{32}P-phosphorylated intermediate (see Section 13.3.5.1). Therefore, prior to its use in the aforementioned procedures, nonradioactive ATP is first purified from contaminant ADP and P_i by ionic exchange chromatography and kept at −20°C in suitable aliquots.

The linearity of ATPase activity with time should be first tested at saturating free Ca^{2+} concentrations (e.g., at 100 μM) in the presence or absence of calcium ionophore, but normally a 10-min incubation at 37°C is within the optimal linear phase for SERCA1, SERCA2, and SERCA3 isoforms studied so far [34, 35, 136]. In the linear phase, the ATPase activity proceeds at constant rates and, in these steady-state conditions, the dependence of enzyme activity on the substrate (Ca^{2+}, ATP, and Mg^{2+}) or inhibitor (e.g., vanadate) concentration, temperature, and pH can be determined.

Irrespective of the method used to measure Ca^{2+}-ATPase activity, the molecular turnover rate (s^{-1}) for ATP hydrolysis can be calculated as the ratio between the ATP hydrolysis rate (nmol ATP hydrolyzed per mg of microsomal protein per second) and the maximum capacity for phosphorylation (active site concentration) with [γ-^{32}P]ATP (nmol ATPase per mg of microsomal protein). Addition of ionophore usually results in a two- to threefold increase in the maximal ATPase turnover rate of SERCA1 and SERCA2 isoforms [35]. Lack of any activating effect of calcium ionophore on the ATPase activity at neutral pH has also been documented for SERCA3 isoforms [34], for some of the studied SERCA1a mutants (e.g., Lys^{758} $\longrightarrow$ Ile and Tyr^{763} $\longrightarrow$ Gly) [136, 161], and for the Darier disease (SERCA2b) mutant Ser^{920} $\longrightarrow$ Tyr^{35}. For mutant Tyr^{763} $\longrightarrow$ Gly, its insensitivity to ionophore can be ascribed to the lack of Ca^{2+} accumulation into the microsomes (i.e., as a result of the mutation, the Ca^{2+} translocation and ATPase activities are uncoupled). Since SERCA3 isoforms and the other mutant SERCA proteins mentioned above are able to accumulate Ca^{2+} in the microsomes, the difference from SERCA1a with respect to ionophore insensitivity can be explained either by an enhanced efflux of lumenal Ca^{2+} through the pump in the E_2 form or by insensitivity to inhibition exerted by lumenal Ca^{2+} on the lumenally facing transport sites [34, 35, 136, 161]. The lack of ionophore effect as well as the modulatory effect of ATP on the molecular turnover rate of the ATPase activity can also be simulated with the kinetic software SimZyme developed in the laboratory led by coauthor Dr Andersen [95].

The analysis of ligand concentration dependences is based on the Hill function (Equation 13.3). Furthermore, the data for the pH dependence of the ATPase activity can be fitted by a classical Michaelis pH function using the equation below:

$$v = \frac{V}{1 + \dfrac{[H^+]}{K_1} + \dfrac{K_2}{[H^+]}} \tag{13.5}$$

where K_1 and K_2 are the molecular dissociation constants (mol·l^{-1}) of the intermediate forms of a dibasic acid. For instance, the pH-dependence of the steady-state ATPase activity of SERCA1a and the SERCA3 isoforms determined at 100 μM free [Ca^{2+}] and at 37°C, in the presence or absence of 1 μM calcium ionophore has allowed the authors to first extract the ionization constants K_1 and K_2 from the regression curves and second to calculate the optimum pH corresponding to each isoform. This analysis has demonstrated that the pH optimum for SERCA3a,

SERCA3b, or SERCA3c is alkaline-shifted by 0.5–0.6 pH units relative to SER-CA1a both in the presence and absence of calcium ionophore [34].

If the measurements of Ca^{2+} uptake and Ca^{2+}-dependent ATPase activity are performed in identical conditions (except for the $^{45}Ca^{2+}$ present in the uptake solution only) including the presence of 5 mM potassium oxalate to trap Ca^{2+} inside the microsomes, then it is possible to calculate the apparent coupling ratio between Ca^{2+} translocation and ATP hydrolysis [136]. The molar ratio is 2 only in the initial phase of the reaction as explained above in Section 13.3.4 [142].

13.3.5 STEADY STATE AND TRANSIENT STATE KINETIC ANALYSES OF PARTIAL REACTIONS

Several partial reactions of the overall catalytic cycle (Figure 13.2) can be studied separately in steady state and transient state conditions by means of specific methods that rely on the intrinsic propensity of any Ca^{2+}-ATPase to become phosphorylated in the presence of Ca^{2+} and ATP (forward reaction 3) or in the absence of Ca^{2+} and presence of inorganic phosphate (reverse reaction 6). Such analyses are needed especially when the functional consequences of amino acid substitutions are examined in mutant Ca^{2+}-ATPases and compared to their corresponding wild-type enzymes in order to gain information with respect to the relationship between enzyme structure and function [28, 32–35].

13.3.5.1 Phosphorylation Approach

As mentioned above (Section 13.2.2.2), the phosphorylated intermediate $E_1{\sim}P$ $(Ca^{2+})_2$ obtained upon the simultaneous incubation of enzyme with Ca^{2+} and $[\gamma\text{-}^{32}P]ATP$ is stable in acidic conditions, thereby allowing quenching of the phosphorylation reaction with an ice-cold acid solution containing 25% (w/v) TCA and 100 mM H_3PO_4. Formation of the $E_1{\sim}P{\cdot}(Ca^{2+})_2$ is stimulated by the presence of saturating $[Ca^{2+}]$ and millimolar potassium, neutral pH, and low temperatures (0°C). If the phosphorylation from ATP is performed at pH 8.35, in the absence of potassium, but presence of high $[Mg^{2+}]/[Ca^{2+}]$ ratio, then the formation of $E_2\text{-}P$ is favored (forward reactions 3–5 in Figure 13.2). The acid-precipitated microsomal protein is further centrifuged at maximum speed in a refrigerated centrifuge (14,000 rpm at 4°C), washed with an ice-cold solution containing 7% (w/v) TCA and 1 mM H_3PO_4, centrifuged, and subjected to SDS-polyacrylamide acid-gel electrophoresis (pH 6.0) to separate the acid-stable phosphorylated Ca^{2+}-ATPases from other proteins [82, 83]. The radioactivity of the dried gels associated with the bands corresponding to the Ca^{2+}-ATPase is quantified by imaging using a Packard Cyclone Storage Phosphor System (Figure 13.1). Background phosphorylation levels are determined in the presence of excess EGTA (mM) and absence of Ca^{2+}.

These practical steps also apply to the phosphorylation reaction with $^{32}P_i$ ("back door" phosphorylation or reverse reaction 6 in Figure 13.2), leading to the formation of the $E_2\text{-}P$ intermediate, but the reaction conditions (pH 6.0, presence of Mg^{2+} and dimethyl sulfoxide, absence of K^+) are different from those mentioned above for ATP phosphorylation and the background phosphorylation level is measured in the presence of excess Ca^{2+} (mM).

When estimation of active site concentration is intended, microsomes (~10 to 30 μg of total protein) are incubated either in a reaction medium containing 40 mM MOPS/Tris (pH 7.0), 80 mM KCl, 5 mM $MgCl_2$, and 100 μM $CaCl_2$ prior to phosphorylation in the presence of 5 μM (final concentration) of [γ-^{32}P]ATP for 15 s at 0°C or in a reaction mixture containing 100 MES/Tris (pH 6.0), 10 mM $MgCl_2$, 2 mM EGTA, 30% (v/v) dimethyl sulfoxide (to increase the pump's affinity for P_i) followed by phosphorylation with 0.5 mM (final concentration) of $^{32}P_i$ for 10 min at 25°C. In both cases, phosphorylation is terminated by acid quenching as described above.

The steady-state kinetics of the partial reactions involving phosphorylation procedures can be studied by determining the phosphorylation levels as a function of a given substrate (Ca^{2+}, ATP, or P_i) or inhibitor (vanadate, TG, cyclopiazonic acid (CPA), or 2,5-di-(tert-butyl)-1,4-hydroquinone (tBHQ) are discussed in Section 13.4) concentration using ~0.5 pmol of expressed Ca^{2+}-ATPase for each experimental point. Ca^{2+} dependence of phosphorylation by [γ-^{32}P]ATP can also be followed at 25°C, but the phosphorylation is acid quenched immediately after 5 s. Taking as 100% the level of maximum phosphorylation reached, all data from the ligand titration of phosphorylation are normalized and fitted to the Hill equation (Equation 13.3), thereby allowing the extraction of the values for $K_{0.5}$ and Hill coefficient displayed by various SERCAs [28, 34, 35, 136].

The transient state kinetics of the partial reactions can be analyzed by measuring the levels of ^{32}P-phosphorylated intermediates accumulated or disappeared at serial time intervals and fitting the data for the time-courses of phosphorylation and dephosphorylation by first-order kinetic equations, or by using the kinetic simulation software SimZyme [95]. These nonlinear regression analyses allow the extraction of the rate constants for phosphorylation and dephosphorylation reactions. The protocols designed to enable the measurements of phosphorylation and dephosphorylation rates can be carried out either at 0°C by means of a manual mixing procedure (with an experimental limit set at 1 s for the first data point collected) [136, 161] or at 25°C by means of the rapid Bio-Logic quench-module QFM-5 (Bio-Logic Science Instruments, Claix, France), thus allowing essential kinetic measurements to be performed with microsomal Ca^{2+}-ATPases (0.05–0.2 pmol/experimental point) on a millisecond time scale [95].

The decay (dephosphorylation) rate of ADP-sensitive $E_1{\sim}P{\cdot}(Ca^{2+})_2$ phosphoenzyme accumulated with [γ-^{32}P]ATP under conditions discussed above can be examined at 0°C with 1 mM ADP and 1 mM EGTA (ADP chase) which promotes the dephosphorylation in the backward direction with yield of ATP and in the forward direction with 0.6 mM cold ATP (ATP chase; 1 mM EGTA can also be part of the chase solution) [83, 94, 136].

Implementation of the rapid quench-flow methodology (Figure 13.6) allows accurate transient kinetic measurements of the phosphorylation rate with enzyme initially accumulated in either Ca^{2+}-deprived E_2 form (Figure 13.6A) or Ca^{2+}-saturated $E_1{\cdot}(Ca^{2+})_2$ form (in Figure 13.6B), thereby providing unique information

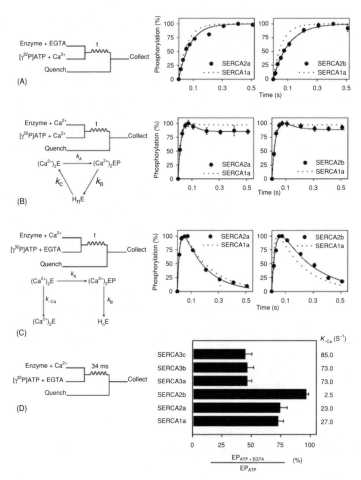

FIGURE 13.6 Rapid quench-flow methodology. Rapid kinetic phosphorylation measurements (at 25°C with [γ-^{32}P]ATP) were conducted with enzyme in both Ca^{2+}-deprived (diagram in panel A) and Ca^{2+}-saturated (diagram in panel B) forms to study the rate of E_2 to $E_1(Ca^{2+})_2$ transition. Diagram in panel (C) illustrates the protocol for monitoring a single turnover cycle, whereas diagram in panel (D) shows the protocol used for the determination of Ca^{2+} dissociation rates. Modified examples from our recent studies [34, 35] are shown at the right of each diagram. Simplified catalytic cycle schemes (below diagrams in panels B and C) were used to compute experimental data with SimZyme simulation software as previously described [95]. Panel (A), the lines show the best fits of a monoexponential function giving the rate constants (s^{-1}) of 22, 13, and 9.8 for SERCA1a, SERCA2a, and SERCA2b, respectively. Panel B, the lines were generated by computer simulation using the following rate constants (s^{-1}): SERCA1a ($k_A = 35$; $k_B = 5$; $k_C = 25$), SERCA2a ($k_A = 35$; $k_B = 6$; $k_C = 13$), and SERCA2b ($k_A = 35$; $k_B = 3$; $k_C = 10$). Panel C, the lines were generated by computer simulation using the following rate constants (s^{-1}): SERCA1a ($k_A = 35$; $k_B = 5$; $k_{-Ca} = 27$), SERCA2a ($k_A = 35$; $k_B = 6$; $k_C = 23$), and SERCA2b ($k_A = 35$; $k_B = 4$; $k_C = 2.5$). Panel (D), rate constants (s^{-1}) for Ca^{2+} dissociation (k_{-Ca}) shown at the right for the indicated SERCA1 [95], SERCA2 [35], and SERCA3 [34] isoforms were determined at 25°C as described [95].

concerning the E_2 to E_1 (or Ca^{2+} binding) transition rate (forward reaction 1 in Figure 13.2). Simultaneous addition of excess Ca^{2+} and [γ-^{32}P]ATP to the E_2 form (Figure 13.6A) initiates its transition to $E_1 \cdot (Ca^{2+})_2$ (reactions 1 and 2) followed by phosphorylation (reaction 3) to form $E_1 \sim P \cdot (Ca^{2+})_2$. Usually, the $E_2 \longrightarrow E_1 \cdot (Ca^{2+})_2$ transition is rate limiting in SERCA1a due to the conformational changes associated with enzyme deprotonation [97]. Therefore, the phosphorylation rate will be affected by any amino acid substitutions that slow or accelerate this transition in SERCA1a mutants. Furthermore, the experimental design shown in Figure 13.6B provides additional information on the $E_2 \longrightarrow E_1 \cdot (Ca^{2+})_2$ transition, because the enzyme recycles after the first formation of $E_1 \sim P \cdot (Ca^{2+})_2$, so that the overshoot reflects the accumulation of E_2 form. Hence, the overshoot becomes larger when $E_2 \longrightarrow E_1 \cdot (Ca^{2+})_2$ is slow.

Information on the rates of dephosphorylation (forward reactions 4–6 in Figure 13.2) and Ca^{2+} dissociation (reverse reaction 2) can be further obtained by monitoring of a single turnover cycle (Figure 13.6C). In addition, the Ca^{2+} dissociation rate can be independently determined using a slightly modified protocol illustrated in Figure 13.6D. In Figure 13.6C, the enzyme is only phosphorylated once (single turnover measurement), because Ca^{2+}-chelating EGTA is added simultaneously with [γ-^{32}P]ATP to the enzyme initially present in its Ca^{2+}-saturated $E_1 \cdot (Ca^{2+})_2$ form, thus preventing enzyme rephosphorylation after the dephosphorylation has occurred. The slope of the decay in Figure 13.6C reflects the rate of dephosphorylation, which is normally limited at neutral pH by the $E_1 \sim P \cdot (Ca^{2+})_2$ to E_2-P conversion (E_2-P to E_2 is rather fast). This type of experiment performed at 25°C provides a more physiological estimation of the dephosphorylation rate than that measured at 0°C using a manual mixing protocol. The ascendant part of the peak is determined by the rate constants for phosphorylation (k_A, measured according to Figure 13.6B) and Ca^{2+} dissociation (k_{-Ca}). The experimental data describing the disappearance of phosphoenzyme (Figure 13.6C) and the occurrence of the phosphorylation overshoot (Figure 13.6B) can be analyzed by computer simulation of the corresponding simplified catalytic cycle schemes (shown in Figure 13.6) using the kinetic simulation software SimZyme as described [95].

The experimental setup in Figure 13.6D is designed to determine independently the rate of Ca^{2+} dissociation from $E_1 \cdot (Ca^{2+})_2$ toward the cytoplasmic side. When EGTA is added together with [γ-^{32}P]ATP to Ca^{2+}-saturated enzyme, the $E_1 \cdot (Ca^{2+})_2$ species partitions between phosphorylation (forming $E_1 \sim P \cdot (Ca^{2+})_2$) and dissociation of the first Ca^{2+}, leading to the nonphosphorylatable $E_1 \cdot Ca^{2+}$ intermediate. Therefore, an increase of the rate of Ca^{2+} dissociation reduces the phosphorylation level detected after 34 ms and from this level the rate constant (k_{-Ca}) for Ca^{2+} dissociation can be calculated as described [95], by numerically solving the equation:

$$\frac{EP_{ATP+EGTA}}{EP_{ATP}} = \frac{\frac{k_A}{k_A + k_{-Ca}} \cdot (1 - e^{-(k_A + k_{-Ca})t})}{(1 - e^{-k_A t})} \tag{13.6}$$

where $EP_{ATP+EGTA}$ is the amount of phosphoenzyme measured at $t = 34$ ms after the simultaneous addition of [γ-^{32}P]ATP and EGTA to Ca^{2+}-saturated enzyme,

EP_{ATP} is the amount of phosphoenzyme measured after a 34-ms incubation of the Ca^{2+}-saturated enzyme with $[\gamma\text{-}^{32}P]ATP$ only, and k_A represents the pseudo-first-order rate constant for phosphorylation of $E_1 \cdot (Ca^{2+})_2$ (k_A determined as in Figure 13.6B).

All the above-mentioned methods for transient kinetic characterization using the rapid quench-flow methodology have been previously established in connection with studies on SERCA1a mutants [95], and we have directly applied them to the study of both SERCA2 (wild-type and Darier disease mutants) and SERCA3 isoforms [34, 35]. From such analyses, it became obvious that the kinetic differences existing between the various SERCA isoforms can have consequences for the Ca^{2+} binding both at the cytoplasmic and lumenal sites. For instance, SERCA3a, SERCA3b, and SERCA3c splice variants (characterized by fast rates of Ca^{2+} dissociation) in the E_1 form exhibit true low affinity for cytosolic Ca^{2+} [34], whereas the SERCA2b enzyme manifests a true high affinity for Ca^{2+} at cytosolic sites [35]. Several differences between SERCA1a and SERCA3 isoforms such as the pH dependence of ATPase activity, increased turnover rate at the alkaline-shifted pH optimum, increased sensitivity to vanadate inhibition at steady state and ionophore insensitivity could now be explained for SERCA3 isoforms by an enhancement of their rate of dephosphorylation from E_2-P phosphoenzyme intermediate to E_2, which is most pronounced at alkaline pH and increased with the length of the alternatively spliced C terminus (highest for the longest SERCA3b isoform) [34]. This partial reaction step is critically important for the regulation of the lumenal Ca^{2+} content by SERCA pumps. Relative to SERCA1a, SERCA2b is characterized by reduced dephosphorylation rates from $E_1 \sim P \cdot (Ca^{2+})_2$ to E_2 and from E_2-P to E_2, which can explain the reduced catalytic turnover rate in SERCA2b [35].

13.3.5.2 Ca^{2+} Occlusion Approach

The Ca^{2+} occlusion assay allows direct measurements of the Ca^{2+} binding properties of SERCAs and their mutants independently of their phosphorylation properties. This is unambiguously achieved through the use of the nonphosphorylating MgATP analog CrATP (β,γ-bidentate chromium(III) ATP) [186–188]. CrATP forms a very stable complex with the enzyme that probably mimics the transition state for phosphorylation. The ionic radius of Cr^{3+} (0.58 Å) is not significantly different from that of Mg^{2+} (0.65 Å). Following incubation in the presence of $^{45}Ca^{2+}$ and CrATP, the Ca^{2+}-ATPase slowly becomes inactivated due to the formation of the stable complex with the enzyme in the Ca^{2+} occluded state, i.e., the $E_1 \cdot (Ca^{2+})_2 \cdot CrATP$ complex. Following separation by size-exclusion chromatography of the Ca^{2+}-ATPase from other Ca^{2+} binding proteins, the occluded $^{45}Ca^{2+}$ can be quantified in the eluate containing the Ca^{2+}-ATPase [188, 189]. Using this method with mutants, Vilsen and Andersen have identified the amino acid residues making up the Ca^{2+} occlusion sites [189, 190]. Except for one residue (Glu^{908}), which is not critical for Ca^{2+} occlusion, the residues pinpointed in the studies with CrATP are the same as those seen to interact with bound Ca^{2+} ions in the Ca^{2+}-ATPase crystal structures obtained more recently [7]. The CrATP-induced $^{45}Ca^{2+}$ occlusion approach can also be used to indirectly assess the nucleotide-binding affinity, since a correlation between defective nucleotide binding and deficient $^{45}Ca^{2+}$ occlusion in the presence of CrATP could be demonstrated [139, 140].

13.3.5.3 Nucleotide-Binding Approach

The ATP-analogue TNP-8N$_3$-ATP (abbreviation for 2′,3′-O-(2,4,6-trinitrophenyl)-8-azidoadenosine 5′-triphosphate) is employed as photoaffinity probe to directly assess the true ATP affinity of Ca^{2+}-ATPases [191], because it can bind covalently to Lys^{492} in the active site of SERCA1a (nucleotide-binding domain or domain N) via its azido group upon exposure to light [192]. Ca^{2+}-ATPases bind TNP-8N$_3$-ATP with much higher affinity than ATP due to the presence of the trinitrophenyl moiety [193]. Interestingly, the nucleotide covalently bound to Lys^{492} is still able to perform the phosphorylation of the Ca^{2+}-pump at Asp^{351}. In order to measure the equilibrium binding of ATP to SERCA1a and its mutants, an ATP binding assay based on the specific photolabeling of Lys^{492} with [γ-^{32}P]TNP-8N$_3$-ATP and its inhibition by competing ATP has been developed by McIntosh et al. (1996) [194]. This approach was used to monitor the importance of specific residues in the catalytic active site on the binding of ATP [139–141, 194]. Lys^{492} can also be labeled by a non-nucleotide affinity label, pyridoxal 5′-phosphate, in a reaction that is competitively inhibited by ATP binding, but not by inorganic phosphate or acetyl phosphate [195].

Residue Lys^{684} of SERCA1a, which is located in the phosphorylation domain (domain P), can be labeled by adenosine 5′-triphosphopyridoxal (an ATP-analogue) in the presence of Ca^{2+}. Interestingly, in the absence of Ca^{2+} adenosine 5′-triphosphopyridoxal is able to equally label both Lys^{492} and Lys^{684} residues [196, 197]. These results imply that that the two residues are closer to each other when the enzyme is in the Ca^{2+}-deprived form (E_2) than in the Ca^{2+}-saturated form $E_1 \cdot (Ca^{2+})_2$. Analyses of the high-resolution models for the atomic structure obtained by x-ray crystallography of SERCA1a crystals in both Ca^{2+}-bound and Ca^{2+}-free forms have confirmed this conclusion [7, 57, 58].

There are many compounds that have been used as non-nucleotide affinity labels because they bind with high affinity to the cytoplasmic catalytic sites (nucleotide-binding and phosphorylation domains), but it is beyond the scope of this review to present them all.

13.4 CALCIUM PUMP INHIBITION AS A TOOL TO STUDY FUNCTION

13.4.1 INHIBITION OF SERCA BY ORGANIC COMPOUNDS

Specific inhibitory compounds are known for SERCA but not for the PMCA and SPCA Ca^{2+}-transport ATPases.

The most potent and specific inhibitor of SERCA pumps is TG. TG binds with high affinity to the Ca^{2+}-free E_2 state of all SERCA isoforms. Because its maximum effect is obtained at subnanomolar concentrations, it can be used to titrate the

SERCA molecules present in a sample (see Section 13.3.3) [137]. TG titration has been used, e.g., to quantitate SERCA levels in cardiac ventricular muscle [198]. Application of TG to cells has become a popular tool to empty intracellular Ca^{2+} stores in studies on the role of the ER in cytosolic Ca^{2+} homeostasis and on the role of ER lumenal Ca^{2+}, e.g., in protein secretion or in the regulation of capacitative Ca^{2+} entry. The importance of specific SERCA inhibition for obtaining new insights in the biology of Ca^{2+} has recently been reviewed [199, 200]. An interesting recent development is the synthesis of various TG analogues that can be conjugated to a peptide [201]. By coupling to a peptide that is a substrate for prostate-specific antigen protease, TG is preferentially liberated in close proximity to prostate cancer cells, effectively inhibiting the growth of these cells [202].

CPA and tBHQ are also effective inhibitors of SERCA-type Ca^{2+} pumps, although with a lower potency than TG. SERCA1 is half-maximally inhibited by tBHQ at 0.4 μM [203]. CPA has a higher affinity for SERCAs but only at ATP concentrations in the micromolar range. ATP protects the enzyme in a competitive manner [204]. At 3 mM ATP, Ma et al. (1999) observed half-maximal inhibition at 0.12 μM CPA [205]. The binding sites for TG and CPA partially overlap [205]. CPA and particularly tBHQ may be less specific for SERCAs than TG, especially at higher concentrations. For example, tBHQ and CPA produced an inhibition of the Ca^{2+} entry pathway in rat thymic lymphocytes [206]. tBHQ blocks N-, P- and Q-type Ca^{2+} currents in peripheral neurons, while CPA or TG does not affect these currents [207]. These compounds also differ from TG in that they are active on a broader range of SERCAs and SERCA-related pumps. For example, the TcSCA Ca^{2+}-ATPase of *Trypanosoma cruzi* is blocked by CPA but not by TG [208].

13.4.2 INHIBITION BY ANALOGUES OF INORGANIC PHOSPHATE

Various substrate analogues have been used to stabilize reaction intermediates of the enzymatic cycle. In addition to analogues of ATP, analogues of inorganic phosphate are particularly important. Analogues of natural ligands have extensively been used in the characterization of intermediate stages of the catalytic cycle. These compounds have also been important in the stabilization of different intermediate conformations for which the crystal structures have been determined (see Table 13.1).

13.4.2.1 Inhibition by Vanadate

Vanadate was first identified as an ATPase inhibitor when it was detected as a contaminant of ATP purified from muscle that potently inhibited Na^+-K^+-ATPase activity [209]. The K_i for inhibition was between 0.01 and 0.1 μM. Subsequently, it was found that vanadate at higher concentrations (in the micromolar range) also inhibits Ca^{2+}-transport ATPases. The Ca^{2+} pumps of the PMCA family are the most sensitive, with half-maximal inhibition at about 1 μM [210]. Inhibition of SERCA1 [211] or SPCA [125] requires tens of micromoles of vanadate. However, both SERCA2 and SERCA3 are about fivefold more sensitive to vanadate than SERCA1 [34, 35].

TABLE 13.1
Intermediate Conformational States of SERCA1a and Their Analogous State Crystal Structures Recently Obtained in the Presence of Various Stabilizing Ligands

Intermediate State[a]	Crystallized Analogous State[a]	Protein Data Bank (PDB) Accession Code[c]	Resolution (Å)	References
$E_1 \cdot (Ca^{2+})_2$	$E_1 \cdot (Ca^{2+})_2$	1EUL	2.6	57
		1SU4	2.5	7
$E_1 \cdot ATP \cdot (Ca^{2+})_2$	$E_1 \cdot AMPPCP \cdot (Ca^{2+})_2^b$	1T5S	2.6	59
		1VFP	2.7	61
$E_1 \cdot ADP \cdot (Ca^{2+})_2$	$E_1 \cdot AIF \cdot ADP \cdot (Ca^{2+})_2^b$	1T5T	2.9	59
		1WPE	2.8	60
$E_2 \cdot P_i$	$E_2 \cdot [MgF_4^{2-}]$	1WPG	2.3	60
E_2	$E_2 \cdot$Thapsigargin	1IWO	3.1	58

[a] Noncovalent binding of a ligand is represented by a dot in all intermediate and analogous states throughout the table.

[b] Crystallization was performed in the presence of Mg^{2+}, which is assumed to bind to the phosphorylated states or to the states with bound ATP or ADP or their analogues.

[c] The $E_1 \cdot (Ca^{2+})_2$ crystal structure was recently refined from 2.6 to 2.5 Å and, consequently, its initial accession code 1EUL replaced by 1SU4.

At neutral pH the orthovanadate ion exists predominantly as $H_2VO_4^-$, which is structurally similar to phosphate. It binds with high affinity to the phosphate acceptor site, thereby inhibiting the enzyme turnover. Whereas phosphate usually has a tetrahedral conformation, vanadate is presumed to have a pentacoordinated bipyramidal structure [212]. Therefore, vanadate most likely mimicks a transition state in which the phosphate is pentacoordinated, and which presents intermediate characteristics between the E_2-P and E_2-P_i states. SERCA also binds vanadate in the form of decavanadate [213]. The formation of decavanadate is favored at acid pH, while monovanadate is the major constituent at neutral and alkaline pH [214, 215]. Decavanadate binds to two sites: like monovanadate to the phosphate acceptor site, and in addition to a site that is part of the ATP-binding site [213, 216]. Both monovanadate and decavanadate induce the formation of crystalline arrays of Ca^{2+}-ATPase, with decavanadate more effective than monovanadate in this respect [213].

13.4.2.2 Inhibition by Fluoride Complexes of Aluminum and Beryllium

Fluoride at millimolar concentrations binds to SERCA Ca^{2+} pumps in the presence of magnesium ions [217]. The crystal structure of SERCA1a with bound magnesium fluoride shows that the compound is bound as MgF_4^{2-} and that an additional Mg^{2+} is present next to the phosphate analogue [61]. Interestingly, it has been shown that fluoride is a useful tool to study the uncoupled ATPase activity of the SERCA pump. In this mode of operation, the energy liberated by ATP hydrolysis is not used to transport Ca^{2+}, but it is instead dissipated as heat. The uncoupled pathway is mainly used after formation of a Ca^{2+} gradient. Fluoride specifically impairs the uncoupled ATPase activity, as shown by the fact that its inhibiting effect on ATPase activity increases with the Ca^{2+} load of the vesicles, and that concomitantly Ca^{2+} uptake is stimulated. Thus fluoride enhances the coupling between ATP consumption and Ca^{2+} transport [145].

In solutions of NaF or KF, aluminum and beryllium form ionic bonds with fluoride resulting in the formation of various complexes. The equilibrium between the different complexes depends on the F^- concentration and on the pH. In contrast to vanadate, phosphate and aluminum, beryllium is strictly tetrahedral. At neutral pH and at millimolar fluoride, the major species is BeF_3^- with a polarized water molecule, whereas aluminum forms species such as AlF_3, $AlF_3(OH)$, and [212]. Beryllium fluoride and aluminum fluoride inhibit Ca^{2+}-ATPases by forming stable analogues of the phosphoenzyme intermediate by binding to Ca^{2+}-free conformations of the ATPase. Beryllium fluoride stabilizes a state that is most similar to the E_2-P ground state in which the catalytic site is strongly hydrophobic and the Ca^{2+}-binding sites are accessible from the lumen. The complex with aluminum fluoride has lost these properties. It mimics a transition state of E_2-P hydrolysis. The hydrolyzed state (E_2-P_i) is represented by the protein containing magnesium fluoride [218], the crystal structure (Table 13.1) of which has been recently determined [61]. In the presence of Ca^{2+} and ADP, aluminum fluoride binds to SERCA pumps to form a stable ADP-E_1-P(Ca^{2+})$_2$ analogue with strong occlusion of both calcium ions and whereby the fluoroaluminate simulates the γ-phosphate of ATP [219, 220].

13.4.3 INHIBITION BY LANTHANIDES

Lanthanides have similar ionic radii to Ca^{2+}. Because of their trivalency, these ions bind with high affinity to Ca^{2+} sites. The fluorescent properties of europium and terbium, and the paramagnetic property of gadolinium have made lanthanide ions very useful as spectroscopic probes for exploring the cytosolic Ca^{2+}-binding sites in SERCA [19]. It has been shown that La^{3+} is an effective inhibitor of the steady-state turnover of Ca^{2+}-ATPases [19, 221]. Binding of La^{3+} does not prevent formation of the phosphorylated intermediate, but it tremendously slows down the dephosphorylation rate. Although La^{3+} can bind both to the Ca^{2+}-transport sites and the catalytic site, the latter interaction is actually responsible for the observed inhibition [19, 222, 223]. La^{3+} in the form of LaATP binds to the catalytic site of SERCA where it replaces Mg^{2+} as the catalytic ion required for phosphoryl transfer. As a consequence, the turnover of ATPase activity in the steady state is inhibited by the much slower than normal rate-limiting transition of the lanthanum phosphoenzyme to the ADP-insensitive E_2-P form [19, 221]. A similar mechanism has been demonstrated for PMCA pumps [221]. The fact that the lanthanum phosphoenzyme is characterized by an unusual extended lifetime has been experimentally exploited to enhance the phosphoprotein detection level. It is particularly suitable for PMCAs, which are characterized by a low level of phosphointermediate in steady-state conditions [224].

13.4.4 INHIBITION OF **SERCA3** BY **PL/IM430** ANTIBODY

The PL/IM430 monoclonal antibody is of special interest because, firstly, it is specific for human SERCA3 isoforms, and, secondly, it specifically inhibits their corresponding Ca^{2+} transporting activity. It may therefore be useful as a probe both for the dissection of SERCA3 isoform diversity at the functional level, and for studying mechanistic features of the catalytic cycle. The antibody was raised using intracellular membrane vesicles of human platelets as an immunogen and initially it was shown to inhibit Ca^{2+} uptake in internal vesicles isolated from human platelets because platelets show a high level of SERCA3 expression [225]. It is commercially available from Chemicon. When bound to human SERCA3, PL/IM430 inhibits the overexpressed pump maximally by about 80%. PL/IM430 does not interact with the other SERCA isoforms or with SERCA3 from other species [226, 227]. Earlier claims that the antibody would inhibit Ca^{2+} uptake and not affect the ATPase activity, and thus uncouple the pump [225], have not been confirmed [227]. The antibody recognizes a linearly noncontiguous set of residues within the actuator domain. It presumably prevents the catalytically critical movement of the actuator domain that accompanies the $E_1{\sim}P{\cdot}(Ca^{2+})_2$ to E_2-P transition (reactions 4 and 5 in Figure 13.2) [227].

13.4.5 INHIBITION BY RNAi EXPERIMENTS

RNA interference (RNAi) techniques provide a fast and easy way to knock down the function of genes with the potential for genome-wide reverse genetics [228]. RNAi involves targeted destruction of mRNAs triggered by double stranded RNAs that are cleaved into short 21–23 base pair duplexes by an RNAse III type called Dicer. The short RNAs, termed small interfering RNAs (siRNAs), act as triggers for targeted RNA degradation. One of the two strands is selectively incorporated into a complex of proteins called the RNA induced silencing complex (RISC). The incorporated small RNA guides the complex to the complementary mRNA sequence, which is followed by endonucleolytic cleavage of the target and recycling of RISC [229]. In mammalian cells, long strands of double stranded RNA activate interferon pathway genes, thus generating a strong cytotoxic response. However, it appears that such an effect is absent when synthetic small (21 to 23 nucleotides) siRNAs are used. The siRNAs can be synthesized *in vitro* and transfected in culture cells or alternatively plasmid or viral vectors can be used [230]. The synthetic siRNAs can be transfected in the cell with very high efficiency, but as they are degraded, the silenced genes can recover in time, limiting this approach to analysis for periods up to 6 days following a single transfection. More prolonged expression of short hairpin RNAs (shRNAs) from plasmid vectors can also be used. These vectors contain an inverted repeat of 19 to 29 nt encompassing the desired targeting sequence, separated by a short loop of 6 to 9 nt. Upon synthesis of the nascent RNA, it will form a stem-loop structure that is processed by Dicer to functional siRNAs; these siRNAs can then be taken up by RISC and direct sequence-specific degradation of target mRNA. The SPCA1 Ca^{2+} pump from the Golgi apparatus of HeLa cells has been successfully knocked down with plasmid-mediated RNAi [231].

Subsequently, the comparison between control cells and cells lacking SPCA1 revealed that endogenous SPCA1 is responsible for Ca^{2+} transport in a subfraction of the Golgi compartment; additionally, the baseline Ca^{2+} oscillations generated upon histamine stimulation occurred less frequently in HeLa cells lacking SPCA1 than in control cells, thus pointing to a certain contribution of SPCA1 in the shaping of the cytoplasmic calcium signal in these cells [231].

Morpholino antisense oligonucleotides present yet another novel and interesting tool for gene silencing both in culture, but most interestingly also in whole organisms where it was found extremely valuable for gene knockdown in developmental biology. Morpholino oligos are short chains of morpholino subunits comprised of a nucleic acid base, a morpholine ring, and a nonionic phosphorodiamidate intersubunit linkage (www.gene-tools.com). Morpholinos act via a steric block mechanism (RNA polymerase H-independent) with high mRNA binding affinity and specificity. They either can block the translation initiation complex (by targeting the 5' UTR through the first 25 bases of coding sequence) or they can block the nuclear splicing machinery (by targeting splice junctions in pre-mRNA). By injecting into a one-cell stage zebrafish embryo of a morpholino antisense oligonucleotide to the predicted translational start site of the *serca* gene (fish orthologue of human *ATP2A1*) fish were obtained with the same typical motility dysfunction described earlier as accordion (*acc*) phenotype [232]. Thus, the phenocopying provides a further argument that *serca* disruption causes the *acc* phenotype. Hence, such fish can be used as organism models for Brody disease where human SERCA1 gene (*ATP2A1*) is affected.

13.4.6   Inhibition of the *Plasmodium Falciparum* Ca^{2+} Pump by Artemisinins

Upon expression in *Xenopus* oocytes, the SERCA orthologue of *Plasmodium falciparum* (PfATP6) can be selectively and specifically inhibited by artemisinins, natural compounds (sesquiterpene lactones) present in sweet wormwood (*Artemisia annua*) [233]. Artemisinins are structurally and functionally similar to thapsigargin (see above Section 13.4.1). Activated artemisinins are the most potent antimalarials available, rapidly killing all asexual stages of *P. falciparum*.

13.5   CALCIUM PUMPS: PURIFICATION AND STRUCTURE

13.5.1   SERCA: Crystallization and 3D Structure

Both two- and three-dimensional crystals of SERCA1 can be obtained and analyzed respectively by cryoelectron microscopy and x-ray diffraction. Whereas the former method allows the reconstruction of the three-dimensional electron density maps at relatively lower resolution (6 to 15 Å), x-ray diffraction permits narrowing the limits to an unprecedented 2.6 to 3.1 Å [57–61]. SERCA1a was the first P-type ion pump that has been characterized at atomic resolution. Several different conformations approximating different catalytic intermediates can now be found in the database. The corresponding protein data bank (PDB) codes and the relevant references are given in Table 13.1. However, in order to understand the functional cycle of a molecular machine like the ATP-driven Ca^{2+} pump one needs to model the dynamics of the transconformations between these various static high-resolution crystal structures. This is where analyses of the structural flexibility of the ATPase by normal mode analysis can be of great help [234, 235]. It allows pinpointing the putative highly flexible hinge interfaces between the more rigid different domains (A, N, P, and M), which allow the pump to function as a Brownian molecular machine [236].

13.5.2 PMCA: CALMODULIN AFFINITY-CHROMATOGRAPHY PURIFICATION

One very fortunate property of the PMCA Ca^{2+} pump is its Ca^{2+}-dependent interaction with calmodulin. This provides an easy way to purify this ATPase with only a few easy steps. These essentially involve the preparation of a plasma-membrane-enriched fraction, solubilization of the membrane proteins by detergents, and application of the solubilized intrinsic membrane proteins on a calmodulin-affinity gel in the presence of Ca^{2+}. After extensive washing in the continued presence of Ca^{2+}, a highly enriched fraction of PMCA can be released from the gel by means of Ca^{2+}-chelating agents such as EGTA [184, 210, 237, 238]. One caveat is that this procedure leads to a delipidated and hence catalytically inactive PMCA preparation. This can however be easily remedied by supplementing the wash solutions with the appropriate phospholipids or by the readdition of the desired phospholipids after elution of the enzyme from the affinity gel. Furthermore, this procedure allows studying the effect of different types of lipids on the ATPase activity. Once purified, the ATPase can be reconstituted in artificial lipid vesicles to study its transport properties. For the latter type of studies it may be desirable to increase the size of the vesicles by means of freeze-thawing in order to avoid a rapid increase in lumenal $[Ca^{2+}]$ with consequent hindrance of further net Ca^{2+} uptake [239].

13.6 STUDIES OF GENETICALLY ALTERED MICE

The possibility to alter the expression of Ca^{2+} pumps or their regulators by ablation of one of their encoding genes (gene knockout) or conversely by boosting their expression in a constitutive or inducible manner (transgenic animals) opens new possibilities. A combination of such approaches even allows replacing a gene by its mutant form. For instance, after phospholamban was successfully ablated, the transgenic cardiac expression of mutant forms of phospholamban lacking either the target site for protein-kinase A-mediated phosphorylation (Ser[16]) or the target site for CaM kinase II (Thr[17]) made it possible to address the physiological function of the corresponding signaling pathways [240].

Other more subtle genetic modifications involve the targeted mutation of an allele in order to change the way its primary transcript is processed. This is the case of the SERCA2$^{b/b}$ mice described above (Section 13.2.1.2) [47, 241].

For a recent overview of the phenotypes of SERCA and PMCA knockout mice the reader is referred to a recent review [242].

13.7   CONCLUSIONS

In this chapter, our aim was to give an overview of the different methods available for the study of the Ca^{2+} pumps. This is not, and could not be, a comprehensive overview but we focused on these methods where our research groups gained experience. Recent developments in the field of molecular biology and the advent of high-throughput analysis techniques are now rapidly producing an overwhelming amount of data. These can only be interpreted with the help of the more classical techniques described in this overview. While it is clear that the Ca^{2+} pumps of the different families are targeted to different subcellular compartments, our knowledge on the targeting mechanisms and the specific functional characteristics of the various isoforms is still limited and requires further investigations. There is still a daunting task to unravel the tasks of each of the many Ca^{2+}-transport systems within the extremely complex networks of Ca^{2+} signaling in which Ca^{2+} plays a pivotal role.

ACKNOWLEDGMENTS

L.D. is Senior Research Associate. This work was supported by the Interuniversity Attraction Poles Progamme–Belgian Science Policy P5/05 and by the Fonds voor Wetenschappelijk Onderzoek Vlaanderen G.0166.04 grant. This review is dedicated to the loving memory of my parents, Eugenia Dode and Vasile Cristian Dode.

REFERENCES

1. Berridge, M.J., Lipp, P., and Bootman, M.D., The versatility and universality of calcium signalling, *Nature Rev. Mol. Cell Biol.*, 1, 11, 2000.
2. Berridge, M.J., Bootman, M.D., and Roderick, H.L., Calcium signalling: dynamics, homeostasis and remodelling, *Nature Rev. Mol. Cell Biol.*, 4, 517, 2003.
3. Ferrari, D., et al., Endoplasmic reticulum, Bcl-2 and Ca^{2+} handling in apoptosis, *Cell Calcium*, 32, 413, 2002.
4. de Meis, L. and Vianna, A.L., Energy interconversion by the Ca^{2+}-dependent ATPase of the sarcoplasmic reticulum, *Annu. Rev. Biochem.*, 48, 275, 1979.
5. Inesi, G., Sumbilla, C., and Kirtley, M.E., Relationship of molecular structure and function in Ca^{2+}-transport ATPase, *Physiol. Rev.*, 70, 749, 1990.
6. MacLennan, D.H., Rice, W.J., and Green, N.M., The mechanism of Ca^{2+} transport by sarco(endo)plasmic reticulum Ca^{2+}-ATPase, *J. Biol. Chem.*, 272, 28815, 1997.
7. Toyoshima, C. and Inesi, G., Structural basis of the ion pumping by Ca^{2+}-ATPase of the sarcoplasmic reticulum, *Annu. Rev. Biochem.*, 73, 269, 2004.
8. Stokes, D.L. and Green, N.M., Structure and function of the calcium pump, *Annu. Rev. Biophys. Biomol. Struct.*, 32, 445, 2003.
9. Wuytack, F., Raeymaekers, L., and Missiaen, L., Molecular physiology of the SERCA and SPCA pumps, *Cell Calcium*, 32, 279, 2002.

10. Wuytack, F., Raeymaekers, L., and Missiaen, L., PMR1/SPCA Ca^{2+} pumps and the role of the Golgi apparatus as a Ca^{2+} store, *Pflugers Arch. Eur. J. Physiol.*, 446, 148, 2003.
11. Van Baelen, K., et al., The Ca^{2+}/Mn^{2+} pumps in the Golgi apparatus, *Biochim. Biophys. Acta*, 1742, 103, 2004.
12. Missiaen, L., et al., SPCA1 pumps and Hailey-Hailey disease, *Biochem. Biophys. Res. Commun.*, 322, 1204, 2004.
13. Strehler, E.E. and Treiman, M., Calcium pumps of plasma membrane and cell interior, *Curr. Mol. Medicine*, 4, 323, 2004.
14. Ton, V.-K. and Rao, R., Functional expression of heterologuous proteins in yeast: insights into Ca^{2+} signalling and Ca^{2+}-transporting ATPases, *Am. J. Physiol. Cell Physiol.*, 287, C580, 2004.
15. Pedersen, P.L. and Carafoli, E., Ion motive ATPases. I. Ubiquity, properties, and significance to cell function, *Trends Biochem. Soc.*, 12, 146, 1987.
16. Møller, J.V., Juul, B., and le Maire, M., Structural organization, ion transport, and energy transduction of P-type ATPases, *Biochim. Biophys. Acta*, 1286, 1, 1996.
17. Axelsen, K.B. and Palmgren, M.G., Evolution of substrate specificities in the P-type ATPase superfamily, *J. Mol. Evol.*, 46, 84, 1998.
18. Palmgren, M.G. and Axelsen, K.B., Evolution of P-type ATPases, *Biochim. Biophys. Acta*, 1365, 37, 1998.
19. Bigelow, D.J. and Inesi, G., Contributions of chemical derivatization and spectroscopic studies to the characterization of the Ca^{2+} transport ATPase of sarcoplasmic reticulum, *Biochim. Biophys. Acta*, 1133, 323, 1992.
20. Dupont, Y. and Leigh, J.B., Transient kinetics of sarcoplasmic reticulum $Ca^{2+}+Mg^{2+}$ ATPase studied by fluorescence, *Nature*, 273, 396, 1978.
21. Guillain, F., et al., A direct fluorescence study of the transient steps induced by calcium binding to sarcoplasmic reticulum ATPase, *J. Biol. Chem.*, 255, 2072, 1980.
22. Dupont, Y., Low-temperature studies of the sarcoplasmic reticulum calcium pump. Mechanisms of calcium binding, *Biochim. Biophys. Acta*, 688, 75, 1982.
23. Henderson, I.M., et al., Binding of Ca^{2+} to the (Ca^{2+}-Mg^{2+})-ATPase of sarcoplasmic reticulum: equilibrium studies, *Biochem J.*, 297, 615, 1994.
24. Henderson, I.M., et al., Binding of Ca^{2+} to the (Ca^{2+}-Mg^{2+})-ATPase of sarcoplasmic reticulum: kinetic studies, *Biochem J.*, 297, 625, 1994.
25. Arrondo, J.L., et al., Infrared spectroscopic characterization of the structural changes connected with the E1 — E2 transition in the Ca^{2+}-ATPase of sarcoplasmic reticulum, *J. Biol. Chem.*, 262, 9037, 1987.
26. Barth, A., Kreutz, W., and Mantele, W., Ca^{2+} release from the phosphorylated and the unphosphorylated sarcoplasmic reticulum Ca^{2+}-ATPase results in parallel structural changes. An infrared spectroscopic study, *J. Biol. Chem.*, 272, 25507, 1997.
27. Barth, A., and Zscherp, C., Substrate binding and enzyme function investigated by infrared spectroscopy, *FEBS Lett.*, 477, 151, 2000.
28. Andersen, J.P., Dissection of the functional domains of the sarcoplasmic reticulum Ca^{2+}-ATPase by site-directed mutagenesis, *Biosci. Rep.*, 15, 243, 1995.
29. Wuytack, F., et al., A sarco/endoplasmic reticulum Ca^{2+}-ATPase 3-type Ca^{2+} pump is expressed in platelets, in lymphoid cells, and in mast cells, *J. Biol. Chem.*, 269, 1410, 1994.
30. Wuytack, F., et al., The SERCA3-type of organellar Ca^{2+} pumps, *Biosci. Rep.*, 15, 299, 1995.
31. Verboomen, H., et al., Modulation of SERCA2 activity: regulated splicing and interaction with phospholamban, *Biosci. Rep.*, 15, 307, 1995.

32. Andersen, J.P., and Vilsen, B., Structure-function relationships of cation translocation by Ca^{2+} and Na^+, K^+-ATPases studied by site-directed mutagenesis, *FEBS Lett.*, 359, 101, 1995.

33. Andersen, J.P., and Vilsen, B., Mutagenesis of sarcoplasmic reticulum Ca^{2+}-ATPase, *Trends Cardiovasc. Med.*, 8, 41, 1998.

34. Dode, L., et al., Dissection of the functional differences between sarco(endo)plasmic reticulum Ca^{2+}-ATPase (SERCA) 1 and 3 isoforms by steady-state and transient kinetic analyses, *J. Biol. Chem.*, 277, 45579, 2002.

35. Dode, L., et al., Dissection of the functional differences between sarco(endo)plasmic reticulum Ca^{2+}-ATPase (SERCA) 1 and 2 isoforms and characterization of Darier disease (SERCA2) mutants by steady-state and transient kinetic analyses, *J. Biol. Chem.*, 278, 47877, 2003.

36. Axelsen, K.B., The P-type ATPase Database, at *http://www.patbase.kvl.dk*.

37. Sambrook, J., and Russell, D.W., *Molecular Cloning — A Laboratory Manual*, 3rd Edition, Cold Spring Harbor Laboratory Press, Cold Spring Harbor, New York, 2001.

38. Strehler, E. and Zacharias, D.A., Role of alternative splicing in generating isoform diversity among plasma membrane calcium pumps, *Physiol. Rev.*, 81, 21, 2001.

39. Korczak, B., et al., Structure of the rabbit fast-twitch skeletal muscle Ca^{2+}-ATPase gene, *J. Biol. Chem.*, 263, 4813, 1988.

40. Gélébart, P., et al., Identification of a new SERCA2 splice variant regulated during monocytic differentiation, *Biochem. Biophys. Res. Comm.*, 303, 676, 2003.

41. Dode, L., et al., cDNA cloning, expression and chromosomal localization of the human sarco/endoplasmic reticulum Ca^{2+}-ATPase 3 gene, *Biochem. J.*, 318, 689, 1996.

42. Dode, L., et al., Structure of the human sarco/endoplasmic reticulum Ca^{2+}-ATPase 3 gene, Promoter analysis and alternative splicing of the SERCA3 pre-mRNA, *J. Biol. Chem.*, 273, 13982, 1998.

43. Martin, V., et al., Three novel sarco/endoplasmic reticulum Ca^{2+}-ATPase (SERCA) 3 isoforms, *J. Biol. Chem.*, 277, 24442, 2002.

44. Bobe R., et al., Identification, expression, function, and localization of a novel (sixth) isoform of the human sarco/endoplasmic reticulum Ca^{2+}ATPase 3 gene, *J. Biol. Chem.*, 279, 24297, 2004.

45. Ribadeau-Dumas, A., et al., Sarco(endo)plasmic reticulum Ca^{2+}-ATPase (SERCA2) gene products are regulated post-transcriptionally during rat cardiac development, *Cardiovasc. Res.*, 43, 426, 1999.

46. van den Hoff, M.J.B. and Moorman, A.F.M., Measure is treasure, *Cardiovasc. Res.*, 43, 288, 1999.

47. Ver Heyen, M., et al., Replacement of the muscle-specific sarcoplasmic reticulum Ca^{2+}-ATPase isoform SERCA2a by the nonmuscle SERCA2b homologue causes mild concentric hypertrophy and impairs contraction-relaxation of the heart, *Circ. Res.*, 89, 838, 2001.

48. Braz, J.C., et al., PKC-alpha regulates cardiac contractility and propensity toward heart failure, *Nat Med.*, 10, 248, 2004.

49. Vanoevelen, J., et al., The secretory pathway Ca^{2+}/Mn^{2+} - ATPase 2 is a golge-localized pump with high affinity for Ca^{2+}ions. *J. Biol. Chem.*, 280, 22800, 2005.

50. Baba-Aissa, F., et al., Purkinje neurons express the SERCA3 isoform of organellar-type Ca^{2+} ATPases, *Mol. Brain Res.*, 41, 169, 1996.

51. Fairclough, R.J., et al., Effect of Hailey-Hailey disease mutations on the function of a new variant of human secretory pathway Ca^{2+}/Mn^{2+}-ATPase (hSPCA1), *J. Biol. Chem.*, 278, 24721, 2003.

52. Kuo, T.H., et al., Co-ordinated regulation of the plasma membrane calcium pump and the sarco(endo)plasmic reticular calcium pump gene expression by Ca^{2+}, *Cell Calcium*, 21, 399, 1997.

53. Hu, P., et al., Transcription rates of SERCA and phospholamban genes change in response to chronic stimulation of skeletal muscle, *Biochim. Biophys. Acta*, 1395, 121, 1998.

54. Misquitta, C.M., et al., Control of SERCA2a Ca^{2+} pump mRNA stability by nuclear proteins: role of domains in the 3'-untranslated region, *Cell Calcium*, 37, 17, 2005.

55. Van Den Bosch, L., et al., Regulation of splicing is responsible for the expression of the muscle-specific 2a isoform of the sarco/endoplasmic-reticulum Ca^{2+}-ATPase, *Biochem. J.*, 302, 559, 1994.

56. Mertens, L., et al., Sequence and spatial requirements for regulated muscle-specific processing of the sarco/endoplasmic-reticulum Ca^{2+}-ATPase 2 gene transcript, *J. Biol. Chem.*, 270, 11004, 1995.

57. Toyoshima, C., et al., Crystal structure of the calcium pump of sarcoplasmic reticulum at 2.6 Å resolution, *Nature*, 405, 647, 2000.

58. Toyoshima, C. and Nomura, H., Structural changes in the calcium pump accompanying the dissociation of calcium, *Nature*, 418, 605, 2002.

59. Sørensen, T.L.-M., Møller, J.V., and Nissen, P., Phosphoryl transfer and calcium ion occlusion in the calcium pump, *Science*, 304, 1672, 2004.

60. Toyoshima, C., Nomura, H., and Tsuda, T., Lumenal gating mechanism revealed in calcium pump crystal structures with phosphate analogues, *Nature*, 423, 361, 2004.

61. Toyoshima, C. and Mizutani, T., Crystal structure of the calcium pump with a bound ATP analogue, *Nature*, 430, 529, 2004.

62. Schwede, T., et al., SWISS-MODEL: an automated protein homology-modeling server, *Nucleic Acids Res.*, 31, 3381, 2003.

63. Zubrzycka-Gaarn, E., Monoclonal antibodies to the $Ca^{2+} + Mg^{2+}$ dependent ATPase of sarcoplasmic reticulum identify polymorphic forms of the enzyme and indicate the presence in the enzyme of a classical high-affinity Ca^{2+} binding side, *J. Bioenerg. Biomembr.*, 16, 441, 1984.

64. Jorgensen, A.O., et al., A monoclonal antibody to the Ca^{2+}-ATPase of cardiac sarcoplasmic reticulum cross-reacts with slow type I but not with fast type II canine skeletal muscle fibers: an immunocytochemical and immunochemical study, *Cell Motil. Cytoskel.*, 9, 164, 1988.

65. Leberer, E. and Pette, D., Immunochemical quantification of sarcoplasmic reticulum Ca^{2+}-ATPase, of calsequestrin and of parvalbumin in rabbit skeletal muscles of defined fiber composition, *Eur. J. Biochem.*, 156, 489, 1986.

66. Clarke, D.M., et al., Functional consequences of glutamate, aspartate, glutamine, and asparagines mutations in the stalk sector of the Ca^{2+}-ATPase of sarcoplasmic reticulum, *J. Biol. Chem.*, 264, 11246, 1989.

67. Bishop, J.E., et al., (Iodoacetamido)fluorescein labels a pair of proximal cysteines on the calcium ATPase of sarcoplasmic reticulum, *Biochemistry*, 27, 5233, 1988.

68. Zador, E., unpublished data, 2005.

69. Eggermont, J.A., et al., Expression of endoplasmic reticulum Ca^{2+}-pump isoforms and of phospholamban in pig smooth-muscle tissues, *Biochem. J.*, 271, 649, 1990.

70. Wuytack, F., et al., Antibodies against the non-muscle isoform of the endoplasmic reticulum Ca^{2+}-transport ATPase, *Biochem. J.*, 264, 765, 1989.

71. Kovács, T., et al., All three splice variants of the human sarco/endoplasmic reticulum Ca^{2+}-ATPase 3 gene are translated to proteins: a study of their co-expression in platelets and lymphoid cells, *Biochem. J.*, 358, 559, 2001.

72. Geisler, M., et al., Molecular aspects of higher plant P-type Ca^{2+}-ATPases, *Biochim. Biophys. Acta*, 1465, 52, 2000.

73. Bonza, M.C., et al., *At*-ACA8 encodes a plasma membrane-localized calcium-ATPase of *Arabidopsis* with a calmodulin-binding domain at the N terminus, *Plant Physiol.*, 123, 1495, 2000.

74. Wu, Z., et al., An endoplasmic reticulum-bound Ca^{2+}/Mn^{2+} pump ECA1, supports plant growth and confers tolerance to Mn^{2+} stress, *Plant Physiol.*, 130, 128, 2002.

75. Antebi, A. and Fink, G.R., The yeast Ca^{2+}-ATPase homologue, PMR1, is required for normal Golgi function and localizes in a novel Golgi-like distribution, *Mol. Biol. Cell*, 3, 633, 1992.

76. Van Baelen, K., et al., The Golgi PMR1 P-type ATPase of *Caenorhabditis elegans*. Identification of the gene and demonstration of calcium and manganese transport, *J. Biol. Chem.*, 276, 10683, 2001.

77. Xiang, M., Mohamalawari, D., and Rao, R., A novel isoform of the secretory pathway Ca^{2+}, Mn^{2+}-ATPase, hSPCA2, has unusual properties and is expressed in brain, *J. Biol. Chem.*, 1280, 11608, 2005.

78. Stauffer, T.P., Guerini, D., and Carafoli, E., Tissue distribution of the four gene products of the plasma membrane Ca^{2+} pump. A study using specific antibodies, *J. Biol. Chem.*, 270, 12184, 1995.

79. Filoteo, A.G., et al., Plasma membrane Ca^{2+} pump in rat brain. Patterns of alternative splices seen by isoform-specific antibodies, *J. Biol. Chem.*, 272, 23741, 1997.

80. Spencer, G.G., et al., Expression of isoforms of internal Ca^{2+} pump in cardiac, smooth muscle and non-muscle tissues, *Biochim. Biophys. Acta*, 1063, 15, 1991.

81. Mountian, I., et al., Expression of Ca^{2+} transport genes in platelets and endothelial cells in hypertension, *Hypertension*, 37, 135, 2001.

82. Sarkadi, B., et al., Molecular characterization of the *in situ* red cell membrane calcium pump by limited proteolysis, *J. Biol. Chem.*, 261, 9552, 1986.

83. Andersen, J.P., et al., Functional consequences of mutations in the beta-strand sector of the Ca^{2+}-ATPase of sarcoplasmic reticulum, *J. Biol. Chem.*, 264, 21018, 1989.

84. Schatzmann, H.J., The calcium pump of the surface membrane and of the sarcoplasmic reticulum, *Annu. Rev. Physiol.*, 51, 473, 1989.

85. Vangheluwe, P., Raeymaekers, L., and Wuytack, F., unpublished data, 2005.

86. Hasselbach, W. and Makinose, M., Die calciumpumpe der "Erschlaffungsgrana" des Muscles und ihre Abhängigkeit von der ATP-Spaltung, *Biochem. Z.*, 333, 518, 1961.

87. Ebashi, S., Calcium binding activity of vesicular relaxing factor, *J. Biochem.*, 50, 236, 1961.

88. Ebashi, S. and Lipmann, F., Adenosine trisphosphate-linked concentration of calcium ions in a particulate fraction of rabbit muscle, *J. Cell Biol.*, 14, 389, 1962.

89. MacLennan, D.H., Purification and properties of an adenosine triphosphatase from sarcoplasmic reticulum, *J. Biol. Chem.*, 245, 4508, 1970.

90. Martonosi, A.N., The structure and interactions of Ca^{2+}-ATPase, *Biosci. Rep.*, 15, 263, 1995.

91. de Meis, L., *The Sarcoplasmic Reticulum: Transport and Energy Transduction*, John Wiley and Sons, New York, USA, 1981.

92. Jencks, W.P., The mechanism of coupling chemical and physical reactions by the calcium ATPase of sarcoplasmic reticulum and other coupled vectorial systems, *Biosci. Rep.*, 15, 283, 1995.

93. Jencks, W.P., How does a calcium pump pump calcium?, *J. Biol. Chem.*, 264, 18855, 1989.

94. Andersen, J.P., et al., Importance of transmembrane segment M3 of the sarcoplasmic reticulum Ca^{2+}-ATPase for control of the gateway to the Ca^{2+} sites, *J. Biol. Chem.*, 276, 23312, 2001.

95. Sørensen, T.L.-M., et al., Fast kinetic analyses of conformational changes in mutants of the Ca^{2+}-ATPase of sarcoplasmic reticulum, *J. Biol. Chem.*, 275, 5400, 2000.

96. Inesi, G., et al., Cooperative calcium binding and ATPase activation in sarcoplasmic reticulum vesicles, *J. Biol. Chem.*, 255, 3025, 1980.

97. Forge, V., Mintz, E., and Guillain, F., Ca^{2+} binding to sarcoplasmic reticulum ATPase revisited. I. Mechanism of affinity and cooperative modulation by H^+ and Mg^{2+}, *J. Biol. Chem.*, 268, 10953, 1993.

98. Forge, V., Mintz, E., and Guillain, F., Ca^{2+} binding to sarcoplasmic reticulum ATPase revisited. II. Equilibrium and kinetic evidence for a two-route mechanism, *J. Biol. Chem.*, 268, 10961, 1993.

99. Girardet, J.L. and Dupont, Y., Ellipticity changes of the sarcoplasmic reticulum Ca^{2+}-ATPase induced by cation binding and phosphorylation, *FEBS Lett.*, 296, 103, 1992.

100. Pick, U. and Karlish, S.J., Indications for an oligomeric structure and for conformational changes in sarcoplasmic reticulum Ca^{2+}-ATPase labeled selectively with fluorescein, *Biochim. Biophys. Acta*, 626, 255, 1980.

101. Wakabayashi, S. and Shigekawa, M., Mechanism for activation of the 4-nitrobenzo-2-oxa-1,3-diazole-labeled sarcoplasmic reticulum ATPase by Ca^{2+} and its modulation by nucleotides, *Biochemistry*, 29, 7309, 1990.

102. Makinose, M., Phosphoprotein formation during osmo-chemical energy conversion in the membrane of the sarcoplasmic reticulum, *FEBS Lett.*, 25, 113, 1972.

103. Makinose, M., Possible functional states of the enzyme of the sarcoplasmic calcium pump, *FEBS Lett.*, 37, 140, 1973.

104. Madeira, V.M., Alkalinization within sarcoplasmic reticulum during the uptake of calcium ions, *Arch. Biochem. Biophys.*, 193, 22, 1979.

105. Chiesi, M. and Inesi, G., Adenosine 5'-triphosphate dependent fluxes of manganese and hydrogen ions in sarcoplasmic reticulum vesicles, *Biochemistry*, 19, 2912, 1980.

106. Yamaguchi, M. and Kanazawa, T., Protonation of the sarcoplasmic reticulum Ca^{2+}-ATPase during ATP hydrolysis, *J. Biol. Chem.*, 259, 9526, 1984.

107. Yamaguchi, M. and Kanazawa, T., Coincidence of H^+ binding and Ca^{2+} dissociation in the sarcoplasmic reticulum Ca^{2+}-ATPase during ATP hydrolysis, *J. Biol. Chem.*, 260, 4896, 1985.

108. Levy, D., et al., Evidence for proton countertransport by the sarcoplasmic reticulum Ca^{2+}-ATPase during calcium transport in reconstituted proteoliposomes with low ionic permeability, *J. Biol. Chem.*, 265, 19524, 1990.

109. Cornelius, F. and Møller, J.V., Electrogenic pump current of sarcoplasmic reticulum Ca^{2+}-ATPase reconstituted at high lipid/protein ratio, *FEBS Lett.*, 284, 46, 1991.

110. Yu, X., et al., H^+ countertransport and electrogenicity of the sarcoplasmic reticulum Ca^{2+} pump in reconstituted proteoliposomes, *Biophys. J.*, 64, 1232, 1993.

111. Yu, X., Hao, L., and Inesi, G., A pK change of acidic residues contributes to cation countertransport in the Ca-ATPase of sarcoplasmic reticulum. Role of H^+ in Ca^{2+}-ATPase countertransport, *J. Biol. Chem.*, 269, 16656, 1994.

112. Meissner, G. and Young, R.C., Proton permeability of sarcoplasmic reticulum vesicles, *J. Biol. Chem.*, 255, 6814, 1980.

113. Shigekawa, M. and Dougherty, J.P., Reaction mechanism of Ca^{2+}-dependent ATP hydrolysis by skeletal muscle sarcoplasmic reticulum in the absence of added alkali metal salts. III. Sequential occurrence of ADP-sensitive and ADP-insensitive phosphoenzymes, *J. Biol. Chem.*, 253, 1458, 1978.

114. Shigekawa, M. and Dougherty, J.P., On the mechanism of Ca^{2+}-dependent adenosine triphosphatase of sarcoplasmic reticulum. Occurrence of two types of phosphoenzyme intermediate in the presence of KCl, *J. Biol. Chem.*, 254, 4726, 1979.

115. Kaufmann, R.J., et al., The phosphorylation state of eukaryotic initiation factor 2 alters translational efficiency of specific mRNAs, *Mol. Cell Biol.*, 9, 946, 1989.

116. Tavernier, J., et al., Deletion mapping of the inducible promoter of human IFN-beta gene, *Nature*, 301, 634, 1983.

117. Huylebroek, D., et al., High-level transient expression of influenza virus proteins from a series of SV40 late and early replacement vectors, *Gene*, 66, 163, 1988.

118. Verboomen, H., et al., Functional difference between SERCA2a and SERCA2b Ca^{2+} pumps and their modulation by phospholamban, *Biochem. J.*, 286, 591, 1992.

119. Verboomen, H., et al., The functional importance of the extreme C-terminal tail in the gene 2 organellar Ca^{2+}-transport ATPase (SERCA2a/b), *Biochem. J.*, 303, 979, 1994.

120. Lytton, J. and MacLennan, D.H., Molecular cloning of cDNAs from human kidney coding for two alternatively spliced products of the cardiac Ca^{2+}-ATPase gene, *J. Biol. Chem.*, 263, 15024, 1988.

121. Chen, C. and Okayama, H., High-efficiency transformation of mammalian cells by plasmid DNA, *Mol. Cell Biol.*, 7, 2745, 1987.

122. Shigekawa, K. and Dower, W.J., Electroporation of eukaryotes and prokaryotes: a general approach to the introduction of macromolecules into cells, *BioTechniques*, 6, 742, 1988.

123. Felgner, P.L., et al., Lipofection: a highly efficient, lipid-mediated DNA-transfection procedure, *Proc. Natl. Acad. Sci. U.S.A.*, 84, 7413, 1987.

124. Durocher, Y., Perret, S., and Kamen, A., High-level and high-throughout recombinant protein production by transient transfection of suspension-growing human 293-EBNA1 cells, *Nucleic Acids Res.*, 30, 1, 2002.

125. Strock, C., et al., Direct demonstration of Ca^{2+} binding defects in sarco-endoplasmic reticulum Ca^{2+} ATPase mutants overexpressed in COS-1 cells transfected with adenovirus vectors, *J. Biol. Chem.*, 273, 15104, 1998.

126. Sumbilla, C., et al., Comparison of SERCA1 and SERCA2a expressed in COS-1 cells and cardiac myocytes, *Am. J. Physiol.*, 277, H2381, 1999.

127. Sorin, A., Rosas, G., and Rao, R., PMR1, a Ca^{2+}-ATPase in yeast Golgi, has properties distinct from sarco/endoplasmic reticulum and plasma membrane calcium pumps, *J. Biol. Chem.*, 272, 9895, 1997.

128. Wei, Y., et al., Phenotypic screening of mutations in Pmr1, the yeast secretory pathway Ca^{2+}/Mn^{2+}-ATPase, reveals residues critical for ion selectivity and transport, *J. Biol. Chem.*, 275, 23927, 2000.

129. Mandal, D., Woolf, T.B., and Rao, R., Manganese selectivity of Pmr1, the yeast secretory pathway ion pump, is defined by residue Gln^{783} in transmembrane segment 6. Residue Asp^{778} is essential for cation transport, *J. Biol. Chem.*, 275, 23933, 2000.

130. Mandal, D., Rulli, S.J., and Rao, R., Packing interactions between transmembrane helices alter ion selectivity of the yeast Golgi Ca^{2+}/Mn^{2+}-ATPase PMR1, *J. Biol. Chem.*, 278, 35292, 2003.

131. Ton, V.-K., et al., Functional expression in yeast of the human secretory pathway Ca^{2+}, Mn^{2+}-ATPase defective in Hailey-Hailey disease, *J. Biol. Chem.*, 277, 6422, 2002.

132. Lenoir, G., et al., Overproduction in yeast and rapid efficient purification of the rabbit SERCA1a Ca^{2+}-ATPase, *Biochim. Biophys. Acta*, 1560, 67, 2002.

133. Guerini, D., Pan, B., and Carafoli, E., Expression, purification, and characterization of isoform 1 of the plasma membrane Ca^{2+} pump, *J. Biol. Chem.*, 278, 38141, 2003.

134. Parys, J.B., et al., Subcellular fractionation and intracellular calcium stores, in: *Calcium Signaling (CRC Methods in Signal Transduction Series)*, edited by Putney J.W., CRC Press, Boca Raton, FL, 71, 1999, chap. 3.

135. Maruyama, K. and MacLennan, D.H., Mutation of aspartic acid-351, lysine-352, and lysine-515 alters the Ca^{2+} transport activity of the Ca^{2+}-ATPase expressed in COS-1 cells, *Proc. Natl. Acad. Sci. U.S.A.*, 85, 3314, 1988.

136. Sørensen, T.L.-M., Vilsen, B., and Andersen, J.P., Mutation $Lys^{758} \longrightarrow$ Ile of the sarcoplasmic reticulum Ca^{2+}-ATPase enhances dephosphorylation of E_2P and inhibits the E_2 to E_1Ca_2 transition, *J. Biol. Chem.*, 272, 30244, 1997.

137. Sagara, Y. and Inesi, G., Inhibition of sarcoplasmic reticulum Ca^{2+} transport ATPase by thapsigargin at subnanomolar concentrations, *J. Biol. Chem.*, 266, 13503, 1991.

138. Mintz, E. and Guillain, F., How do Ca^{2+} ions pass through the sarcoplasmic reticulum membrane, *Biosci. Rep.*, 15, 377, 1995.

139. Clausen, J.D., et al., Importance of Thr-353 of the conserved phosphorylation loop of the sarcoplasmic reticulum Ca^{2+}-ATPase in MgATP binding and catalytic activity, *J. Biol. Chem.*, 276, 35741, 2001.

140. Clausen, J.D., et al., Importance of conserved N-domain residues Thr^{441}, Glu^{442}, Lys^{515}, Arg^{560}, and Leu^{562} of sarcoplasmic reticulum Ca^{2+}-ATPase for MgATP binding and subsequent catalytic steps. Plasticity of the nucleotide-binding site, *J. Biol. Chem.*, 278, 20245, 2003.

141. McIntosh, D.B., et al., Roles of conserved P domain residues and Mg^{2+} in ATP binding in the ground and Ca^{2+}-activated states of sarcoplasmic reticulum Ca^{2+}-ATPase, *J. Biol. Chem.*, 279, 32515, 2004.

142. Inesi, G. and de Meis, L., Regulation of steady state filling in sarcoplasmic reticulum. Roles of back-inhibition, leakage, and slippage of the calcium pump, *J. Biol. Chem.*, 264, 5929, 1989.

143. de Meis, L., Uncoupled ATPase activity and heat production by the sarcoplasmic reticulum Ca^{2+}-ATPase. Regulation by ADP, *J. Biol. Chem.*, 276, 25078, 2001.

144. Sumbilla, C., et al., The slippage of the Ca^{2+} pump and its control by anions and curcumin in skeletal and cardiac sarcoplasmic reticulum, *J. Biol. Chem.*, 277, 13900, 2002.

145. Reis, M., et al., Correlation between uncoupled ATP hydrolysis and heat production by the sarcoplasmic reticulum Ca^{2+}-ATPase, Coupling effect of fluoride, *J. Biol. Chem.*, 276, 42793, 2001.

146. Fleischer, S. and Fleischer, B., *Methods in Enzymology, Volume 157. Biomembranes, Part Q. ATP-Driven Pumps and Related Transport: Calcium, Proton, and Potassium Pumps*, Academic Press, San Diego, 1988.

147. Raeymaekers, L., et al., Isolation of a plasma-membrane fraction from gastric smooth muscle. Comparison of the calcium uptake with that in endoplasmic reticulum, *Biochem. J.*, 210, 315, 1983.

148. Fabiato, A., Computer programs for calculating total from specified free or free from specified total ionic concentrations in aqueous solutions containing multiple metals and ligands, *Methods Enzymol.*, 157, 378, 1988.

149. Martell, A.E. and Smith, R.M., *Critical Stability Constants*, Plenum Press, New York, NY, 1974.

150. Bers, D.M., Patton, C.W., and Nuccitelli, R., A practical guide to the preparation of Ca^{2+} buffers, *Methods Cell Biol.*, 40, 3, 1994.

151. Inesi, G. and Scarpa, A., Fast kinetics of adenosine trisphosphate dependent Ca^{2+} uptake by fragmented sarcoplasmic reticulum, *Biochemistry*, 11, 356, 1972.

152. Durham, A.C. and Walton, J.M., A survey of the available colorimetric indicators for Ca^{2+} and Mg^{2+} ions in biological experiments, *Cell Calcium*, 4, 47, 1983.

153. Poenie, M., Fluorescent calcium indicators based on BAPTA, in: *Calcium Signaling (CRC Methods in Signal Transduction Series)*, edited by Putney J.W., CRC Press, Boca Raton, FL, 71, 1999, chap. 1.

154. Simpson, A.W.M., Fluorescent measurement of $[Ca^{2+}]_c$, in: *Calcium Signaling Protocols, (Methods in Molecular Biology, volume 114)* edited by Lambert D.G., Humana Press, Totowa, NJ, 1999, chap. 1.

155. Kargacin, M.E. and Kargacin, G.J., Methods for determining cardiac sarcoplasmic reticulum Ca^{2+} pump kinetics from fura 2 measurements, *Am. J. Physiol.*, 267, C1145, 1994.

156. Kargacin, G.J., Aschar-Sobbi, R., and Kargacin, M.E., Inhibition of SERCA2 Ca^{2+}-ATPases by Cs^+, *Pflügers Archiv Eur. J. Physiol.*, 449, 356, 2005.

157. Haugland, R.P., *Handbook of Fluorescent Probes and Research Products*, 9th Ed Molecular Probes (www.probes.com), U.S.A, 2002.

158. Meissner G. and Fleischer S., Characterization of sarcoplasmic reticulum from skeletal muscle, *Biochim. Biophys. Acta*, 241, 356, 1971.

159. Vilsen, B., et al., Functional consequences of proline mutations in the cytoplasmic and transmembrane sectors of the Ca^{2+}-ATPase of sarcoplasmic reticulum, *J. Biol. Chem.*, 264, 21024, 1989.

160. Dode, L., unpublished data, 2002.

161. Andersen, J.P., Functional consequences of alterations to amino acids at the M5S5 boundary of the Ca^{2+}-ATPase of sarcoplasmic reticulum, Mutation $Tyr^{763} \longrightarrow$ Gly uncouples ATP hydrolysis from Ca^{2+} transport, *J. Biol. Chem.*, 270, 908, 1995.

162. Lytton, J., et al., Functional comparison between isoforms of the sarcoplasmic or endoplasmic reticulum family of calcium pumps, *J. Biol. Chem.*, 267, 14483, 1992.

163. Vanoevelen, J., et al., Inositol trisphosphate producing agonists do not mobilize the thapsigargin-insensitive part of the endoplasmic reticulum and Golgi Ca^{2+} store, *Cell Calcium*, 35, 115, 2004.

164. Orci, L., et al., Heterogeneous distribution of filipin-cholesterol complexes across the cisternae of the Golgi apparatus, *Proc. Natl. Acad. Sci. U.S.A.*, 78, 293, 1981.

165. Burgess, G.M., et al., Calcium pools in saponin-permeabilized guinea pig hepatocytes, *J. Biol. Chem.*, 258, 15336, 1983.

166. Fiskum, G., Intracellular levels and distribution of Ca^{2+} in digitonin-permeabilized cells, *Cell Calcium*, 6, 25, 1985.

167. Champeil, P., et al., Fast kinetics of calcium release induced by myo-inositol trisphosphate in permeabilized rat hepatocytes, *J. Biol. Chem.*, 264, 17665, 1989.

168. Hofer, A.M. and Machen, T.E., Technique for *in situ* measurement of calcium in intracellular inositol 1,4,5-trisphosphate-sensitive stores using the fluorescent indicator mag-fura–2, *Proc. Natl. Acad. Sci. U.S.A.*, 90, 2598, 1993.

169. van de Put, F.H. and Elliott, A.C., Imaging of intracellular calcium stores in individual permeabilized pancreatic acinar cells. Apparent homogeneous cellular distribution of inositol 1,4,5-trisphosphate-sensitive stores in permeabilized pancreatic acinar cells, *J. Biol. Chem.*, 271, 4999, 1996.

170. Marchant, J.S., et al., Rapid kinetic measurements of $^{45}Ca^{2+}$ mobilization reveal that Ins(2,4,5)P$_3$ is a partial agonist at hepatic InsP$_3$ receptors, *Biochem. J.*, 321, 573, 1997.

171. Stout, M.A. and Diecke, F.P., ^{45}Ca distribution and transport in saponin skinned vascular smooth muscle, *J. Pharmacol. Exp. Ther.*, 225, 102, 1983.

172. Casteels, R. and Raeymaekers, L., The action of acetylcholine and catecholamines on an intracellular calcium store in the smooth muscle cells of the guinea-pig taenia coli, *J. Physiol.*, 294, 51, 1979.

173. Verma, A., Hirsch, D.J., and Snyder, S.H., Calcium pools mobilized by calcium or inositol 1,4,5-trisphosphate are differentially localized in rat heart and brain, *Mol. Biol. Cell*, 3, 621, 1992.

174. Watson, W.D., et al., Sarco-endoplasmic reticulum Ca^{2+} ATPase (SERCA) inhibitors identify a novel calcium pool in the central nervous system, *J. Neurochem.*, 87, 30, 2003.

175. Taylor, C.W., Measuring Ca^{2+} fluxes in permeabilized cells, in: *Calcium Signaling (CRC Methods in Signal Transduction Series)*, edited by Putney J.W., CRC Press, Boca Raton, FL, 71, 1999, chap. 4.

176. Pinton, P., Pozzan, T., and Rizzuto, R., The Golgi apparatus is an inositol 1,4, 5-trisphosphate-sensitive Ca^{2+} store, with functional properties distinct from those of the endoplasmic reticulum, *EMBO J.*, 17, 5298, 1998.

177. Brini, M., et al., Targeted recombinant aequorins: tools for monitoring $[Ca^{2+}]$ in the various compartments of a living cell, *Microsc. Res. Tech.*, 46, 380, 1999.

178. Chiesa, A., et al., Recombinant aequorinand green fluorescent protein as valuable tools in study of cell signaling, *Biochem. J.*, 355, 1, 2001.

179. Palmer, A.E., et al., Bcl-2-mediated alterations in endoplasmic reticulum Ca^{2+} analyzed with an improved genetically encoded fluorescent sensor, *Proc. Natl. Acad. Sci. U.S.A.*, 101, 17404, 2004.

180. Møller, J.V., Lind, K.E., and Andersen, J.P., Enzyme kinetics and substrate stabilization of detergent-solubilized and membranous ($Ca^{2+}+Mg^{2+}$)-activated ATPase from sarcoplasmic reticulum. Effect of protein–protein interactions, *J. Biol. Chem.*, 255, 1912, 1980.

181. Saborido, A., Delgado, J., and Megías, A., Measurement of sarcoplasmic reticulum Ca^{2+}-ATPase activity and E-type Mg^{2+}-ATPase activity in rat heart homogenates, *Anal. Biochem.*, 268, 79, 1999.

182. Lecocq, J. and Inesi, G., Determination of inorganic phosphate in the presence of adenosine trisphosphate by the molybdo-vanadate method, *Anal. Biochem.*, 15, 160, 1966.

183. Lin, T.-I. and Morales, M.F., Application of a one-step procedure for measuring inorganic phosphate in the presence of proteins: the actomyosin ATPase system, *Anal. Biochem.*, 77, 10, 1977.

184. Kosk-Kosicka, D., Measurement of Ca^{2+}-ATPase activity (in PMCA and SERCA1), in: *Calcium Signaling Protocols, (Methods in Molecular Biology, volume 114)* edited by Lambert D.G., Humana Press, Totowa, NJ, 1999, chap. 23.

185. Baginski, E.S., Foà, P.P., and Zak, B., Microdetermination of inorganic phosphate, phospholipids, and total phosphate in biologic materials, *Clin. Chem.*, 13, 326, 1967.

186. Vilsen, B., and Andersen, J.P., Occlusion of Ca^{2+} in soluble monomeric sarcoplasmic reticulum Ca^{2+}-ATPase, *Biochim. Biophys. Acta*, 855, 429, 1986.

187. Vilsen, B. and Andersen, J.P., Characterization of CrATP-induced calcium occlusion in membrane-bound and soluble monomeric sarcoplasmic reticulum Ca^{2+}-ATPase, *Biochim. Biophys. Acta*, 898, 313, 1987.

188. Vilsen, B. and Andersen, J.P., Interdependence of Ca^{2+} occlusion sites in the unphosphorylated sarcoplasmic reticulum Ca^{2+}-ATPase complex with CrATP, *J. Biol. Chem.*, 267, 3539, 1992.

189. Vilsen, B. and Andersen, J.P., CrATP-induced calcium occlusion in mutants of the Ca^{2+}-ATPase of sarcoplasmic reticulum, *J. Biol. Chem.*, 267, 25739, 1992.

190. Andersen, J.P. and Vilsen, B., Amino acids Asn^{796} and Thr^{799} of the Ca^{2+}-ATPase of sarcoplasmic reticulum bind Ca^{2+} at different sites, *J. Biol. Chem.*, 269, 15931, 1994.

191. Seebregts, C.J. and McIntosh, D.B., $2',3'-O$-(2,4,6,-trinitrophenyl)-8-azido-adenosine mono-, di-, and triphosphates as photoaffinity probes of the Ca^{2+}-ATPase of sarcoplasmic reticulum. Regulatory/superfluorescent nucleotides label the catalytic site with high efficiency, *J. Biol. Chem.*, 264, 2043, 1989.

192. McIntosh, D.B., Woolley, D.G., and Berman, M.C., $2',3'-O$-(2,4,6,-trinitrophenyl)-8-azido-AMP and –ATP photolabel Lys^{492} at the active site of sarcoplasmic reticulum Ca^{2+}-ATPase, *J. Biol. Chem.*, 267, 5301, 1992.

193. Murphy, A.J., Affinity labeling of the active site of the Ca^{2+}-ATPase of sarcoplasmic reticulum, *Biochim. Biophys. Acta*, 946, 57, 1988.

194. McIntosh, D.B., et al., Mutagenesis of segment ^{487}Phe-Ser-Arg-Asp-Arg-Lys^{492} of sarcoplasmic reticulum Ca^{2+}-ATPase produces pumps defective in ATP binding, *J. Biol. Chem.*, 271, 25778, 1996.

195. Yamagata, K., Daiho, T., and Kanazawa, T., Labeling of lysine 492 with pyridoxal $5'$-phosphate in the sarcoplasmic reticulum Ca^{2+}-ATPase. Lysine 492 residue is located outside the fluorescein 5-isothiocyanate-binding region in or near the ATP binding site, *J. Biol. Chem.*, 268, 20930, 1993.

196. Yamamoto, H., et al., Affinity labeling of the ATP-binding site of Ca^{2+}-transporting ATPase of sarcoplasmic reticulum by adenosine triphosphopyridoxal: identification of the reactive lysyl residue, *J. Biochem. (Tokyo)*, 103, 452, 1988.

197. Yamamoto, H., Ca^{2+}-dependent conformational change of the ATP-binding site of Ca^{2+}-transporting ATPase of sarcoplasmic reticulum as revealed by an alteration of the target-site specificity of adenosine triphospho pyridoxal, *J. Biochem. (Tokyo)*, 106, 1121, 1989.

198. Hove-Madsen, L. and Bers, D.M., Sarcoplasmic reticulum Ca^{2+} uptake and thapsigargin sensitivity in permeabilized rabbit and rat ventricular myocytes, *Circ. Res.*, 73, 820, 1993.

199. Treiman, M., Caspersen, C., and Christensen, S.B., A tool coming of age: thapsigargin as an inhibitor of sarco-endoplasmic reticulum Ca^{2+}-ATPases, *Trends Pharm. Sci.*, 19, 131, 1998.

200. Hussain, A. and Inesi, G., Involvement of sarco/endoplasmic reticulum Ca^{2+} ATPases in cell function and the cellular consequences of their inhibition, *J. Membr. Biol.*, 172, 91, 1999.

201. Jakobsen, C.M., et al., Design, synthesis, and pharmacological evaluation of thapsigargin analogues for targeting apoptosis to prostatic cancer cells, *J. Med. Chem.*, 44, 4696, 2001.

202. Denmeade, S.R. and Isaacs, J.T., The SERCA pump as a therapeutic target: making a "smart" bomb for prostate cancer, *Cancer Biol. Ther.* 4, 14, 2005.

203. Wictome, M., et al., The inhibitors thapsigargin and 2,5-di(*tert*-butyl)-1,4-benzohydroquinone favour the E2 form of the Ca^{2+}, Mg^{2+}-ATPase, *FEBS Lett.*, 304, 109, 1992.

204. Seidler, N.W., et al., Cyclopiazonic acid is a specific inhibitor of the Ca^{2+}-ATPase of sarcoplasmic reticulum, *J. Biol. Chem.*, 264, 17816, 1989.

205. Ma, H., et al., Overlapping effects of S3 stalk segment mutations on the affinity of Ca^{2+}-ATPase (SERCA) for thapsigargin and cyclopiazonic acid, *Biochemistry*, 38, 15522, 1999.

206. Mason, M.J., Garcia-Rodriguez, C., and Grinstein, S., Coupling between intracellular Ca^{2+} stores and the Ca^{2+} permeability of the plasma membrane. Comparison of the effects of thapsigargin, 2,5-di-(*tert*-butyl)-1,4-hydroquinone, and cyclopiazonic acid in rat thymic lymphocytes, *J. Biol. Chem.*, 266, 20856, 1991.

207. Scamps, F., et al., Sarco-endoplasmic ATPase blocker 2,5-di-(*tert*-butyl)-1,4-benzohy-droquinone inhibits N-, P-, and Q- but not T-, L-, or R-type calcium currents in central and peripheral neurons, *Mol. Pharmacol.*, 58, 18, 2000.

208. Furuya, T., et al., TcSCA complements yeast mutants defective in Ca^{2+} pumps and encodes a Ca^{2+}-ATPase that localizes to the endoplasmic reticulum of *Trypanosoma cruzi*, *J. Biol. Chem.*, 276, 32437, 2001.

209. Cantley, L.C., et al., Vanadate is a potent (Na,K)-ATPase inhibitor found in ATP derived from muscle, *J. Biol. Chem.*, 252, 7421, 1977.

210. Niggli, V., et al., Purified $(Ca^{2+}-Mg^{2+})$-ATPase of the erythrocyte membrane. Reconstitution and effect of calmodulin and phospholipids, *J. Biol. Chem.*, 256, 395, 1981.

211. Wang, T., et al., Effects of potassium on vanadate inhibition of sarcoplasmic reticulum Ca^{2+}-ATPase from dog cardiac and rabbit skeletal muscle, *Biochem. Biophys. Res. Commun.*, 91, 356, 1979.

212. Chabre, M., Aluminofluoride and berylofluoride complexes: new phosphate analogs in enzymology, *Trends Biochem. Sci.*, 15, 6, 1990.

213. Csermely. P., Varga, S., and Martonosi, A., Competition between decavanadate and fluorescein isothiocyanate in the Ca^{2+}-ATPase of sarcoplasmic reticulum, *Eur. J. Biochem.*, 150, 455, 1985.

214. Varga, S., Csermely. P., and Martonosi, A., The binding of vanadium (V) oligoanions to sarcoplasmic reticulum, *Eur. J. Biochem.*, 148, 119, 1985.

215. Hua, S., Fabris, D., and Inesi, G., Characterization of calcium, nucleotide, phosphate, and vanadate bound states by derivatization of sarcoplasmic reticulum ATPase with ThioGlo1, *Biophys. J.*, 77, 2217, 1999.

216. Hua, S., Inesi, G., and Toyoshima, C., Distinct topologies of mono- and decavanadate binding and photo-oxidative cleavage in the sarcoplasmic reticulum ATPase, *J. Biol. Chem.*, 275, 30546, 2000.

217. Murphy, A.J. and Coll, R.J., Fluoride is a slow, tight-binding inhibitor of the calcium ATPase of sarcoplasmic reticulum, *J. Biol. Chem.*, 267, 5229, 1992.

218. Danko, S., et al., Distinct natures of beryllium fluoride-bound, aluminum fluoride-bound, and magnesium fluoride-bound stable analogues of an ADP-insensitive phos-phoenzyme intermediate of sarcoplasmic reticulum Ca^{2+}-ATPase. Changes in catalytic and transport sites during phosphoenzyme hydrolysis, *J. Biol. Chem.*, 279, 14991, 2004.

219. Trouillier, A., Girardet, J.-L., and Dupont, Y., Fluoroaluminate complexes are bifunc-tional analogues of phosphate in sarcoplasmic reticulum Ca^{2+}-ATPase, *J. Biol. Chem.*, 267, 22821, 1992.

220. Inesi, G., et al., Ca^{2+} occlusion and gating function of Glu[309] in the ADP-fluoroalumi-nate analog of the Ca^{2+}-ATPase phosphoenzyme intermediate, *J. Biol. Chem.*, 279, 31629, 2004.

221. Herscher, C.J. and Rega, A.F., Pre-steady-state kinetic study of the mechanism of inhibition of the plasma membrane Ca^{2+}-ATPase by lanthanum, *Biochemistry*, 35, 14917, 1996.

222. Asturias, F.J., Fischetti, R.F., and Blasie, J.K., Changes in the profile structure of the sarcoplasmic reticulum membrane induced by phosphorylation of the Ca^{2+} ATPase enzyme in the presence of terbium: a time-resolved x-ray diffraction study, *Biophys. J.*, 66, 1653, 1994.

223. Asturias, F.J., Fischetti, R.F., and Blasie, J.K., Changes in the relative occupancy of metal-binding sites in the profile structure of the sarcoplasmic reticulum membrane induced by phosphorylation of the Ca^{2+} ATPase enzyme in the presence of terbium: a time-resolved, resonance x-ray diffraction study, *Biophys. J.*, 66, 1665, 1994.

224. Wuytack, F., et al., Evidence for the presence in smooth muscle for two types of Ca^{2+}-transport ATPase, *Biochem. J.*, 224, 445, 1984.

225. Hack, N., Wilkinson, J.M., and Crawford, N.A., Monoclonal antibody (PL/IM 430) to human platelet intracellular membranes which inhibits the uptake of Ca^{2+} without affecting the $Ca^{2+}+Mg^{2+}$-ATPase, *Biochem. J.*, 250, 355, 1988.

226. Poch, E., et al., Functional characterization of alternatively spliced human SERCA3 transcripts, *Am. J. Physiol.*, 275, C1449, 1998.

227. Chandrasekera, C.P. and Lytton, J., Inhibition of human SERCA3 by PL/IM430. Molecular analysis of the interaction, *J Biol. Chem.*, 278, 12482, 2003.

228. Poulin, G., Nandakumar, R., and Ahringer, J., Genome-wide RNAi screens in *Caenorhabditis elegans*: impact on cancer research, *Oncogene*, 23, 8340, 2004.

229. Scherer, L. and Rossi, J.J., Recent applications of RNAi in mammalian systems, *Curr. Pharm. Biotechnol.*, 5, 355, 2004.

230. Medema, R.H., Optimizing RNA interference for application in mammalian cells, *Biochem. J.*, 380, 593, 2004.

231. Van Baelen, K., et al., The contribution of the SPCA1 Ca^{2+} pump to the Ca^{2+} accumulation in the Golgi apparatus of HeLa cells assessed via RNA-mediated interference, *Biochem. Biophys. Res. Commun.*, 306, 430, 2003.

232. Gleason, M.R., et al., A mutation in *serca* underlies motility dysfunction in accordion zebrafish, *Dev. Biol.*, 276, 441, 2004.

233. Eckstein-Ludwig, U., et al., Artemisinins target the SERCA of *Plasmodium falciparum*, *Nature*, 424, 957, 2003.

234. Reuter, N., Hinsen, K., and Lacapere, J.J., Transconformations of the SERCA1 Ca^{2+}-ATPase: a normal mode study, *Biophys. J.*, 4, 2186, 2003.

235. Hinsen, K., et al., Normal mode based fitting of atomic structure into electron density maps: application to SR Ca-ATPase, *Biophys. J.*, 88, 818, 2005.

236. Li, G. and Cui, Q., Analysis of functional motions in Brownian molecular machines with an efficient block normal mode approach: myosin-II and Ca^{2+}-ATPase, *Biophys. J.*, 86, 743, 2004.

237. Niggli, V., Penniston, J.T., and Carafoli, E., Purification of the $(Ca^{2+}-Mg^{2+})$-ATPase from human erythrocyte membranes using a calmodulin affinity column, *J. Biol. Chem.*, 254, 9955, 1979.

238. De Schutter, G., et al., Tissue levels and purification by affinity chromatography of the calmodulin-stimulated Ca^{2+}-transport ATPase in pig antrum smooth muscle, *Biochim. Biophys. Acta*, 773, 1, 1984.

239. Verbist, J., et al., Reconstitution of the purified calmodulin-dependent $(Ca^{2+}-Mg^{2+})$-ATPase from smooth muscle, *Cell Calcium*, 5, 253, 1984.

240. Chu, G. and Kranias, E.G., Functional interplay between dual site phospholambam phosphorylation: insights from genetically altered mouse models, *Basic Res. Cardiol.*, 97, Suppl 1, I43, 2002.

241. Antoons, G., et al., Ca^{2+} uptake by the sarcoplasmic reticulum in ventricular myocytes of the SERCA2$^{b/b}$ mouse is impaired at higher Ca^{2+} loads only, *Circ. Res.*, 92, 881, 2003.
242. Prasad, V., et al., Phenotypes of SERCA and PMCA knockout mice, *Biochem. Biophys. Res. Commun.*, 322, 1192, 2004.

14 Measuring Single Cell and Subcellular Ca^{2+} Signals

Sandip Patel, Lawrence D. Gaspers, Nicola Pierobon, Walson Metzger, and Andrew P. Thomas

CONTENTS

14.1   INTRODUCTION

Calcium is widely recognized as one of the most universal of intracellular signals, and it controls a diverse array of cellular activities [1–3]. The commonest form of intracellular Ca^{2+} signaling is triggered by extracellular agonists that bind to cell surface receptors that act through a variety of signal transduction pathways to mobilize calcium either at the level of plasma membrane or intracellular calcium channels. Numerous extracellular stimuli, acting through many classes of receptor and ion channels, lead to increases in the cytosolic free Ca^{2+} concentration. Typically, cytosolic Ca^{2+} rises severalfold from resting values of ~100 nM to values as high as tens of micromolar. Once elevated, Ca^{2+} exerts its effects by influencing the activity of Ca^{2+}-dependent target proteins either directly or through

Ca^{2+}-binding proteins such as calmodulin [4, 5]. Multiple extrusion mechanisms exist within the cell to rapidly decrease Ca^{2+}, preventing the potentially toxic consequences of sustained Ca^{2+} elevations [6–8]. Ca^{2+} signals, thus, arise through the concerted activity of a variety of Ca^{2+} channels, pumps, and exchangers.

One of the major pathways that leads to an increase of cytosolic Ca^{2+} in nonexcitable cells is through production of the second messenger, inositol 1,4, 5-trisphosphate (IP_3) [9–11]. IP_3 activates Ca^{2+} channels located within the membranes of intracellular Ca^{2+} stores [10, 12, 13]. It is now becoming clear that other second messengers also play a role in mobilizing Ca^{2+} from intracellular stores, including cyclic ADPribose and NAADP [14–17]. Stemming from the pioneering studies of Cobbold and colleagues [18], it is now clear that cytosolic Ca^{2+} signals can be remarkably complex at the level of individual cells. Submaximal hormonal stimulation has been shown to elicit oscillatory Ca^{2+} signals in a variety of cell types [19]. Figure 14.1, shows repetitive baseline-separated Ca^{2+} oscillations in single hepatocytes stimulated with the α-adrenergic agonist, phenylephrine, which is coupled to IP_3 formation. Importantly, all cells do not oscillate in synchrony or with the same frequency at a given hormone concentration. Thus, in measurements of [Ca^{2+}] in populations of cells, these complex Ca^{2+} signals will be averaged out and remain undetected. Population averages may even distort relatively simple changes in [Ca^{2+}]. For example, if all cells do not respond at low concentrations of hormone, population averages (which take into account both the responsive and unresponsive cells) will underestimate the peak response and rate of Ca^{2+} rise that actually occurs in the responding cells. Averaging may also distort these parameters if individual cells respond with differing latencies. In addition, recording the average Ca^{2+} signal from a large number of cells can incorporate responses derived from unhealthy or contaminating cell types. Thus, whereas population measurements complement parallel determination of other parameters such as enzyme activity that can only be performed with large numbers of cells, monitoring the Ca^{2+} concentration at the single-cell level offers several advantages in quantitation of changes in [Ca^{2+}].

Monitoring Ca^{2+} at the subcellular level reveals further complexity in the Ca^{2+} signal [19–21] that may be overlooked by measurement of global cytosolic Ca^{2+} changes. Oscillations often originate from a specific cellular locus [22, 23] and may then propagate through the cell in the form of waves [22] or spirals [24]. In other systems, Ca^{2+} elevations may remain relatively local such as at the apical poles of secretory cells [23, 25]. Smaller, localized non-propagating Ca^{2+} elevations have been described in many systems, and recruitment of these elementary events is thought to underlie global Ca^{2+} increases [26]. The advent of chemical and engineered protein fluorescent Ca^{2+}-sensitive indicators to monitor [Ca^{2+}] in living cells has greatly furthered our understanding of the mechanisms involved in Ca^{2+} homeostasis [27–30]. In this article, we discuss some of the practical considerations in measuring Ca^{2+} at the cellular and subcellular level with the use of these fluorescent Ca^{2+} indicator dyes and bioengineered Ca^{2+}-sensitive proteins based on green fluorescent protein (GFP) and related fluorescent proteins. Structural details and molecular properties of the chemical dyes and bioengineered proteins are discussed in earlier chapters of this volume.

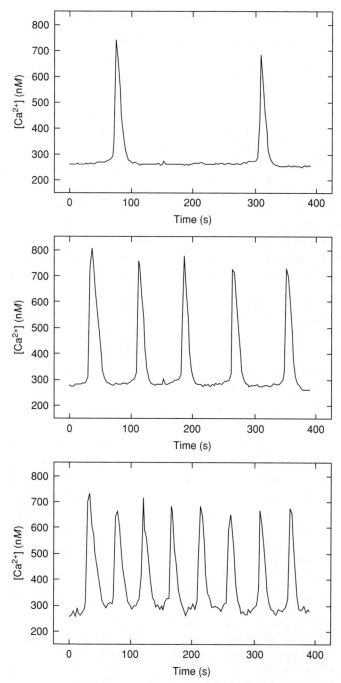

FIGURE 14.1 Oscillations in cytosolic Ca^{2+} in hepatocytes. The traces show [Ca^{2+}] responses of three individual fura-2-loaded hepatocytes from the same imaging field stimulated with phenylephrine (0.5 μM). Note the asynchrony and marked differences in oscillation frequency between cells, even though all were exposed to the same global stimulus.

14.2   FLUORESCENT Ca^{2+}-SENSITIVE INDICATOR DYES

14.2.1   Properties of Fluorescent Ca^{2+} Indicator Dyes

Fluorescence involves the absorption of light to generate an excited state in a molecule, which then emits light upon return of the molecule to the ground state. In single photon excitation, the emitted light is of longer wavelength than that of the absorbed exciting light due to dissipation of some of the energy of the excited state of the molecule prior to relaxation. The fluorescent output of a particular molecule is determined by the efficiency of light absorption (characterized by its molar extinction coefficient) and the amount of light emitted relative to what was absorbed (characterized by its quantum yield). All fluorescent Ca^{2+} indicators change their fluorescence properties upon binding Ca^{2+}, although the form of this change varies, ranging from intensity changes to shifts in spectra and alterations in the efficiency of intermolecular fluorescence resonance energy transfer (FRET). Whereas several other methods exist for monitoring changes in [Ca^{2+}], including the use of Ca^{2+}-sensitive microelectrodes and Ca^{2+}-sensitive photoproteins, fluorescent Ca^{2+} probes provide an easy and versatile approach for quantitating [Ca^{2+}] in living cells, especially when rapid single cell and subcellular information is required. Moreover, despite the advent of a range of genetically engineered fluorescent protein-based Ca^{2+} indicators (see below), chemically synthesized Ca^{2+} indicator dyes remain the most commonly used indicators of Ca^{2+}. The primary advantage of the chemical indicator dyes is their relative ease of loading into almost all cell types, ready availability, and large signal changes on binding Ca^{2+}. The primary disadvantages are that the dyes may become compartmentalized in intracellular organelles, that they can be pumped out of cells, they are difficult to target to specific subcellular compartments, and they can buffer the Ca^{2+} signals under study.

Most of the available chemical fluorescent Ca^{2+} indicator dyes are based on the Ca^{2+}-selective chelators, EGTA and BAPTA. These compounds have a high selectivity for Ca^{2+} over other ions such as Mg^{2+} and in the case of BAPTA are relatively insensitive to pH in the physiological range [31]. However, as with the parent compounds, heavy metals bind many of these fluorescent Ca^{2+} indicator dyes with high affinity and can cause fluorescence quenching or elevate basal fluorescence to interfere with Ca^{2+} measurements.

Single wavelength Ca^{2+} indicator dyes change their fluorescence output intensity upon binding calcium without a marked shift in their excitation or emission spectra. An example is fluo-3, an indicator based on the core structure of fluorescein. This visible wavelength dye has excitation and emission peaks of ~500 and ~525 nm, respectively. Increasing Ca^{2+} results in a progressive increase in emission intensity; fluo-3 fluorescence is thus directly proportional to [Ca^{2+}]. Fluorescence of Ca^{2+}-bound fluo-3 is almost 100 times that of the Ca^{2+}-free form, making this dye particularly sensitive to changes in [Ca^{2+}]. This dye is readily excited with visible light lasers and is thus suitable for confocal microscopy (see below). A whole family of fluo-based indicators is now available, covering a broad range of Ca^{2+}

sensitivities that are useful for a variety of applications, including sensing large and small [Ca^{2+}] changes, rapid kinetics, and the high concentrations of Ca^{2+} that can be encountered in intracellular compartments such as the endoplasmic reticulum (ER). In addition to the fluorescein-based Ca^{2+} indicator dyes, there are a number of other options that offer alternative wavelengths of excitation and emission (from ultraviolet to red light), which can be useful when recording Ca^{2+} in parallel with other fluorescent probes. These are discussed in detail in Chapter 1.

Ratiometric indicators respond to the binding of Ca^{2+} with a shift in the excitation or emission spectra of the molecule. For fura-2, which remains the most popular dye for single cell studies, the excitation maximum is shifted from ~360 nm, in the absence of Ca^{2+} to ~340 nm in the presence of saturating Ca^{2+} levels, without an appreciable change in the wavelength of the fluorescence emission peak (~510 nm). Thus, reciprocal changes occur in emission intensity around the isosbestic (or Ca^{2+}-insensitive wavelength), which are most pronounced when the dye is excited at 340 and 380 nm. Fluorescence increases with [Ca^{2+}] at 340 nm excitation and is inversely related to [Ca^{2+}] at 380 nm. The [Ca^{2+}] is directly related to the ratio of emitted intensity at the two excitation wavelengths, as described below under calibration. Indo-1 is another example of a ratiometric dye; however, this dye differs from fura-2 in that the predominant effect of Ca^{2+} binding is to induce a shift in its emission, as opposed to its excitation spectrum. The Ca^{2+}-free and Ca^{2+}-bound forms of the dye have emission peaks of ~400 and ~480 nm, respectively (excitation at 340 nm).

Ratiometric Ca^{2+} indicator dyes offer considerable advantages over single wavelength indicators for measuring [Ca^{2+}]. Fluorescence signals from single wavelength dyes are dependent not only on [Ca^{2+}] but also on the concentration of dye within the cell. Loss of dye from the cell (see below), inhomogeneities in cell thickness, or changes in cell shape (e.g., contracting cardiomyocytes) can yield differences in fluorescence signals that are difficult to distinguish from real changes in [Ca^{2+}]. These issues are less of a problem with ratiometric dyes, because the ratio of fluorescence signals measured at two wavelengths is essentially independent of dye concentration. Nonreciprocal changes in fluorescence intensity at the two wavelengths (i.e., Ca^{2+} independent effects) can be identified, and with appropriate calibration procedures these changes can be eliminated from the Ca^{2+}-dependent fluorescence signals under study. This important property makes ratiometric recording the method of choice, particularly for Ca^{2+} imaging studies.

In order to optimize signal changes, it is important to choose a dye with an affinity that falls within the range of the anticipated changes in [Ca^{2+}]. As a general guideline, the [Ca^{2+}] range that a Ca^{2+} indicator dye can accurately report will be 0.1–10 times its dissociation constant. Fura-2, for example, has a dissociation constant for Ca^{2+} of about 200 nM, which is well suited for monitoring global changes in cytosolic Ca^{2+} during cell stimulation. Lower affinity dyes, such as fura-2FF (K_d for Ca^{2+} of 35 μM), are appropriate for measurement of larger Ca^{2+} changes such as those that may occur in localized cytosolic domains or in compartments where the free [Ca^{2+}] is relatively high (e.g., the lumen of the ER, see below). Lower affinity Ca^{2+} indicator dyes also typically have faster kinetics for association and dissociation of Ca^{2+}, which makes them better suited for following fast $[Ca^{2+}]$ changes, such as those occurring in nerve and muscle (see Chapter 1 for kinetic constants of some key Ca^{2+} indicator dyes).

14.2.2 LOADING FLUORESCENT Ca^{2+} INDICATOR DYES INTO CELLS

A major advantage of using fluorescent Ca^{2+} indicators to monitor $[Ca^{2+}]$ is the ease with which these dyes can be introduced into cells. Many dyes are available as acetoxymethyl (AM) esters [32]. Esterification of the hydrophilic carboxyl groups renders the dye cell permeable. Once within the cell, the ester bonds are cleaved by endogenous esterases freeing the negatively charged Ca^{2+} indicator dye, essentially trapping it within the cell. Thus, incubating cells with the AM form of the dye in a physiological medium for a defined period of time results in significant concentration of the dye within the cell. In general, a target range of 30–100 μM intracellular active dye concentration is suitable for monitoring Ca^{2+} changes. However, a broad range of intracellular dye concentrations can be obtained, depending on extracellular concentration of AM dye, time, and temperature of incubation, and the density of the cells to be loaded. As most AM esters are fluorescent but Ca^{2+} insensitive, it is important to allow sufficient time to effect complete hydrolysis of the esterified dye.

Loss of dye from the cell is often a problem when using fluorescent Ca^{2+} indicators. Although the dye loss can be due to leakage under some conditions (mostly when cells are damaged), the predominant pathway by which these lipophilic and negatively charged polycarboxylates are lost from the cell is through anion transport proteins. This can be reduced by inclusion of organic anion transport inhibitors such as bromosulfophthalein, probenecide, or sulfinpyrazone in media during both loading and subsequent data acquisition. Lowering the temperature may also slow dye loss. Dye loading can be enhanced by premixing of the indicator with Pluronic F-127 (a surfactant) to aid dispersion. It should be noted, however, that high levels of dye loading may have adverse effects on the cell. An important consideration is buffering of Ca^{2+} by the dye itself. Whereas steady-state cytosolic $[Ca^{2+}]$ will be relatively unaffected by additional Ca^{2+} buffering, the kinetics of Ca^{2+} changes may be significantly slowed. Also, the hydrolysis of AM esters results in the generation of potentially toxic byproducts (acetate and formaldehyde). For these reasons, the level of dye within the cell should be minimized. On the other hand, sufficient dye level must be attained in order to achieve adequate signal-to-noise ratio to monitor the Ca^{2+} fluctuations of interest. This in turn is governed by the magnitude of fluorescence changes that the dye undergoes upon binding Ca^{2+} and the detection efficiency of the instrumentation used (see below).

14.2.3 COMPARTMENTALIZATION OF Ca^{2+} INDICATOR DYES

In most studies, the target of $[Ca^{2+}]$ measurements with fluorescent Ca^{2+} indicator dyes is the cytosolic compartment. Unfortunately, the esterases that liberate the active indicator from the cell-permeant AM form are not exclusive to the cytoplasm, and this can result in accumulation of Ca^{2+}-sensitive and Ca^{2+}-insensitve dye forms in other intracellular compartments. This is because AM esters pass through membranes of intracellular organelles such as mitochondria and ER, in addition to the plasma membrane. Depending on the activity and specificity of esterases in these organelles, significant and even predominant compartmentalization of the dye can occur. The final intracellular distribution of the fluorescent dyes following AM loading is dependent on the particular dye used, the cell type, and the loading conditions employed.

In many cases, the majority of the dye is found in the cytosol, but even a relatively small proportion of compartmentalized dye can pose problems for measurement of cytosolic Ca^{2+}. For example, ER accumulation of high affinity indicators used to monitor cytosolic Ca^{2+} such as fura-2, results in an increase in the basal signal. This is because the dye is Ca^{2+}-saturated in the ER where the Ca^{2+} concentration is much higher than the cytosol. As long as the compartmentalized Ca^{2+} does not change, it is usually possible to account for it as part of the calibration of cytosolic Ca^{2+} signals (see below). For the most part, signals from high-affinity dyes in the ER remain saturated unless this compartment is extensively depleted of Ca^{2+}, as may occur with Ca^{2+} pump inhibitors such as thapsigargin [33, 34]. Compartmentalization of dyes into mitochondria can be more problematic, because the affinity of dyes used for cytosolic measurements are also in the range of mitochondrial matrix Ca^{2+}. Since mitochondrial Ca^{2+} changes do not necessarily mirror cytosolic Ca^{2+} changes [35], "mixed" signals can result from cells when there is significant mitochondrial compartmentalization [36]. Dye compartmentalization can often be reduced by decreasing the dye concentration or temperature during loading. Alternatively, direct delivery of the free acid form of the dye (normally cell impermeant) into the cytosol can be achieved through microinjection or through patch pipettes. These methods have the added advantage of being able to load indicators conjugated to large inert dextrans, thereby largely eliminating the problems of dye leakage and redistribution.

The accumulation of fluorescent Ca^{2+} indicator dyes (loaded in the AM form) into other compartments is problematic for monitoring Ca^{2+} changes in the cytosol, but it can be exploited to allow measurements of Ca^{2+} changes in the particular organelle where the dye resides. The primary requirement for exploiting compartmentalized dye signals in this way is to have a means to eliminate the signals from other compartments, particularly the cytosol. Low-affinity fluorescent Ca^{2+} indicator dyes have been used successfully to measure fluctuations in Ca^{2+} in the lumen of the ER in both intact [37–40] and permeabilized cells [41–43]. Figure 14.2 shows recordings of ER Ca^{2+} changes in primary cultured hepatocytes loaded with the low affinity indicator, fura-2FF (K_d for Ca^{2+} of 35 μM) and subsequently permeabilized with digitonin to release cytosolic indicator. Addition of ATP resulted in an increase in the fluorescence ratio of the dye, reflecting loading of the intracellular stores with Ca^{2+} (Figure 14.2A). Subsequent addition of IP$_3$ caused a dose-dependent decrease in ER luminal [Ca^{2+}], and the residual stored Ca^{2+} could then be released with the Ca^{2+} ionophore ionomycin. In contrast to Figure 14.2A, where EGTA was used to buffer Ca^{2+} in the cytoplasmic compartment and prevent Ca^{2+} feedback effects on the IP$_3$ receptor Ca^{2+} channel, the experiment in Figure 14.2B was carried out in the absence of Ca^{2+} buffer. Under these conditions, local positive

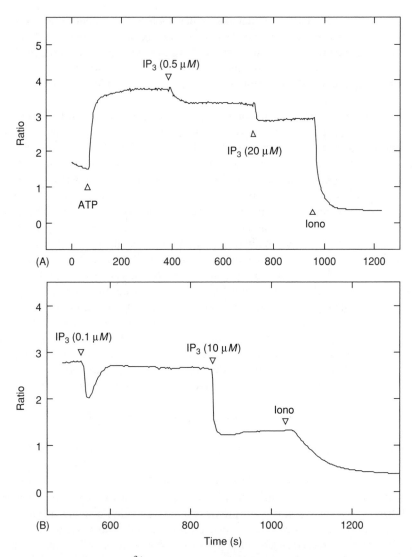

FIGURE 14.2 Monitoring Ca^{2+} changes in the endoplasmic reticulum with a low-affinity luminal fluorescent Ca^{2+} indicator dye. Fura-2FF-loaded hepatocytes were permeabilized with digitonin in Ca^{2+}-free intracellular-like buffer. The increase in the fluorescence ratio of fura-2FF (340 nm/380 nm) following Mg^{2+}–ATP (2 mM) addition indicates loading of intracellular stores. Subsequent addition of the indicated concentrations of IP_3 and ionomycin results in depletion of these stores. The two panels emphasize the different responses to submaximal doses of IP_3 and in the presence of Ca^{2+} buffer (500 μM EGTA, 150 μM $CaCl_2$; Panel A) and in the absence of EGTA (Panel B).

and negative Ca^{2+} feedback serves to amplify and then terminate the Ca^{2+} release at low levels of IP_3. Indeed, these conditions can give rise to IP_3-induced Ca^{2+} oscillations in the absence of plasma membrane signaling and allow spatial characterization of the location and activity of IP_3-sensitve Ca^{2+} stores in imaging measurements of single cells [34, 41–45].

Mitochondrial Ca^{2+} in intact cells can be monitored using fluorescent Ca^{2+} indicator dyes, with rhod-2 having received the most attention in this regard [35, 46–49]. The AM ester of this dye accumulates into mitochondria, probably because its positive charge and membrane permeability allow the mitochondrial membrane potential to drive uptake. Cytosolic "contamination" can be reduced by allowing cells to extrude cytosolic dye by incubating for a period following loading, thereby increasing the degree of mitochondrial compartmentalized dye [35, 47]. Figure 14.3 compares the distribution of rhod-2 in a single COS cell with that of MitoTracker green, a mitochondrial marker [50]. Addition of the purinergic agonist ATP rapidly increases rhod-2 fluorescence in these cells, indicating elevated mitochondrial Ca^{2+}. When studies are carried out using permeabilized cells or isolated mitochondria, a broader range of Ca^{2+} indicator dyes can be used, including fura-2 and its ratiometric derivatives [51–53].

14.2.4 CALIBRATION OF FLUORESCENCE SIGNALS

The signals from Ca^{2+}-sensitve fluorescent indicator dyes can be calibrated to yield absolute or relative values for the $[Ca^{2+}]$ in the compartment of interest. Most of the calibration approaches are designed for use with cytosolic dyes, but these approaches can also be adapted for use in other compartments. Moreover, similar calibration approaches can also be utilized with bioengineered fluorescent protein-based Ca^{2+} indicators of both single and dual wavelength types (see Chapter 4). Importantly, calibration procedures can be applied to either a single spatially averaged signal (e.g., from a photomultiplier) or on a pixel-by-pixel basis to fluorescence image data, yielding a spatially resolved image of $[Ca^{2+}]$. In all cases, fluorescence changes can be followed through time to give a time-course of $[Ca^{2+}]$ change. Conversion of fluorescence intensity data into $[Ca^{2+}]$ is subject to various uncertainties and errors that are discussed below, but calibration is useful to provide information on the relative and absolute change in Ca^{2+} levels that occur within the cell.

The nonlinear relationship between fluorescence intensity (F) and $[Ca^{2+}]$ necessitates the use of some form of calibration to obtain either an index of the relative change in $[Ca^{2+}]$, or preferably an absolute measure of $[Ca^{2+}]$. For single wavelength dyes, the following equation applies [54]:

$$[Ca^{2+}] = \frac{(F - F_{min})}{(F_{max} - F)} K_d \tag{14.1}$$

where F_{min} and F_{max} are the fluorescence intensities of the Ca^{2+}-free and Ca^{2+}-saturated dye, respectively, and K_d is the dissociation constant of the dye. As

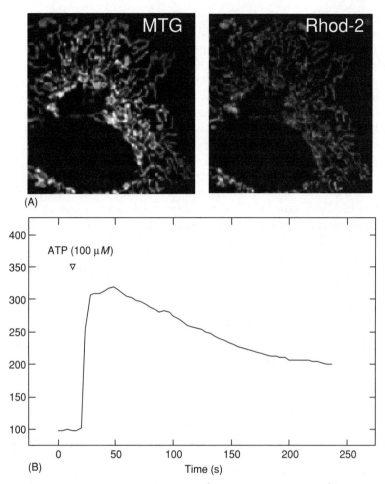

(A)

(B)

FIGURE 14.3 (A) Monitoring mitochondria Ca^{2+} with a fluorescent Ca^{2+} indicator dye. Confocal images of COS cells co-loaded with MitoTracker green (MTG; excitation = 514 nm, emission = 525–545 nm) and rhod-2 (excitation 514 nm, emission > 600 nm) showing a similar subcellular distribution of the mitochondrial marker with rhod-2 fluorescence. (B) Mean rise in mitochondrial Ca^{2+} in response to ATP stimulation (5 cells).

mentioned previously, since the fluorescence of this class of dye is dependent upon both the $[Ca^{2+}]$ and dye concentration it is imperative that F_{min} and F_{max} be determined under the same conditions of dye loading as those employed for determination of F. Thus, use of single wavelength indicators, necessitates *in situ* calibration of the dye. This is achieved by first equilibrating the $[Ca^{2+}]$ of the cytosol with that of the extracellular medium with a divalent ionophore such as ionomycin. Addition of EGTA and excess Ca^{2+} to the outside of the cell then allows determination of F_{min} and F_{max}, respectively. The Ca^{2+} concentration is then calculated from the known affinity of the dye. This assumes that the measured affinity of the dye in solution is similar to that within the cell. The dissociation constant of the dye may be determined *in situ* by Ca^{2+} titration in the presence of ionophore, but it is difficult to ensure true equilibration of Ca^{2+} across the cell membrane, and the ionophore may introduce pH changes that alter the apparent Ca^{2+} dissociation constant. Compartmentalized dye may further confound the calibration, but can be corrected for as described below.

The underlying principle for calibration of fluorescence signals from ratiometric dyes is essentially the same as that for single wavelength indicators. The fluorescence ratio, R (fluorescence intensity where the Ca^{2+}-bound dye predominates divided by the fluorescence intensity where the Ca^{2+}-free form predominates) is related to $[Ca^{2+}]$ as follows [55]:

$$[Ca^{2+}] = \frac{(R - R_{min})}{(R_{max} - R)} K_d \beta \qquad (14.2)$$

where R_{min} and R_{max} are the fluorescence ratio values of the dye under Ca^{2+}-free and Ca^{2+}-saturating conditions, respectively, K_d is the dissociation constant of the dye and β is the fluorescence intensity of the Ca^{2+}-free dye divided by the Ca^{2+}-bound form at the wavelength where the Ca^{2+}-free form of the dye predominates [55]. R_{min}, R_{max}, and β can be determined *in situ* as described above. However, since fluorescence ratios are independent of dye concentration, it is also possible to obtain these values in free solution for the particular instrument configuration used. For these measurements the free acid form of the indicator is added to a medium designed to mimic the ionic composition of the intracellular compartment (high K, low Na) and the fluorescence of the Ca^{2+}-free and Ca^{2+}-bound dye determined with added EGTA and excess Ca^{2+}, respectively. Again, it should be noted that the conditions of *in vitro* calibrations may not faithfully represent the environment of the dye within the cell, particularly with respect to interactions with higher molecular weight components and the microviscosity of intracellular environments (see Chapter 1).

Under conditions where it is not necessary to calculate absolute $[Ca^{2+}]$, or it is difficult to obtain appropriate calibration values, it is widely accepted that the ratio (R) alone is a reasonable index of the relative magnitude and kinetics of $[Ca^{2+}]$ changes in cells and tissues. A similar approach is frequently taken with single wavelength Ca^{2+} indicator dyes, where a self-ratio can be calculated by taking the ratio of fluorescence at a given time after treatment (F_t) to the fluorescence under resting conditions at time zero (F_0). Thus, relative $[Ca^{2+}]$ changes are commonly reported as F_t/F_0 or F/F_0. This is particularly useful for imaging studies with single wavelength fluorescent Ca^{2+} indicators, where measurement of appropriate calibration parameters on a spatially resolved basis is extremely difficult.

In order to accurately calibrate fluorescence intensities in terms of absolute $[Ca^{2+}]$, it is essential that any fluorescence not emanating from the Ca^{2+}-sensitive form of the dye in the appropriate compartment be subtracted from the measured signal. Autofluorescence represents the intrinsic cell fluorescence, which can derive from a variety of cellular constituents such as lipofuscin, reduced pyridine nucleotides (see below), and flavoproteins. The amounts of these components vary greatly depending on the cell type and excitation wavelength used, but typically contribute most significantly to the measured fluorescence in primary cells. Autofluorescence is likely to be most problematic for cells loaded with relatively low levels of fluorescent Ca^{2+} indicators. Another source of extraneous fluorescence is Ca^{2+}-insensitive dye, such as certain unhydrolyzed AM esters and dye trapped in other compartments.

Autofluorescence in single cell studies is determined by measuring the residual signal remaining after elimination of the fluorescence due to the reporter dye. For indicators such as fura-2, this can be achieved with Mn^{2+} that quenches fura-2 fluorescence. Incubating cells with millimolar levels of $MnCl_2$ in the presence of ionomycin (to facilitate Mn^{2+} entry) will result in rapid quench of cellular fura-2 (Figure 14.4A). However, this method is not suitable for cells where there is significant compartmentalized dye, since ionomycin will equilibrate divalent ions across most intracellular membranes. Furthermore, not all indicators are completely quenched by Mn^{2+} (e.g., fluo-3). An alternative approach is to selectively permeabilize the plasma membrane with detergents such as digitonin or saponin to effect release of just the cytosolic dye, which can then be easily washed away (Figure 14.4B). However, this is a rather disruptive treatment that may lead to significant movement artifacts in imaging studies, and may not account for fluorescence due to residual unhydrolyzed dye in the cytosol. Microinjection of the free acid form of fluorescent Ca^{2+} indicator dyes reduces errors associated with incomplete hydrolysis and dye compartmentalization. Note that autofluorescence need not be determined independently in the calibration of single wavelength fluorescence data when *in situ* calibration parameters are obtained, since this quantity cancels out in the denominator and numerator of Equation (14.1). For ratiometric dyes, autofluorescence must be subtracted at each wavelength prior to calculation of the ratio.

The procedures discussed above to evaluate autofluorescence are normally carried out at the completion of an experiment, because they are irreversible and disruptive to the cell. Therefore, it is assumed that the autofluorescence was constant throughout the preceding experiment. However, this is not always the case, particularly in cells where mitochondrial pyridine nucleotide (excitation maximum 360 nm) and flavoprotein (excitation maximum 470 nm) fluorescence signals are substantial. This is because the mitochondrial components of these signals respond to changes in mitochondrial Ca^{2+} due to stimulation of Krebs cycle dehydrogenases; giving an increase in pyridine nucleotide fluorescence and a decrease in flavoprotein fluorescence. These mitochondrial redox signals can actually be used as a real-time measure of the metabolic activity of intact cells during Ca^{2+} signals [35, 56, 57]. Nevertheless, the changes in autofluorescence are usually a relatively minor signal relative to the loaded Ca^{2+} indicator dye. The extent and magnitude of these or other dynamic autofluorescence changes should be estimated by carrying out parallel experiments with unloaded cells using the same protocols and data acquisition parameters used with the cells loaded with the Ca^{2+} indicator.

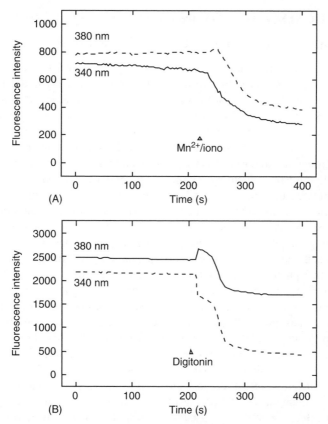

FIGURE 14.4 Determination of autofluorescence. (A) Quenching of fluorescence of a fura-2-loaded COS cell by addition of 2 mM MnCl$_2$ and 4 μM ionomycin. (B) Release of cytosolic fura-2 from a single hepatocyte by digitonin (25 μg/ml). In both cases, emitted fluorescence (>400 nm) was measured after excitation of cells at 340 nm (solid line) and 380 nm (dashed line). The residual fluorescence is subtracted from acquired signals before calculation of the fluorescence ratio.

Another important issue that must be considered in calibration of fluorescence data is loss of dye due to photobleaching or active extrusion of dye from the cell. For single wavelength indicators, reduction in dye concentration is difficult to distinguish from actual decreases in [Ca^{2+}] and will result in underestimation of F_{min} and F_{max}. In some cases, it is possible to correct for changes in dye loss if the kinetics of decay can be mathematically fitted. Although ratio measurements are relatively unaffected by changes in intracellular dye concentration, it is wise to minimize photobleaching at all times due to the concomitant production of damaging free radicals.

14.3   FLUORESCENT PROTEIN-BASED Ca^{2+} INDICATORS

14.3.1   ADVANTAGES OF PROTEIN-BASED Ca^{2+} INDICATORS

One of the biggest drawbacks of the fluorescent Ca^{2+} indicator dyes is the uncertainty in their actual distribution within a cell, and the difficulty in localizing them to a specific subcellular compartment. One approach to overcome this problem has been to inject fluorescent dyes conjugated with an appropriate targeting peptide, such as the nuclear-targeted Ca^{2+} green dextran described by Allbritton et al. [58]. There are also a series of fluorescent Ca^{2+} indicator dyes linked to lipophilic aliphatic groups designed to target them to membranes (described in Chapter 1), but these do not have selectivity for specific membrane locations.

In contrast to the limitations of chemical indicators, Ca^{2+} indicators based on bioengineered proteins offer enormous flexibility in subcellular targeting. This is because it is possible to exploit the myriad native cell protein targeting machinery, by generating chimeric protein constructs with an appropriate Ca^{2+} sensor and a location-specific targeting sequence. This methodology was first established for the Ca^{2+}-sensitive photoprotein aequorin by Rizzuto and coworkers (see Chapter 5), and has since found many applications in the Ca^{2+} signaling field. Molecular targeting of this type can lead to very selective localization within a cell (see below). Other advantages of protein-based Ca^{2+} indicators include that they tend to be more stable in cells than chemical indicators and they can be used to record over longer time periods. In addition, they may be less perturbing to the signaling pathways they are designed to measure. Aequorin constructs, for example, have very little buffering effect on the Ca^{2+} pools they are designed to measure. However, calmodulin-based fluorescent Ca^{2+} indicator proteins may have more significant effects on Ca^{2+}, as discussed below. It should also be noted that the GFP-based fluorescent Ca^{2+} indicator proteins typically have relatively small fluorescence signal changes on Ca^{2+} binding compared to chemical fluorescent Ca^{2+} indicator dyes. Nevertheless, there is continuing improvement in the signal-to-noise ratio and in reducing sensitivity to other factors such as pH, and it is clear that these highly versatile protein-based indicators will find increasing use in the future.

14.3.2   IMAGING OF PHOTOPROTEIN SIGNALS

Although most of this section will focus on fluorescent protein-based Ca^{2+} indicator constructs, a considerable amount of important work has been done using the naturally occurring Ca^{2+}-sensitve photoprotein, aequorin. Aequorin is a 21-kDa photoprotein isolated from the jellyfish *Aequorea*. Aequorin is not fluorescent, but yields luminescence in a Ca^{2+}-sensitive fashion [59]. Expression of aequorin fused to appropriate organellar targeting peptides has allowed measurements of Ca^{2+} in mitochondria [60], nucleus [61], ER [62], inner leaflet of the plasma membrane [63], secretory vesicles [64], and golgi apparatus [65]. In general, these measurements have relied on averaging signals over relatively large cell populations, because the signals from expressed aequorin are very small, usually requiring photon counting. Despite this limitation, this technique can yield important information about the subcellular organization of calcium signals due to the targeting technology that reports Ca^{2+} signals from discrete compartments (Chapter 5).

As noted above, the light output from expressed aequorin constructs is generally too low for imaging studies. However, some of the earliest recordings of Ca^{2+} images in intact cells were obtained with aequorin injected into large cells such as eggs [66, 67]. Despite advances in the sensitivity of imaging devices, single cell and subcellular imaging of expressed aequorins remains at the threshold of detection at present, with significant trade-offs of temporal and spatial resolution necessary compared to fluorescent Ca^{2+} indicators. Nevertheless, it is possible to record images of localized Ca^{2+} signals with expressed aequorin in small mammalian cells using multistage intensified cameras in a photon-counting mode [68, 69]. The main advantage of using aequorin in such measurements is that it is likely to be the least perturbing of all available reagents for measuring Ca^{2+}, because only a small proportion of the protein molecules are actually bound to Ca^{2+} at any time. However, the technical limitation of low light output greatly restricts the usefulness of this approach for single cell and subcellular imaging in most applications.

14.3.3 GFP-Based Ca^{2+} Indicators

The structure and properties of the many fluorescent protein-based Ca^{2+} indicators are described in detail in Chapter 4 of this volume. Most of these indicators rely on the GFP, which like aequorin is derived from jellyfish (it partners with aequorin by converting the blue luminescence to green fluorescence). Mutagenesis of GFP and other fluorescent proteins has yielded a wide array of different colored fluorescent proteins with improved properties for use as biological markers and indicators [71–73]. The primary prototype for the genetically engineered Ca^{2+}-sensing proteins is the "cameleon," developed by Miyawaki et al. [70]. The cameleons and a number of related Ca^{2+} indicators are fusion proteins composed of calmodulin and a calmodulin-binding peptide (M13 derived from the CaM-binding peptide of myosin light chain kinase M13), surrounded by two distinct spectrally shifted mutants of GFP [70].

The original cameleon used the GFP mutant CFP with GFP, but subsequent developments have relied on more spectrally separated mutant forms of GFP, most often CFP and YFP. The spectral properties of these proteins allow fluorescence resonance energy transfer (FRET) from the CFP (donor), which is excited at a lower wavelength, to the YFP (acceptor), which is excited at a higher wavelength. Binding of Ca^{2+} to calmodulin, induces interaction of the complex with the M13 peptide causing a large conformational change that brings the CFP and YFP closer together and therefore enhances FRET between the two fluorophores. Thus, when CFP is excited, fluorescence decreases at the emission peak of the donor CFP and increases at the emission peak of the acceptor YFP in response to Ca^{2+} binding. This is the principal of the Ca^{2+} sensing of this class of proteins. These cleverly designed molecules are therefore, dual emission ratiometric Ca^{2+} indicators. The Ca^{2+} sensitivity of cameleons is determined by the affinity of calmodulin for Ca^{2+}, which can be altered by mutation of the calmodulin sequence.

A number of important improvements have been made in the molecular properties of GFP-based Ca^{2+} indicators that have been introduced over the last few years. These include reducing the sensitivity to pH (still a problem with many of these indicators), increasing the brightness and dynamic range of the fluorescent signal changes, and modifying the Ca^{2+}-sensitivty [74–77]. In addition, circular permutation of the calmodulin Ca^{2+}-sensor component has been used to generate novel Ca^{2+}-binding proteins that undergo appropriate conformational changes for both dual and single fluorescent protein sensing of Ca^{2+} changes [76, 78, 79]. As a result of these biotechnology advances in the properties of fluorescent protein-based Ca^{2+} indicators, these probes are finding increasing utilization for studies of cytosolic Ca^{2+} signals. Moreover, they remain the technique of choice for studies of organelle Ca^{2+}, and Ca^{2+} in the immediate vicinity of discrete proteins such as Ca^{2+} channels, because of the ability to use fusion protein constructs to achieve selective targeting within cells (see Chapter 4). The primary limitation of the fluorescent protein-based Ca^{2+} indicators remains their relatively limited dynamic range and consequent lower signal-to-noise ratio compared to the best chemical dye Ca^{2+} indicators.

14.3.4 USING GFP-BASED Ca^{2+} INDICATORS

The measurement of signaling events with Ca^{2+}-sensitive biosensors poses a different set of problems from the fluorescent indicator dyes. The first hurdle is to express the protein at sufficient levels to allow detection of Ca^{2+} changes without significantly buffering the Ca^{2+} changes under study or causing cellular toxicity. Commercially available liposome-based transfection reagents have been utilized successfully to express a variety of transgenes in commonly studied cell lines (e.g., HEK 293, CHO, COS-7). However, these protocols do not mediate efficient gene expression in most primary cell types. For example, even with the strong CMV promoters typically found in mammalian expression vectors, transfection rates rarely exceed 1.0% in cultured hepatocytes. Viral-mediated gene transfer, either lenti- or adenoviruses, can circumvent this limitation. At relatively high viral titers (500–1000 virus particles/cell) we can attain 100% gene expression in liver cells without overt toxicity. However, viral infection does not have an entirely benign effect on cell physiology; a decrease in cell number can be observed with high viral loads. Moreover, the existence of Toll-like receptors that mediate the innate immune responses to viral DNA exemplifies some of the possible obstacles that can be encountered with viral-mediated gene transfer. It should also be pointed out that working with viral vectors and propagating the virus particles introduces additional procedures and requires precautions beyond those usually necessary for plasmid-based transfections.

Electroporation technology offers an alternative to viral transfection gene transfer. The confocal images in Figure 14.5 and Figure 14.6 depict genetically engineered Ca^{2+}-sensitive fluorescence proteins targeted to the nucleus or mitochondrial matrix in primary cultured hepatocytes. For each cDNA plasmid, the electroporation protocol did not modify the targeting of the fluorescent proteins to its appropriate subcellular location (Figure 14.5 and Figure 14.6). Cells that are

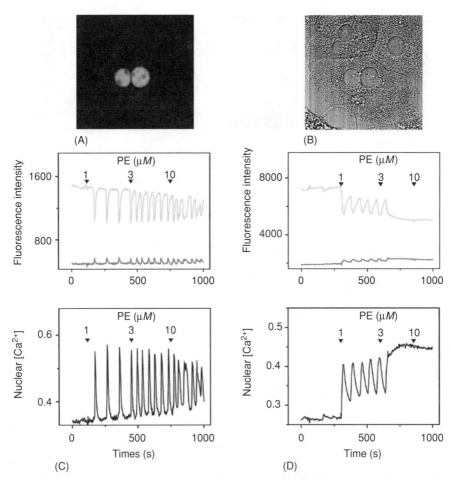

FIGURE 14.5 Measurement of nuclear calcium with genetically targeted Ca^{2+}-sensitive fluorescent proteins. Primary cultured hepatocytes were electroporated with 5 μg of plasmid for ratiometric-pericam targeted to the nucleus [79] using Nucleofection technology (Amaxa), then cultured for 16–18 h as previously described [57]. (A) Confocal z-stack of ratiometric-pericam-nu fluorescence acquired with 488-nm excitation. (B) Reflective image of the cell acquired at midsection. (C, D) The top panels show the time courses of fluorescence intensity changes observed at 405 nm and 495 nm excitation with an emission wavelength of 515–600 nm. The bottom panels show the corresponding changes in the ratio of 495/405 nm. Cells were challenged with the indicated phenylephrine (PE) concentrations.

successfully transfected using this electroporation procedure are generally more responsive to hormones than cultures transfected with liposome-based reagents. Nevertheless, some cells expressing the biosensor have three or more nuclei suggesting that cells may fuse during the electroporation protocol. These cells are easily identified and should be excluded from analysis.

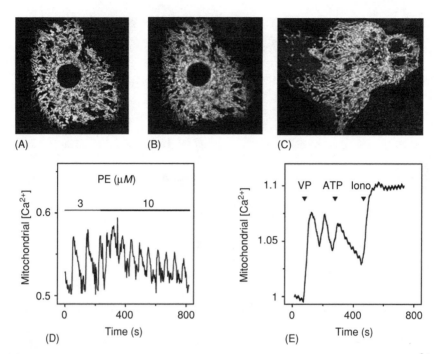

FIGURE 14.6 Measurement of mitochondrial calcium with genetically engineered Ca^{2+}-sensitive fluorescent proteins targeted to the mitochondrial matrix. Primary cultured hepatocytes were electroporated with 5 μg of plasmid for ratiometric-pericam-mt [79] (A, B, D) or 2mt yellow-cameleon-2 [81] (C, E) using Nucleofection technology. (A) shows a single confocal section and B a z-stack through a hepatocyte expressing ratiometric-pericam-mt, while (C) shows a z-stack of a hepatocyte expressing 2mt yellow-cameleon-2. All confocal images were acquired with the 488-nm argon laser line. Panel (D) shows the time course of changes in the 495/405 nm fluorescence ratio observed after phenylephrine (PE) challenge. In panel (E), the trace shows the change in FRET between CFP and YFP in a hepatocyte expressing 2mt yellow-cameleon-2 (436 nm excitation; >520 nm emission). Where indicated the culture was sequentially challenged with vasopressin (2 nM, VP), ATP (100 μM) or ionomycin (5 μM).

One of the key attributes of genetically engineered fluorescence proteins is the ability to target the probe and measure Ca^{2+} responses exclusively in subcellular compartments. The traces in Figure 14.5 and Figure 14.6 show agonist-stimulated Ca^{2+} signals in the nucleus (Figure 14.5) or mitochondrial matrix (Figure 14.6). Addition of a submaximal dose of the α-adrenergic agonist phenylephrine evoked baseline separated oscillations in nuclear Ca^{2+}. These nuclear Ca^{2+} responses have similar properties to cytosolic Ca^{2+} signals reported by us and others in fura-2 loaded hepatocytes [18–20, 22], namely the oscillation frequency increases with agonist concentration, while the spike amplitude is relatively insensitive to the strength of the stimulus (Figure 14.5C). Figure 14.5 also illustrates another potential pitfall when overexpressing Ca^{2+}-binding proteins. The fluorescence intensity of the

Ca^{2+}indicator in Figure 14.5D is fivefold higher than in the cell depicted in Figure 14.5C. These higher levels of protein expression dramatically slow the kinetics of agonist-evoked Ca^{2+} responses and decrease the magnitude of the nuclear Ca^{2+} spikes, consistent with an increase in Ca^{2+} buffering. A similar phenomenon also occurs with cameleons targeted to the mitochondrial matrix (Figure 14.6). Note the relatively sluggish rate of Ca^{2+} rise and decay back to baseline in Figure 14.6E compared to Figure 14.6D, where a low loading of ratiometric pericam was used. In hepatocytes, agonist-evoked mitochondrial calcium signals measured with aequorin have time-to-peak of 10–15 s [57], significantly shorter than observed with a high load of mitochondrial-targeted cameleon in Figure 14.6E. At very high loading levels (which of course give the best signal-to-noise) it is not possible to detect any measurable Ca^{2+} increase after maximal agonist challenge.

The intracellular targeting sequences can also have unexpected effects on the Ca^{2+}-sensing properties of fluorescent protein-based Ca^{2+} indicators. For example, the dissociation constant for Ca^{2+} has been reported to increase from 2 to 11 μM when ratiometric-pericam is targeted to the mitochondrial matrix with a signal peptide derived from yeast cytochrome c oxidase subunit 4 (COX IV) [80]. The underlying mechanism for the shift in Ca^{2+} affinity is not currently understood, but may arise from an inability of the matrix proteases to cleave the leader peptide [81]. Pozzan and collaborators have shown that tandem repeats of the leader sequence derived from cytochrome c oxidase subunit 8 is efficiently cleaved in the matrix and does not alter the probes' affinity for calcium. A decrease in Ca^{2+}-binding affinity can be advantageous in cell types where agonist challenge increases mitochondrial Ca^{2+} to high micromolar levels, but not in cells such as hepatocytes, where mitochondrial Ca^{2+} responses peak at 2 μM [57]. The trace in Figure 14.6D illustrates mitochondrial Ca^{2+} responses measured with COX IV-targeted ratiometric pericam, where mitochondrial Ca^{2+} spikes are relatively noisy, in part due to the low loading and in part due to the dynamic range of the protein indicator.

14.4 INSTRUMENTATION

14.4.1 EPIFLUORESCENCE MICROSCOPE

While it is possible to measure Ca^{2+} signals from cells in populations of cells using readily available fluorometer systems, studies of [Ca^{2+}] in single cells or subcellular regions typically require some form of digital imaging fluorescence microscopy to visualize the individual cells. In general, adherent cells loaded with fluorescent indicator dyes or transfected with fluorescent protein Ca^{2+} sensors are placed in an appropriate incubation chamber on the stage of an epifluorescence microscope. Excitation light of the desired wavelength is directed to the cells via the objective through which cells are imaged. The resulting emitted fluorescence passes back through the objective and is discriminated from the exciting light by means of a dichroic mirror. Emitted light of the appropriate wavelength is then selected, focused on to a detector, digitized, and stored to computer. Ideally, the illumination of the cells is limited to the period of image acquisition through opening of shutters under computer control. An important innovation in fluorescence Ca^{2+} imaging is the use of total internal reflection fluorescence microscopy [82], which images fluorescence signals only from the immediate vicinity (~200 nm) of the plasma membrane in contact with the coverslip through which the specimen is viewed. This technique offers the opportunity to measure Ca^{2+} signals from only this limited subcellular domain, but cannot be extended to other, less accessible subcellular domains.

A standard research-grade epifluorescence microscope is sufficient to record Ca^{2+} signals from single cells loaded with fluorescent Ca^{2+} indicators, but it is important to optimize the efficiency of the light path because fluorescent signals tend to be relatively small, and short acquisition times must be used to follow Ca^{2+} dynamics. The most important single element is the objective, since in an epifluorescence microscope this serves to both deliver the excitation light and collect the emitted fluorescence. The objective lens must transmit light efficiently at the wavelengths to be used. For example, conventional objectives do not transmit in the UV range and are thus not suited for excitation of fura-2. For these measurements, quartz or special Fluor objectives are required. Regardless of the dye, in order to maximize fluorescence signals, lenses with high numerical aperture (an index of the light collecting capacity) should be used. Optimal numerical aperture is usually obtained with lenses designed for use with immersion fluids such as glycerol, oil, or water. In most cases, single cell studies are performed with an inverted microscope as opposed to an upright system, because this configuration is suitable for use with cells in culture and is amenable to facile perfusion and other manipulations of the extracellular medium.

Fluorescence excitation requires a high-intensity illumination source. Although lasers are used in some applications, xenon and mercury arc lamps are the standard excitation source for most epifluorescence microscope imaging studies of Ca^{2+} signals. A xenon arc lamp is preferred over a mercury arc lamp, because it gives a more even illumination over the broad range of wavelengths required for excitation of the commonly used Ca^{2+} indicators. A 75 W xenon lamp is sufficient for most applications, and will usually have to be attenuated with neutral-density filters to minimize photobleaching and photodamage.

The selection of excitation wavelength is usually achieved using an optical interference filter placed in the illumination path. For multiexcitation measurements such as with ratiometric Ca^{2+} indicator dyes, it is necessary to have an automated device for changing filters. This can be achieved by placing the appropriate filters in a rotating disk or other mechanically switched systems [83]. Monochromator systems, similar in design to those used in fluorometers, can also be used for selection of the excitation wavelength. This approach is more flexible than using filter changers, in that the number of available wavelengths is essentially unlimited. With a galvanometer-driven monochromator, rapid rates of wavelength change can be achieved. To minimize photobleaching, it is necessary to block illumination when data is not acquired, usually with an electronically controlled shutter in front of the light source. In the epifluorescence configuration, a dichroic mirror is used to reflect the excitation light into the objective, while allowing the returning emission light to pass to the detector (Figure 14.7). Dichroic mirrors are designed to reflect

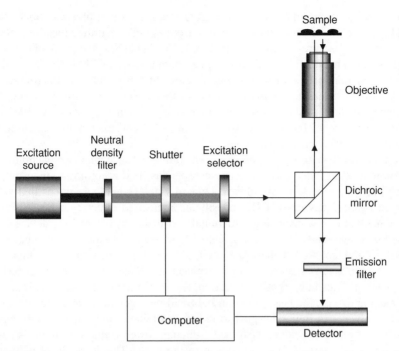

FIGURE 14.7 Schematic representation of a typical setup for monitoring single cell fluorescence.

light only in a specific wavelength range and must be selected for the specific fluorophore under study. For example, a dichroic mirror with a reflection cutoff of 400 nm would be appropriate for excitation of fura-2 at 340 and 380 nm, since this would pass the emission light which peaks at about 500 nm.

It is generally necessary to eliminate stray excitation light by means of an emission filter placed after the dichroic mirror. An ideal emission filter passes a relatively broad range of wavelengths in order to maximize the fluorescence signal arriving at the detector. However, when multiple wavelengths of emission light must be measured, a narrower bandpass filter may be necessary. In addition, for dual emission fluorescent Ca^{2+} indicator dyes such as indo-1 and FRET-based fluorescent protein Ca^{2+} indicators, the emitted fluorescence must be further discriminated. This can be achieved by using a second dichroic mirror to split the two emissions to individual detectors, or the emitted light can be alternately selected with a filter changer placed between the exit port of the microscope and the detector in a manner similar to the selection of excitation wavelengths for dual-excitation dyes.

14.4.2 DETECTORS AND IMAGE ACQUISITION

It is possible to obtain dynamic fluorescence Ca^{2+} recordings from single cells and even subcellular regions without an imaging device, by masking the region of interest and recording the fluorescence from this region to a photomultiplier tube (PMT). While this can provide information from a single identified location, it does not provide a "picture" of the distribution of Ca^{2+} changes, as can be obtained in imaging measurements. Nevertheless, signals captured with a PMT can be acquired rapidly, and may provide better signal-to-noise ratio than imaging measurements because of the inherent spatially averaging of this detection method. It should also be noted that laser scanning confocal microscopy systems typically use PMTs to detect fluorescence, but in this case the image is constructed by scanning the illumination beam across the sample.

Fluorescence imaging systems based on epifluorescence microscopy commonly use some form of video camera. These can be classified according to their output, which may be in the form of either a standard video signal (e.g., intensified video charged coupled devices [CCD] and silicon intensified target cameras [SIT]) or digital readout following analogue to digital conversion by the imaging device (e.g., cooled or slow-scan CCD cameras). Better signal-to-noise ratios are obtained with cooled CCD cameras but these are often slower than video cameras. Important parameters in determining camera performance are sensitivity (often described by the quantum efficiency), spatial resolution, and dynamic range. The more sensitive the imaging device, the less excitation light will be required for image acquisition, which is always an important goal to reduce photobleaching. Imaging formats of 1024×1024 pixels are commonly used, although lower spatial resolution is sufficient for some types of Ca^{2+} imaging measurements. This lower level of spatial resolution may also be achieved by "binning" multiple pixels for readout. Binning involves integrating the signal from several adjacent pixels within the camera prior to digitization to improve signal strength and enhance the readout time at the expense of spatial resolution. In order to carry out quantitative measurements of Ca^{2+} changes, a minimum digitization range of 8 bits ($2^8 = 256$ intensity levels) is required. However, 12- to 16-bit cameras give greater dynamic range, improving accuracy and reducing the potential for parts of the image to be saturated with excess light. See Ref. [83] for further details on detector selection.

Rates of data acquisition need to be optimized to the experimental system under study. In some cases, the $[Ca^{2+}]$ changes occur relatively slowly, such as the Ca^{2+} oscillations observed in nonexcitable cells such as those shown in Figure 14.1 and Figure 14.5. Studies such as these can be carried out with acquisition rates of one image per second or less. On the other hand, more rapid Ca^{2+} changes such as those that occur in muscle and nerve cells necessitate faster acquisition times, which can involve PMT-based systems (including confocal microscopy line scans), high speed cameras, and hybrid systems where image time-series are stored on the same CCD camera imaging chip where they are acquired [84]. It should also be noted that the kinetics of Ca^{2+} binding to the fluorescent Ca^{2+} indicator can become a limiting factor in fast measurements, leading to the use of more rapid low affinity Ca^{2+} indicator dyes (see Chapter 1). Even with relatively slow global events, a higher spatial and temporal resolution can reveal additional organization of Ca^{2+} signals at the subcellular level, such as intracellular Ca^{2+} waves. Figure 14.8 shows fura-2 loaded hepatocytes responding to vasopressin assessed on a more rapid time scale

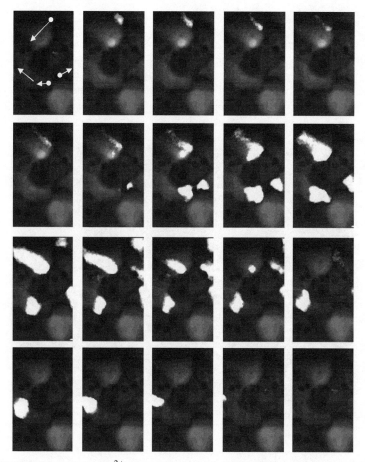

FIGURE 14.8 Intracellular Ca^{2+} waves. Images of fura-2 loaded hepatocytes (excitation 380 nm, emission > 420 nm) stimulated with vasopressin (10 nM). The overlaid white image is a difference image representing the change in fura-2 intensity at that time point. The top left panel shows the initiation locus (circle) and direction (arrow) of the Ca^{2+} wave in three responding cells. Data were acquired at 650 ms intervals (left to right).

than used for the whole cell recordings of Figure 14.1. In these imaging experiments, it becomes clear that the hormone-evoked cytosolic Ca^{2+} signal starts at a specific subcellular region and spreads through the cell in the form of a Ca^{2+} wave [22]. Another example where the subcellular organization of Ca^{2+} signals gives fundamentally different information from global Ca^{2+} measurements are the elemental events known as Ca^{2+} sparks, that have been observed in both excitable and non-excitable cells [26, 52, 85–87].

14.4.3 CONFOCAL MICROSCOPY

A detailed description of confocal microscopy is beyond the scope of this chapter, but a series of good reviews can be found in [88, 89]. Confocal microscopy differs from conventional epifluorescence microscopy in that the excitation light is scanned over the sample to build up the image. The scanned beam is typically derived from a laser source, although other configurations are possible. The key

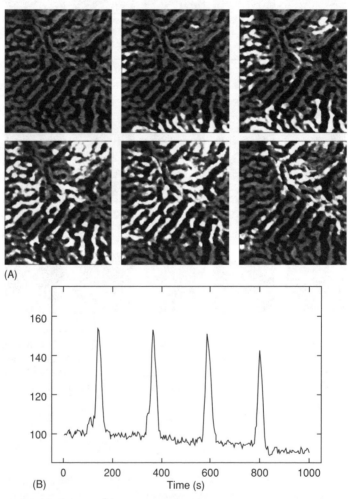

(A)

(B)

FIGURE 14.9 Intercellular Ca^{2+} waves in an intact organ. (A) Confocal images of an intact perfused rat liver loaded with fluo-3 (excitation 488 nm, argon laser, emission > 515 nm) and then perfused with vasopressin (50 p*M*). The overlaid white image is a difference image representing the change in fluo-3 intensity at that time point. Images were acquired at 20 s intervals (left to right). (B) Typical measured change in fluo-3 fluorescence of a representative individual hepatocyte within the liver showing Ca^{2+} oscillations in response to vasopressin.

element of confocal microscopy is the arrangement of optical pinholes that allow exclusion of out of focus light. This arrangement provides excellent spatial resolution and can yield optical sections and three-dimensional reconstruction of subcellular structures (see Figure 14.3, Figure 14.5, and Figure 14.6). Among other advantages, confocal microscopy is particularly well suited for monitoring fluorescence changes in thick specimens such as large cells and tissue slices. Indeed, considerable success has been achieved in monitoring single cell Ca^{2+} changes even in intact organs [91–93]. Figure 14.9 shows a confocal image of an intact liver loaded with fluo-3. Perfusion with the IP$_3$-forming agonist, vasopressin, induces Ca^{2+} oscillations in cells throughout the lobe (Figure 14.9A) that can be resolved individually (Figure 14.9B). Indeed, highly coordinated intercellular waves traveling the entire length of the liver lobule have been characterized [90]. Scan times for most confocal systems are relatively slow (1–2 s); however, higher temporal resolution can be obtained by scanning the laser line repeatedly on a single axis, to generate a line-scan image with temporal resolution of 1–2 ms and high spatial resolution in a single dimension [52, 85–87].

An important innovation in confocal microscopy is two-photon or multiphoton excitation [94–96], in which coincident excitation of a fluorophore by more than one photon of longer wavelength gives rise to the fluorescence signal. Advantages of this approach include excitation in a diffraction-limited spatial spot without the need for a pinhole in the fluorescence emission pathway, less tissue damage due to the use of longer wavelength light, and the ability to penetrate more deeply into tissue. The most suitable fluorescent Ca^{2+} indicator dyes for use with two-photon confocal microscopy are discussed in Chapter 1.

ACKNOWLEDGMENTS

This study was supported by NIH grants DK38422 and AA014980 (to A.P.T.).

REFERENCES

1. Berridge, M.J., Bootman, M.D., and Lipp, P. Calcium — a life and death signal. *Nature* 395, 645, 1998.
2. Carafoli, E. The ambivalent nature of the calcium signal. *J Endocrinol Invest* 27, 134, 2004.
3. Berridge, M.J., Bootman, M.D., and Roderick, H.L. Calcium signalling: dynamics, homeostasis and remodelling. *Nat Rev Mol Cell Biol* 4, 517, 2003.
4. Means, A.R., Vanberkum, M.F.A., Bagchi, I., Lu, K.P., and Rasmussen, C.D. Regulatory functions of calmodulin. *Pharmac Ther* 50, 255, 1991.
5. Stull, J.T. Ca^{2+}-dependent cell signaling through calmodulin-activated protein phosphatase and protein kinases minireview series. *J Biol Chem* 276, 2311, 2001.
6. Carafoli, E. and Brini, M. Calcium pumps: structural basis for and mechanism of calcium transmembrane transport. *Curr Opin Chem Biol* 4, 152, 2000.
7. Orrenius, S. and Bellomo, G. Toxicological implications of perturbation of Ca^{2+} homeostasis in hepatocytes, in *Calcium and Cell Function*. Volume 6, Ed., Cheung, W.Y., Academic Press Inc., New York, 185, 1986.

8. Philipson, K.D. and Nicoll, D.A. Sodium–calcium exchange: a molecular perspective. *Annu Rev Physiol* 62, 111, 2000.

9. Berridge, M.J. Inositol trisphosphate and calcium signalling. *Nature* 361, 1993, 315.

10. Patterson, R.L., Boehning, D., and Snyder, S.H. Inositol 1,4,5-trisphosphate receptors as signal integrators. *Annu Rev Biochem* 73, 437, 2004.

11. Nowycky, M.C. and Thomas, A.P. Intracellular calcium signaling. *J Cell Sci* 115, 3715, 2002.

12. Taylor, C.W. and Richardson, A. Structure and function of inositol trisphosphate receptors. *Pharmac Ther* 51, 97, 1991.

13. Patel, S., Joseph, S.K., and Thomas, A.P. Molecular properties of inositol 1,4,5-trisphosphate receptors. *Cell Calcium* 25, 247, 1999.

14. Lee, H.C. Physiological functions of cyclic ADP-ribose and NAADP as calcium messengers. *Annu Rev Pharmacol Toxicol* 41, 317, 2001.

15. Patel, S. NAADP-induced Ca^{2+} release — a new signalling pathway. *Biol Cell* 96, 19, 2004.

16. Cancela, J.M., Charpentier, G., and Petersen, O.H. Co-ordination of Ca^{2+} signalling in mammalian cells by the new Ca^{2+}-releasing messenger NAADP. *Pflug. Arch* 446, 322, 2003.

17. Galione, A. and Churchill, G.C. Interactions between calcium release pathways: multiple messengers and multiple stores. *Cell Calcium* 32, 343, 2002.

18. Woods, N.M., Cuthbertson, K.S.R., and Cobbold, P.H. Repetitive transient rises in cytoplasmic free calcium in hormone-stimulated hepatocytes. *Nature* 319, 600, 1986.

19. Thomas, A.P., Bird, G.St.J., Hajnóczky, G., Robb-Gaspers, L.D., and Putney, J.W., Jr. Spatial and temporal aspects of cellular calcium signaling. *FASEB J* 10, 1505, 1996.

20. Rooney, T.A., Sass, E.J., and Thomas, A.P. Characterization of cytosolic calcium oscillations induced by phenylephrine and vasopressin in single fura-2 loaded hepatocytes. *J Biol Chem* 264, 17131, 1989.

21. Petersen, O.H. Local and global Ca^{2+} signals: physiology and pathophysiology. *Biol Res* 37, 661, 2004.

22. Rooney, T.A., Sass, E.J., and Thomas, A.P. Agonist-induced cytosolic calcium oscillations originate from a specific locus in single hepatocytes. *J Biol Chem* 265, 10792, 1990.

23. Thorn, P., Lawrie, A.M., Smith, P.M., Gallacher, D.V., and Petersen, O.H. Local and global cytosolic Ca^{2+} oscillations in exocrine cells evoked by agonists and inositol trisphosphate. *Cell* 74, 661, 1993.

24. Lechleiter, J., Girard, S., Peralta, E., and Clapham, D. Spiral calcium wave propagation and annihilation in *Xenopus laevis* oocytes. *Science* 252, 123, 1991.

25. Kasai, H. and Augustine, G.J. Cytosolic Ca^{2+} gradients triggering unidirectional fluid secretion from exocrine pancreas. *Nature* 348, 735, 1990.

26. Berridge, M.J. Elementary and global aspects of calcium signalling. *J Exp Biol* 200, 315, 1997.

27. Tsien, R.Y. Fluorescent indicators of ion concentration. *Methods Cell Biol* 30, 127, 1989.

28. Miyawaki, A., Griesbeck, O., Heim, R., and Tsien, R.Y. Dynamic and quantitative Ca^{2+} measurements using improved cameleons. *Proc Natl Acad Sci U.S.A* 96, 2135, 1999.

29. Zhang, J., Campbell, R.E., Ting, A.Y., and Tsien, R.Y. Creating new fluorescent probes for cell biology. *Nat Rev Mol Cell Biol* 3, 906, 2002.

30. Miyawaki, A. Fluorescence imaging of physiological activity in complex systems using GFP-based probes. *Curr Opin Neurobiol* 13, 591, 2003.

31. Tsien, R.Y. New calcium indicators and buffers with high selectivity against magnesium and protons: design, synthesis, and properties of prototype structures. *Biochemistry* 19, 2396, 1980.

32. Tsien, R.Y. A non-disruptive technique for loading calcium buffers and indicators into cells. *Nature* 290, 527, 1981.

33. Thastrup, O., Cullen, P.J., Drobak, B.K., Hanley, M.R., and Dawson, A.P. Thapsigargin, a tumour promoter, discharges intracellular stores by specific inhibition of the endoplasmic reticulum Ca^{2+}-ATPase. *Proc Natl Acad Sci U.S.A.* 87, 2466, 1990.

34. Glennon, M.C., Bird, G.S., Takemura, H., Thastrup, O., Leslie, B.A., and Putney, J.W. Jr. *In situ* imaging of agonist-sensitive calcium pools in AR4–2J pancreatoma cells. Evidence for an agonist- and inositol 1,4,5-trisphosphate-sensitive calcium pool in or closely associated with the nuclear envelope. *J Biol Chem* 267, 25568, 1992.

35. Hajnóczky, G., Robb-Gaspers, L.D. Seitz, M.B., and Thomas, A.P. Decoding of cytosolic calcium oscillations in the mitochondria. *Cell* 82, 415, 1995.

36. Quintana, A. and Hoth, M. Apparent cytosolic calcium gradients in T-lymphocytes due to fura-2 accumulation in mitochondria. *Cell Calcium* 36, 99, 2004.

37. Tse, F.W., Tse, A., and Hille, B. Cyclic Ca^{2+} changes in intracellular stores of gonadotropes during gonadotropin-releasing hormone-stimulated Ca^{2+} oscillations. *Proc Natl Acad Sci U.S.A* 91, 9750, 1994.

38. Golovina, V. and Blaustein, M. Spatially and functionally distinct Ca^{2+} stores in sarcoplasmic and endoplasmic reticulum. *Science* 275, 1643, 1997.

39. Mogami, H., Tepikin, A.V., and Petersen, O.H. Termination of cytosolic Ca^{2+} signals: Ca^{2+} reuptake into intracellular stores is regulated by free Ca^{2+} concentration in the store lumen. *EMBO J* 17, 435, 1998.

40. Hofer, A.M., Landolfi, B., Debellis, L., Pozzan, T., and Curci, S. Free [Ca^{2+}] dynamics measured in agonist-sensitive stores of single living intact cells: a new look at the refilling process. *EMBO J* 17, 1986, 1998.

41. Hofer, A.M. and Machen, T.E., Technique for *in situ* measurement of calcium in intracellular inositol 1,4,5-trisphosphate-sensitive stores using the fluorescent indicator mag-fura-2. *Proc Natl Acad Sci U.S.A* 90, 2598, 1993.

42. Tanimura, A. and Turner, R.J., Inositol 1,4,5-trisphosphate-dependent oscillations of luminal [Ca^{2+}] in permeabilized HSY cells. *J Biol Chem* 271, 30904, 1996.

43. Hajnóczky, G. and Thomas, A.P., Minimal requirements for calcium oscillations driven by the IP$_3$ receptor. *EMBO J* 16, 3533, 1997.

44. Hajnóczky, G., Hager, R., and Thomas, A.P. Mitochondria suppress local feedback activation of inositol 1,4,5-trisphosphate receptors by Ca^{2+}. *J Biol Chem* 274, 14157, 1999.

45. Park, M.K., Petersen, O.H., and Tepikin, A.V. The endoplasmic reticulum as one continuous Ca^{2+}-pool: visualization of rapid Ca^{2+} movements and equilibration. *EMBO J* 19, 5729, 2000.

46. Simpson, P.B. and Russell, J.T. Mitochondria support inositol 1,4,5-trisphosphate-mediated Ca^{2+} waves in cultured oligodendrocytes. *J Biol Chem* 271, 33493, 1996.

47. Smaili, S.S., Stellato, K.A., Burnett, P., Thomas, A.P., and Gaspers, L.D. Cyclosporin A inhibits Inositol 1,4,5-trisphosphate-dependent Ca^{2+} signals by enhancing Ca^{2+} uptake into the ER and mitochondria. *J Biol Chem* 276, 23329, 2001.

48. Voronina, S., Sukhomlin, T., Johnson, P.R., Erdemli, G., Petersen, O.H., and Tepikin, A. Correlation of NADH and Ca^{2+} signals in mouse pancreatic acinar cells. *J Physiol* 539, 41, 2002.

49. Rakhit, R.D., Mojet, M.H., Marber, M.S., and Duchen, M.R. Mitochondria as targets for nitric oxide-induced protection during simulated ischemia and reoxygenation in isolated neonatal cardiomyocytes. *Circulation* 103, 2617, 2001.

50. Haughland, R.P. *Handbook of Fluorescent Probes and Research Chemicals.* Ed., Spence, M.T.Z., Eugene, Oregon, 1992.

51. Csordás, G., Thomas, A.P., and Hajnóczky, G. Quasi-synaptic calcium signal transmission between endoplasmic reticulum and mitochondria. *EMBO J* 18, 96, 1999.

52. Pacher, P., Thomas, A.P., and Hajnóczky, G. Miniature calcium signals in single mitochondria driven by ryanodine receptors. *Proc Natl Acad Sci U.S.A* 99, 2380, 2002.

53. Chalmers, S. and Nicholls, D.G. The relationship between free and total calcium concentrations in the matrix of liver and brain mitochondria. *J Biol Chem* 278, 19062, 2003.

54. Tsien, R.Y., Pozzan, T., and Rink, T.J. Calcium homeostasis in intact lymphocytes: cytoplasmic free calcium monitored with a new, intracellularly trapped fluorescent indicator. *J Cell Biol* 94, 325, 1982.

55. Grynkiewicz, G., Poenie, M., and Tsien, R.Y. A new generation of Ca^{2+} indicators with greatly improved fluorescence properties. *J Biol Chem* 260, 3440, 1985.

56. Pralong, W.-F., Spat, A., and Wollheim, C.B. Dynamic pacing of cell metabolism by intracellular Ca^{2+} transients. *J Biol Chem* 269, 27310, 1994.

57. Robb-Gaspers, L.D., Burnett, P., Rutter, G.A., Denton, R.M., Rizzuto, R., and Thomas, A.P. Integrating cytosolic calcium signals into mitochondrial metabolic responses. *EMBO J* 1, 49877, 1998.

58. Allbritton, N.L., Oancea, E., Kuhn, M.A., and Meyer, T. Source of nuclear calcium signals. *Proc Natl Acad Sci U.S.A.* 91, 12458, 1994.

59. Cobbold, P.H. and Rink, T.J. Fluorescence and bioluminscence measurement of cytoplasmic free calcium. *Biochem J* 248, 313, 1987.

60. Rizzuto, R., Simpson, A.W., Brini, M., and Pozzan, T. Rapid changes of mitochondrial Ca^{2+} revealed by specifically targeted recombinant aequorin. *Nature* 267, 10939, 1992.

61. Brini, M., Murgia, M., Pasti, L., Picard, D., Pozzan, T., and Rizzuto, R. Nuclear Ca^{2+} concentration measured with specifically targeted recombinant aequorin. *EMBO J* 12, 4813, 1993.

62. Montero, M., Brini, M., Marsault, R., Alvarez, J., Sitia, R., Pozzan, T., and Rizzuto, R. Monitoring of dynamic changes in free Ca^{2+} concentration in the endoplasmic reticulum of intact cells. *EMBO J* 14, 5467, 1995.

63. Marsault, R., Murgia, M., Pozzan, T., and Rizzuto, R. Domains of high Ca^{2+} beneath the plasma membrane of living A7r5 cells. *EMBO J* 16, 1575, 1997.

64. Pouli, A.E., Karagene, N., Wasmeier, C., Hutton, J.C, Bright, N., Arden, S., Schofield, J.G., and Rutter, G.A. A phogrin-aequorin chimaera to image free Ca^{2+} in the vicinity of secretory granules. *Biochem J* 330, 1339, 1998.

65. Pinton, P., Pozzan, T., and Rizzuto, R. The golgi apparatus is an inositol 1,4,5-trisphosphate-sensitive store, with functional properties distinct from those of the endoplasmic reticulum. *EMBO J* 17, 5298, 1998.

66. Miyazaki, S., Hashimoto, N., Yoshimoto, Y., Kishimoto, T., Igusa, Y., and Hiramoto, Y. Temporal and spatial dynamics of the periodic increase in intracellular free calcium at fertilization of golden hamster eggs. *Dev Biol* 118, 259, 1986.

67. Speksnijder, J.E., Sardet, C., and Jaffe, L.F. The activation wave of calcium in the ascidian egg and its role in ooplasmic segregation. *J Cell Biol* 110, 1589, 1990.

68. Rutter, G.A., Burnett, P., Rizzuto, R., Brini, M., Murgia, M., Pozzan, T., Tavare, J.M., and Denton, R.M. Subcellular imaging of intramitochondrial Ca^{2+} with recombinant

targeted aequorin: significance for the regulation of pyruvate dehydrogenase activity. *Proc Natl Acad Sci U.S.A.* 93, 5489, 1996.

69. Ainscow, E.K. and Rutter, G.A. Mitochondrial priming modifies Ca^{2+} oscillations and insulin secretion in pancreatic islets. *Biochem J* 353, 175, 2001.

70. Miyawaki, A., llopis, J., Heim, R., McCaffery, J.M., Adams, J.A., and Tsien, R.Y. Fluorescent indicators for Ca^{2+} based on green fluorescent proteins and calmodulin. *Nature* 388, 882, 1997.

71. Tsien, R.Y. The green fluorescent protein. *Annu Rev Biochem* 67, 509, 1998.

72. Lippincott-Schwartz, J. and Patterson, G.H. Development and use of fluorescent protein markers in living cells. *Science* 300, 87, 2003.

73. Miyawaki, A. Fluorescent proteins in a new light. *Nat Biotechnol* 22, 1374, 2004.

74. Miyawaki, A., Griesbeck, O., Heim, R., and Tsien, R.Y. Dynamic and quantitative Ca^{2+} measurements using improved cameleons. *Proc Natl Acad Sci U.S.A.* 96, 2135, 1999.

75. Griesbeck, O., Baird, G.S., Campbell, R.E., Zacharias, D.A., and Tsien, R.Y. Reducing the environmental sensitivity of yellow fluorescent protein. Mechanism and applications. *J Biol Chem* 276, 29188, 2001.

76. Nagai, T., Yamada, S., Tominaga, T., Ichikawa, M., and Miyawaki, A. Expanded dynamic range of fluorescent indicators for Ca^{2+} by circularly permuted yellow fluorescent proteins. *Proc Natl Acad Sci U.S.A* 101, 10554, 2004.

77. Nagai, T., Ibata, K., Park, E.S., Kubota, M., Mikoshiba, K., and Miyawaki, A. A variant of yellow fluorescent protein with fast and efficient maturation for cell-biological applications. *Nat Biotechnol* 20, 87, 2002.

78. Baird, G.S., Zacharias, D.A., and Tsien, R.Y. Circular permutation and receptor insertion within green fluorescent proteins. *Proc Natl Acad Sci U.S.A.* 96, 11241, 1999.

79. Nagai, T., Sawano, A., Park, E.S., and Miyawaki, A. Circularly permuted green fluorescent proteins engineered to sense Ca^{2+}. *Proc Natl Acad Sci USA* 98, 3197, 2001.

80. Filippin, L., Magalhaes, P.J., Di Benedetto, G., Colella, M., and Pozzan, T. Stable interactions between mitochondria and endoplasmic reticulum allow rapid accumulation of calcium in a subpopulation of mitochondria. *J Biol Chem* 278, 39224, 2003.

81. Filippin, L., Abad, M.C., Gastaldello, S., Magalhaes, P.J., Sandona, D., and Pozzan, T. Improved strategies for the delivery of GFP-based Ca^{2+} sensors into the mitochondrial matrix. *Cell Calcium* 37, 129, 2005.

82. Demuro, A. and Parker, I. Imaging the activity and localization of single voltage-gated Ca^{2+} channels by total internal reflection fluorescence microscopy. *Biophys J* 86, 3250, 2004.

83. Thomas, A.P. and Delaville, F. The use of fluorescent indicators for measurement of cytosolic-free calcium concentration in cell populations and single cells, in *Cellular Calcium. A Practical Approach.* Eds., McCormack, J.G. and Cobbold, P.H., Oxford University Press, Oxford, 1, 1991.

84. O'Rourke, B., Reibel, D.K., and Thomas, A.P. High speed digital imaging of cytosolic Ca^{2+} and contraction in single cardiomyocytes. *Am J Physiol* 259, H230, 1990.

85. Guatimosim, S., Dilly, K., Santana, L.F., Saleet Jafri, M., Sobie, E.A., and Lederer, W.J. Local Ca^{2+} signaling and EC coupling in heart: Ca^{2+} sparks and the regulation of the [Ca^{2+}](i) transient. *J Mol Cell Cardiol* 34, 941, 2002.

86. Wellman, G.C. and Nelson, M.T. Signaling between SR and plasmalemma in smooth muscle: sparks and the activation of Ca^{2+}-sensitive ion channels. *Cell Calcium* 34, 211, 2003.

87. Zima, A.V. and Blatter, L.A. Inositol-1,4,5-trisphosphate-dependent Ca^{2+} signalling in cat atrial excitation-contraction coupling and arrhythmias. *J Physiol* 555, 607, 2004.

88. Pawley, J.B., Ed., *Handbook of Biological Confocal Microscopy*. Plenum Press, New York, 459, 1995.
89. Matsumoto, B., Ed., Cell biological applications of confocal microscopy. *Methods in Cell Biology*. vol. 70, 2002.
90. Robb-Gaspers, L.D. and Thomas, A.P. Coordination of Ca^{2+} signaling by intercellular propagation of Ca^{2+} waves in the intact liver. *J Biol Chem* 270, 8102, 1995.
91. Nathanson, M.H., Burgstahler, A.D., Mennone, A., Fallon, M.B., Gonzalez, C.B., and Saez, J.C. Ca^{2+} waves are organized among hepatocytes in the intact organ. *Am J Physiol* 269, G1667, 1995.
92. Minamikawa, T., Cody, S.H., and Williams, D.A. *In situ* visualization of spontaneous calcium waves within the perfused whole rat heart by confocal imaging. *Am J Physiol* 272, H236, 1997.
93. Patel, S., Robb-Gaspers, L.D., Stellato, K.A., Shon, M., and Thomas, A.P. Coordination of Ca^{2+} signaling by endothelial-derived nitric oxide in the intact liver. *Nature Cell Biol* 1, 467, 1999.
94. Denk, W., Strickler, J.H., and Webb, W.W. Two-photon laser scanning fluorescence microscopy. *Science* 248, 73, 1990.
95. Williams, R.M., Piston, D.W., and Webb, W.W. Two-photon molecular excitation provides intrinsic 3-dimensional resolution for laser-based microscopy and microphoto-chemistry. *FASEB J* 8, 804, 1994.
96. Zipfel, W.R., Williams, R.M., and Webb, W.W. Nonlinear magic: multiphoton microscopy in the biosciences. *Nat Biotechnol* 21, 1369, 2003.

15 Subcellular Compartmentalization of Calcium Signaling

Nicholas J. Dolman, Michael C. Ashby,
Myoung K. Park, Ole H. Petersen,
and Alexei V. Tepikin

CONTENTS

15.1 INTRODUCTION

This chapter is about the methods used to investigate the localization of calcium signals. Local or localized will be among the most frequently used adjectives in the chapter. We will start by describing the technology of local stimulation of receptors in polarized cells. This will be followed by the methods of observing calcium signals with respect to the location of cellular organelles and a description of the powerful technology of local uncaging. The last section will deal with localized extrusion of calcium. And when we finish, we will go to the local pub and celebrate with local ales.

15.2 INTRA-PATCH PIPETTE UNCAGING — TECHNIQUE FOR STUDYING THE DISTRIBUTION OF RECEPTORS IN POLARIZED CELLS

The idea behind this technique was to limit the stimulation to a small area encompassed by lipid–glass contact formed between the plasma membrane and patch pipette. This allows monitoring of downstream events following activation of receptors in the small selected region of a cell. The technique has the potential to reveal cellular events induced by stimulation of a single activated receptor. A schematic drawing of the setup for intra-patch pipette uncaging is shown in Figure 15.1. The caged precursor of a neurotransmitter was added to the patch pipette solution. A pulse of UV light was applied at the selected time point to illuminate the pipette resulting in the photorelease of the agonist. The cell-attached configuration of the patch-clamp technique has been used in these experiments [1]. To limit the spread of the photoreleased agonist the uncaging is triggered only after the formation of a gigaseal (tight electrical contact between the pipette and the cell membrane). The possibility of applying the pulses of UV light locally, illuminating the pipette but avoiding the patched cell, is advantageous as this allows the experimenter to start optical monitoring of postuncaging events earlier and avoids numerous artifacts that can be induced by illuminating a cell with intense UV light. Ideally, the technique should be combined with videoimaging (confocal or two-photon microscopy add further advantages). To achieve better illumination of the intrapipette solution with UV light, it is beneficial to use a relatively shallow angle of approach of the patch pipette (angle between cover glass and the pipette); in our experiments the angle was less than 45°. During uncaging we usually applied UV light to the region of interest surrounding 20–50 μm of the length of the patch pipette but avoiding a few micrometers immediately adjacent to the area of the seal

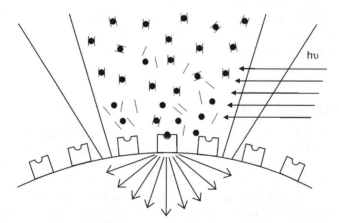

FIGURE 15.1 Schematic drawing of intra-patch pipette uncaging technique. Following the formation of a gigaseal, the neurotransmitter (shown as small filled circles) is photoreleased from the caged precursor by a pulse of UV light.

with the plasma membrane. Initially, caged fluorescein and caged fluorescein dextran were used for visualization and preliminary calibration of the technique. Images obtained after uncaging of these compounds could also be used to estimate the area of the cell occluded by the patch pipette.

We developed and applied intra-patch pipette uncaging to investigate the mechanism of formation of local calcium signals in pancreatic acinar cells [1] using caged carbachol as a photoactivatable precursor for local stimulation. We employed a relatively high concentration of caged carbachol (usually 100–1000 μM). This is advantageous since a lower intensity of light (or shorter pulses) is required to release enough neurotransmitter to stimulate the cell. One of the limitations on the concentration of caged compound is the level of contamination by the active compound (present even before uncaging), another restraint is the price of the caged agonist.

In our study, we found that local calcium signals were formed on the opposite side of the cell with regard to the place of stimulation. The pipette was placed on the basal part of the cell while local calcium signals were detected approximately 15–20 μm away — in the granular region. These results suggested long-distance communication between the receptors and calcium release sites. The robustness of this technique was illustrated by experiments with atropine (an inhibitor of muscarinic ACh receptors). A very high concentration of atropine (100 μM) was added to the extracellular solution (outside the cell–pipette system), after the formation of a gigaseal, which failed to prevent calcium signals induced by the uncaging. Another surprising finding was that stimulation of a very small proportion of the plasma membrane (1–2% of the cell area) was able to produce all forms of calcium signals — local calcium signals (induced with short pulses of uncaging) or global calcium signals (triggered by longer uncaging), and even calcium waves propagating from the stimulated cell to other cells of the cluster. This finding indicates that the receptor cascade operates with a remarkable degree of amplification. The pancreatic acinar cell is an example of a polarized epithelial cell. It is possible that the technology of intrapipette uncaging could become useful in studies of other types of polarized cells with well-defined specialized regions of the plasma membrane (e.g., other types of exocrine secretory cells and neurons). Other techniques of "probing" the receptor distribution employ local iontophoretic stimulation, two-photon uncaging of caged compounds in extracellular solution (not limited to the patch pipette), and the recently developed technique of local stimulation with a pipette containing glycerol and an agonist [2]. The advantage of these techniques is the possibility to stimulate multiple regions of the cell during a single experiment, however these techniques do not provide a physical barrier for diffusion of the agonist and therefore are probably less stringent (with regard to the limiting of the stimulation area).

The technique of localized intrapipette uncaging could become a useful tool in biophysical studies of the downstream signaling cascades induced by local uncaging/stimulation of single receptors. Carbachol, glutamate, GABA, NMDA, and ATP are already commercially available in the caged form (e.g., from Molecular Probes). The list of caged neurotransmitters and neurotransmitter analogs is likely to grow. The downstream signaling cascades that could be investigated by localized intrapipette uncaging in addition to calcium include other second messengers with available fluorescent probes (e.g., IP_3, PKC, cAMP).

15.3 LOCAL CALCIUM SIGNALS AND THE GEOGRAPHY OF CELLULAR ORGANELLES OR HOW TO PUT CALCIUM SIGNALS ON THE MAP

To achieve signal specificity, the cell encodes calcium signals in terms of both their spatial (where) and temporal (when) properties [3]. The characteristics of the calcium signal can be determined or influenced by intracellular organelles [4, 5]. Intracellular organelles act as a source and a sink for calcium, thereby regulating calcium signaling. Conversely, the concentration of calcium in the cytoplasm around the organelle as well as in the lumen of the organelle itself regulates the functioning of the organelle. Luminal calcium regulates processing of enzymes within compartments of both the endoplasmic reticulum (ER) (reviewed in Ref. [6]) and the Golgi (reviewed in Ref. [7]). Mitochondrial metabolism is regulated by intra-matrix calcium levels [5, 8]. Cytosolic calcium elevations are also needed for membrane trafficking throughout the secretory pathway [9, 10].

At rest, the concentration of calcium is higher in organelles such as the ER and Golgi, giving rise to calcium gradients between the cytosol and organelles. Agonist stimulation leads to calcium release from the ER [11], the Golgi [12], nuclear envelope [13, 14], and granule/lysosomal structures [15–17], elevating cytosolic calcium. Spatially restricted calcium release and calcium uptake results in the formation of complex calcium gradients in the cytosol of a cell. The visualization of cytosolic calcium changes, using high-resolution microscopy, has provided information about the spatial and temporal characteristics of these gradients [18–21]. One productive approach to study local calcium signaling was to correlate the positions of calcium transients with the location of cellular organelles. This section will consider experimental studies in which calcium signals and calcium gradients were studied simultaneously with imaging of intracellular organelles.

15.3.1 SOME CONSIDERATIONS ABOUT THE INSTRUMENTS

Local calcium signals are usually short-lived events [22–24], therefore high temporal resolution of at least a few frames per second (for some cell types tens or even hundreds of frames per second) is required. As indicators with different properties have to be used in these experiments, employing multiple excitation and emission wavelengths is highly advantageous. Confocal microscopy is most commonly used in this type of experiment. Nonconfocal digital imaging can also be used and indeed may be beneficial where spatial resolution can be sacrificed for increased temporal resolution (particularly in systems where the localized signals are very short-lived). It should however be noted that relatively high optical resolution (e.g., axial resolution of 1 μm) is required to determine the boundaries of the organelles. Such a resolution is difficult to achieve in nonconfocal systems.

15.3.2 COMBINATIONS OF FLUOROPHORES FOR SIMULTANEOUS ORGANELLAR LABELING AND STUDIES OF CALCIUM SIGNALING

Fluorophores used to localize organelles and measure calcium must have sufficiently different fluorescent properties. Fluorescent markers for organelles can be
either small organic fluorescent dyes, loaded into cells or organelle-targeted
fluorescent proteins that are expressed in the cells. It is important to note that
popular calcium indicators (for confocal microscopy) from the fluo family of
dyes (e.g., fluo-4) are not ideal for simultaneous staining with many fluorescent
proteins (e.g., CFP/GFP/YFP) because of partial spectral overlap. Useful information on the compatibility of fluorescent proteins and calcium indicators can be
found in the work by S. Bolsover and colleagues [25]. We found that Fura Red
and X-Rhod-1 are suitable for simultaneous imaging with EYFP [26, 27]. Fura
Red was also compatible with small organic indicators NBD C_6 Ceramide (for
the Golgi) and NADH autofluorescence (for mitochondria) (see Figure 15.2).
Nuclei usually accumulate Fura Red and can be distinguished from the cytosol
because of the slightly increased fluorescence (Figure 15.2). Staining with Hoechst
33342 can be done at the end of the experiment for more detailed nuclear
staining [21].

Investigations of calcium signals with simultaneous organellar labeling have
been conducted for most subcellular compartments. A few prominent examples are
considered here:

Wahl and colleagues performed a comprehensive study of calcium gradients
and the localization of organelles [19]. Fura-2 was used in combination with either
NBD C_6 Ceramide (Golgi), Rhodamine 123 (Mitochondria), or acridine orange
(acidic structures) to map calcium gradients with respect to these cellular compartments. Paemaeleire and colleagues used fluo-3 in combination with ER tracker-
blue/white to image the generation of cytosolic calcium signals with respect to the
ER [28].

Differences in the kinetics of the cytosolic and nuclear calcium signals have
been systematically investigated in a number of studies. In particular, a clear
difference in the shape of local calcium signals (calcium puffs) between the cytosol
and the cell nucleus has been documented [29]. Exclusion of the nucleoplasm from
some types of calcium signals [21] and attenuation of nuclear responses have also
been reported [30]. The nucleus is a large organelle and in some cell types, it is
visible in the transmitted light and does not require further staining. The nucleus can
also be identified by a higher than cytosolic fluorescence when the cell is stained
with calcium indicators. This has been seen for Fura Red [21] and for fluo-3 [30]. In
the study employing fluo-3, nuclear staining with ethidium bromide was used to
verify the nuclear localization at the end of some experiments [30]. Marangos and
colleagues investigated the regulation of calcium oscillations by the formation of
the pronuclei [31]. The authors used BSA conjugated nuclear localization signal
(NLS) coupled to the fluorescent molecule FITC. In some experiments, this nuclear
probe, FITC-NLS–BSA, was imaged simultaneously with Fura-2 dextran or with
Fura Red [31].

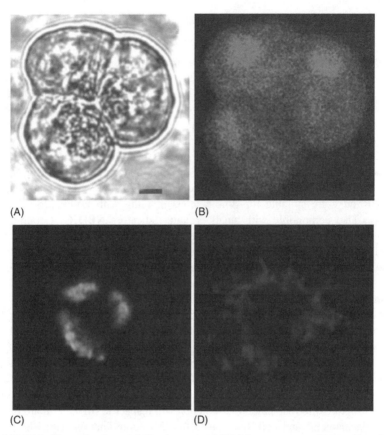

FIGURE 15.2 (See color insert following page 140.) Measurements of cytosolic calcium and organelle location. Transmitted light image (A) of a triplet of freshly isolated pancreatic acinar cells loaded with the calcium-zsensitive dye Fura Red (B, note the slightly higher fluorescence in the nuclei) and NBD C_6 Ceramide (C) to visualize the Golgi. UV excitation of NADH autofluorescence reveals the location of the mitochondria (D).

Radding and colleagues successfully imaged the release of acidic pools simultaneously with elevations in cytosolic calcium by using fluo-3 and lysotracker red [32].

There are a substantial number of studies in which calcium signals were measured simultaneously with positions and movements or responses of mitochondria (e.g., Ref. [33]). A particularly useful technical study was recently published by Quintana and Hoth [34]. These authors correlated the position of mitochondria (labeled with Mito Tracker Green) with calcium gradients (measured with Fura-2). This approach indicated that the apparent calcium gradients were in fact due to the accumulation of Fura-2 in mitochondria. In the same study, the authors suggested a method of Fura-2 loading that prevented mitochondrial accumulation of this indicator and consequently the formation of the artefactual calcium transients [34]. The use of genetically encoded calcium indicators [35] is becoming increasingly common. Cells expressing mitochondrially targeted Cam-Garoo were loaded with Fura-2 and the spread of a calcium wave across the cell in both the cytosol and in the mitochondria visualized [36]. In a study by Nagai and colleagues, cells were transfected with plasmids encoding nuclear and mitochondrially targeted pericams that allowed simultaneous imaging of calcium in these organelles [37].

15.3.3 MEASURING LOCAL CALCIUM GRADIENTS AND ORGANELLE LOCALIZATION IN POLARIZED SECRETORY CELLS

Many primary cell types are difficult to transfect and therefore small organic fluorescent probes are the only option for labeling organelles. In our experiments, we measured cytosolic calcium gradients with respect to the Golgi and mitochondria. The experiments were conducted on freshly isolated pancreatic acinar cells. The cells were loaded first with NBD C_6 Ceramide (Molecular Probes, 2.5 μM, for 10 min at 4°C) then with the calcium-sensitive dye Fura Red (5 μM for 45 min at room temperature); after both loading procedures the cells were washed twice in dye-free buffer. Both dyes were excited at 488 nm, and the emission collected between 500 and 530 nm for NBD C_6 Ceramide and above 580 nm for Fura Red (light collection above 580 nm was sufficient to largely avoid the contamination of the Fura Red signal by NBD C_6 Ceramide fluorescence). The position of mitochondria could be visualized in the same cells using NADH autofluorescence (excitation 351 nm, emission 420–470 nm).

A combination of Fura Red and the membrane-permeable nuclear stain Hoechst 33342 (excitation 363 nm, emission above 400 nm) was used in our laboratory for correlation of calcium gradients and the boundary of the nucleus [21]. We found that the best way to generate long-lasting local calcium signals in our cell type is to apply the calcium-releasing neurotransmitter (ACh) via a patch pipette located close to the cell surface (the application can be abruptly terminated when the calcium signal is formed in the apical part of the cell but has not yet spread to the basal region). After recording calcium signals and simultaneously imaging positions of cellular organelles, the calcium gradient along a specific line of interest (e.g., line selected on the two-dimensional image of the cell in such a way that it crosses the organelle of interest) can be calculated and plotted. Plotting the distribution of fluorescence of an organelle-specific probe (for the Golgi, nucleus, or mitochondria) and the calcium gradient along the same line allows us to examine the interrelationships between calcium signals and the organelle of interest. During calculation of the intracellular calcium gradient, it is important to take into account the resting distribution of the calcium probe and to make the appropriate corrections (e.g., by subtracting appropriate resting levels of fluorescence).

We found that putting calcium signaling on the map of cellular organelles provides useful and sometimes unexpected additional information. The possibilities for such combined measurements have improved substantially with the recent developments in fluorescent probes and confocal microscopy.

15.4 PROBING CALCIUM SIGNALING WITH CALCIUM UNCAGING (IN THE CYTOSOL AND THE LUMEN OF CELLULAR ORGANELLES)

The caged second messengers IP_3, cADP-ribose, NAADP, and Calcium were developed in the 1980s and 1990s [38–44] and have played an important role in clarifying the mechanisms of local calcium signaling processes (e.g., [45–50]). In this section, we will concentrate on applications of caged calcium to studies of calcium signaling.

Photosensitive calcium buffers (caged calcium) include derivatives of BAPTA- nitr-2, nitr-5, and nitr-7 synthesized by R.Y. Tsien and colleagues [41, 42]. DM-nitrophen, a derivative of EDTA (this probe has a relatively high-affinity for Mg^{2+}), and NP-EGTA were produced by G.C. Ellis-Davies and J.H. Kaplan [43, 44]. All these compounds display a decrease in their affinity to calcium as a result of absorption of UV light, which results in release of Ca^{2+} into solution. Nitr-5, nitr-7, DM-nitrophen, and NP-EGTA are commercially available. During the last few years NP-EGTA has become the most frequently used caged calcium probe. Calcium uncaging has been used to study mechanisms of calcium homeostasis, e.g., calcium buffering [51], propagation of intracellular and intercellular calcium waves [49, 52], and the formation of fundamental calcium release events [22]. Furthermore, calcium uncaging has been extensively used to study downstream reactions induced by calcium signaling. These applications include studies into the kinetics of calcium-dependant exocytosis [53], translocation of calcium-sensitive proteins [26], activation of calcium-dependant channels [45, 54], and the kinetics of muscle contraction [55, 56].

Usually calcium uncaging involves illumination of caged calcium with high intensity UV light (at approximately 340–360 nm). More recent technology involving two-photon calcium uncaging by infrared (IR) light (at around 700–720 nm) from a femtosecond Ti:S laser [57] has been developed. An interesting recent innovation is the application of frequency doubled Ti:S output to produce femtosecond pulses of 430–460 nm suitable for very efficient uncaging of NPE groups (which are difficult to photorelease using IR light). This technology has so far been applied to localized uncaging of NPE-IP_3 and NPE-ATP.

Uncaging of caged calcium in the cytosol is an excellent tool for probing the local sensitivity to calcium-induced calcium release (CICR). UV uncaging could be achieved using a mercury lamp supplied with an electronic shatter or a xenon flash lamp [58]. Normally, the light is delivered to the sample via the objective of a microscope. Local delivery of UV light using a small diameter quartz fiber has also been described [59]. A specialized laser, added to the imaging or confocal system [57], is probably the most frequently used instrument for UV uncaging. Global or local calcium uncaging of caged calcium can now be made routinely using Leica or Zeiss confocal microscopes equipped with UV lasers. The software of modern confocal systems allows one to select a region (or regions) for application of UV light on the image of the cell, adjust the duration of uncaging, apply the UV light, and monitor the results of uncaging in real time (without interrupting confocal or two-photon imaging during UV light application). It is important to note that UV uncaging targeted into a specific point occurs in reality within a volume of an "hour-glass figure" formed by the light rays, therefore the targeting from drawing the region of interest on the image recorded from the focal plane determines only approximately the space for uncaging. In our experience with UV-based uncaging, selecting an uncaging area less than 1 μm in diameter is not productive, this does not result in further improvement of targeting.

Individual local uncaging allows one to characterize calcium signaling in a specific region of the cell. Using such a technique, we were able to determine that CICR could be triggered only in the apical part of a pancreatic acinar cell [49]. In our study, fluo-4 was used as a cytosolic calcium indicator and NP-EGTA as the caged calcium. Another interesting example of local calcium uncaging activating CICR has been reported in a study on developing dendrites [60]. In this work, local uncaging of NP-EGTA triggered an extended calcium transient (monitored using Oregon Green 488 BAPTA-1), which could be eliminated by inhibition of calcium release from internal stores. The uncaging induced CICR was shown to locally stabilize dendritic processes [60].

Modern optical instruments also allow production of multiple simultaneous uncaging events. This is a potentially interesting development as it makes it possible to directly compare calcium signals in different parts of a cell produced in the same imaging/uncaging experiment (for an example of this application see Ref. [49]). In separate calibration experiments, caged fluorescein could be used to probe the efficiency of light delivery and the efficiency of uncaging in different parts of a cell. Dextran-bound calcium indicators could be employed in such experiments to decrease spillover of calcium originating from multiple uncaging regions.

The combination of calcium-imaging, electrophysiology, and local calcium uncaging in the cytosol of a cell provides an interesting opportunity for testing the distribution of calcium-dependant channels in the plasma membrane. This possibility was utilized in experiments conducted in our laboratory [54]. In these experiments, both caged calcium and the calcium indicator fluo-4 were delivered to the cytosol of the cell via the patch pipette (whole-cell configuration). Local calcium uncaging in the apical, lateral, and basal regions of a cell was used to determine the localization of calcium-dependant chloride channels. Only apical calcium uncaging was able to induce a chloride current (recorded in the whole-cell patch-clamp configuration). This technology demonstrated the specific apical localization of chloride channels without the need for complicated patching of small regions of the apical membrane.

Local calcium uncaging can be used in combination with organelle-specific calcium probes to investigate the role of specific organelles in this local calcium regulation. In our studies, we used local calcium uncaging in the nucleoplasm to probe the ability of perinuclear mitochondria to accumulate calcium and to segregate nuclear and cytosolic calcium signals. In these experiments, mitochondria were loaded with rhod-2 by incubation with the AM form of this probe, the indicator preferentially accumulates in mitochondria due to its positive charge (see http://www.probes.com/handbook/sections/1903.html). UV uncaging of NP-EGTA was used to generate restricted nuclear calcium signals, which resulted in calcium sequestration by perinuclear mitochondria [61]. These and other experiments showed that mitochondria could indeed play an important role in the regulation of perinuclear calcium in pancreatic acinar cells [61]. Preferential perinuclear clustering of mitochondria and their strong influence on calcium signaling around the nucleus was also found in parotid acinar cells [62].

Intraorganellar calcium uncaging is still a relatively novel technology. In our studies, we used intra-ER uncaging for monitoring calcium movement through the lumen of this organelle [63]. The protocol involved the following steps (Figure 15.3): (1) Loading of both cytosol and the ER lumen with caged calcium (NP-EGTA) and an ER — calcium probe (Mag-fluo-4-AM) using the membrane permeable (AM) form of these substances. (2) Formation of whole-cell patch-clamp configuration using a pipette containing a high concentration of a calcium buffer (based on 10 mM of BAPTA). This allows us to remove the caged calcium and the calcium indicator from the cytosol and additionally to clamp the cytosolic calcium concentration, preventing possible contamination with cytosolic calcium signals. (3) We found that the best results (largest proportional changes of ER calcium) are obtained when ER calcium is partially depleted. We used the calcium-releasing neurotransmitter ACh to deplete the ER prior to uncaging. (4) Local application of UV light resulted in a localized intra-ER calcium elevation. The process of ER calcium diffusion could then be monitored following calcium uncaging. From these experiments, we reached the conclusion that the ER is luminally connected and that calcium can diffuse rapidly in the lumen of the ER [63]. The ER-free regions of cytosol could be used to verify the absence of cytosolic contamination (the inner region of the nucleus was used in our experiments). The recent development of low-affinity calcium indicators could make the intra-ER calcium measurements more convenient. Particularly important are genetically engineered indicators — low-affinity cameleons [64, 65]. It is important to note that, while intra-ER calcium uncaging is relatively rare, the local photobleaching of fluorescent proteins expressed in the lumen of cellular organelles is an established technique used for studies into the connectivity of organelles and the diffusibility of organelle-targeted proteins [14, 66–68].

NP-EGTA is not suitable for two-photon uncaging with IR light [69]. The development of caged calcium probes suitable for two-photon uncaging; azid-1 [69] and DMNPE-4 [57] hold the promise of local calcium release in smaller volumes than possible for UV uncaging; this should allow more precise probing of the local calcium-dependent reactions in different regions of a cell. It would of course be advantageous to have these caged calcium reagents commercially available. This technology will hopefully develop simultaneously with instrumentation for local two-photon uncaging.

15.5 MEASURING LOCAL CALCIUM EXTRUSION

Extracellular calcium concentration can be monitored using ion-selective electrodes [70] or fluorescent calcium probes [71–74]. One of the problems of such measurements is the fast diffusion of calcium in the extracellular solution; this decreases the

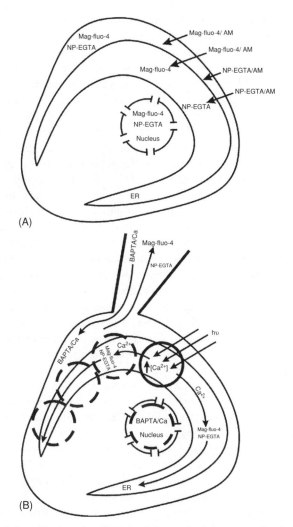

FIGURE 15.3 Schematic drawing of the technique of calcium uncaging in ER lumen. (A) Mag-fluo-4 and NP-EGTA are loaded into the cytosol and the ER lumen by incubation with the AM forms of these probes. (B) A patch pipette is used to remove the NP-EGTA and the Mag-fluo-4 from the cytosol, at the same time BAPTA/Ca buffer is delivered into the cytosol by the duffusion from the patch pipette. UV light is applied to the region containing both the cytosol and the ER (area indicated as a solid circle) but only in the ER lumen can the calcium rise occur (unlike cytosol, ER lumen contains NP-EGTA) and only in ER lumen can the calcium rise be detected (the ER lumen contains the calcium indicator, which is removed from the cytosol). Strong calcium buffer in the cytosol serves to suppress unwanted cytosolic calcium changes. Regions of interest selected for measurement of fluorescent changes contain both ER and cytosol (indicated by dashed circles) but only ER calcium changes are recorded. One of the regions of interest is placed in the nucleoplasm, which, because of high permeability of the nuclear pore complexes to small molecules, serves to verify the absence of cytosolic calcium changes.

near-membrane calcium changes and makes any change difficult to resolve. One approach is to restrict calcium diffusion by placing a cell in a very small (approximately 10^{-10} L) droplet of extracellular solution so that the calcium extruded from the cell causes a large change in the calcium concentration of the droplet. This technique is very sensitive; it allows direct measurement and precise quantification of calcium extrusion as a result of a single calcium transient in a small mammalian cell. The disadvantage of this method of measurement is that it does not provide information on the distribution of calcium extrusion sites. The lipophilic calcium-sensitive dyes (e.g., Calcium Green C_{18}, Fura C_{18}), with appropriate loading, become connected to the outer leaflet of the plasma membrane and report near-membrane extracellular calcium changes [71], thus allowing imaging of calcium extrusion. In particular, Fura C_{18} has been employed recently to monitor (and image) calcium extrusion from HEK cells stimulated with calcium-releasing agonists such as carbachol, ATP, or spermine (used in this study as an activator of the calcium-sensing receptor) [72]. Calcium Green C_{18} was used to measure changes of extracellular calcium near the plasma membrane of cardiac myocytes [75]. Introduction of lipophilic, low-affinity calcium indicators (Fura-FF-C_{18} and mag-fura-C_{18}) by Molecular Probes could result in further progress of this technology. In our laboratory, we used a different method for the measurement of calcium extrusion [74, 76, 77]. Instead of a lipophilic indicator, we used calcium indicators attached to heavy dextrans. The indicator-linked dextran serves in such experiments as both a calcium indicator and as a calcium buffer. It is important to note that this calcium buffer is much less mobile than the calcium ions in water-based solutions. This technique allows the slowing down of calcium diffusion and consequently the localization of calcium extrusion sites can be resolved at the level of a single cell. We found considerable regional variations in the intensity of calcium extrusion in different specialized cellular regions [74, 77]. The technique employing heavy calcium-binding dextrans appeared sufficiently sensitive to resolve small extracellular calcium transients occurring due to secretion of calcium from a single-fused secretory granule [77]. In our experiments, we used Calcium Green 1 bound to Dextran (Mr 500,000, dextrans with the molecular weight of 70,000 are also suitable for this type of measurement); we found that Fura Red could be conveniently used for simultaneous intracellular measurement. Both indicators can be excited by a 488-nm laser line; the emission spectra of the probes are significantly different to prevent cross-talk. The concentration of calcium-binding sites (Ca-Green 1 molecules) has to be relatively high to slow down calcium diffusion. We used 30–100 μM of Ca-Green-1 (bound to dextran). Please note that the concentration of the indicator is different from the concentration of the dextran since each molecule of dextran is linked to a number of molecules of the indicator. The degree of labeling will depend on the molecular weight of the dextran and can vary from batch to batch; for the availability and properties of the dextran-conjugated calcium-indicators, see http://www.probes.com/handbook/sections/1904.html. Imaging of extracellular calcium changes with dextran-conjugated probes could be combined with simultaneous recording of cytosolic calcium signals and with calcium photorelease from caged calcium (e.g., NP-EGTA) [74].

REFERENCES

1. Ashby, M.C. et al. Long distance communication between muscarinic receptors and calcium release channels revealed by carbachol uncaging in cell-attached patch pipette. *J. Biol. Chem.*, 278, 20860, 2003.
2. Li, Q., Luo, X., and Muallem, S. Functional mapping of calcium signaling complexes in plasma membrane microdomains of polarized cells. *J. Biol. Chem.*, 279, 27837, 2004.
3. Berridge, M.J, Bootman, M.D., and Roderick, H.L. Calcium signalling: dynamics, homeostasis and remodelling. *Nat. Rev. Mol. Cell. Biol.*, 4, 517, 2003.
4. Sorrentino, V. and Rizzuto, R. Molecular genetics of calcium stores and intracellular calcium signalling. *Trends Biochem. Sci.*, 22, 459, 2001.
5. Rutter, G.A. and Rizzuto, R. Regulation of mitochondrial metabolism by ER calcium release: an intimate connection. *Trends Biochem. Sci.*, 25, 215, 2000.
6. Corbett, E.F. and Michalak, M. Calcium, a signalling molecule in the endoplasmic reticulum? *Trends Biochem. Sci.*, 25, 307, 2000.
7. Wuytack, F.L., Raeymaekers, L., and Missiaen, L. PMR1/SPCA calcium pumps and the role of the Golgi apparatus as a calcium store. *Pflugers Arch.*, 446, 148, 2003.
8. Hajnoczky, G. et al. Decoding of cytosolic calcium oscillations in the mitochondria. *Cell*, 82, 415, 1995.
9. Porat, A. and Elazar, Z. Regulation of intra-Golgi membrane transport by calcium. *J. Biol. Chem.*, 275, 29233, 2000.
10. Ahluwalia, J.P. et al. A role for calcium in stabilizing transport vesicle coats. *J. Biol. Chem.*, 276, 34148, 2001.
11. Streb, H. et al. Release of calcium from a non mitochondrial intracellular store in pancreatic acinar cells by Inositol 1,4,5-risphosphate. *Nature*, 306, 67, 1983.
12. Pinton, P., Pozzan, T., and Rizzuto, R. The Golgi apparatus is an inositol 1,4,5, trisphosphate-sensitive Ca^{2+} store, with functional properties distinct from those of the endoplasmic reticulum. *EMBO J.*, 17, 5298, 1998.
13. Gerasimenko, O.V. et al. ATP-dependent accumulation and inositol trisphosphate- or cyclicADP-ribose-mediated release of Ca^{2+} from the nuclear envelope. *Cell*, 80, 439, 1995.
14. Subramanian, K. and Meyer, T. Calcium induced restructuring of nuclear envelope and endoplasmic reticulum calcium stores. *Cell*, 89, 963, 1997.
15. Gerasimenko, O.V. et al. Inositol trisphosphate and cyclicADP-ribose mediated Ca^{2+} release from single isolated pancreatic zymogen granules. *Cell*, 84, 473, 1996.
16. Mitchell, K.J. et al. Dense core secretory granules revealed as a dynamic calcium store in neuroendocrine cells with a vesicle associated membrane protein aequorin chimera. *J. Cell. Biol.*, 155, 41, 2001.
17. Churchill, G.C. et al. NAADP mobilizes Calcium from reserve granules, lysosome related organelles, in sea urchin eggs. *Cell*, 111, 703, 2002.
18. Williams, D.A. et al. Calcium gradients in single smooth muscle cells revealed by the digital imaging microscope using Fura-2. *Nature*, 318, 558, 1985.
19. Wahl, M., Sleight, R.G., and Gruenstein, E. Association of cytoplasmic free calcium gradients with subcellular organelles. *J. Cell. Physiol.*, 150, 593, 1992.
20. Bolsover, S.R., Kater, S.B., and Guthrie. P.B. Spatial gradients of cytosolic calcium concentration in neurones during paradoxical activation by calcium. *Cell Calcium*, 20, 373, 1996.

21. Gerasimenko, O.V. et al. Short pulses of acetylcholine stimulation induce cytosolic Ca^{2+} signals that are excluded from the nuclear region in pancreatic acinar cells. *Pflugers Arch.* 432, 1055, 1996.

22. Lipp, P. and Niggli, E. Fundamental calcium release events revealed by two-photon excitation photolysis of caged calcium in guinea-pig cardiac myocytes. *J. Physiol.,* 508, 801, 1998.

23. Yao, Y., Choi, J., and Parker, I. Quantal puffs of intracellular calcium evoked by inositol trisphosphate in *Xenopus* oocytes. *J. Physiol.,* 482, 533, 1995.

24. Thorn, P. et al. Local and global cytosolic calcium oscillations in exocrine cells evoked by agonists and inositol trisphosphate. *Cell,* 74, 661, 1993.

25. Bolsover, S. et al. Use of fluorescent calcium dyes with green fluorescent protein and its variants: problems and solutions. *Biochem. J.,* 356, 345, 2001.

26. O'Callaghan, D.W., Tepikin, A.V., and Burgoyne, R.D. Dynamics and calcium sensitivity of the calcium/myristoyl switch protein hippocalcin in living cells. *J. Cell. Biol.,* 163, 715, 2003.

27. Haynes, L.P., Tepikin, A.V., and Burgoyne, R.D. Calcium-binding protein 1 is an inhibitor of agonist-evoked, inositol 1,4,5-trisphosphate-mediated calcium signaling. *J. Biol. Chem.,* 279, 547, 2004.

28. Paemeleire, K. et al. Intercellular calcium waves in HeLa cells expressing GFP-labeled connexin 43, 32, or 26. *Mol. Biol. Cell.,* 5, 1815, 2000.

29. Lipp, P. et al. Nuclear calcium signalling by individual cytoplasmic calcium puffs. *EMBO J.,* 16, 7166, 1997.

30. al-Mohanna, F.A., Caddy, K.W., and Bolsover, S.R. The nucleus is insulated from large cytosolic calcium ion changes. *Nature,* 367, 745, 1994.

31. Marangos, P., FitzHarris, G., and Carroll, J. Calcium oscillations at fertilization in mammals are regulated by the formation of pronclei. *Development,* 130, 1461, 2003.

32. Radding, W. et al. Intracellular calcium puffs in osteoclasts. *Exp. Cell. Res.,* 253, 689, 1999.

33. Yi, M., Weaver, D., and Hajnoczky, G. Control of mitochondrial motility and distribution by the calcium signal: a homeostatic circuit. *J. Cell. Biol.,* 167, 661, 2004.

34. Quintana, A. and Hoth, M. Apparent cytosolic calcium gradients in T-lymphocytes due to fura-2 accumulation in mitochondria. *Cell Calcium,* 36, 99, 2004.

35. Miyawaki, A., et al. Fluorescent indicators for calcium based on green fluorescent proteins and calmodulin. *Nature,* 388, 882, 1997.

36. Rapizzi, E. et al. Recombinant expression of the voltage-dependent anion channel enhances the transfer of calcium microdomains to mitochondria. *J. Cell. Biol.,* 159, 613, 2002.

37. Nagai, T. et al. Circularly permuted green fluorescent proteins engineered to sense calcium. *Proc. Natl. Acad. Sci. U.S.A,* 98, 3197, 2001.

38. Walker, J.W., Feeney, J., and Trentham, D.R. Photolabile precursors of inositol phosphates. Preparation and properties of 1-(2-nitrophenyl)ethyl esters of myo-inositol 1,4,5-trisphosphate. *Biochemistry,* 28, 3272, 1989.

39. Aarhus, R., Gee, K., and Lee, H.C. Caged cyclic ADP-ribose. Synthesis and use. *J. Biol. Chem.,* 270, 7745, 1995.

40. Lee, H.C., et al. Caged nicotinic acid adenine dinucleotide phosphate. Synthesis and use. *J. Biol. Chem.,* 272, 4172, 1997.

41. Tsien, R.Y. and Zucker, R.S. Control of cytoplasmic calcium with photolabile tetracarboxylate 2-nitrobenzhydrol chelators. *Biophys. J.,* 50, 843, 1986.

42. Gurney, A.M., Tsien, R.Y., and Lester, H.A. Activation of a potassium current by rapid photochemically generated step increases of intracellular calcium in rat sympathetic neurons. *Proc. Natl. Acad. Sci. U.S.A*, 84, 3496, 1987.

43. Kaplan, J.H. and Ellis-Davies, G.C. Photolabile chelators for the rapid photorelease of divalent cations. *Proc. Natl. Acad. Sci. U.S.A*, 85, 6571, 1988.

44. Ellis-Davies, G.C. and Kaplan, J.H. Nitrophenyl-EGTA, a photolabile chelator that selectively binds calcium with high affinity and releases it rapidly upon photolysis. *Proc. Natl. Acad. Sci. U.S.A*, 91, 187, 1994.

45. Wang, S.S. and Augustine, G.J. Confocal imaging and local photolysis of caged compounds: dual probes of synaptic function. *Neuron.*, 15, 755, 1995.

46. Lee, H. C. and Aarhus, R. Functional visualization of the separate but interacting calcium stores sensitive to NAADP and cyclic ADP-ribose. *J. Cell. Sci.*, 113, 4413, 2000.

47. Santella, L. et al. Nicotinic acid adenine dinucleotide phosphate-induced calcium release. Interactions among distinct calcium mobilizing mechanisms in starfish oocytes. *J. Biol. Chem.*, 275, 8301, 2000.

48. Straub, S.V., Giovannucci, D.R., and Yule, D.I. Calcium wave propagation in pancreatic acinar cells: functional interaction of inositol 1,4,5-trisphosphate receptors, ryanodine receptors, and mitochondria. *J. Gen. Physiol.*, 116, 547–560, 2000.

49. Ashby, M.C. et al. Localized calcium uncaging reveals polarized distribution of calcium-sensitive calcium release sites: mechanism of unidirectional calcium waves. *J. Cell. Biol.*, 158, 283, 2002.

50. Echevarria, W. et al. Regulation of calcium signals in the nucleus by a nucleoplasmic reticulum. *Nat. Cell Biol.*, 5, 440, 2003.

51. Fleet, A., Ellis-Davies, G., and Bolsover, S. Calcium buffering capacity of neuronal cell cytosol measured by flash photolysis of calcium buffer NP-EGTA. *Biochem. Biophys. Res. Commun.*, 250, 786, 1998.

52. Leybaert, L. et al. Inositol-trisphosphate-dependent intercellular calcium signaling in and between astrocytes and endothelial cells. *Glia.*, 24, 398, 1998.

53. Heinemann, C. et al. Kinetics of the secretory response in bovine chromaffin cells following flash photolysis of caged calcium. *Biophys. J.*, 67, 2546, 1994.

54. Park, M.K. et al. Local uncaging of caged calcium reveals distribution of calcium-activated chloride channels in pancreatic acinar cells. *Proc. Natl. Acad. Sci. U.S.A*, 98, 10948, 2001.

55. Somlyo, A.V. et al. Cross-bridge kinetics, cooperativity, and negatively strained cross-bridges in vertebrate smooth muscle. A laser-flash photolysis study. *J. Gen. Physiol.*, 91, 165, 1988.

56. Lea, T.J. and Ashley, C.C. Calcium release from the sarcoplasmic reticulum of barnacle myofibrillar bundles initiated by photolysis of caged calcium. *J. Physiol.*, 427, 435 1990.

57. DelPrincipe, F. et al. Two-photon and UV-laser flash photolysis of the calcium cage, dimethoxynitrophenyl-EGTA-4. *Cell Calcium.*, 25, 85, 1999.

58. Kishimoto, T. et al. Ion selectivities of the calcium sensors for exocytosis in rat phaeochromocytoma cells. *J. Physiol.*, 533, 627, 2001.

59. Eberius, C. and Schild, D. Local photolysis using tapered quartz fibres. *Pflugers Arch.*, 443, 323–330, 2001.

60. Lohmann, C., Myhr, K.L., and Wong, R.O. Transmitter-evoked local calcium release stabilizes developing dendrites. *Nature*, 418, 177, 2002.

61. Park, M.K. et al. Perinuclear, perigranular and sub-plasmalemmal mitochondria have distinct functions in the regulation of cellular calcium transport. *EMBO J.*, 20, 1863, 2001.

62. Bruce, J.I. et al. Modulation of [Calcium]i signaling dynamics and metabolism by perinuclear mitochondria in mouse parotid acinar cells. *J. Biol. Chem.*, 279, 12909, 2004.

63. Park, M.K., Petersen, O.H., and Tepikin, A.V. The endoplasmic reticulum as one continuous calcium pool: visualization of rapid calcium movements and equilibration. *EMBO J.*, 19, 5729, 2000.

64. Nagai, T. et al. Expanded dynamic range of fluorescent indicators for calcium by circularly permuted yellow fluorescent proteins. *Proc. Natl. Acad. Sci. U.S.A.*, 101, 10554, 2004.

65. Palmer, A.E. et al. Bcl-2-mediated alterations in endoplasmic reticulum calcium analyzed with an improved genetically encoded fluorescent sensor. *Proc. Natl. Acad. Sci. USA*, 101, 17404, 2004.

66. Snapp, E. L., The fusome mediates intercellular endoplasmic reticulum connectivity in *Drosophila* ovarian cysts. *Mol. Biol. Cell.*, 15, 4512, 2004.

67. Collins, T.J. et al. 2002. Mitochondria are morphologically and functionally heterogeneous within cells. *EMBO J.*, 21, 1616–1627.

68. Szabadkai, G. et al. Drp-1-dependent division of the mitochondrial network blocks intraorganellar calcium waves and protects against calcium-mediated apoptosis. *Mol. Cell.*, 16, 59, 2004.

69. Brown, E.B., et al. Photolysis of caged calcium in femtoliter volumes using two-photon excitation. *Biophys. J.*, 76, 489, 1999.

70. Smith, P.J. et al. Self-referencing, non-invasive, ion selective electrode for single cell detection of trans-plasma membrane calcium flux. *Microsc. Res. Tech.*, 46, 398, 1999.

71. Lloyd, Q.P., Kuhn, M.A., and Gay, C.V. Characterization of calcium translocation across the plasma membrane of primary osteoblasts using a lipophilic calcium-sensitive fluorescent dye, calcium green C_{18}. *J. Biol. Chem.*, 270, 22445, 1995.

72. De Luisi, A. and Hofer, A.M. Evidence that calcium cycling by the plasma membrane Calcium ATPase increases the 'excitability' of the extracellular calcium-sensing receptor. *J. Cell Sci.*, 116, 1527, 2003.

73. Tepikin, A.V. et al. Acetylcholine-evoked increase in cytoplasmic calcium concentration and calcium extrusion measured simultaneously in single mouse pancreatic acinar cells. *J. Biol. Chem.*, 267, 3569, 1992.

74. Belan, P.V. et al. Distribution of calcium extrusion sites on the mouse pancreatic acinar cell surface. *Cell Calcium*, 22, 5, 1997.

75. Blatter, L.A. and Niggli, E. Confocal near-membrane detection of calcium in cardiac myocytes. *Cell Calcium*, 23, 269, 1998.

76. Belan, P.V. et al. A new technique for assessing the microscopic distribution of cellular calcium exit sites. *Pflugers Arch.*, 433, 200, 1998.

77. Belan, P.V. et al. Isoproterenol evokes extracellular calcium spikes due to secretory events in salivary gland cells. *J. Biol. Chem.*, 273, 4106, 1998.

16 Apoptosis

Clark W. Distelhorst

CONTENTS

16.1 INTRODUCTION

The aim of this chapter is to summarize how established methods of Ca^{2+} signaling research are used to elucidate the involvement of Ca^{2+} in apoptosis. Selected examples of apoptosis are used to illustrate methodological principles and applications. For a comprehensive review of the role of Ca^{2+} in a wide range of apoptotic processes, the reader is referred to a number of excellent in-depth reviews [1–5]

This chapter is directed at two major audiences. First are those who have considerable experience in Ca^{2+} signaling but want to know more about apoptosis and how it is triggered by Ca^{2+} signals. Second are those who work in the area of apoptosis research and are interested in learning more about Ca^{2+} and its measurement because of the increasing number of reports implicating Ca^{2+} in apoptosis. The author formally belongs to neither of the camps and therefore has the unique perspective of one who was drawn into apoptosis research by the desire to understand a physiological process of great clinical relevance (i.e., the killing of lymphocytes by glucocorticosteroid hormones) and was lured to the realm of Ca^{2+} signaling by unexpected findings implicating Ca^{2+} release from the endoplasmic reticulum (ER) in this process [6, 7] and the regulation of ER Ca^{2+} release by the apoptosis regulatory protein Bcl-2 [8]. This avenue of research has led to recent evidence that the anti-apoptotic protein Bcl-2 interacts functionally with inositol 1,4,5-trisphospate receptors and thereby regulates Ca^{2+} release from the ER in lymphocytes [9]. But applying the techniques of Ca^{2+} measurement in the context of apoptosis has been sobering and therefore one goal of this chapter is to alert newcomers to some of the pitfalls unique to this area of research.

16.2 APOPTOSIS AND METHODS FOR ITS DETECTION

16.2.1 APOPTOSIS AND NECROSIS

Cell death is generally considered to be by either necrosis or apoptosis. Necrosis and apoptosis are distinguished by a variety of criteria. Perhaps the most useful are morphological features. Apoptosis is characterized by condensation of chromatin around the periphery of the nucleus accompanied by the preservation of intact nuclear and plasma membranes, whereas in necrosis nuclear chromatin remains dispersed throughout the nucleus in a typical heterochromatin pattern, while the continuity of the plasma membrane is disrupted. From a molecular standpoint, apoptosis is characterized by the systematic destruction of intracellular structure and function "from within," mediated by specifically regulated endogenous proteases and endonucleases, whereas necrosis is characterized by destruction of the plasma membrane "from without," generally by toxic chemicals or proteases. Moreover, apoptosis is genetically programmed (hence the term "programmed cell death," which is used interchangeably with apoptosis), meaning that a cell encodes its tools of self-destruction (i.e., proteases, endonucleases), whereas necrosis is an accidental form of cell death utilizing tools foreign to the cell itself (e.g., complement-mediated destruction of a red cell membrane, toxin-induced destruction of liver or bone marrow cells). These definitions are useful guidelines, but in reality the distinction between necrosis and apoptosis is often blurred, and there are many cases in which cells display features of both processes. An example is the induction of cell death in immature lymphocytes by glucocorticosteroid hormones, which displays morphological features of both apoptosis and necrosis [10].

16.2.2 CELL DEATH INITIATION AND EXECUTION: A BRIEF OVERVIEW

Progress in understanding fundamental mechanisms of apoptosis has expanded at an extraordinary rate during the past decade (for review, see Refs. [11–14]). It is useful to divide the process of apoptosis into two successive stages: initiation and execution. A simplified view of apoptosis is illustrated in Figure 16.1, which captures the essence of our current knowledge of this process and avoids specific details about which there may still be some disagreement. Initiation is considered to be the process by which diverse cellular death signals set off a cascade of events that trigger the execution phase of apoptosis, which is considered here to be the activation of endonucleases and proteases that mediate cellular destruction, and the engulfment of corpses by macrophages. The Bcl-2 family of proteins plays a central role in the initiation stage of apoptosis. Members of this protein family share regions of sequence homology (i.e., BH domains). Certain members of this protein family (e.g., Bcl-2, Bcl-xL) share all four of the BH domains and have primarily a pro-survival function, whereas members sharing only three of the BH domains (e.g., Bax, Bak) or only the BH3 domain (e.g., Bim, Bad, Bid, Noxa, Puma) play critical roles in promoting cell death. The BH3-only proteins are distributed strategically throughout the cell and act as sentinels of cellular dysfunction [15]. Their activation by cellular stress, including disruption of Ca^{2+} homeostasis, initiates the cell death

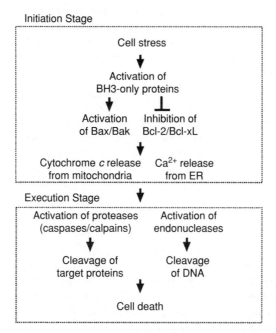

FIGURE 16.1 A simplified model of apoptosis signaling separated into initiation and execution stages. A wide range of apoptotic signals activate BH3-only proteins that serve as sentinels of cellular stress. These in turn activate the proapoptotic proteins Bax and Bak, or inhibit the pro-survival activity of the anti-apoptotic proteins Bcl-2 and Bcl-xL. At least two consequences are release of pro-apoptotic factors from mitochondria, including cytochrome c, and release of Ca^{2+} from the ER. These factors trigger the execution phase of apoptosis in which proteases and endonucleases play an active part.

process by either activating Bax or Bak, or by inhibiting a survival function provided by Bcl-2 or Bcl-xL. Bax and Bak then mediate release of cytochrome c and other pro-apoptotic factors from mitochondria and Ca^{2+} from the ER (this effect on Ca^{2+} homeostasis is described in greater detail below). The release of these factors from organelles in turn activates proteases (caspases and calpains), leading to the degradation of critical structural elements of the cell and activation of endonucleases that cleave DNA.

16.2.3 DIAGNOSTIC TESTS AND QUANTITATIVE MEASURES OF APOPTOSIS

A variety of techniques are commonly employed to detect apoptosis and to distinguish apoptosis from necrosis (for details of particular assays, see Refs. [16, 17]). The application of these techniques varies widely among investigators and the selection of which technique to use is governed mostly by the specific goals of the experiment (e.g., to determine if cells are alive versus dead, by whatever mechanism; or to precisely distinguish between apoptosis and necrosis). Based on convenience and simplicity, cell viability is often defined as the ability of the cell to exclude a cell impermeable dye, such as trypan blue or acridine orange. Dead cells are stained with these dyes due to reduced plasma membrane integrity. Hence, the uptake of one of these dyes is more a marker of necrosis than apoptosis. Perhaps the simplest and most informative assay for apoptosis is the detection of nuclear condensation by fluorescence microscopy using one of a number of cell permeable dyes that bind to DNA (e.g., acridine orange, Hoechst 33342). The typical morphology of apoptosis is illustrated by electron microscopic images of cells induced to undergo apoptosis by the Ca^{2+}-ATPase inhibitor thapsigargin (Figure 16.2A). These same morphological changes are readily detected by epifluorescence microscopy of cells stained with acridine orange, where apoptotic nuclei appear as punctate condensations of the fluorescent dye, often referred to as "apoptotic bodies" (Figure 16.2B). Since apoptosis was originally defined and distinguished from necrosis by morphological criteria, this method provides tangible evidence of apoptosis versus necrosis. One precaution, however, is that it is important to assay cells for apoptosis over a range of time points following treatment with the suspected apoptotic stimulus. This is because apoptotic cells *in vivo* are quickly removed and destroyed by macrophages, and apoptotic cell *in vitro* (i.e., in tissue culture) progress from apoptosis to necrosis and typical nuclear morphological changes of apoptosis may disappear over time. This precaution pertains to other types of apoptosis assays where positive findings typically peak during a particular time frame following treatment with the apoptosis-inducing agent and then decline, rather than maintaining a stable plateau.

Assays that detect inter-nucleosomal DNA fragmentation typical of apoptosis are also useful. These include analysis of DNA by agarose gel electrophoresis to detect a typical ladder pattern of DNA fragments (Figure 16.2C), analysis of propidium iodide-stained DNA by flow cytometry to detect a typical sub-G_1 DNA peak (i.e., cells having DNA content less than that of cells in the G_1 phase of the cell cycle) (Figure 16.2D), or the TUNEL assay to label the terminal

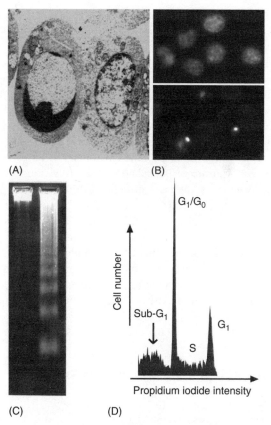

(A)   (B)

(C)   (D)

FIGURE 16.2 Diagnostic markers of apoptosis. Panel (A), electron microscopy of thapsigargin-treated lymphocytes, with the condensed chromatin pattern of apoptosis on the left,
and the normal heterochromatin pattern of a nonapoptotic cell on the right. Panel (B),
epifluorescence microscopy of lymphocytes stained with acridine orange. Top panel, untreated cells displaying heterochromatin pattern typical of nonapoptotic cells; bottom panel,
dexamethasone treated cells displaying chromatin condensation, or apoptotic bodies, typical
of apoptosis. Panel (C), agarose gel electrophoresis of DNA from untreated (left lane) and
dexamethasone-treated (right lane) lymphocytes, illustrating the typical ladder pattern of
DNA fragments from apoptotic cells. Panel (D), flow cytometry analysis of propidium iodide
stained lymphocytes following treatment with dexamethasone. The sub-G_1 peak represents
DNA fragments and is indicative of apoptosis.

$3'$-hydroxyl exposed by DNA strand breaks. Caution needs to be used in interpreting the results of these assays since some degree of DNA fragmentation is often
detected in cells undergoing necrosis. Typically, DNA is fragmented at random
sites in cells undergoing necrosis in contrast to the inter-nucleosomal DNA cleavage characteristic of apoptosis. Therefore, resolution of DNA from necrotic cells
on agarose gel electrophoresis gives a smear rather than the ladder pattern typical of
apoptosis.

Members of the caspase family of proteases are typically involved in mediating apoptosis and can be detected by Western blotting or immunofluorescence techniques. Alternatively, a commonly applied method is to analyze the integrity of poly-ADP-ribose polymerase by Western blotting. Caspase-mediate cleavage of this DNA repair enzyme is frequently observed in cells undergoing apoptosis and detection of appropriately sized cleavage fragments is a useful marker of apoptotic cell death. These assays are increasingly employed but some degree of caution is warranted as some forms of apoptosis are caspase-independent [18, 19].

Another test for apoptosis detects phospholipid asymmetry on the plasma membrane [20]. Loss of plasma membrane asymmetry is an early event in apoptosis, resulting in the exposure of phosphatidylserine residues at the outer plasma membrane leaflet. Annexin V interacts strongly with the exposed phosphatidylserine and therefore labeled annexin V can be used in flow cytometry to detect and quantify cells in early stages of apoptosis. However, annexin V is a Ca^{2+}-binding protein and brief elevation of cytoplasmic Ca^{2+} may be sufficient to induce phosphatidylserine exposure in the absence of apoptosis [21, 22]. Therefore, it is important not to rely solely on annexin V binding as a measure of apoptosis in those settings where elevated Ca^{2+} is thought to play a role in apoptosis.

In summary, there are multiple assays that are commonly employed to detect and quantify apoptosis. In view of the known limitations of each assay, it is wise to apply several of these assays rather than relying on a single assay.

16.3 ALTERATIONS OF Ca^{2+} HOMEOSTASIS THAT SIGNAL APOPTOSIS

16.3.1 Ca^{2+} Homeostasis and Signaling

The term Ca^{2+} homeostasis refers to the normal distribution of Ca^{2+} ions in excitable and nonexcitable cells under normal resting conditions. Under normal physiological conditions, resting cytoplasmic Ca^{2+} concentration is low (≤ 100 nM), whereas extracellular Ca^{2+} is high (~1.5 mM). A primary goal of Ca^{2+} homeostasis is to maintain a low cytoplasmic Ca^{2+} concentration so that intended elevations of cytoplasmic Ca^{2+} in the form of oscillations and waves can be used to transmit information throughout the cell. Various Ca^{2+}-sensitive effector molecules (e.g., kinases, phosphatases) convert the information encoded by the Ca^{2+} elevation into cellular responses which regulate transcription, proliferation, differentiation, contraction, secretion, or axis formation (for in-depth review, see Refs. [23–28]). Among the effector molecules are Ca^{2+}-regulated metabolic enzymes located within mitochondria. Hence, a critical component of intracellular Ca^{2+} homeostasis is the maintenance of mitochondrial Ca^{2+} at relatively low levels so that transient elevations of mitochondrial Ca^{2+}, produced by transient elevations of cytoplasmic Ca^{2+}, serve a signaling function. Another major goal of Ca^{2+} homeostasis is to maintain the Ca^{2+} concentration within the ER lumen at high levels (estimated to be several orders of magnitude higher that that of the surrounding cytoplasm). Maintenance of high luminal Ca^{2+} is necessary for a variety of essential ER-associated functions, including protein processing within the ER lumen and translation on adjacent ribosomes. A myriad of Ca^{2+} pumps, channels, and buffers, considered to be components of a "Ca^{2+} signaling toolkit," work in a coordinated fashion to regulate and maintain Ca^{2+} homeostasis (for an in-depth review, see Ref. [26]).

Dynamic changes in cytoplasmic Ca^{2+} levels that constitute Ca^{2+} signals are generated either by uptake of extracellular Ca^{2+} through channels in the plasma membrane or by release of Ca^{2+} from the ER lumen through channels in the membrane of the sarcoplasmic reticulum or the ER (i.e., ryanodine receptors or inositol 1,4,5-trisphosphate (IP_3) receptors). Importantly, the amplitude and duration of cytoplasmic Ca^{2+} elevation is controlled by a variety of factors including pumps (Ca^{2+}-ATPases) on the plasma membrane that extrude Ca^{2+} from the cell, pumps (SERCA pumps, sarcoplasmic endoplasmic reticulum Ca^{2+}-ATPases) that pump Ca^{2+} from the cytoplasm into the ER lumen, cytoplasmic Ca^{2+}-binding proteins that buffer Ca^{2+}, and mitochondria that are capable of taking up large amounts of Ca^{2+}. As a consequence, the cell imposes limits on the level and duration of physiologic Ca^{2+} signals and prevents depletion of ER Ca^{2+} or mitochondrial Ca^{2+} overload.

16.3.2 PHYSIOLOGICAL AND NONPHYSIOLOGICAL Ca^{2+} CHANGES THAT INITIATE CELL DEATH PATHWAYS

Marked alterations in Ca^{2+} homeostasis are a well-recognized cause of necrosis (e.g., toxin-treated hepatocytes) [29]. Alterations in Ca^{2+} homeostasis can also induce the typical hallmarks of apoptosis. These alterations in Ca^{2+} homeostasis include sustained elevation of cytoplasmic Ca^{2+}, depletion of ER Ca^{2+}, or elevation of mitochondrial Ca^{2+}. Calcium ionophores (e.g., ionomycin) or SERCA pump inhibitors (e.g., thapsigargin) produce gross disruption of Ca^{2+} homeostasis and consequently induce both apoptosis and necrosis [8, 30]. But physiological and pharmacological forms of apoptosis are often mediated by more selective alterations in intracellular Ca^{2+} homeostasis. In this regard, Ca^{2+} activates both the extrinsic or receptor-mediated death pathways (e.g., induced by engagement of Fas receptors or tumor necrosis factor receptors on the plasma membrane by their respective ligands) and intrinsic or mitochondria-mediated death pathways (e.g., induced by growth factor withdrawal, DNA damage, or steroid hormones) [2]. Ca^{2+} plays a role in both the initiation and execution phases of apoptosis (Figure 16.3). Ca^{2+} acts in the initiation phase by activating BH3-only proteins. An example of the action of Ca^{2+} signaling in the initiation phase of apoptosis is the induction of apoptosis following T-cell receptor activation. This process is initiated by IP_3-mediated release of Ca^{2+} from the ER, which elevates cytoplasmic Ca^{2+} and thereby activates the Ca^{2+}-sensitive protein phosphatase calcineurin (reviewed in Refs. [31, 32]). IP_3 receptor-mediated elevation of cytoplasmic Ca^{2+} also triggers apoptosis in T cells following treatment with glucocorticosteroids or ionizing radiation [33, 34] and in B cells following surface IgM ligation [35]. Ca^{2+} is also released from the ER following growth factor deprivation, leading to an intracellular redistribution of Ca^{2+} from the ER lumen into mitochondria, although the specific role of IP_3 receptors in this process has not yet been delineated [36].

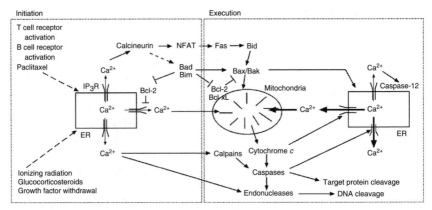

FIGURE 16.3 Model illustrating the roles of Ca^{2+} in both the initiation and execution of apoptosis. Multiple signals induce Ca^{2+} release from the ER, which in turn activates downstream effectors of apoptosis (e.g., calcineurin, calpains, endonucleases) or alters mitochondrial function. Ca^{2+} release from the ER is also induced downstream, during the execution stage, by the actions of Bax/Bak, by protease-mediated cleavage of IP_3 receptors, and by cytochrome c-mediated enhancement of Ca^{2+} release via IP_3 receptors. Bcl-2 may in part oppose apoptosis by inhibiting IP_3-mediated release of Ca^{2+} from the ER.

16.3.3 DOWNSTREAM EFFECTORS THAT MEDIATE THE EXECUTION PHASE OF Ca^{2+}-INDUCED APOPTOSIS

Alterations in Ca^{2+} homeostasis trigger the execution phase of apoptosis by activating apoptosis effector molecules or by disrupting organelle function (Figure 16.3). Ca^{2+} activates a variety of apoptosis effectors either directly or indirectly (reviewed in Refs. [2, 3]). For example, the elevation of cytoplasmic Ca^{2+} induced by T-cell receptor activation increases expression of the death receptor ligand Fas through calcineurin-mediated activation of the transcription factor, nuclear factor of activated T cells (NFAT) [37, 38]. Calcineurin-mediated elevation of Fas expression not only mediates cell death following T-cell receptor activation [39, 40], but also following treatment with the chemotherapeutic agent paclitaxel [41]. In addition, T-cell receptor stimulation, via IP_3-mediated Ca^{2+} elevation and calcineurin activation, sensitizes resting T cells to Fas-mediated cell death by decreasing the steady-state level of FLIP (FLICE-like inhibitory protein), an inhibitor of the Fas signaling pathway [42]. Moreover, calcineurin overexpression sensitizes cells to apoptosis induction following growth factor withdrawal [43]. Ca^{2+} also activates intrinsic death pathways by activating calcineurin, which dephosphorylates and thereby activates the pro-apoptotic Bcl-2 family member Bad [44, 45] or mediates induction of other pro-apoptotic Bcl-2 family members (e.g., Bik, Bim) [46]. Cells from Bim knockout mice are resistant to Ca^{2+}-induced apoptosis, substantiating the role of Bim as a downstream mediator in this process [47]. Thus, the Ca^{2+}/calmodulin-activated protein phosphatase calcineurin is a key downstream effector involved in mediating apoptosis in response to cytoplasmic Ca^{2+} elevation. In addition to calcineurin, another effector of Ca^{2+} signaling related to cell cycle entry is calmodulin kinase (CAMK). This kinase has been implicated in the re-entry pathway as it phosphorylates the CREB (Ca^{2+}-cyclic AMP response element binding factor) transcription factor and regulates expression of early response genes such as *c-fos* [48, 49].

Proteases, including calpains and caspases, are also important downstream effectors of Ca^{2+}-mediated apoptosis (for review, see Ref. [2]). Calpains are cytosolic Ca^{2+}-activated neutral cysteine proteases that are expressed in all animal cells. Calpains are implicated in mediating apoptosis induction by glucocorticosteroids [50], vitamin D [51], etoposide [52], cisplatin [53], ischemia/reperfusion injury [54], platelet activation [55], and ionomycin [56]. Moreover, there is strong evidence for crosstalk between caspases and calpains. For example, mu-calpain cleaves the ER-associated procaspase-12 to generate active caspase-12 [57]. Also, calpain has been shown to activate caspases during etoposide-induced apoptosis in T cells [52]. A number of proteins are calpain cleavage targets during apoptosis, including the pro-apoptotic Bcl-2 family members Bid [53, 54], Bax [58, 59], DNA ligase III [60], p53 [61], and the endogenous calcineurin inhibitor cain/cabin [62]. Moreover, destruction of the nuclear structural protein lamin, an event that appears to precede endonucleolytic DNA cleavage during glucocorticosteroid-induced apoptosis, is blocked by calpain inhibition [63, 64].

While the preceding discussion has emphasized molecular mediators of Ca^{2+}-induced apoptosis, there is also a wealth of evidence that alterations in Ca^{2+} homeostasis directly trigger apoptosis by disrupting normal organelle function (reviewed in Refs. [3–5, 65]). Major emphasis in this regard is placed on mitochondria. Mitochondria play critical roles in both cell life and cell death. Mitochondria play a central role in cell death by releasing apoptotic factors (e.g., cytochrome *c*) that mediate caspase activation [66–69]. In life, increases of mitochondrial matrix Ca^{2+} evoked by Ca^{2+} mobilizing agonists play a fundamental role in cellular energy metabolism [70–72]. Whereas transient elevation of mitochondrial Ca^{2+} serves a physiological role in regulating energy metabolism, sustained elevation of cytosolic Ca^{2+} leads to mitochondrial Ca^{2+} overload and cell death by apoptosis or necrosis (reviewed in Refs. [2, 3]). IP_3-mediated Ca^{2+} signals activate cell intrinsic death pathways by direct and indirect effects on mitochondria. Apoptotic stimuli induce a switch in mitochondrial Ca^{2+} signaling at the beginning of the apoptotic process by facilitating Ca^{2+}-induced opening of the mitochondrial permeability transition pore [73, 74]. Thus, IP_3-induced Ca^{2+} elevation, under certain circumstances, triggers mitochondrial permeability transition and, in turn, cytochrome *c* release. Pro-apoptotic BH3-only members of the Bcl-2 protein family (e.g., Bid) contribute to this process by selectively permeabilizing the outer mitochondrial membrane, thereby increasing the magnitude of mitochondrial Ca^{2+} elevation [75].

Disturbances in ER Ca^{2+} homeostasis associated with ER stress also induce apoptosis [76, 77]. Activation of caspase-12, located on the ER membrane, is a critical step in ER stress-mediated apoptosis. Ca^{2+} mediates ER stress-induced apoptosis by activating calpain, which in turn cleaves and thereby activates caspase-12. Apoptosis can also be a consequence of decreased luminal Ca^{2+} [78, 79].

Recent findings indicate that pro-apoptotic members of the Bcl-2 protein family (i.e., Bax, Bak) induce Ca^{2+} release from the ER, thereby activating ER stress and inducing apoptosis [80–82]. Two additional mechanisms amplify the release of Ca^{2+} from the ER during apoptosis. One involves the binding of cytochrome c to IP$_3$ receptors [83] and another involves caspase-mediated cleavage of IP$_3$ receptors [84, 85]. Paradoxically, a decrease in luminal Ca^{2+} is also proposed to play an anti-apoptotic role by decreasing the transfer of Ca^{2+} from ER to mitochondria [86]. Accordingly, it has been reported that decreased expression of Bax and Bak reduces luminal Ca^{2+}, thereby protecting mitochondria by inhibiting Ca^{2+} release from the ER [87]. Moreover, findings differ regarding the action of the anti-apoptotic protein Bcl-2 on ER Ca^{2+}, with some suggesting that Bcl-2 causes a leakage of Ca^{2+} from the ER and thereby decreases luminal Ca^{2+}, while others suggest that Bcl-2 inhibits IP$_3$-mediated Ca^{2+} release from the ER without decreasing luminal Ca^{2+} (for review, see Ref. [88]). Because of their role in mediating Ca^{2+} elevation in response to many apoptotic signals, IP$_3$ receptors are emerging as key targets for regulation by pro- and anti-apoptotic members of the Bcl-2 family [89]. Bcl-2 may also suppress pro-apoptotic Ca^{2+} signaling by sequestering calcineurin to intracellular membranes [41, 90] or by docking calcineurin to IP$_3$ receptors [91, 92].

16.4   METHODS TO DETECT ALTERATIONS IN Ca^{2+} HOMEOSTASIS ASSOCIATED WITH APOPTOSIS

The purpose of this section is to illustrate how both conventional and unconventional methods of Ca^{2+} measurement have been used to investigate the role of Ca^{2+} in apoptosis. Methodological details are provided in a number of the preceding chapters and therefore not repeated here. Special emphasis is placed on unique aspects of Ca^{2+} signaling during apoptosis and special challenges presented to those measuring Ca^{2+} in the context of apoptosis.

16.4.1   MEASURING CYTOPLASMIC Ca^{2+} ELEVATION INDUCED BY AN APOPTOTIC SIGNAL

In assessing whether or not Ca^{2+} is involved in mediating a particular apoptotic process the simplest first step is to determine if cytoplasmic Ca^{2+} concentration is elevated in response to the apoptotic signal, and if this elevation of Ca^{2+} precedes overt signs of apoptosis (e.g., caspase activation, DNA fragmentation, nuclear chromatin condensation). Although documenting an elevation of cytoplasmic Ca^{2+} is relatively straightforward using Ca^{2+}-sensitive dyes (e.g., Fura-2 AM) described in detail in the first two chapters, there are potential caveats depending primarily on the nature of the apoptotic signal and the cell system selected for study. Apoptotic signals that rapidly induce Ca^{2+} elevation are most amenable to study because the agonist-induced Ca^{2+} elevation is detected soon after agonist addition. Thus, Ca^{2+} is measured soon after dye loading and de-esterification. An example is from our work where we have measured the effect of Bcl-2 over-expression on the elevation of cytoplasmic Ca^{2+} induced by T-cell receptor activation [9]. But many apoptotic signals cause a delayed, rather than immediate, elevation of cytoplasmic Ca^{2+}. A typical example is the elevation of Ca^{2+} induced in lymphocyte cell lines by the glucocorticosteroid hormone dexamethasone [7, 93]. This problem is not unique to steroid hormone-induced cell death and pertains to any form of cell death in which Ca^{2+} changes are delayed following the apoptotic stimulus (e.g., withdrawal of growth factors or cytokines, treatment with chemotherapeutic agents). In these situations, it is not feasible to load cells with the fluorescent Ca^{2+} indicator before treatment with the apoptotic signal inducer (i.e., dexamethasone) and then monitor Ca^{2+} over a prolonged period (12–18 h) until Ca^{2+} becomes elevated because the fluorescent dye may leak from the cells or be taken up into organelles, a problem that increases over time. To circumvent this problem, one approach is to initiate the apoptotic signal (e.g., treatment with an apoptosis-inducing agent, growth factor withdrawal) before loading with the Ca^{2+} indicator dye. Ca^{2+} measurements are taken immediately after loading is complete. However, several precautions need to be taken when using this approach. First, the amount of dye taken up by cells should be monitored to insure that it is similar in treated and untreated populations (i.e., the fluorescent emission when dye is saturated with Ca^{2+}, F_{max}, should be similar when equal numbers of treated and untreated cells are compared). Second, the degree of dye de-esterification should be monitored to insure that it is complete in treated and untreated cells. An esterified dye (e.g., Fura-2 AM) undergoes a spectral shift after it is taken up by cells and de-esterified. Thus, it is wise to compare the spectral scan of treated and untreated cells to insure that it is the same, and shifted by comparison to the esterified form of the dye. Third, it is necessary to monitor organelle uptake of Ca^{2+}-sensitive dyes to insure that the dye is indeed cytoplasmic in location. It may be necessary to empirically determine optimal dye loading conditions (e.g., time, temperature) for each cell type under investigation to maximize cytoplasmic dye uptake and retention and to reduce organelle uptake. Moreover, it is necessary to exclude possible increases in organelle dye uptake when cells are treated with an apoptosis-inducing agent. As discussed above, the ER has a very high luminal Ca^{2+} concentration. Hence, sequestration of Ca^{2+} indicator by the ER could be mistakenly interpreted as an elevation of cytoplasmic Ca^{2+}. One convenient method to monitor organelle dye uptake is to treat cells sequentially with the detergents digitonin and Triton X-100 while continuously monitoring dye fluorescence. Typically, digitonin at relatively low concentrations permeabilizes the plasma membrane but not the membranes of organelles. Permeabilization of the plasma membrane by digitonin will release dye from the cytoplasm, reducing the fluorescence signal. The subsequent addition of a strong detergent (e.g., Triton X-100) will then permeabilize the organelle membranes. A decrease in fluorescence signal following addition of Triton X-100 is an indication that dye has been sequestered in the ER or some other high Ca^{2+} compartment.

Ca^{2+} elevation in the setting of apoptosis can be measured either on a population of cells or on single cells, with each approach having advantages and disadvantages. The simplest and most readily applicable technique involves recording the fluorescent output of Ca^{2+}-sensitive dyes using a cuvette-based fluorometer capable of continuous recording over time. Since most systems are cuvette based, serial additions of agonists can be made, and careful calibrations to determine R_{max} and R_{min} can be taken on each sample. In this method, however, Ca^{2+} is averaged over the cell population in the cuvette. In most cases this is sufficient. But transient elevations of Ca^{2+} in the form of spikes and oscillations may not be detected by population-based methods as such changes in Ca^{2+} are likely to occur asynchronously in the cell population under investigation. Ca^{2+} can be measured at a single cell level by digital imaging using an epifluoresence microscope or by flow cytometry. The former is preferred, as it permits continuous recording of cytoplasmic Ca^{2+} over a prolonged period of time.

16.4.2 DETERMINING WHETHER CYTOPLASMIC Ca^{2+} ELEVATION MEDIATES APOPTOSIS

Once it has been demonstrated that the apoptotic signal of interest induces an elevation of cytoplasmic Ca^{2+}, the next step is to determine whether or not Ca^{2+} elevation contributes to apoptosis or is secondary to metabolic changes (e.g., decreased ATP levels) that accompany cell death. Methods used to document that Ca^{2+} elevation contributes to glucocorticosteroid-induced apoptosis are illustrative of the range of experimental approaches that can be employed:

(a) One approach is to determine the effect of reducing extracellular Ca^{2+} on apoptosis induction. This is illustrated by studies demonstrating that reduction of extracellular Ca^{2+} inhibits the induction of apoptosis in thymocytes by glucocorticosteroid hormones [94–96]. The applicability of this approach, however, is dependent upon the cell system under investigation and the degree to which cytoplasmic Ca^{2+} elevation depends on extracellular Ca^{2+} on uptake. In the case of thymocytes this approach works well since the onset of apoptosis in thymocytes following glucocorticosteroid hormone treatment is relatively rapid (on the order of 6–12 h). However, the onset of apoptosis in transformed lymphoid cell lines is delayed by over 24 h after hormone treatment [7, 93]. In the case of lymphoid cell lines, reducing extracellular Ca^{2+} concentration leads to a diminution of intracellular Ca^{2+} stores (e.g., ER luminal Ca^{2+}) and this alone is sufficient to trigger apoptosis [78]. Thus, experimental approaches involving depletion of extracellular Ca^{2+} are only appropriate for cells that do not require extracellular Ca^{2+} for cell viability during the duration of the experiment. Thus, it is important to first determine that the cell type under investigation does not require extracellular Ca^{2+} for maintenance of cell viability before using extracellular Ca^{2+} depletion to test the dependency of apoptosis induction on extracellular Ca^{2+} uptake.

(b) A more direct approach to document the contribution of cytoplasmic Ca^{2+} elevation to apoptosis is to determine if buffering the Ca^{2+} elevation inhibits apoptosis. Two approaches are commonly employed. One is to use an intracellular Ca^{2+} chelator (e.g., BAPTA-AM) [97] and the other is to express a cytoplasmic Ca^{2+}-binding protein (e.g., calbindin) [98]. Both these measures have documented the role of cytoplasmic Ca^{2+} elevation in glucocorticosteroid-induced apoptosis. However, caution must be applied in applying these experimental approaches. In our experience, the uptake and retention of BAPTA-AM varies with cell type. Also, the use of BAPTA-AM to buffer Ca^{2+} is most effective in those situations where the elevation of cytosolic Ca^{2+} is acute and transient, rather than delayed or sustained for prolonged periods of time. There are also several caveats with transient or stable over-expression of Ca^{2+}-binding proteins. First, careful control experiments need to be included to exclude possible artifacts due to the transfection procedure itself or secondary gene expression changes (e.g., loss of cell viability and induction of stress responses, respectively). Second, it is important to document by careful Ca^{2+} measurements that expression of a Ca^{2+}-binding protein has indeed suppressed Ca^{2+} elevation following the apoptotic signal under investigation. Third, it is important to insure experimentally that an effect of over-expressing a Ca^{2+}-binding protein on apoptosis is indeed due to the Ca^{2+}-binding activity of the protein, rather than due to an unanticipated action of the protein. Ideally, this could be accomplished by expressing a mutant form of the protein that fails to bind Ca^{2+} as a negative control.

16.4.3   DETECTING ALTERATIONS IN INTRACELLULAR Ca^{2+} DISTRIBUTION

As already discussed, alterations in Ca^{2+} homeostasis associated with apoptosis may be more complex than a simple elevation in cytoplasmic Ca^{2+}. In response to certain apoptotic signals, Ca^{2+} is redistributed from organelles that normally have a high concentration of Ca^{2+} (i.e., the ER) into organelles that normally have a low concentration of Ca^{2+} (i.e., mitochondria). An example is the response of cells to growth factor withdrawal (i.e., interleukin-3 withdrawal from hematopoietic cells), in which Ca^{2+} was found to redistribute from the ER into mitochondria preceding apoptosis [36]. Paradoxically, there are a number of reports suggesting that Bcl-2, an inhibitor of apoptosis, also decreases luminal Ca^{2+} concentration (see above). Thus, methods to detect changes in ER Ca^{2+} levels are playing an increasing role in studies of the role of Ca^{2+} in signaling apoptosis.

(a) A common but indirect method of estimating ER luminal Ca^{2+} concentration is to measure cytoplasmic Ca^{2+} concentration before and after addition of thapsigargin, using a standard cytoplasmic Ca^{2+} indicator (e.g., Fura-2 AM). This method has been used to document a decrease in luminal Ca^{2+} concentration associated with apoptosis induction by glucocorticosteroid treatment in lymphocytes [7, 93] and by growth factor withdrawal [36]. Moreover, findings using this technique have suggested an alteration in luminal Ca^{2+} concentration in cells that over-express the anti-apoptotic protein Bcl-2 (reviewed in Ref. [88]). Thapsigargin is a selective inhibitor of the ER-associated Ca^{2+}-ATPase that is required to maintain a high concentration of Ca^{2+} within the ER lumen. Under normal physiological conditions, the action of this Ca^{2+}-ATPase counterbalances a continuous leakage of Ca^{2+} from the ER lumen into the cytoplasm. Hence, when the Ca^{2+}-ATPase is inhibited by thapsigargin, Ca^{2+} leaks down a steep concentration gradient from ER lumen into the surrounding cytoplasm, inducing an elevation of cytoplasmic Ca^{2+}. The magnitude of the cytoplasmic Ca^{2+} elevation provides an indirect measure of luminal Ca^{2+}. However, the elevation of cytoplasmic Ca^{2+} that follows addition of thapsigargin is influenced by many factors in addition to luminal Ca^{2+} content (e.g., the buffering capacity of luminal and cytoplasmic proteins, the activity of plasma membrane channels and pumps that serve to reduce cytoplasmic Ca^{2+} concentration, and the buffering capacity of mitochondria). Also, in employing this technique, one must account for the contribution of extracellular Ca^{2+} to the thapsigargin-induced Ca^{2+} elevation. Typically, Ca^{2+} release activated channels (CRAC) located in the plasma membrane open and permit influx of extracellular Ca^{2+} in response to a diminution of ER luminal Ca^{2+} concentration. Hence, the Ca^{2+} elevation induced by thapsigargin has two components, one due to the efflux of Ca^{2+} from the ER lumen into the cytoplasm and the other due to influx of extracellular Ca^{2+}. To distinguish between these two components, it is necessary to either suspend cells in nominally Ca^{2+}-free buffer or to add EGTA in quantities sufficient to chelate extracellular Ca^{2+} before adding thapsigargin. When these steps are taken, the elevation of cytoplasmic Ca^{2+} following thapsigargin addition will be due to efflux of Ca^{2+} from the ER without significant contribution from extracellular Ca^{2+}. One caveat, however, is that depending on cell type, ER luminal Ca^{2+} concentration may be heavily dependent upon extracellular Ca^{2+}, and will decline when extracellular Ca^{2+} concentration is lowered. Hence, it is wise to measure thapsigargin-induced Ca^{2+} elevation immediately after resuspending cells in nominally Ca^{2+}-free buffer or chelating extracellular Ca^{2+} with EGTA.

(b) The use of genetically encoded fluorescent proteins as Ca^{2+} indicators has been a major advance (see Chapter 4), but these indicators have not yet received widespread use in studying apoptosis. However, two instances where they have been targeted to the ER lumen and used to measure luminal Ca^{2+} provide an example of their use. In both situations investigators employed cameleons, which are composed of two fluorescent proteins linked together by calmodulin and a calmodulin-binding peptide. Fluorescence resonance energy transfer (FRET) between the two fluorescent proteins occurs upon binding of Ca^{2+} to the binding interface of calmodulin and the calmodulin-binding peptide. The cameleon is targeted to the ER lumen by incorporating an ER uptake and retention sequence similar to that found on resident ER proteins. Demaurex and colleagues used an original version of a cameleon to measure luminal

Ca^{2+} in cells that overexpress Bcl-2, providing evidence that Bcl-2 decreases luminal Ca^{2+} concentration [99]. However, the K_d for Ca^{2+} in this cameleon was not well suited to monitor changes in ER Ca^{2+} and therefore had low sensitivity. A new cameleon with improved reaction kinetics and a K_d ideal for imaging Ca^{2+} in the ER was recently engineered by Tsien and colleagues and used to confirm a reduction of luminal Ca^{2+} concentration in cells over-expressing Bcl-2 [100].

(c) ER-targeted aequorin probes (see Chapter 5) can also be genetically targeted to the ER to measure ER Ca^{2+} directly, but the aequorin probes are luminescent and therefore not bright enough for single-cell imaging without specialized equipment. Perhaps more problematic is that for aequorin to be active it must be reconstituted with the coelenterazine cofactor, and to do so the ER must initially be completely depleted of Ca^{2+}. However, perturbation of ER Ca^{2+} can lead to secretion of luminal Ca^{2+}-binding proteins [101]. Nevertheless, ER-targeted aequorin has been used to investigate the effect of Bcl-2 protein family members on luminal Ca^{2+}. The findings of these studies indicate that luminal Ca^{2+} is reduced in cells that over-express Bcl-2 [102] and in cells lacking both of the pro-apoptotic Bcl-2 family members, Bax and Bak [87].

(d) Compartmentalized fluorescent dyes (e.g., Mag-Fura-2 and Fura-2 FF) have also been used to measure Ca^{2+} but, unlike either fluorescent proteins or aequorins, cannot be selectively targeted to the ER lumen [103]. Cells are loaded with these dyes under empirically derived conditions that favor uptake and retention of the dye in the ER or mitochondria (e.g., 37° incubation temperature) and cytoplasmic dye is released (e.g., by digitonin permeabilization) or quenched by manganese before measurements are recorded. Compartmentalized fluorescent dyes have been used to investigate the effect of Bcl-2 on ER luminal Ca^{2+} concentration. One group, using Mag-Fura-2, found that Bcl-2 overexpression reduced luminal Ca^{2+} [104]. On the other hand, we have used Fura-2 FF to measure luminal Ca^{2+} in T cells and have not detected a significant difference in Bcl-2 positive and negative cells [9]. In both cases, there was complete consistency between observations made with compartmentalized dyes and measurements of thapsigargin-releasable Ca^{2+} (see above), suggesting that differences in measured effects of Bcl-2 on luminal Ca^{2+} were more likely due to differences in the experimental system than the technique of Ca^{2+} measurement.

In summary, a variety of techniques are available to measure luminal Ca^{2+} and each has distinct advantages and disadvantages. Aequorins and fluorescent proteins can be targeted selectively into organelles, which is a tremendous advantage over other techniques. However, a major caveat with their use is that they must be expressed either transiently or stably in cells by means of transfecting cells with expression vectors. This calls for careful controls to insure that the transfections procedure itself does not alter Ca^{2+} homeostasis. Those studies intended to measure effects of Bcl-2 on ER Ca^{2+} serve as a useful example. The fact that over-expression of the anti-apoptotic protein Bcl-2 can induce cell death is so unexpected that it is easily overlooked [105, 106]. Hence, cell viability must be carefully monitored when the effect of transiently over-expressing Bcl-2 on luminal Ca^{2+} is under investigation. One way to avoid loss of viability is to selectively target Bcl-2 to the ER, since it is the insertion of Bcl-2 into the outer mitochondrial membrane that is toxic to cells [106]. In experiments where Bcl-2 is stably over-expressed, one must be cautious to control for altered expression of other Ca^{2+} regulatory proteins. The need for this precaution is illustrated by the work of Vanden Abeele, cited above, who used Mag-Fura-2 to measure luminal Ca^{2+} and found that decreased luminal Ca^{2+} in cells that over-express Bcl-2 was due to decreased SERCA pump levels and decreased levels of the luminal Ca^{2+}-binding protein calreticulin [104].

16.5   METHODS TO INVESTIGATE THE INVOLVEMENT OF IP$_3$ RECEPTORS IN APOPTOSIS

The involvement of IP$_3$ receptors in mediating Ca^{2+} signals associated with apoptosis has been described above. A wide variety of standard methods have been used to investigate the role of IP$_3$ receptors in apoptosis, including immunoblotting to detect caspase-mediated cleavage [84, 85] antisense-mediated repression of IP$_3$ receptor expression to test the essential role of IP$_3$ receptors in apoptosis induction [33, 34], and both co-immunoprecipitation and blue native gel electrophoresis to detect interactions of IP$_3$ receptors with other proteins (i.e., Bcl-2) [9]. But one approach that is particularly unique is the use of the chicken DT40 B lymphocyte line. This cell line is increasing in popularity due to the ease with which it can be manipulated genetically [107]. All three subtypes of IP$_3$ receptors have been knocked out in this cell line, significantly inhibiting apoptosis induction through the B-cell antigen receptor [35].

16.6   CONCLUDING REMARKS

The role of Ca^{2+} in signaling apoptosis is far from completely understood. Methods of measuring Ca^{2+} are not applied as easily as are methods of protein chemistry and molecular biology. Thus, the role of Ca^{2+} in apoptosis is not as well understood as are many other facets of this process. Improved technologies to measure Ca^{2+}, particularly in organelles, will help advance understanding of the role of Ca^{2+} in apoptosis and clarify some of the confusion that has arisen regarding the effect of Bcl-2 family members on Ca^{2+}. But Ca^{2+} measuring technologies are only part of the equation. The model system under investigation is critical. As already discussed, many factors contribute to regulation of Ca^{2+} signaling and an alteration in one factor can cause secondary changes in other factors. Indeed, it is not always easy to insure that changes in Ca^{2+} attributed to altered expression of an apoptosis regulatory protein (e.g., a Bcl-2 family member) are mediated directly through the protein of interest or alternatively are mediated by secondary changes in expression of one or more members of the Ca^{2+}

signaling toolkit. Thus, it is important in all experiments to control for multiple variables that may influence changes in Ca^{2+} signaling or homeostasis.

ACKNOWLEDGMENTS

George Dubyak, Martin Bootman, Llewelyn Roderick, and Michael Berridge have exercised both diligence and patience in their attempts to teach the author of this chapter the complexities of Ca^{2+} and its measurement. For this the author is forever grateful. The author also thanks Michael Malone for providing data for figures.

REFERENCES

1. Krebs, J., The role of calcium in apoptosis. *Bio Metals*, 11, 375, 1998.
2. Orrenius, S., Zhivotovsky, B., and Nicotera, P., Regulation of cell death: the calcium-apoptosis link. *Nature Rev Mol Cell Biol*, 4, 552, 2003.
3. Hajnoczky, G., Davies, E., and Madesh, M., Calcium signaling and apoptosis. *Biochem Biophys Res Commun*, 304, 445, 2003.
4. Rizzuto, R., Pinton, P., Ferrari, D., Chami, M., Szabadkai, G., Magalhaes, P.J., Di Virgilio, F., and Pozzan, T., Calcium and apoptosis: facts and hypotheses. *Oncogene*, 22, 8619, 2003.
5. Szabadkai, G. and Rizzuto, R., Participation of endoplasmic reticulum and mitochondrial calcium handling in apoptosis: more than just neighborhood? *FEBS Lett*, 567, 111, 2004.
6. Lam, M., Vimmerstedt, L.J., Schlatter, L.K., Hensold, J.O., and Distelhorst, C.W., Preferential synthesis of the 78-Kd glucose-regulated protein in glucocorticoid-treated S49 mouse lymphoma cells. *Blood*, 79, 3285, 1992.
7. Lam, M., Dubyak, G., and Distelhorst, C.W., Effect of glucocorticosteroid treatment on intracellular calcium homeostasis in mouse lymphoma cells. *Mol Endocrinol*, 7, 686, 1993.
8. Lam, M., Dubyak, G., Chen, L., Nuñez, G., Miesfeld, R.L., and Distelhorst, C.W., Evidence that BCL-2 represses apoptosis by regulating endoplasmic reticulum-associated Ca^{2+} fluxes. *Proc Natl Acad Sci U.S.A*, 91, 6569, 1994.
9. Chen, R., Valencia, I., Zhong, F., McColl, K.S., Roderick, H.L., Bootman, M.D., Berridge, M.J., Conway, S.J., Holmes, A.B., Mignery, G.A., Velez, P., and Distelhorst, C.W., Bcl-2 functionally interacts with inositol 1,4,5-trisphosphate receptors to regulate calcium release from the ER. *J Cell Biol*, 166, 193, 2004.
10. Wyllie, A.H., Glucocorticoid-induced thymocyte apoptosis is associated with endonuclease activation. *Nature*, 284, 555, 1990.
11. Cory, S. and Adams, J.M., The Bcl-2 family: regulators of the cellular life-or-death switch. *Nat Rev Cancer*, 2, 647, 2002.
12. Bouillet, P. and Strasser, A., BH3-only proteins — evolutionarily conserved proapoptotic Bcl-2 family members essential for initiating programmed cell death. *J Cell Sci*, 115, 1567, 2002.
13. Willis, S., Day, C.L., Hinds, M.G., and Huang, D.C., The Bcl-2-regulated apoptotic pathway. *J Cell Science*, 116, 4053, 2003.
14. Danial, N.N.and Korsmeyer, S.J., Cell death: critical control points. *Cell*, 116, 205, 2004.

15. Puthalakath, H. and Strasser, A., Keeping killers on a tight leash: transcriptional and post-translational control of the pro-apoptotic activity of BH3-only proteins. *Cell Death Differ*, 9, 505, 2002.

16. Schwartz, L.M.and Osborne, B.A., Cell death, in *Methods in Cell Biology*, Vol. 46, Wilson, L. and Matsudaira, P., eds., Academic Press, New York, 1995, 453.

17. Reed, J.C., (Ed.) Apoptosis, in *Methods in Enzymology*, Vol. 322, Academic Press, San Diego, 2000, 569.

18. Nicotera, P., Apoptosis and age-related disorders: role of caspase-dependent and caspase-independent pathways. *Toxicol Lett*, 127, 189, 2002.

19. Cande, C., Cecconi, F., Dessen, P., and Kroemer, G., Apoptosis-inducing factor (AIF): key to the conserved caspase-independent pathways of cell death? *J Cell Sci*, 115, 4727, 2002.

20. van Engeland, M., Nieland, L.J., Ramaekers, F.C., Schutte, B., and Reutelingsperger, C.P., Annexin V-affinity assay: a review on an apoptosis detection system based on phosphatidylserine exposure. *Cytometry*, 31, 1, 1998.

21. Li, W. and Tait, J. F., Regulatory effect of CD9 on calcium-stimulated phospatidylserine exposure in Jurkat T lymphocytes. *Arch Biochem Biophys*, 351, 89, 1998.

22. Tait, J. F. and Gibson, D., Phospholipid binding of annexin V: effects of calcium and membrane phosphatidylserine content. *Arch Biochem Biophys*, 298, 187, 1992.

23. Berridge, M.J., Lipp, P., and Bootman, M.D., The versatility and universality of calcium signalling. *Nature Rev. Mol Cell Biol*, 1, 11, 2000.

24. Putney, J. W., Jr., Broad, L.M., Braun, F.J., Lievremont, J. P., and Bird, G.S., Mechanisms of capacitative calcium entry. *J Cell Sci*, 114, 2223, 2001.

25. Petersen, O.H., Calcium signal compartmentalization. *Biol Res*, 35, 177, 2002.

26. Berridge, M.J., Bootman, M.D., and Roderick, H.L., Calcium signaling: dynamics, homeostasis and remodelling. *Nature Rev Mol Cell Biol*, 4, 517, 2003.

27. Lewis, R.S., Calcium oscillations in T-cells: mechanisms and consequences for gene expression. *Biochem Soc Trans*, 31, 925, 2003.

28. Winslow, M.M., Neilson, J. R., and Crabtree, G.R., Calcium signaling in lymphocytes. *Curr Opinion Immunol*, 15, 299, 2003.

29. Schanne, F.A.X., Kane, A.B., Young, E.E., and Farber, J. L., Calcium dependence of toxic cell death: a final common pathway. *Science*, 206, 700, 1979.

30. Jiang, S., Chow, S.C., Nicotera, P., and Orrenius, S., Intracellular Ca^{2+} signals activate apoptosis in thymocytes: studies using the Ca^{2+}-ATPase inhibitor thapsigargin. *Exp Cell Res*, 212, 84, 1994.

31. Alberola-Ila, J., Takaki, S., Kerner, J.D., and Perlmutter, R.M., Differential signaling by lymphocyte antigen receptors. *Annu Rev Immunol*, 15, 125, 1997.

32. Crabtree, G.R., Generic signals and specific outcomes: signaling through Ca^{2+}, calcineurin, and NFAT. *Cell*, 96, 611, 1999.

33. Khan, A.A., Soloski, M.J., Sharp, A.H., Schilling, G., Sabatini, D.M., Li, S.-H., Ross, C.A., and Snyder, S.H., Lymphocyte apoptosis: mediation by increased type 3 inositol 1,4,5,-trisphosphate receptor. *Science*, 273, 503, 1996.

34. Jayaraman, T. and Marks, A.R., T cells deficient in inositol 1,4,5-trisphosphate receptor are resistant to apoptosis. *Mol Cell Biol*, 17, 3005, 1997.

35. Sugawara, H., Kurosaki, M., Takata, M., and Kurosaki, T., Genetic evidence for involvement of type 1, type 2 and type 3 inositol 1,4,5-trisphosphate receptors in signal transduction through the B-cell antigen receptor. *EMBO J*, 16, 3078, 1997.

36. Baffy, G., Miyashita, T., Williamson, J. R., and Reed, J. C., Apoptosis induced by withdrawal of interleukin-3 (IL-3) from an IL-3-dependent hematopoietic cell line is

associated with repartitioning of intracellular calcium and is blocked by enforced Bcl-2 oncoprotein production. *J Biol Chem*, 268, 6511, 1993.

37. Hildeman, D.A., Zhu, Y., Mitchell, T.C., Kappler, J., and Marrack, P., Molecular mechanisms of activated T cell death *in vivo*. *Curr Opinion Immunol*, 14, 354, 2002.

38. Green, D.R., Droin, N., and Pinkoski, M., Activation-induced cell death in T cells. *Immunol Rev*, 193, 70, 2003.

39. Zheng, L., Trageser, C.L., Willerford, D.M., and Lenardo, M.J., T cell growth cytokines cause the superinduction of molecules mediating antigen-induced T lymphocyte death. *J Immunol*, 160, 763, 1998.

40. Parijs, L.V., Refaeli, Y., Lord, J. D., Nelson, B.H., Abbas, A.K., and Baltimore, D., Uncoupling IL-2 signals that regulate T cell proliferation, survival, and Fas-mediated activation-induced cell death. *Immunity*, 11, 281, 1999.

41. Srivastava, R.K., Sasaki, C.Y., Hardwick, J. M., and Longo, D.L., Bcl-2-mediated drug resistance: inhibition of apoptosis by blocking nuclear factor of activated T lymphocytes (NFAT)-induced Fax ligand transcription. *J Exp Med*, 190, 253, 1999.

42. Algeciras-Schimnich, A., Griffith, T.S., Lynch, D.H., and Paya, C.V., Cell cycle-dependent regulation of FLIP levels and susceptibility to Fas-mediated apoptosis. *J Immunol*, 162, 5205, 1999.

43. Shibasaki, F. and McKeon, F., Calcineurin functions in Ca^{2+}-activated cell death in mammalian cells. *J Cell Biol*, 131, 735, 1995.

44. Wang, H.G., Pathan, N., Ethell, I.M., Krajewski, S., Yamaguchi, Y., Shibasaki, F., McKeon, F., Bobo, T., Franke, T.F., and Reed, J. C., Ca^{2+}-induced apoptosis through calcineurin dephosphorylation of BAD. *Science*, 284, 339, 1999.

45. Saito, S., Hiroi, Y., Zou, Y., Aikawa, R., Toko, H., Shibasaki, F., Yazaki, Y., Nagai, R., and Komuro, I., β-Adrenergic pathway induces apoptosis through calcineurin activation in cardiac myocytes. *J Biol Chem*, 275, 34528–34533, 2000.

46. Jiang, A. and Clark, E.A., Involvement of Bik, a proapoptotic member of the Bcl-2 family, in surface IgM-mediated B cell apoptosis. *J Immunol*, 166, 6025, 2001.

47. Bouillet, P., Metcalf, D., Huang, D.C., Tarlinton, D.M., Kay, T.W., Kontgen, F., Adams, J. M., and Strasser, A., Proapoptotic Bcl-2 relative Bim required for certain apoptotic responses, leukocyte homeostasis, and to preclude autoimmunity. *Science*, 286, 1735, 1999.

48. Means, A.R., Calcium, calmodulin and cell cycle regulation. *FEBS Lett*, 347, 1, 1994.

49. Santella, L., The role of calcium in the cell cycle: facts and hypotheses. *Biochem Biophys Res Commun*, 244, 317, 1998.

50. Squier, M.K. and Cohen, J. J., Calpain, an upstream regulator of thymocyte apoptosis. *J Immunol*, 158, 3690, 1997.

51. Mathiasen, I.S., Sergeev, I.N., Bastholm, L., Elling, F., Norman, A.W., and Jaattela, M., Calcium and calpain as key mediators of apoptosis-like death induced by vitamin D compounds in breast cancer cells. *J Biol Chem*, 277, 30738, 2002.

52. Varghese, J., Radhika, G., and Sarin, A., The role of calpain in caspase activation during etoposide induced apoptosis in T cells. *Eur J Immunol*, 31, 2035, 2001.

53. Mandic, A., Viktorsson, K., Starndberg, L., Heiden, T., Hansson, J., Linder, S., and Shoshan, M.C., Calpain-mediated Bid cleavage and calpain-independent Bak modulation: two separate pathways in cisplatin-induced apoptosis. *Mol Cell Biol*, 22, 3003, 2002.

54. Chen, M., Won, D.-J., Krajewski, S., and Gottlieb, R.A., Calpain and mitochondria in ischemia/reperfusion injury. *J Biol Chem*, 277, 29181, 2002.

55. Wolf, B.B., Goldstein, J. C., Stennicke, H.R., Beere, H., Amarante-Mendes, G.P., Salvesen, G.S., and Green, D.R., Calpain functions in a caspase-independent

manner to promote apoptosis-like events during platelet activation. *Blood*, 94, 1683, 1999.

56. Gil-Parrado, S., Fernandez-Montalvan, A., Assfalg-Machleidt, I., Popp, O., Bestvater, F., Holloschi, A., Knoch, T.A., Auerswald, E.A., Welsh, K., Reed, J. C., Fritz, H., Fuentes-Prior, P., Spiess, E., Salvesen, G.S., and Machleidt, W., Ionomycin-activated calpain triggers apoptosis. A probable role for Bcl-2 family members. *J Biol Chem*, 277, 27217, 2002.

57. Nakagawa, T. and Yuan, J., Cross-talk between two cysteine protease families: activation of caspase-12 by calpain in apoptosis. *J Cell Biol*, 150, 887, 2000.

58. Wood, D.E., Thomas, A., Devi, L.A., Berman, Y., Beavis, R.C., Reed, J. C., and Newcomb, E.W., Bax cleavage is mediated by calpain during drug-induced apoptosis. *Oncogene*, 17, 1069, 1998.

59. Choi, W.S., Lee, E.H., Chung, C.W., Jung, Y.K., Jin, B.K., Kim, S.U., Oh, T.H., Saido, T.C., and Oh, Y.J., Cleavage of Bax is mediated by caspase-dependent or -independent calpain activation in dopaminergic neuronal cells: protective role of Bcl-2. *J Neurochem*, 77, 1531, 2001.

60. Bordone, L. and Campbell, C., DNA ligase III is degraded by calpain during cell death induced by DNA-damaging agents. *J Biol Chem*, 277, 26673, 2002.

61. Kubbutat, M.H. and Vousden, K.H., Proteolytic cleavage of human p53 by calpain: a potential regulator of protein stability. *Mol Cell Biol*, 17, 460, 1997.

62. Kim, M.-J., Jo, D.-G., Hong, G.-S., Kim, B.-J., Lai, M., Cho, D.-H., Kim, K.-W., Bandyopadhyay, A., Hong, Y.-M., Kim, D.H., Cho, C., Liu, J. O., and Snyder, S.H., Calpain-dependent cleavage of cain/cabin 1 activates calcineurin to mediate calcium-triggered cell death. *Proc Natl Acad Sci U.S.A*, 9870, 2002.

63. Neamati, N., Fernanadez, A., Wright, S., Kiefer, J., and McConkey, D.J., Degradation of lamin B1 precedes oligonucleosomal DNA fragmentation in apoptotic thymocytes and isolated thymocyte nuclei. *J Immunol*, 154, 3788, 1995.

64. McConkey, D.J., Calcium-dependent, interleukin 1β-converting enzyme inhibitor-insensitive degradation of lamin B1 and DNA fragmentation in isolated thymocyte nuclei. *J Biol Chem*, 271, 22398, 1996.

65. Polster, B.M. and Fiskum, G., Mitochondrial mechanisms of neural cell apoptosis. *J Neurochemistry*, 90, 1281, 2004.

66. Desagher, S. and Martinou, J. C., Mitochondria as the central control point of apoptosis. *Trends Cell Biol*, 10, 369, 2000.

67. Ferri, K.F. and Kroemer, G., Organelle-specific initiation of cell death pathways. *Nat Cell Biol*, 3, E255, 2001.

68. Wang, X., The expanding role of mitochondria in apoptosis. *Genes Dev*, 15, 2922, 2001.

69. Nieminen, A.-L., Apoptosis and necrosis in health and disease: role of mitochondria. *Int Rev Cytol*, 224, 29, 2003.

70. Smaili, S.S., Hsu, Y.T., Youle, R.J., and Russell, J. T., Mitochondria in Ca^{2+} signaling and apoptosis. *J Bioenerg Biomembr*, 32, 35, 2000.

71. Hajnoczky, G., Csordas, G., Krishnamurthy, R., and Szalai, G., Mitochondrial calcium signaling driven by the IP3 receptor. *J Bioenerg Biomembr*, 32, 15, 2000.

72. Rutter, G.A. and Rizzuto, R., Regulation of mitochondrial metabolism by ER Ca^{2+} release: an intimate connection. *Trends Biochem Sci*, 25, 215, 2000.

73. Szalai, G., Krishnamurthy, R., and Hajnoczky, G., Apoptosis driven by IP(3)-linked mitochondrial calcium signals. *EMBO J*, 18, 6349, 1999.

74. Pacher, P. and Hajnoczky, G., Propagation of the apoptotic signal by mitochondrial waves. *EMBO J*, 20, 4107, 2001.

75. Csordas, G., Madesh, M., Antonsson, B., and Hajnoczky, G., tcBid promotes Ca^{2+} signal propagation to the mitochondria: control of Ca^{2+} permeation through the outer mitochondrial membrane. *EMBO J*, 21, 2198, 2002.

76. Szegezdi, E., Fitzgerald, U., and Samali, A., Caspase-12 and ER stress-mediated apoptosis. *Ann NY Acad Sci*, 1010, 186, 2003.

77. Rutkowski, D.T. and Kaufman, R.J., A trip to the ER: coping with stress. *Trends Cell Biol*, 14, 20, 2004.

78. He, H., Lam, M., McCormick, T.S., and Distelhorst, C.W., Maintenance of calcium homeostasis in the endoplasmic reticulum by Bcl-2. *J Cell Biol*, 138, 1219, 1997.

79. Wertz, I.E. and Dixit, V.M., Characterization of calcium release-activated apoptosis in LNCaP prostate cancer cells. *J Biol Chem*, 275, 11470, 2000.

80. Nutt, L.K., Chandra, J., Pataer, A., Fang, B., Roth, J., Swisher, S.G., O'Neil, R.G., and McConkey, D.J., Bax-mediated Ca^{2+} mobilization promotes cytochrome *c* release during apoptosis. *J Biol Chem*, 277, 20301, 2002.

81. Nutt, L.K., Pataer, A., Pahler, J., Fang, B., Roth, J., McConkey, D.J., and Swisher, S. G., Bax and Bak promote apoptosis by modulating endoplasmic reticular and mitochondrial Ca^{2+} stores. *J Biol Chem*, 277, 9219, 2002.

82. Zhong, W.X., Li, C., Hatzivassiliou, G., Lindsten, T., Yu, Q.-C., Yuan, J., and Thompson, C.B., Bax and Bak can localize to the endoplasmic reticulum to initiate apoptosis. *J Cell Biol*, 162, 59, 2003.

83. Boehning, D., Patterson, R.L., Sedaghat, L., Glebova, N.O., Kurosaki, T., and Snyder, S.H., Cytochrome *c* bind to inositol (1,4,5) trisphosphate receptors, amplifying calcium-dependent apoptosis. *Nat Cell Biol*, 5, 1051, 2003.

84. Hirota, J., Furuichi, T., and Mikoshiba, K., Inositol 1,4,5-trisphosphate receptor Type 1 is a substrate for caspase-3 and is cleaved during apoptosis in a caspase-3-dependent manner. *J Biol Chem*, 274, 34433, 1999.

85. Assefa, Z., Bultynck, G., Szlufcik, K., Kasri, N.N., Vermassen, E., Goris, J., Missiaen, L., Callewaert, G., Parys, J. B., and De Smedt, H., Caspase-3-induced truncation of Type 1 inositol trisphosphate receptor accelerates apoptotic cell death and induces inositol trisphosphate-independent calcium release during apoptosis. *J Biol Chem*, 279, 43227, 2004.

86. Pinton, P., Ferrari, D., Rapizzi, E., De Virgilio, F., Pozzan, T., and Rizzuto, R., The Ca^{2+} concentration of the endoplasmic reticulum is a key determinant of ceramide-induced apoptosis: significance for the molecular mechanism of Bcl-2 action. *EMBO J*, 20, 2690, 2001.

87. Scorrano, L., Oakes, S.A., Opferman, J. T., Cheng, E.H., Sorcinelli, M.D., Pozzan, T., and Korsmeyer, S.J., Regulation of endoplasmic reticulum calcium dynamics by Bax and Bak: a control point for apoptosis. *Science*, 300, 135139, 2003.

88. Distelhorst, C.W. and Shore, G.C., Bcl-2 and calcium: controversy beneath the surface. *Oncogene*, 23, 2875, 2004.

89. Hanson, C.J., Bootman, M.D., and Roderick, H.L., Cell signalling: IP3 receptors channel calcium into cell death. *Curr Biol*, 14, R933, 2004.

90. Shibasaki, F., Kondo, E., Akagi, T., and McKeon, F., Suppression of signalling through transcription factor NF-AT by interactions between calcineurin and Bcl-2. *Nature*, 386, 728, 1997.

91. Erin, N., Lehman, R.A.W., Boyer, P.J., and Billingsley, M.L., *In vitro* hypoxia and excitotoxicity in human brain induce calcineurin-Bcl-2 interactions. *Neuroscience*, 117, 557, 2003.

Calcium Signaling

92. Erin, N., Bronson, S.K., and Billingsley, M.L., Calcium-dependent interaction of calcineurin with Bcl-2 in neuronal tissue. *Neuroscience*, 117, 541, 2003.
93. Bian, X., Hughes, F.M., Huang, Y., Cidlowski, J. A., and Putney, J. W., Roles of cytoplasmic Ca^{2+} and intracellular Ca^{2+} stores in induction and suppression of apoptosis in S49 cells. *Am J Physiol*, 272 (*Cell Physiol*, 41), C1241, 1997.
94. Kaiser, N. and Edelman, I.S., Calcium dependence of glucocorticoid-induced lymphocytolysis. *Proc Natl Acad Sci U.S.A*, 74, 632, 1977.
95. Wyllie, A.H., Morris, R.G., Smith, A.L., and Dunlop, D., Chromatin cleavage in apoptosis: association with condensed chromatin morphology and dependence on macromolecular synthesis. *J Pathol*, 142, 67, 1984.
96. Cohen, J.J. and Duke, R.C., Glucocorticoid activation of a calcium-dependent endonuclease in thymocyte nuclei leads to cell death. *J Immunol*, 132, 38, 1984.
97. Zhivotovsky, B., Cedervall, B., Jiang, S., Nicotera, P., and Orrenius, S., Involvement of Ca^{2+} in the formation of high molecular weight DNA fragments in thymocyte apoptosis. *Biochem Biophys Res Commun*, 202, 120, 1994.
98. Dowd, D.R., MacDonald, P.N., Komm, B.S., Haussler, M.R., and Miesfeld, R.L., Stable expression of the calbindin-D28K complementary DNA interferes with the apoptotic pathway in lymphocytes. *Mol Endo*, 6, 1843, 1992.
99. Foyouzi-Youssefi, R., Arnaudeau, S., Borner, C., Kelley, W.L., Tschopp, J., Lew, D.P., Demaurex, N., and Krause, K.H., Bcl-2 decreases the free Ca^{2+} concentration within the endoplasmic reticulum. *Proc Natl Acad Sci U.S.A*, 97, 5723, 2000.
100. Palmer, A.E., Jin, C., Reed, J. C., and Tsien, R.Y., Bcl-2-mediated alterations in endoplasmic reticulum Ca^{2+} analyzed with an improved genetically encoded fluorescent sensor. *Proc Natl Acad Sci U.S.A*, 101, 17404, 2004.
101. Booth, C. and Koch, G.L.E., Perturbation of cellular calcium induces secretion of luminal ER proteins. *Cell*, 59, 729, 1989.
102. Pinton, P., Ferrari, D., Magalhaes, P., Schulze-Osthoff, K., Di Virgilio, F., Pozzan, T., and Rizzuto, R., Reduced loading of intracellular Ca(2+) stores and downregulation of capacitative Ca(2+) influx in Bcl-2-overexpressing cells. *J Cell Biol*, 148, 857, 2000.
103. Hofer, A.M., Measurement of free $[Ca^{2+}]$ changes in agonist-sensitive internal stores using compartmentalized fluorescent indicators. *Methods Mol Biol*, 114, 249, 1999.
104. Vanden Abeele, F., Skryma, R., Shuba, Y., Van Coppenolle, F., Slomianny, C., Roudbaraki, M., Mauroy, B., Wuytack, F., and Prevarskaya, N., Bcl-2-dependent modulation of Ca^{2+} homeostasis and store-operated channels in prostate cancer cells. *Cancer Cell*, 1, 169, 2002.
105. Uhlmann, E.J., Subramanian, T., Vater, C.A., Lutz, R., and Chinnadurai, G., A potent cell death activity associated with transient high level expression of BCL-2. *J Biol Chem*, 273, 17926, 1998.
106. Wang, N.S., Unkila, M., Reineks, E.Z., and Distelhorst, C.W., Transient expression of wild-type or mitochondrially targeted Bcl-2 induced apoptosis, whereas transient expression of endoplasmic reticulum-targeted Bcl-2 is protective against Bax-induced cell death. *J Biol Chem*, 276, 44117, 2001.
107. Winding, P. and Berchtold, M.W., The chicken B cell line DT40: a novel tool for gene disruption experiments. *J Immunol Meth*, 249, 1, 2001.

17 Calcium Entry Channels and Drug Discovery

Su Li, Anne Dodge, Chris T. Poll, and Martin Gosling

CONTENTS

17.1 DRUG DISCOVERY AND CALCIUM ENTRY CHANNELS: A HISTORICAL PERSPECTIVE

In all cell types Ca^{2+} influx through plasma membrane channels is of fundamental importance in both physiology and pathophysiology, e.g., contraction of muscle cells, neurotransmitter release from nerve terminals, secretion by epithelial cells, and leukocyte activation. Therefore, modulation of cell function by targeting these channels represents a potentially effective approach for therapeutic intervention.

Ca^{2+} influx pathways can be broadly divided into Ca^{2+} channels that are voltage-activated and those that are not. Both types of channels have been studied intensely over several decades. Although their function is similar i.e., regulated sources of one of the most important cytosolic signaling messengers in the body — Ca^{2+} ions, the path of drug discovery for these two types of Ca^{2+} channel is distinctly different. Both areas of research actually began from similar starting points, i.e., observations that certain pharmacological agents had an effect on physiological responses either *in vitro* or *in vivo* that led to a proposed mode of action, i.e., inhibition of Ca^{2+} influx [1]. While therapeutically useful drugs have resulted from the combined efforts of academics and their industry colleagues in developing blockers of voltage-activated Ca^{2+} channels (i.e., blockers of the L-type voltage-operated Ca^{2+} channel, L-VOCC), the therapeutic potential of blockers of non-voltage-gated Ca^{2+} channels is yet to be realized.

17.1.1 DISCOVERY OF L-TYPE VOLTAGE-OPERATED Ca^{2+} CHANNEL BLOCKERS

Blockers of L-VOCC Ca^{2+} channels present in vascular smooth muscle are widely used and are very effective drugs for the treatment of cardiovascular diseases such as hypertension, and in particular angina. The very earliest compounds, verapamil (Knoll AG) and prenylamine (Hoescht AG), were developed as coronary vasodilators working as β-adrenoceptor antagonists. However, both compounds had unexpected cardio-suppressant side effects that Fleckenstein investigated and discovered were due to Ca^{2+} channel blockade in 1964. These studies demonstrated that these compounds mimicked the cardiac effect of withdrawal of Ca^{2+} ions, which inhibits the cardiac excitation coupling, resulting in diminished contractile force (reviewed in Fleckenstein 1983 [2]). Cardio-suppressant effects of the compounds could be rapidly overcome by addition of Ca^{2+} ions, β catecholamines, or cardiac glycosides. The concept of Ca^{2+} antagonism was introduced in 1969, disproving the initial assumption that both these compounds were acting as β-adrenoceptor antagonists. Additional Ca^{2+} antagonists were also identified such as D600, a more potent methoxy derivative of verapamil, (Figure 17.1) which had a similar profile to verapamil although it was more potent, and the dihydropyridines including nifedipine (Figure 17.1), (reviewed by Fleckenstein 1983 [2]). The dihydropyridines, which share a similar mechanism of action to verapamil and D600, were also strong coronary vasodilators and decreased the force of muscle contraction (i.e., negative ionotropic effects). These observations led Fleckenstein to propose a new distinct pharmacological class of potent inhibitors of contraction–excitation coupling, the "calcium antagonists." The biochemical isolation and identification for the site of action of the compounds came from the availability of highly radioactive ([^{3}H]-labeled) and potent (low nanomolar binding affinity) ligands such as nitrendipine (Figure 17.1) that demonstrated high-affinity binding

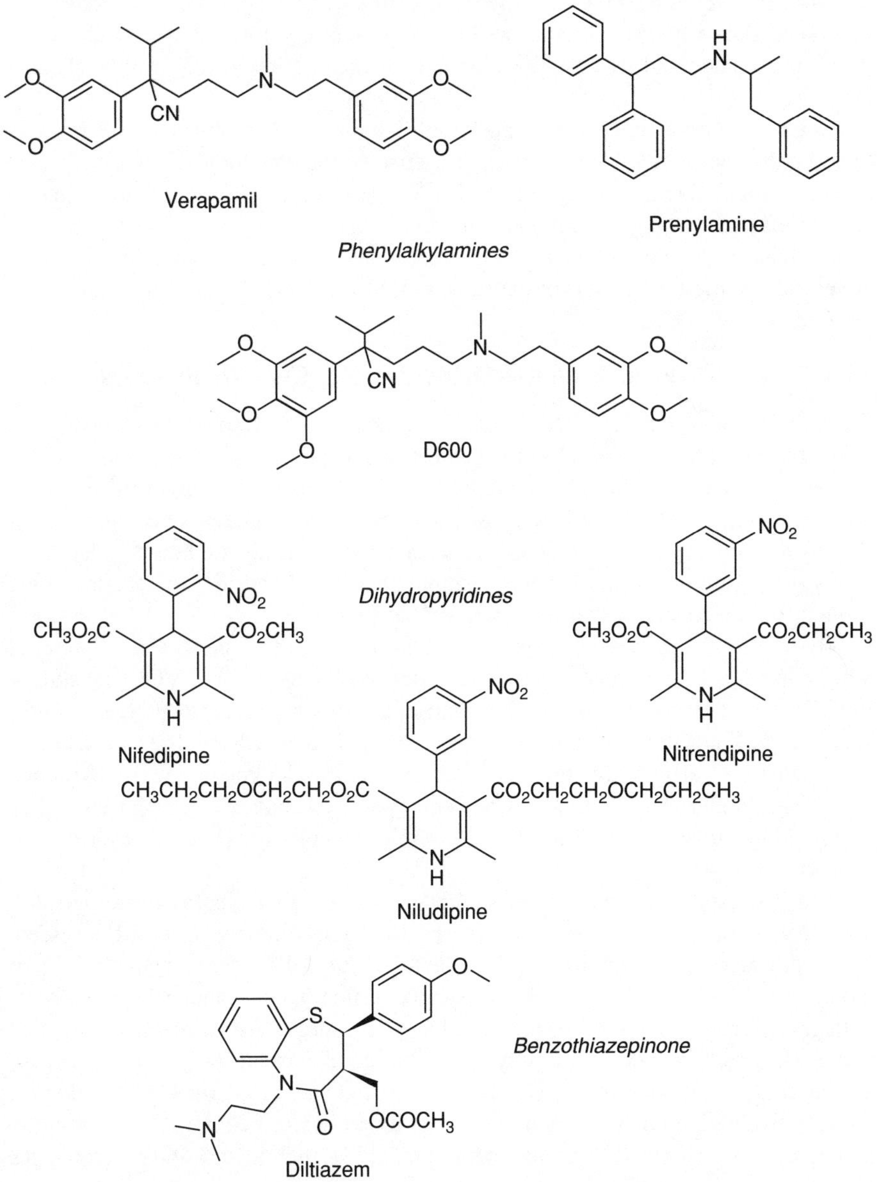
CN
N
O
Verapamil
H
N
Prenylamine
Phenylalkylamines
CN
D600
NO2
Dihydropyridines
CH3O2C
CO2CH3
N
H
Nifedipine
CO2CH2CH3
Nitrendipine
CH3CH2CH2OCH2CH2O2C
CO2CH2CH2OCH2CH2CH3
Niludipine
S
Benzothiazepinone
OCOCH3
Diltiazem

FIGURE 17.1 Structures of L-VOCC channel blockers.

to protein extracts from heart, coronary arteries, and aorta [3–6]. This work led to the identification of the α subunit of the led to (L-VOCC) and subsequently the additional accessory subunits of the native channel [7]. The results of these biochemical experiments also suggested that the dihydropyridines are Ca^{2+} antagonists which are chemically distinct from verapamil or diltiazem (Figure 17.1) and do not have identical binding sites on the channel [8].

Molecular cloning of the L-VOCC subunits came in the late 1980s [9] and functional data obtained with these cloned proteins confirmed the biochemical, pharmacological, and electrophysiological characteristics of the native channel [10, 11].

Selective high-affinity ligands contributed significantly to the advances made in the drug discovery of L-VOCCs, e.g., they enabled the development of specific and robust biological assays (which are still in use in the drug discovery environment) and helped generate significant understanding of the biology of this ion channel and its modulators. As a consequence, pharmaceutical companies were able to develop novel therapeutic treatments for cardiovascular disorders.

17.1.2 Discovery of Receptor-Operated Ca^{2+} Channel Blockers

Receptor-operated Ca^{2+} influx channels are not voltage-activated and open as a result of the activation of plasma membrane receptors, e.g., G-protein coupled receptors (GPCRs). Although the signal transduction pathways involved in receptor-operated Ca^{2+} channel (ROCC) opening are not well understood, intracellular second messengers e.g., arachidonic acid and the depletion of intracellular Ca^{2+} stores (i.e., store-operated channels, commonly abbreviated to SOC channels) are some of the mechanisms that may be involved [12–14].

Bolton first described receptor-operated Ca^{2+} influx pathways in smooth muscle in 1979 [15], almost two decades after the discovery of L-VOCC channels. ROCCs have a broader, cellular distribution than their voltage-operated counterparts, as they are present in both electrically excitable cells such as muscle and nerve cells and electrically unexcitable cells such as leukocytes and endothelial cells. These non-voltage-activated Ca^{2+} entry pathways have been proposed to play an important functional role in all the unexcitable cells in which they have been studied.

ROCCs have been intensively studied *in vitro*, either electrophysiologically, using patch-clamp techniques pioneered by Hamill et al. [16], or with fluorescent Ca^{2+} indicators that first became available in the 1980s, e.g., Quin 2 [17] to monitor intracellular Ca^{2+} levels in isolated primary cells and cell lines. Non-voltage-activated Ca^{2+} influx pathways have been described in platelets [18] and leukocytes, and include the highly Ca^{2+} selective current I_{CRAC} (Ca^{2+}-release activated Ca^{2+} current) in mast cells and lymphocytes [19, 20]. Ca^{2+}-permeable nonselective cation channels have also been found in mast cells [21], the promyelocytic cell line HL60 [22], human neutrophils [23], monocytes, and macrophages [24, 25].

TABLE 17.1
Summary of Receptor-Operated Calcium Channel (ROCC) Channel Inhibitors Originating from Pharmaceutical Companies

Compound	Company	Activity	References
SC 38249	SmithKline Beecham	Inhibition of platelet aggregation	28
SK&F 96365	SmithKline Beecham	Inhibiton of neutrophil activation	26
		Inhibition of lymphocyte activation	27
LOE 908	Boehringer Ingelheim	Inhibition of non-selective cation channel currents in HL60 cells	22
		Inhibition of non-selective cation channel currents in smooth muscle	104
CAI	Merck & Co	Anti-proliferative and inhibitor of calcium influx in cancer cells	34, 105
LU52396	BASF	Inhibitor of store-operated calcium influx	31
BTP2	Abbott Laboratories	Inhibitors of lymphocyte activation and store-operated calcium influx/I_{CRAC}	45, 47
YM-58483	Yamanouchi	Inhibitor of store-operated calcium influx/I_{CRAC}	46

The availability of fluorescent intracellular Ca^{2+} indicators that could be used in a wide variety of native cells in the 1980s enabled pharmaceutical companies to undertake research to identify blockers of these ROCCs (see Table 17.1). These efforts resulted in the identification of compounds of different structural types that inhibited receptor-mediated Ca^{2+} entry (Figure 17.2).

17.1.3 ROCC CHANNEL BLOCKERS

SK&F 96365, probably the best-known synthetic ROCC-blocker compound, inhibits receptor-activated Ca^{2+} influx and activation of platelets [26], neutrophils [26], and lymphocytes [27]. SK&F 96365 was an optimized compound with improved potency over the prototypical compound SC 38249 (Figure 17.2) that was originally designed as a thromboxane synthetase inhibitor [28]. SK&F 96365 also has relatively low potency and selectivity for ROCC channels versus other channels e.g., L-VOCC channels [26], K^+ currents [29], and Cl^- channels [21]. In addition, the molecular identity of the channels responsible for Ca^{2+} influx in the target cells was, and still is, unknown.

LOE 908 (an isoquinoline derivative) developed as an inhibitor of human neutrophil ROCC channels [22] has also proved to be a useful experimental tool especially when used in combination with other pharmacological tools. For instance, LOE 908 was used together with SK&F 96365 to demonstrate the presence of at least three pharmacologically distinct Ca^{2+} influx pathways in rat aortic smooth muscle cells [30].

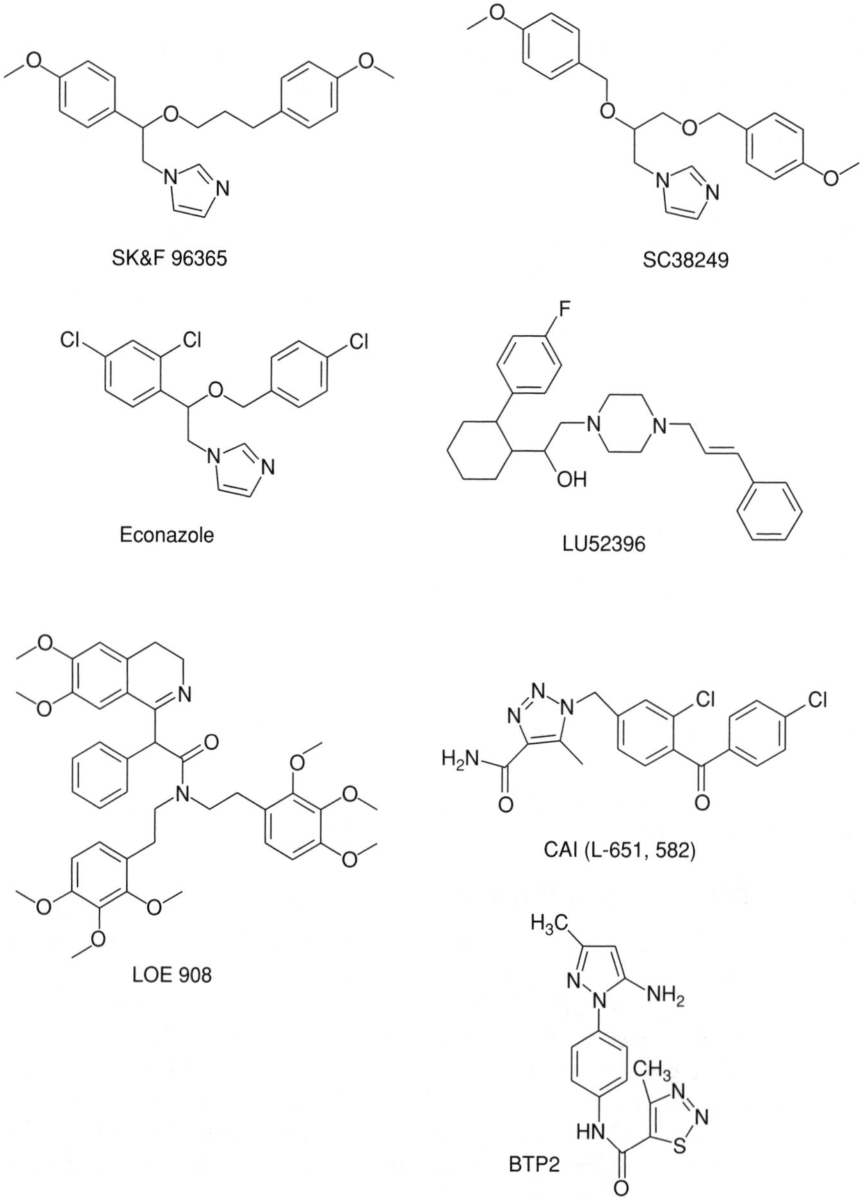

FIGURE 17.2 Structures of ROCC channel blockers.

LU52396 (Figure 17.2), a cycloalkyl-2-piperazineinoethanol, like SC 38249 was designed as a thromboxane synthetase inhibitor and found to inhibit thromboxane-independent agonist-stimulated Ca^{2+} influx in platelets [31].

CAI (also known as L-651,582, a carboxyamidotriazole) was initially developed as an anticoccidiostat [32] (an agent to combat coccidiosis, an endemic parasitic disease of poultry caused by protozoa of the genus *Eimeria*). This compound was also shown to have anti-proliferative effects in ovarian cancer disease models [33]. Further work suggested that the anti-proliferative effect was due to its ability to inhibit receptor-mediated Ca^{2+} entry [34]. This compound was shown to inhibit receptor-stimulated Ca^{2+} entry in several cell types, including neutrophils and inhibited voltage-operated Ca^{2+} channels at similar concentrations [34].

However, all the above compounds have characteristics that have limited their utility in a drug discovery environment including low potency as inhibitors of Ca^{2+} influx in their target primary cell types. For instance, the ROCC channel-blocking activity is typically only observed between 1 and 50 μM, unlike the many of the L-VOCC channel blockers that have low nanomolar potency, e.g., dihydropyridines [3–6]. Thus, none of these ROCC channel blockers are potent or selective enough to be used as pharmacological tools to isolate and identify the molecular target. Without a selective and potent ligand, i.e., blocker or activator, developing assays for the target channel in order to develop more selective potent compounds is problematic. (This subject is discussed in more detail in the drug discovery approaches section of this chapter, Section 17.2.)

The ROCC channel blockers also lack selectivity versus other ion channels e.g., VOCCs and K^+ channels, Cl^- channels [22, 26, 29]. The variable potency with which these compounds block ROCCs in different cell types and different stimuli in any given cell type suggests that there is heterogeneity between ROCCs in different types. Without the knowledge of the molecular identity of the ROCC channels it is unclear whether the same channels mediate Ca^{2+} influx in different cell types — although this is likely to be the case in at least some instances. Without a defined target for drug discovery and using only native cell-based assays or phenotypic animal models for functional readouts, the so-called "black-box assays," ROCC blocker drug discovery in general has made limited progress. The exception is CAI which has progressed into patients (tested in Phase II clinical trials for relapsed ovarian cancer) [35]. As far as we are aware, none of the other compounds achieved therapeutic utility. Although several of these compounds have been patented as having therapeutic potential [36–44], the identities of their primary molecular targets still remain to be elucidated. Even in recent years this approach is still evident e.g., the pyrazole derivatives BTP2 and YM-58483 that have been proposed to be inhibitors of store-operated CRAC Ca^{2+} channels [45–47]. To our knowledge, the molecular identity of the CRAC channel is still unknown, and some of the properties of these compounds described are not consistent with these compounds acting directly as channel blockers i.e., prolonged incubation times of up to 24 h for demonstration of inhibitory effects [47].

In summary, circumstantial experimental evidence generated using the currently available pharmacological tools suggests that each cell type can possess multiple Ca^{2+} influx pathways, and that there is heterogeneity between cell types (reviewed by Li et al. [1]). To enable the full exploration and exploitation of the therapeutic potential of ROCC channels requires, of the molecular components of these channels and the development of more discriminating pharmacological tools with properties that allow their development as useful therapeutic agents are required.

17.1.4 EVOLUTION OF DRUG DISCOVERY APPROACHES: IMPLICATIONS FOR Ca²⁺ CHANNELS

The completion of the sequencing of the human genome [48] has provided molecular information about the candidate genes encoding Ca^{2+} channel proteins. The escalation and miniaturization of drug discovery technologies in the last decade suggests that never before has there been a better chance for identifying novel Ca^{2+} channel targets for drug discovery. The combination of molecular information and increases in compound screening capabilities available now, allows the use of a rational approach to systematically identify these elusive channels in disease-relevant cells and to identify modulators of these channels that could fully exploit their therapeutic potential.

17.2 CURRENT DRUG DISCOVERY APPROACHES

17.2.1 OVERVIEW

Over the last 10–12 years, traditional physiology-based drug discovery, as typified by the discovery of L-VOCC channels described earlier (17.1.1), has largely been replaced by molecular target-based approaches [49]. Using these newer approaches, it can take up to 15 years for a new drug to complete the drug discovery and development process (Figure 17.3). Furthermore, under this regimen, it is estimated that only 1 out of 5000 originally synthesized compounds will become a drug approved by the US Food and Drug Administration (FDA) according to information provided by the Pharmaceutical Research and Manufacturers of America association (PhRMA) in 2003 [50]. These statistics underlie the fact that the current approaches in drug discovery (as shown in Figure 17.3) are complex involving multiple scientific disciplines at different phases of the process. These phases are briefly discussed in the remainder of this section.

17.2.2 TARGET SELECTION

The first step of the drug discovery process clearly requires selecting a target with which to progress. Potential targets may arise from the identification of genes expressed in a target cell or tissue type, e.g., the pore-forming subunit of the T-type VOCC channel was isolated with the help of sequence information originating from the Human Genome Project associations [51, 52] or where the defect in a specific gene is strongly associated with the disease, e.g., mutations in the cystic fibrosis transmembrane conductance regulator (CFTR) gene encoding a dysfunctional CFTR chloride channel underlying cystic fibrosis [53]. Clearly, a significant understanding of the cause or mechanism underlying the disease state is required for target selection.

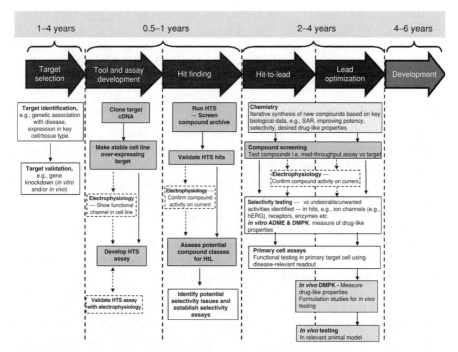

FIGURE 17.3 (See color insert following page 140.) The main phases of drug discovery. Additional activities required for ion channel targets are indicated by boxes with dashed outlines. Data source for duration of drug discovery phases is Bannerjee, P. and Baker, A. (2001): http://www.accenture.com/xdoc/en/industries/hls/pharma/hpdd.pdf. HTS (high-throughput screen), HtL (hit to lead), ADME (absorption, distribution, metabolism, elimination), DM (drug metabolism and pharmacokinetics).

Once a suitable potential target has been identified, the process of target validation begins. Validating drug targets assesses whether modulating the function of putative targets will have the desired biological effect. Molecular approaches to target validation include using specific gene knockdown approaches either *in vitro* or *in vivo*, e.g., using RNA interference (RNAi) [54] or using transgenic animals (gene knockout or gene knock-in). RNAi and generating gene knockout animals [55] are ways of reducing or preventing the expression of the gene of interest to determine the functional importance of the protein encoded by the gene. With gene knock-in animals, genes of interest are expressed in an animal that does not express the target and the biological consequences of the heterologous gene expression is determined [56].

In addition to molecular approaches to target validation, pharmacological tools can also be useful at this early stage of drug discovery, e.g., natural ligands such as capsaicin (the pungent chemical found in hot chilli peppers) both helped to identify and validate the transient receptor potential V1 cation channel (TRPV1) as a drug target for chronic pain [57] — see the following section on emerging targets (17.3.1) for further details.

Biological information gained at this early stage may reap benefits further down the line of drug discovery and development. For instance, the functional readouts used to validate a target may turn out to be disease-relevant functional readouts (biomarkers) that can be used in humans in clinical studies to monitor the short-term efficacy of a compound and predict the long-term potential of treatment [58–59].

17.2.3 HIT-FINDING

Once a target has been selected, the work begins to identify a compound that modulates the function of the target as a chemical starting point for developing a new drug. The most common method used to provide a chemical starting point, a "hit," for a synthetic chemistry program in a drug discovery process is to run a high throughput screen (HTS). A HTS campaign requires an *in vitro* assay to measure the activity of compounds on the target. The assay will be used to test large numbers of compounds (up to one or more million) to identify those compounds that exhibit the desired biological activity against the target. The current assay technologies available for screening Ca^{2+} channel targets and HTS assay development are described in detail in the section entitled "Assay technologies for Ca^{2+} channel drug discovery." For ion channel targets including Ca^{2+} channels, the functional activity of any putative Ca^{2+} channel modulator, i.e., a "hit" needs to be confirmed using electrophysiological techniques (Figure 17.3). If there are many hits to be tested in electrophysiology, this process can be time consuming and be a bottleneck to progress ion channel targets through this stage of the drug discovery process. There have been many recent advances in electrophysiology technologies that enhance the utility and speed of electrophysiological compound testing in "hit-finding" activities for Ca^{2+} channels (see section 17.4 on technological advances for Ca^{2+} channels for more detail).

17.2.4 HIT TO LEAD

During the hit to lead phase, a series of modifications of the chemical structure of the original HTS "hits" are performed and the biological activity of these new compounds determined in order to optimize the activity against the target, e.g., by increasing potency. For all ion channel targets including Ca^{2+} channels, this requires the compounds to be tested in electrophysiology assays to confirm any improvements in biological activity (Figure 17.3). The link between the chemical structure of a series of similar compounds derived from the original hit is often referred to as the structure–activity relationship (SAR). The SAR is used for the rational design of additional compounds to optimize activity against the target in order to identify a lead compound with the desired biological activity, e.g., potency for the target but also increasing selectivity against undesirable properties. In this phase of the drug discovery process, biological activity on the target protein is not the only property of the "hit" under investigation [60].

Key additional information required at this stage includes identification of any undesirable biological activity not associated with the target, i.e., those selectivity issues that may result in potential side effects for the patients. Such biological activities need to be minimized for compounds before they can be considered potential drug molecules and suitable for progressing into the drug development phase. At this stage of the drug discovery process, these potential selectivity issues are highlighted and assays to measure selectivity are established, e.g., the screening of compounds for inhibitory activity at the hERG (human ether-a-go-go related gene) K^+ channel. Blockade of hERG channels is the main cause of drug-induced long QT syndrome (LQTS), which is associated with Torsades de Pointes (TdP) arrhythmias, potentially leading to sudden cardiac death [61]. Although many types of drugs are known to cause LQTS, e.g., class IA, IC, and III anti-arrhythmic drugs, many antibiotics, neurotropic, antifungal, and antimalarial drugs, several L-VOCC blockers also block hERG K^+ channels [61–63]. Therefore, determining any potential hERG channel-blocking activity is a key selectivity issue to address for novel Ca^{2+} channel drugs.

In summary, chemists design new chemical compound molecules in order to produce lead compound that has the required level of activity on the target and a sufficiently low amount of the undesirable properties. The stringency of the criteria used to define a lead compound, e.g., potency on the target and the acceptable level of undesirable activities can vary from target to target and therapeutic indication.

17.2.5 LEAD OPTIMIZATION

Once a lead chemical structure has been identified (a lead class), the lead optimization phase focuses on modifying the chemical structure to optimize drug-like properties of the compounds [64]. This means optimizing biological activity *in vivo*, i.e., in a disease-relevant animal model to choose a compound that is suitable for use in the first human clinical trials. The parameters that are considered include the physicochemical properties for dosing, which impacts on drug formulation, and the drug metabolism and pharmacokinetics (DMPK), i.e., how much and how long does the compound stay in the body in order to give the desired biological effect and how long does this last.

A lead chemical candidate needs to possess adequate bioactivity, appropriate physical and chemical properties to enable formulation development, the ability to cross crucial membranes, have metabolic stability, and have appropriate safety and efficacy in humans. The use of earlier developed biomarkers would be of benefit at this stage to give a functional readout of predictive activity. The definitive goal is to gain a window within which a therapeutic dose can be achieved while simultaneously minimizing side effects. A fast proof of concept in man is desirable, which may be achieved with the use of disease-relevant tissue to give an improved readout, and an increased confidence when aiming for the clinic [58].

17.2.6 SUMMARY

To date, the main targets for drug discovery have been GPCRs and enzymes, while ion channels have generally been considered more difficult to pursue as drug targets. The lack of assay technologies for many ion channel targets suitable for drug discovery purposes and the slow and resource-intensive nature of electrophysiology, the "gold standard" for measuring ion channel activity but an absolute requirement for ion channel drug discovery, have been major drawbacks. Additional activities and infrastructure needed to support ion channel drug targets through the different phases of the current drug discovery process are illustrated in Figure 17.3. However, recent advances in the technologies for ion channel assays and electrophysiology have led to a renaissance for ion channels as drug targets.

17.3 EMERGING Ca^{2+} CHANNEL TARGETS

Here, we highlight two of the most recently emerged Ca^{2+} channels that have been pursued as drug targets in the last decade of gene-targeted drug discovery approaches.

17.3.1 TRPV1 CHANNELS

TRPV1 is the founding member of the TRPV subfamily of cation channels. The alternative names for this channel, i.e., capsaicin receptor and vanilloid receptor give clues about the original discovery of this channel. Capsaicin, the pungent chemical found in hot chilli peppers was shown to directly activate Ca^{2+}-permeable nonselective cation channels present in nociceptors (sensory nerve fibers involved in pain sensation), resulting in a burning pain sensation. The channel was proposed to be a target for the treatment of pain [65]. Subsequently, the molecular identity was determined by reverse pharmacology where the capsaicin-activated channel was identified by expression cloning of dorsal root ganglion genes [57]. This comprehensive study also pharmacologically characterized this capsaicin-activated channel with other known ligands including the antagonist capsazepine and the high-affinity agonist resiniferotoxin and showed that TRPV1 was selectively expressed in the Aδ and C sensory nerve fibers involved in nociception. The availability of specific ligands for TRPV1 channel facilitated drug discovery efforts for the development of TRPV1 channel blockers in two important ways. First of all, even before the TRPV1 channel had been cloned capsaicin was known to be a channel activator and provided an obvious chemical starting point for chemists to modify the compound structure and change its biological properties from an agonist to an antagonist, leading to the development of the first blockers, e.g., capsazepine [66] which have proved to be useful pharmacological tools for target validation and assay development. Second, having a specific channel activator enabled the development of a highly selective functional assay for screening for low molecular weight blockers that would be novel and diverse in structure. This has formed the basis of most companies' efforts with fluorescence based Ca^{2+} assays for high throughput screening to develop novel blockers of TRPV1 channels for the treatment for chronic pain. TRPV1 channel blockers may also be useful for the treatment of cough and bladder instability and hyperreactivity. Several patents have now been filed and companies are publishing data on their respective chemical approaches [67]. Although all RPV1 channel blockers are still at the preclinical stage, TRPV1 is currently the most advanced "novel" Ca^{2+} channel target and the closest to fulfilling its therapeutic potential.

17.3.2 N-Type Voltage-Operated Ca^{2+} Channels

N-type voltage-operated Ca^{2+} channels (N-VOCC channels) have a different distribution to the other voltage-operated Ca^{2+} channels, and are expressed in specific populations of central and peripheral neurones. Pharmacologically, they can be discriminated from the L-VOCC channels by virtue of their insensitivity (at therapeutic concentrations) to the L-VOCC blockers namely the dihydropyridines, phenylalylamines, and benzothiazepines and high sensitivity to the certain omega-conotoxins (ω-conotoxins). The N-VOCC channel was purified from brain using ω-conotoxin GVIA (an irreversible binder of N-VOCC channels) as a marker and the purified channel re-constituted into lipid bilayers and shown to be functional [68]. Another type of omega-conotoxin, ω-conotoxin MVIIA, was found to be a reversible blocker, and a synthetic peptide corresponding to its peptide sequence folded to form four cysteinyl disulfide bridges was made (SNX-111, Neurex Corporation, USA), which competed with ω-conotoxin MVIIA. Subsequently SNX-111 was found to have therapeutic potential as it was a potent and reversible N-VOCC blocker that was neuroprotective, antihypertensive, and analgesic in a range of animal models [69]. SNX-111 (USAN generic name *ziconitide*) entered clinical trials for prevention of ischemic neurodegeneration (but discontinued in phase 2 clinical trials) and analgesia for severe chronic pain, where it is in the final stages of clinical development having completed phase 3 clinical trials [70]. The main disadvantages of SNX-111 are that it is a peptide and therefore not suitable for oral administration and when dosed intravenously induces bradycardia as a side effect. It has been proposed that channel subtypes and variants of N-VOCCs exist and that SNX-111 exhibits a lack of selectivity between N-VOCC channel subtypes. Therefore the N-VOCC channels on sympathetic nerve endings mediating the cardiac effect of SNX-111 may be a different N-VOCC channel variant (either a splice variant or a multimolecular channel complex with different auxiliary subunits) to those on the nociceptor terminals involved in pain sensation [70].

The search for orally active selective small molecule inhibitors for N-VOCC channels intensified after demonstration of the clinical efficacy of SNX-111 and several approaches have been adopted. Companies have made use of the high potency and selectivity of the omega-toxins for N-VOCC channels to aid their drug discovery efforts, e.g., developing a ^{125}I-ω-conotoxin MVIIA binding assay for high-throughput screening to look for novel structures using the three-dimensional solution crystal structure of ω-conotoxin MVIIA to determine the key binding amino acids to design novel compounds [71]. Another approach has been to modify known drugs that also inhibit N-VOCC channels, e.g., the L-VOCC channel blockers verapamil (a phenylalkylamine) and cilnipidine (a dihydropyridine) as chemical starting points [72]. The success of these new strategies to develop the next generation of N-VOCC channel blockers with improved selectivity and drug-like properties over SNX-111 remains to be determined.

17.4 ASSAYS AND TECHNOLOGIES FOR Ca^{2+} CHANNEL DRUG DISCOVERY

Although the ability to discover lead chemical templates for therapeutic targets remains the first critical step in any drug discovery program, the approaches adopted by most pharmaceutical companies in all target areas, including ion channels, have evolved substantially in recent decades. Historically, an empirical or pharmacologically driven approach involving the front line screening in either an animal model or *in vitro* preparation was utilized to identify compounds with efficacy/affinity (e.g., L-VOCC inhibitors such as the dihydropyridines) [2, 73]. Today in the postgenomic era, rich with an explosion of molecular information regarding ion channel structure, *in vitro* and *in vivo* models, although still employed, are commonly used in the later stages of drug discovery flowcharts (see Figure 17.3). This has been the result of the adoption of HTS as the primary mechanism for identifying compounds that interact with target channels, commonly termed "hits." HTS, driven by a requirement to increase throughput in the discovery process to both ameliorate high compound attrition rates and exploit the numerous potential target opportunities offered by modern genomic approaches, aims to maximize the number of chemical starting points (templates) for any given target. This paradigm shift has been further fueled by developments in combinatorial chemistry that have massively increased the numbers of compounds available for screening [74–77]. This has occurred to such an extent that large-scale pharmaceutical industry HTS campaigns commonly utilize between 500,000 and >1,000,000 structurally distinct compounds [78].

17.4.1 Cell Lines

Obviously, the throughput of *in vivo* and *ex vivo* assays precludes their usage in the economic screening of large (>1000) compound numbers from both a time and financial perspective. Instead primary screening assays in the ion channel arena are commonly based upon heterologous cell lines over-expressing the cloned channel of interest. These engineered cell lines offer a number of advantages over native cells that are compatible with the requirements and rigors of HTS campaigns [79]. Engineered cell lines, using backgrounds such as human embryonic kidney cell line HEK293 and Chinese hamster ovary cell line CHO-K1, proliferate rapidly and do not dedifferentiate, providing the huge cell numbers and temporal phenotypic stability required for large compound number HTS campaigns that may take months to complete. These lines are easy to maintain, requiring uncomplicated culture and feeding regimes, and are thus readily amenable to both automated cell culture systems and the liquid handling automation commonly employed by HTS groups. By expressing the target channel of interest at high, and commonly supra-physiological levels, the engineered lines offer further advantages over native cells. The low expression levels of endogenous channels in the commonly utilized background cell lines (e.g., CHO-K1, HEK293) provide a lack of contaminating conductances, which is an important consideration with certain assay formats, e.g., membrane potential. Coupled with high expression levels, this makes it much more straight-forward to develop an assay whereby the readout is dominated by the target channel of interest. However, it is also worth highlighting the possible downsides of engineered high-level expressing cell lines. Some Ca^{2+}-permeable channels when over-expressed in heterologous systems adopt a non-native conformation — this is a not an uncommon finding with the TRP channel family (e.g., a subpopulation of TRPV6 channels become constitutively active when expressed in HEK293 back-ground) [80]. The underlying cause of these phenomena are unknown; however, we now know that many channels have accessory proteins, with diverse functions ranging from direct modulation of channel function to chaperoning channels to the appropriate cellular compartment during synthesis (e.g., voltage-operated ion channel β subunits) [81]. The lack of expression of these proteins, particularly in non-human background lines such as CHO-K1, may be a contributory factor to non-native channel behavior. Channel expression level can also contribute profoundly to pharmacological sensitivity in certain formats, particularly membrane potential. Depending upon the background cell's electrical characteristics (e.g., membrane resistance), the number of channels required to affect changes in membrane potential may vary. In a cell with high resistance, only a few open channels may be required to drive membrane potential to the reversal potential for the expressed channel type. If the channel is over-expressed at a level which is orders of magnitude higher, then blocker sensitivity will be severely abrogated as a high number of channels will need to be blocked before a change in membrane potential is observed.

Today, the number of assay formats on offer to the Ca^{2+} channel drug discov-erer looking to identify new chemotypes is unparalleled. In some cases, the major challenge posed lies in translating these into cost-effective, miniaturized, high density plate-based formats (96-, 384-, or 1536-well). Each of these formats is briefly reviewed below highlighting the considerations in choosing a format for the development of a Ca^{2+} channel screening assay and summarized in Table 17.2.

17.4.2 RADIOLIGAND BINDING

Radioligand binding was historically employed in the discovery and characteriza-tion of the L-VOCC channel inhibitors such as nifedipine, diltiazem, and verapamil, and used to demonstrate that L-VOCCs have distinct bindings sites for different classes of blocker [82]. These well-established assays are still employed today, predominantly for lead optimization [83] or selectivity testing (www.mds.com). Although a high throughput technique, there are a number of limitations of radi-oligand binding that in many cases render this an unsuitable approach for ion channel HTS campaigns.

The basic requirement for the development of a radioligand binding assay is a high-affinity ligand (K_D commonly in the low nM to pM range), which binds to the site of interest, that can be radiolabeled with a high specific activity (>20 Ci/mmol as an approximate estimate). For the VOCCs these ligands are readily available (e.g., ^{3}H-PN200–110) but for novel Ca^{2+} entry channels such as the TRPs where the pharmacology is as yet ill-defined, this is problematic. The binding sites involved in

TABLE 17.2
Summary of Ion Channel Essay Methods Suitable for Screening of Ca^{2+} Channels

Assay format	Information content	Throughput	Sensitivity	Response time	Comments
Radioligand binding (e.g., ^{3}H-PN200-110)	Low	High (10^5)	Low–medium	N/A	Requires high affinity/high-energy radioligand probe
Radiotracer flux (e.g., ^{45}Ca^{2+})	Medium	Medium–high (10^4–10^5)	Medium	Slow (seconds–minutes)	Often requires high level of channel expression plus pharmacological modification of channel gating
Ca^{2+} sensitive dyes (e.g., Fluo-4, Fura-2)	Medium	High (10^5–10^6)	Medium–high	Medium (seconds commonly)	Can be prone to artefacts from fluorescent compounds/quenchers
Membrane potential:					
Oxonol dye	Medium	High (10^5–10^6)	Medium	Slow–medium	Can be prone to dye artefacts
FRET-based sensors	High	High (10^5–10^6)	High	Fast (subsecond)	Ratiometric so less prone to dye artefacts
Oocyte electrophysiology	High–very high	Low (10^2–10^3)	High	Fast (millisecond)	Require microinjection; lipid-rich yolk sac can sequester compounds
Patch clamp electrophysiology	Very high	Low (10^1–10^3)	High	Fast (millisecond)	"Gold standard"; washout of cell contents

Information content is assessed upon the assay technology's ability to discriminate channel state; throughput is defined as number of datapoints per normal 8 hour day; sensitivity is based upon determination of potency compared to patch clamp electrophysiology and reciprocal of the number of channels per cell required to give a robust signal-to-noise ratio.

the required modulatory action are unknown as is the identity of a suitable ligand — a chicken and egg situation. Even when suitable ligands are identified they provide a readout from only a single binding domain and may not be able to detect allosteric inhibitors, nor compounds with voltage- or state-dependent binding. Additionally, displacement of radioligand only signifies affinity, with no indication of the functional consequences of compound binding, and as such is considered a low information content assay format.

17.4.3 RADIOTRACER FLUX

Radiotracers have long been used to assess flux through channels in cells, and even intact tissues. In the case of Ca^{2+} channels, $^{45}Ca^{2+}$ is the tracer of choice [84], a β-emitter, with energy approximately ten times that of ^{3}H. The assay protocol is relatively straightforward for Ca^{2+}-permeable channels, opening the channel with an appropriate stimuli (KCl-induced depolarization for VOCCs, ligand or store depletion for ROCCs and SOCCs) and assessing the amount of $^{45}Ca^{2+}$ influx. Generally, assays with channels that permit influx are more time efficient as they avoid the long cell loading times associated with efflux experiments (e.g., $^{86}Rb^{+}$ efflux for K^{+} channel screening). Quantification of $^{45}Ca^{2+}$ influx requires that the excess tracer be removed and cells lysed, commonly by addition of the scintillant required for counting. These additional steps can add a significant amount of time to the overall assay and hence affect throughput, an obvious consideration for an HTS campaign. The advent of scintillating microplates such as the Cytostar T (Amersham International) and FlashPlate (PerkinElmer Life Sciences) in the late 1990s has dramatically reduced the need for separation and washing. As $^{45}Ca^{2+}$ accumulates within the cell cytosol, by virtue of flux through the open channel, it is brought into near proximity with the scintillant that is incorporated into the base of the plate (cells must form a confluent monolayer to keep nonincorporated radiotracer from generating a signal). Quantification can be undertaken using a standard scintillation counter that measures the emitted photons using photomultiplier tubes or a more rapid counting charge-couple device (CCD) image-based system (e.g., Amersham Biosciences LEADSeeker). Figure 17.4 illustrates the basic principles of these radioactive assay formats. These advances have greatly improved the practicality of using radiotracer flux for ion channel HTS campaigns, however there are a number of disadvantages and limitations.

The magnitude of the assay window for $^{45}Ca^{2+}$ uptake experiments is dependent upon a number of variables that limit its use for screening. The background cell type chosen for Ca^{2+} channel screening may, and often does, have a number of nonspecific background fluxes (e.g., Ca^{2+} exchangers) that may produce assay interference. Often these effects are accounted for by subtraction from the screening data.

In the same way that channel number is an important consideration, so is the amount of time that the channels spend in their open configuration. To maximize open time, sometimes to the order of minutes or even hours, pharmacological modifiers, if available, are used. For the L-VOCCs, the dihydropyridine agonist BayK8644 or the benzoyl pyrrole agonist FPL64176 have been employed. This

(A) Radioligand binding using SPA   (B) Radiotracer flux using scintiplate

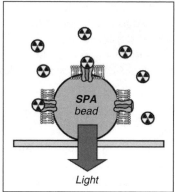 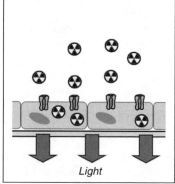

FIGURE 17.4 Ion channel assay formats commonly employed in drug discovery. Radioactive formats: (A) Radioligand binding employing a wheatgerm agglutinin scintillation proximity assay (SPA) bead, which encompasses the scintillant allows binding events to be detected without the requirement of removing unbound ligand. (B) Radiotracer flux measures the flow of labeled permeation into or out of the channel-expressing cell. Influx assays are suited to microtitre plates with the scintillant incorporated into the base (e.g., Cytostar T, Flashplate), which avoids the need for removal of nonincorporated tracer and cell lysis for quantitation.

strategy does improve assay signal but introduces an alternative mechanism of action (i.e., antagonism of the agonist site) that may produce false-positives if screening for channel blockers. However, despite these manipulations, signal strength remains a challenge when miniaturizing flux assays. This issue, along with a general trend in the reduction of using potentially harmful and environmentally unfriendly radioisotopes, means that flux assays are not commonly a frontline choice for high throughput screening.

17.4.4   OPTICAL READOUTS — INTRACELLULAR Ca^{2+} CONCENTRATION AND MEMBRANE POTENTIAL

Assays based upon fluorescence readouts are widely employed as screens for Ca^{2+} channel activity. These assays utilize fluorescent probes that quantify ion channel induced changes in either intracellular free Ca^{2+} ion concentration ($[Ca^{2+}]_i$)or less commonly membrane potential. As such these assays offer high sensitivity, report functional channel activity, often in real time, and are readily amenable to miniaturization and automation.

The opening of Ca^{2+} channels can elicit a substantial increase in intracellular Ca^{2+} concentrations, typically 100– to 1000-fold over basal. This can be easily detected by a wide range of commercially available fluorescent Ca^{2+}-sensitive dyes such as Fura-2, Fluo-4, and Calcium green-1 (Molecular Probes, USA; www.probes.com). The high-throughput measurement of the rapid, and often transient, changes that occur in $[Ca^{2+}]_i$ has been enabled by the widespread availability of CCD-based fluorescence plate readers. The prototypical instrument, the Fluorometric Imaging

Plate Reader (FLIPR, Molecular Devices, USA) [85], has a 488-n*M* argon laser and is compatible with any fluophore with an excitation spectrum in this range — the FLIPR emission filters are readily interchangeable. The laser can rapidly scan the assay plate and acquire datapoints from dye loaded cells at a frequency of approximately 1 per sec, allowing real time kinetic measurement. With integrated fluidics and the ability to self load both pipette tips as well as cell and compound plates, the FLIPR is an ideal HTS instrument that can form the basis of an entire screening platform (Figure 17.5A). In its current guise, the FLIPR^{TETRA} (www. moleculardevices.com), the FLIPR is compatible with 96-, 384-, and 1536-well plates enabling the screening of many thousands of compounds per day. Standard protocols using membrane permeant Ca^{2+}-sensitive fluophores (e.g., Fluo-4-AM), require the cells to be washed subsequent to loading to remove any unincorporated dye that would yield extraneous fluorescence, abrogating the stimulated signal. Assay preparation time has been further reduced by the introduction of "no wash" fluorescence Ca^{2+} dye kits (e.g., Calcium Plus, Calcium 3 kits; Molecular Devices), which utilize a nonmembrane permeant quench molecule to bind the unincorporated dye [86].

Although not routinely used as an HTS format for Ca^{2+} channels, the Ca^{2+} sensitive photoprotein aequorin has been validated by Euroscreen (www. euroscreen.com) for ion channel screening [87]. Euroscreen has engineered mitochondrially expressed aequorin directly into cell lines enabling the utilization of the low background, high sensitivity, and large dynamic range associated with this indicator. The readout, changes in luminescence, can be readily measured by a number of HTS compatible readers.

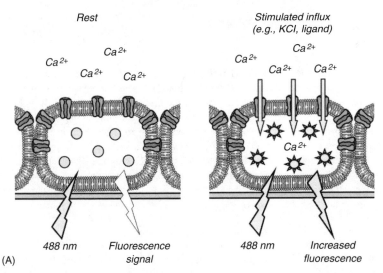

FIGURE 17.5A (See color insert following page 140.) Ion channel assay formats commonly employed in drug discovery. Optical formats: (A) Ca^{2+} sensitive dyes such as Fura-2 and Fluo-4 allow direct real-time quantitation of Ca^{2+} flux into cells and are compatible with a variety of high throughput plate readers such as FLIPR. (*Continued*)

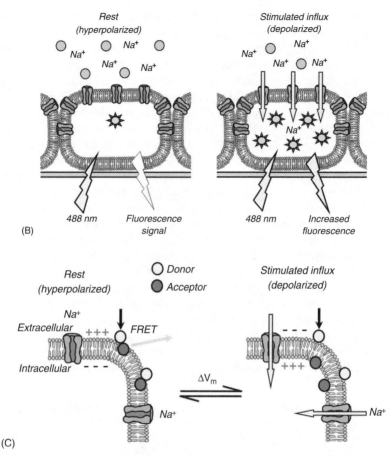

FIGURE 17.5B,C (See color insert following page 140.) Ion channel assay formats commonly employed in drug discovery. Optical formats: (B) Potentiometric dyes are sensitive to charges on the inner and outer leaflets of cell membranes. Redistribution dyes such as the oxonol dyes partition into the intracellular compartment upon cellular depolarization, binding to intracellular proteins and increasing their fluorescence emission. This is easily detected by standard fluorescence plate readers. (C) The FRET-based voltage-sensor probes (VSP dyes) utilize two dyes, a donor and an acceptor to produce a FRET signal. Upon cellular depolarization the FRET donor migrates to the inner leaflet of the cell membrane away from the acceptor and reduces the FRET signal. As the acceptor only has to move a short distance to elicit a change in signal, VSP dyes are very rapid to respond to changes in membrane potential.

For nonselective cation channels, such as the majority of the TRP channel family (see Ref. [88] for review), an alternative to Ca^{2+}-sensitive probes are measurements of membrane potential. This approach has recently been applied to a TRPC6 overexpressing HEK cell line that did not differentiate itself from a null transfected line when assessed using a Ca^{2+}-sensitive probe [89]. Exploiting the nonselective cation permeability profile of TRPC6, while simultaneously chelating intracellular Ca^{2+} using BAPTA-AM (to enhance TRPC6-mediated currents and reduce contamination from any Ca^{2+}-sensitive conductances), agonist-induced membrane potential changes were clearly evident in the transfected line. These changes were mediated by Na^+ influx as ionic replacement with the large impermeant organic cation, N-methyl-D-glucamine (NMDG), abolished the membrane potential changes (Figure 17.5B). Membrane potential is a sensitive and versatile format for ion channel assays as the high electrical resistance of biological membranes means that relatively small currents can induce substantial changes in potential. Historically, membrane potential indicator dyes with an acceptable voltage sensitivity have had slow response times (in the order of seconds to minutes) and are strongly temperature-sensitive. These dyes, exemplified by the lipophilic, negatively charged oxonol dyes such as bis-(1,2-dibutylbarbituric acid) trimethine oxonol (DiBAC$_4$), have a low fluorescence in aqueous solution but significantly increase their quantum yield upon binding to intracellular proteins [90]. The basis of the assay is straightforward — redistribution of the dye. As membrane potential depolarizes cells slowly accumulate the negatively charged dye that increases its fluorescence when intracellular. The slow response time means that dyes of this type can only be used in assays configured to detect steady-state changes in membrane potential which limits their usage to channels that do not rapidly inactivate or desensitize, or those that can be pharmacologically modified to remain in their open state. More recently dyes with faster response times have become available such as the proprietary FLIPR membrane potential dye (FMP) that is "oxonol-like" but equilibrates completely within seconds [91, 92]. It is worth noting that the FLIPR system was originally developed to quantify DiBAC$_4$ associated fluorescence. The fastest responding dyes are the fluorescence resonance energy transfer (FRET)-based voltage sensor dyes pioneered for ion channel screening by Aurora Biosciences, now Vertex (www.vpharm.com). Quantification of membrane potential by FRET utilizes two dyes, a coumarin-linked phospholipid (CC2-DMPE) in addition to an oxonol dye. The CC2-DMPE partitions into the outer membrane leaflet where it acts as a fixed FRET donor to the mobile voltage-sensitive oxonol acceptor. Because FRET can occur over only short distances (<100 nm), excitation of the coumarin donor results in specific monitoring of the movement of the oxonol dye within the membrane (Figure 17.5C). Using a ratiometric measure of the responses of both the donor and the acceptor, the result is a rapidly responding system (<500 ms) with a high degree of sensitivity (1% ratio change per mV) [93]. Aurora developed a kinetic plate reader, the voltage/ion probe reader (VIPR), specifically for usage with the FRET dyes; however, the dyes are compatible with most instruments capable of FRET measurement [94]. A head to head comparison of the available potentiometric dyes for screening has been recently performed by Wolff and colleagues (2003) [95].

Although well suited to HTS, fluorescent dyes, in common with most other approaches, can produce both false-positives and false-negatives. A common problem is the actual compounds to be screened. Compounds with an emission or absorption spectra that coincides with that of the fluophore can produce strong assay interference. If suspected this can be overcome by establishing the wavelengths at which compounds absorb and emit, or by the use of an appropriately designed control assay that may be as simple as adding the fluophore to the compound directly. Compounds with unwanted actions such as increasing cell membrane fluidity can also provide misleading results.

17.4.5 ELECTROPHYSIOLOGY — PATCH CLAMP AND OOCYTE RECORDING

Voltage clamp techniques remain the "gold standard" in ion channel recording, providing direct, real time, mechanistic information on ion channel function. However, the most commonly industry-used techniques, whole-cell patch clamp (Figure 17.6B) and twin-electrode voltage clamp (TEVC, Figure 17.6A), do not easily lend themselves to the screening paradigm. Although both techniques provide data that is information rich, they are technically challenging methods to employ, requiring highly trained operatives. Perhaps more crucially the throughput using traditional equipment of around 10–20 datapoints per day per operator does not make it viable to undertake large compound screening. This has led a number of companies to design automated systems to improve throughput while maintaining the clear advantages in data quality and information that these approaches confer.

Several systems are now marketed that have automated the TEVC process either with serial (e.g., Robocyte, Multichannel Systems) [96] or parallel recording (e.g., Opus Xpress, Axon/Molecular Devices). These systems have dramatically increased throughput, and in the case of the Robocyte have automated the microinjection of the channel encoding RNA or cDNA. Even though these systems have been validated for channel screening [97] there is a general reluctance by the pharmaceutical industry to use oocytes *per se* as they are non-mammalian in nature and compound IC_{50} values do not always correlate well with those determined with patched mammalian cells. This is particularly evident with lipophilic molecules that may concentrate in the lipid-rich yolk of the oocytes.

Increasing throughput in the patch clamp process has been a long-term aim for many manufacturers and researchers alike, not only providing the promise of direct electrophysiology-based screening campaigns but also removing bottlenecks within lead optimization or pharmacological safety testing programs (e.g., hERG). The earliest attempts to achieve this involved automation of a conventional rig by replacing the human operator with computerized image analysis and automated manipulators while still employing traditional patch clamp pipettes [98]. The Apatchi system, developed by Sophion Biosciences (www.sophion.dk), was however a serial system and required substantial user input in its operation and was therefore described as semi-automated. Subsequent and more recent developments (reviewed by Wood et al.) [99] have moved away from using a pipette as the interface between the cell and the amplifier electrode and instead utilized a planar substrate (Figure 17.6C). Bringing the cell to the substrate has circumvented the need for accurate micropositioning and de-skilled the patch clamp process, but infers that systems adopting this approach can only use cells in a nonadherent format. Today there are several planar patch based systems either marketed or in late stage development and although this approach is the most common, it is not the only one, e.g., Flyion uses conventional patch pipettes but

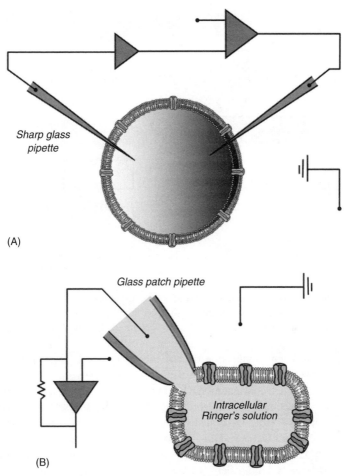

FIGURE 17.6 Ion channel assay formats commonly employed in drug discovery. Electrophysiological formats: (A) Twin-electrode voltage clamp of oocytes is commonly used for channels with multiple subunits that can be difficult to stably express in mammalian cells in the correct stoichiometry. Even with automation, oocyte recording is not high throughput and requires each oocyte be injected with DNA or RNA encoding the channel of interest. (B) Conventional patch clamp electrophysiology uses glass pipettes to form high-resistance seals capable of recording small currents (even single channels). This technique is considered "gold standard" as it is a sensitive and direct functional measure of channel activity that can be assessed in a millisecond time frame. This technique can provide information regarding channel transition states and kinetic information on compound binding but is very low throughput.

injects the cells into the pipette [100]. The instruments with the most data to support their utilization are the PatchXpress 7000A (developed by Axon Instruments, now Molecular Devices) and the IonWorks HT (Molecular Devices/Essen Instruments), both planar patch systems. The PatchXpress records whole-cell current from 16 individual chambers of a planar electrode chip (manufactured by Aviva Bioscience;

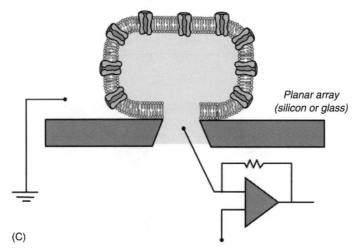

FIGURE 17.6 *(Cont'd)* (C) Planar patch clamp is the basis of several commercialized automated electrophysiology systems (e.g., Patch Xpress). This approach relies upon replacing the patch pipette with a "hole" in a flat substrate, commonly glass or plastic. The cells are added to the array and this avoids the requirement for micropositioning and additionally vibration isolation. Systems based upon this concept are capable of generating thousands of datapoints per day, several orders of magnitude higher than a conventional rig.

www. avabio.com). The application of individually controlled negative pressure to each chamber enables the formation of true GΩ seals and is used to rupture the patch, producing the whole cell configuration. Continuous voltage control, measurement, and compensation produce high-quality recordings comparable to those obtained by conventional techniques [101]. Integrated fluidics allow simultaneous add and read in addition to washout protocols and this system is suited to all ion channel classes.

Molecular Devices IonWorks HT system uses a 384-well patch plate in combination with a 48-channel amplifier [102]. This system differs substantially from the PatchXpress in that the seals obtained are as commonly as low as 100–200 MΩ and the system relies upon a perforating antibiotic (amphotericin B) to gain electrical access to the cell interior. The 48-channel amplifier head, which records from eight different positions on the patch plate, physically moves out of each well to allow addition from 12-channel integrated fluidics; thus simultaneous add and read is not possible. This would preclude the use of the IonWorks HT for many ligand-gated channel applications. Despite the marked differences from the classical patch clamp technique, particularly the seal resistance, this system has been able to reproduce the pharmacological response for several channel types including hERG [103].

Both systems described above exploit the high levels of data fidelity offered by patch clamp electrophysiology combined with an increased throughput. However utilizing a low-volume well with microliter volumes of compound addition, rather than a larger (250 μl) bath with continuous superfusion, characteristic of a traditional rig, may make the automated systems vulnerable to some of the unwanted effects associated with miniaturization. Compound depletion, resulting from nonspecific binding to either the instrument dispensing tips or the planar arrays themselves, has been observed and may contribute to differences in compound potency between automated and conventional patch systems. Manufacturers are attempting to minimize this occurrence by offering alternative planar array substrates (glass and plastic; see www.axon.com).

Undoubtedly, these systems represent a substantial advance in throughput, with the PatchXpress reported to deliver ~2000 datapoints per day and the IonWorks HT ~3000 datapoints [www.moleculardevices.com], however this comes at a significant price. The consumable, the planar patch arrays, are expensive to produce which equates to a high cost per usable datapoint. Realistic estimates place the cost per datapoint for the PatchXpress as high as US \$25 per datapoint and \$3 per datapoint for IonWorks HT. Costs at this level do not compare favorably with other assay formats for ion channel screening that commonly have costs as low as 5 cents per well. Recently, Molecular Devices have launched their second-generation instrument, the IonWorks Quattro. This has four times the throughput of the Ion-Works HT and claims a near 100% success rate for recordings. Currently, the exact details of the system are sparse but it is reported by Molecular Devices to utilize "population patch clamp (PPC)" technology that involves sealing multiple cells per well and thus generating a "well-averaged" recording.

Cost and throughput currently preclude the use of automated patch clamp systems for true high throughput screening. These systems are finding a niche in secondary screening, downstream to HTS campaigns performed using the other formats considered above, or during lead optimization. When used as a hit-finding approach it is only economically viable to use a selected compound set, either a "core set," which covers the chemical diversity of a larger set but with fewer compounds, or alternatively a "focused set," which has been synthesized around a known channel pharmacophore. However, the pace at which these systems are evolving is rapid and many anticipate that true HTS electrophysiology systems with an associated low consumable cost will be reality in the next 2–5 years.

17.5   CONCLUDING REMARKS

Drug discovery in the field of Ca^{2+} channel targets is undergoing a renaissance with the increasing availability of functional assay technologies suitable for Ca^{2+} channels. The plethora of advances in innovative new technologies has revolutionized the development of higher throughput functional assays for Ca^{2+} channels that are vital for current drug discovery HTS-based approaches. The pharmaceutical industry has never been so well equipped to fully explore and exploit the therapeutic potential of all types of Ca^{2+} channels, and this represents an exciting challenge for all drug hunters.

ACKNOWLEDGMENTS

We thank Dr. Pamela Tranter, Novartis Institutes for Biomedical Research (Horsham, U.K.) for her critical review of the manuscript.

APPENDIX

How Good Is an Assay?

Assay development for ion channels, and indeed any biological target, can involve the optimization of multiple variables with the end goal of producing an assay with high sensitivity and reproducibility, that can easily discriminate between molecules which do and do not interact with the target protein. Assessing assay performance in response to changes such as miniaturisation or automation, is commonly undertaken mathematically. Historically, the 'quality' of an assay has been loosely quantified by the calculation of a signal-to-noise (S/N) ratio or signal-to-background (S/B) ratio:

$$S/B = \frac{\text{Mean signal}}{\textit{Mean background}}$$

$$S/N = \frac{\text{Mean signal} - \text{mean background}}{\textit{Standard deviation of background}}$$

These two closely related terms give an indication of the assay window and the ability of an assay to discern a true signal from baseline noise, respectively. However, neither equation fully takes into account data variation and thus neither is ideal for assay evaluation. Zhang $et\ al$ [106] reported a new parameter, the Z or Z' factor which has become widely accepted as a term for evaluating the quality of an HTS assay:

$$Z' = 1 - \frac{(3\sigma s + 3\sigma c)}{(\mu s - \mu c)}$$

where σs and σc are the standard deviations of the sample and the control respectively; μs and μc are the means of the samples and controls respectively. If positive controls are used as samples to define the upper and lower limits of the assay then the calculation reveals the Z'. This is a very useful term to aid in assay development by quantitatively identifying the assay conditions which produce the highest 'quality' assay. This may however not be the assay with the biggest window but rather one which combines a good window with low data variance. For normal HT screening the samples will be unknown test compounds and hence the equation yields the Z factor. For an ideal assay the Z or Z' factor will equal 1, <1–0.5 is a good assay, <0.5–0.1 is a moderate assay and 0 or below is an assay with poor discrimination and should not be employed for HTS.

REFERENCES

1. Li, S. et al. (2004) TRP channels as drug targets. *Novartis Found Symp* 258:204–213.
2. Fleckenstein, A. (1983) History of calcium antagonists. *Circ Res* 52:13–16.
3. Bellemann, P. et al. (1982) [^{3}H]-Nimodipine and [^{3}H]-nitrendipine as tools to directly identify the sites of action of 1,4-dihydropyridine calcium antagonists in guinea-pig tissues. Tissue-specific effects of anions and ionic strength. *Arzneimittelforschung* 32:361–363.
4. DePover, A. et al. (1982) Specific binding of [^{3}H]nitrendipine to membranes from coronary arteries and heart in relation to pharmacological effects. Paradoxical stimulation by diltiazem. *Biochem Biophys Res Commun* 108:110–117.
5. Triggle, C.R. et al. (1982) Calcium-channel antagonist binding to isolated vascular smooth muscle membranes. *Can J Physiol Pharmacol* 60:1738–1741.
6. Williams, L.T. and Tremble, P. (1982) Binding of a calcium antagonist, [^{3}H]nitrendipine, to high affinity sites in bovine aortic smooth muscle and canine cardiac membranes. *J Clin Invest* 70:209–212.
7. Curtis, B.M. and Catterall, W.A. (1984) Purification of the calcium antagonist receptor of the voltage-sensitive calcium channel from skeletal muscle transverse tubules. *Biochemistry* 23:2113–2118.
8. Glossmann, H. et al. (1984) Molecular pharmacology of the calcium channel: evidence for subtypes, multiple drug-receptor sites, channel subunits, and the development of a radioiodinated 1,4-dihydropyridine calcium channel label, [^{125}I]iodipine. *J Cardiovasc Pharmacol* 6 Suppl 4:S608–S621.
9. Ellis, S.B. et al. (1988) Sequence and expression of mRNAs encoding the alpha 1 and alpha 2 subunits of a DHP-sensitive calcium channel. *Science* 241:1661–1664.
10. Tanabe, T. et al. (1987) Primary structure of the receptor for calcium channel blockers from skeletal muscle. *Nature* 328:313–318.
11. Mikami, A. et al. (1989) Primary structure and functional expression of the cardiac dihydropyridine-sensitive calcium channel. *Nature* 340:230–233.
12. Putney, J. W., Jr. (1986) A model for receptor-regulated calcium entry. *Cell Calcium* 7:1–12.
13. Mignen, O. and Shuttleworth, T.J. (2000) I(ARC), a novel arachidonate-regulated, noncapacitative Ca(2+) entry channel. *J Biol Chem* 275:9114–9119.
14. Gill, D.L. and Patterson, R.L. (2004) Toward a consensus on the operation of receptor-induced calcium entry signals. *Sci STKE* 2004:e39.
15. Bolton, T.B. (1979) Mechanisms of action of transmitters and other substances on smooth muscle. *Physiol Rev* 59:606–718.
16. Hamill, O.P. et al. (1981) Improved patch-clamp techniques for high-resolution current recording from cells and cell-free membrane patches. *Pflugers Arch* 391:85–100.
17. Tsien, R.Y., Pozzan, T., and Rink, T.J. (1982) Calcium homeostasis in intact lymphocytes: cytoplasmic free calcium monitored with a new, intracellularly trapped fluorescent indicator. *J Cell Biol* 94:325–334.
18. Mahaut-Smith, M.P., Sage, S.O., and Rink, T.J. (1990) Receptor-activated single channels in intact human platelets. *J Biol Chem* 265:10479–10483.
19. Hoth, M. and Penner, R. (1992) Depletion of intracellular calcium stores activates a calcium current in mast cells. *Nature* 355:353–356.
20. Zweifach, A. and Lewis, R.S. (1995) Rapid inactivation of depletion-activated calcium current (I_{CRAC}) due to local calcium feedback. *J Gen Physiol* 105:209–226.

21. Franzius, D., Hoth, M., and Penner, R. (1994) Non-specific effects of calcium entry antagonists in mast cells. *Pflugers Arch* 428:433–438.

22. Krautwurst, D. et al. (1993) Novel potent inhibitor of receptor-activated nonselective cation currents in HL-60 cells. *Mol Pharmacol* 43:655–659.

23. von Tscharner, V. et al. (1986) Ion channels in human neutrophils activated by a rise in free cytosolic calcium concentration. *Nature* 324:369–372.

24. Malayev, A. and Nelson, D.J. (1995) Extracellular pH modulates the Ca^{2+} current activated by depletion of intracellular Ca^{2+} stores in human macrophages. *Membr Biol* 146:101–111.

25. Naumov, A.P. et al. (1995) ATP-operated calcium-permeable channels activated via a guanine nucleotide-dependent mechanism in rat macrophages. *J Physiol* 486:339–347.

26. Merritt, J. E. et al. (1990) SK&F 96365, a novel inhibitor of receptor-mediated calcium entry. *Biochem J* 271:515–522.

27. Chung S.C., McDonald T.V., and Gardner P. (1994) Inhibition by SK&F 96365 of Ca^{2+} current, IL-2 production and activation in T lymphocytes. *Br J Pharmacol* 113:861–868.

28. Howson, W. et al. (1990) Design and synthesis of a series of glycerol-derived receptor mediated calcium entry (RCME) blockers. *Euro J Med Chem* 25:595–602.

29. Schwarz, G., Droogmans, G., and Nilius, B. (1994) Multiple effects of SK&F 96365 on ionic currents and intracellular calcium in human endothelial cells. *Cell Calcium* 15:45–54.

30. Iwamuro, Y. et al. (1999) Activation of three types of voltage-independent Ca^{2+} channel in A7r5 cells by endothelin-1 as revealed by a novel Ca^{2+} channel blocker LOE 908. *Br J Pharmacol* 126:1107–1114.

31. Clementi, E. et al. (1995) LU52396, an inhibitor of the store-dependent (capacitative) Ca^{2+} influx. *Eur J Pharmacol* 289:23–31.

32. Bochis, R. J. et al. (1991) Benzylated 1,2,3-triazoles as anticoccidiostats. *J Med Chem* 34:2843–2852.

33. Kohn, E.C. and Liotta, L.A. (1990) L651582: a novel antiproliferative and antimetastasis agent. *J Natl Cancer Inst* 82:54–60.

34. Hupe, D.J. et al. (1991) The inhibition of receptor-mediated and voltage-dependent calcium entry by the antiproliferative L-651,582. *J Biol Chem* 266:10136–10142.

35. Hussain, M.M. et al. (2003) Phase II trial of carboxyamidotriazole in patients with relapsed epithelial ovarian cancer. *J Clin Oncol* 21:4356–4363.

36. Hupe, D. et al. (1989) 5-Amino-1,2,3-Triazoles Useful as Antiproliferative Agents. EP 1988-307418:1–21.

37. Kohn, E.C. and Liotta, L.A. (1989) Therapeutic Potential of Compound L651582, an Antitumor and Antiinvasion Agent. US 1989-355744:1–27.

38. Kohn, E.C. and Liotta, L.A. (1991) Triazole Derivatives, Especially L 651582, for Inhibition of Tumour Proliferation, Invasion, and Metastasis. US 1991-637145:1–49.

39. Arndts, D., Loesel, W., and Roos, O. (1993) Preparation of Carbocyclically- or Heterocyclically-fused dihydropyridines as Drugs for Treatment of Ulcerative Colitis and Crohn's Disease. DE-92-4220319:1–37.

40. Desantis, L.J. (1993) Topical Ophthalmic Compositions Comprising a Combination of Calcium Antagonists with Known Antiglaucoma Agents. WO 1993-US4505:1–20.

41. (1991) Preparation of 5-Amino-1,2,3-triazole Derivatives for Treating Tumors with High Metastasis Probability. JP-1990-117000:1–14.

42. Ensinger, H. et al. (2000) Dihydroisoquinoline LOE 908 for Reduction of Motor Dysfunction Resulting from Traumatic Brain Injury. GB 1999-3524:1–9.

43. Delorenzo, R.J. (2001) Inhibition of a Novel Calcium Injury Current that Forms in neurons During Injury Prevents Neuronal Cell Death. WO 2001-US9516:1–49.

44. Kim, T.W., Tanzi, R.E., and Yoo, A.S. (2001) Method for treatment of neurodegenerative diseases. WO 2000-US20138:1–68.

45. Trevillyan, J.M. et al. (2001) Potent inhibition of NFAT activation and T cell cytokine production by novel low molecular weight pyrazole compounds. *J Biol Chem* 276:48118–48126.

46. Ishikawa, J. et al. (2003) A pyrazole derivative, YM-58483, potently inhibits store-operated sustained Ca^{2+} influx and IL-2 production in T lymphocytes. *J Immunol* 170:4441–4449.

47. Zitt, C. et al. (2004) Potent inhibition of Ca^{2+} release-activated Ca^{2+} channels and T-lymphocyte activation by the pyrazole derivative BTP2. *J Biol Chem* 279:12427–12437.

48. Venter, J.C. et al. (2001) The sequence of the human genome. *Science* 291:1304–1351.

49. Sams-Dodd, F. (2005) Target-based drug discovery: is something wrong? *Drug Discov Today* 10:139–147.

50. Pharmaceutical Research and Manufacturers of America (PhRMA) (2003) Pharmaceutical Industry Profile 2003 (Washington, DC: PhRMA, 2003). http://www.wcuppd.org/compete/sld011.htm.

51. Tsien, R.W. (1998) Molecular physiology. Key clockwork component cloned. *Nature* 391:839, 841.

52. Perez-Reyes, E. et al. (1998) Molecular characterization of a neuronal low-voltage-activated T-type calcium channel. *Nature* 391:896–900.

53. Riordan, J.R. et al. (1989) Identification of the cystic fibrosis gene: cloning and characterization of complementary DNA. *Science* 245:1066–1073.

54. Hannon, G.J. and Rossi, J.J. (2004) Unlocking the potential of the human genome with RNA interference. *Nature* 431:371–378.

55. Zambrowicz, B.P. and Sands, A.T. (2003) Knockouts model the 100 best-selling drugs — will they model the next 100? *Nat Rev Drug Discov* 2:38–51.

56. Gines, S. et al. (2003) Specific progressive cAMP reduction implicates energy deficit in presymptomatic Huntington's disease knock-in mice. *Hum Mol Genet* 12:497–508.

57. Caterina, M.J. et al. (1997) The capsaicin receptor: a heat-activated ion channel in the pain pathway. *Nature* 389:816–824.

58. Frank, R. and Hargreaves, R. (2003) Clinical biomarkers in drug discovery and development. *Nat Rev Drug Discov* 2:566–580.

59. Barnes, P.J. (2004) COPD: is there light at the end of the tunnel? *Curr Opin Pharmacol* 4:263–272.

60. Bleicher, K.H. et al. (2003) Hit and lead generation: beyond high-throughput screening. *Nat Rev Drug Discov* 2:369–378.

61. Abriel, H. et al. (2004) Molecular and clinical determinants of drug-induced long QT syndrome: an iatrogenic channelopathy. *Swiss Med Wkly* 134:685–694.

62. Zhang, S. et al. (1999) Mechanism of block and identification of the verapamil binding domain to HERG potassium channels. *Circ Res* 84:989–998.

63. Fermini, B. and Fossa, A.A. (2003) The impact of drug-induced QT interval prolongation on drug discovery and development. *Nat Rev Drug Discov* 2:439–447.

64. Pritchard, J.F. et al. (2003) Making better drugs: decision gates in non-clinical drug development. *Nat Rev Drug Discov* 2:542–553.
65. Dray, A. (1995) Inflammatory mediators of pain. *Br J Anaesth* 75:125–131.
66. Bevan, S. et al. (1992) Capsazepine: a competitive antagonist of the sensory neurone excitant capsaicin. *Br J Pharmacol* 107:544–552.
67. Krause, J.E., Chenard, B.L., and Cortright, D.N. (2005) Transient receptor potential ion channels as targets for the discovery of pain therapeutics. *Curr Opin Investig Drugs* 6:48–57.
68. De Waard, M., Witcher, D.R., and Campbell, K.P. (1994) Functional properties of the purified N-type Ca^{2+} channel from rabbit brain. *J Biol Chem* 269:6716–6724.
69. Bowersox, S.S. and Luther, R. (1998) Pharmacotherapeutic potential of omega-conotoxin MVIIA (SNX-111), an N-type neuronal calcium channel blocker found in the venom of Conus magus. *Toxicon* 36:1651–1658.
70. Miljanich, G.P. (2004) Ziconotide: neuronal calcium channel blocker for treating severe chronic pain. *Curr Med Chem* 11:3029–3040.
71. Cox, B. and Denyer, J.C. (1998) N-type calcium channel blockers in pain and stroke. *Exp Opin Ther Patents* 8:1237–1250.
72. Cox, B. (2000) Calcium channel blockers and pain therapy. *Curr Rev Pain* 4:488–498.
73. Li, S. et al. (2005) Therapeutic scope of modulation of non-voltage-gated cation channels. *Drug Discov Today* 10:129–137.
74. Houghten, R.A. (2000) Parallel array and mixture-based synthetic combinatorial chemistry: tools for the next millennium. *Annu Rev Pharmacol Toxicol* 40:273–282.
75. Gray, N.S. (2001) Combinatorial libraries and biological discovery. *Curr Opin Neurobiol* 11:608–614.
76. Geysen, H.M. et al. (2003) Combinatorial compound libraries for drug discovery: an ongoing challenge. *Nat Rev Drug Discov* 2:222–230.
77. Young, S.S. and Ge, N. (2004) Design of diversity and focused combinatorial libraries in drug discovery. *Curr Opin Drug Discov Devel* 7:318–324.
78. Krall, R. (2001) Consolidating to Compete. West Chester University Pharmaceutical Product Development Program. CMR International 2003 R&D Compendium. http://www.wcuppd. org/compete/sld011.htm.
79. Horrocks, C. et al. (2003) Human cell systems for drug discovery. *Curr Opin Drug Discov Devel* 6:570–575.
80. Vennekens, R. et al. (2000) Permeation and gating properties of the novel epithelial Ca(2+) channel. *J Biol Chem* 275:3963–3969.
81. Hanlon, M.R. and Wallace, B.A. (2002) Structure and function of voltage-dependent ion channel regulatory beta subunits. *Biochemistry* 41:2886–2894.
82. Glossmann, H. et al. (1983) Identification of voltage operated calcium channels by binding studies: differentiation of subclasses of calcium antagonist drugs with ^{3}H-nimodipine radioligand binding. *J Recept Res* 3:177–190.
83. Lebsack, A.D. et al. (2004) Identification and synthesis of [1,2,4]triazolo[3,4-a]phthalazine derivatives as high-affinity ligands to the alpha 2 delta-1 subunit of voltage gated calcium channel. *Bioorg Med Chem Lett* 14:2463–2467.
84. Tammela, P. and Vuorela, P. (2004) Miniaturisation and validation of a cell-based assay for screening of Ca^{2+} channel modulators. *J Biochem Biophys Methods* 59:229–239.
85. Schroeder, K.S. and Neagle, B.D. (1996) FLIPR: A new instrument for accurate, high throughput optical screening. *J Biomol Screen* 1:75–80.
86. Zhang, Y. et al. (2003) Evaluation of FLIPR Calcium 3 Assay Kit — a new no-wash fluorescence calcium indicator reagent. *J Biomol Screen* 8:571–577.

87. Dupriez, V.J. et al. (2002) Aequorin-based functional assays for G-protein-coupled receptors, ion channels, and tyrosine kinase receptors. *Receptors Channels* 8:319–330.

88. Vazquez, G. et al. (2004) The mammalian TRPC cation channels. *Biochim Biophys Acta* 1742:21–36.

89. Estacion, M. et al. (2004) Activation of human TRPC6 channels by receptor stimulation. *J Biol Chem* 279:22047–22056.

90. Waggoner, A. (1976) Optical probes of membrane potential. *J Membr Biol* 27:317–334.

91. Whiteaker, K.L. et al. (2001) Validation of FLIPR membrane potential dye for high throughput screening of potassium channel modulators. *J Biomol Screen* 6:305–312.

92. Baxter, D.F. et al. (2002) A novel membrane potential-sensitive fluorescent dye improves cell-based assays for ion channels. *J Biomol Screen* 7:79–85.

93. Gonzalez, J.E. et al. (1999) Cell-based assays and instrumentation for screening ion-channel targets. *Drug Discov Today* 4:431–439.

94. Gonzalez, J.E. and Maher, M.P. (2002) Cellular fluorescent indicators and voltage/ion probe reader (VIPR) tools for ion channel and receptor drug discovery. *Receptors Channels* 8:283–295.

95. Wolff, C., Fuks, B., and Chatelain, P. (2003) Comparative study of membrane potential-sensitive fluorescent probes and their use in ion channel screening assays. *J Biomol Screen* 8:533–543.

96. Schnizler, K. et al. (2003) The roboocyte: automated cDNA/mRNA injection and subsequent TEVC recording on *Xenopus* oocytes in 96-well microtiter plates. *Receptors Channels* 9:41–48.

97. Pehl, U. et al. (2004) Automated higher-throughput compound screening on ion channel targets based on the *Xenopus laevis* oocyte expression system. *Assay Drug Dev Technol* 2:515–524.

98. Asmild, M. et al. (2003) Upscaling and automation of electrophysiology: toward high throughput screening in ion channel drug discovery. *Receptors Channels* 9:49–58.

99. Wood, C., Williams, C., and Waldron, G. J. (2004) Patch clamping by numbers. *Drug Discov Today* 9:434–441.

100. Lepple-Wienhues, A. et al. (2003) Flip the tip: an automated, high quality, cost-effective patch clamp screen. *Receptors Channels* 9:13–17.

101. Tao, H. et al. (2004) Automated tight seal electrophysiology for assessing the potential hERG liability of pharmaceutical compounds. *Assay Drug Dev Technol* 2:497–506.

102. Schroeder, K. et al. (2003) Ionworks HT: a new high-throughput electrophysiology measurement platform. *J Biomol Screen* 8:50–64.

103. Kiss, L. et al. (2003) High throughput ion-channel pharmacology: planar-array-based voltage clamp. *Assay Drug Dev Technol* 1:127–135.

104. Krautwurst, D. et al. (1994) The isoquinoline derivative LOE 908 selectively blocks vasopressin-activated nonselective cation currents in A7r5 aortic smooth muscle cells. *Naunyn Schmiedebergs Arch Pharmacol* 349:301–307.

105. Kohn, E.C. et al. (1994) Structure-function analysis of signal and growth inhibition by carboxyamido-triazole, CAI. *Cancer Res* 54:935–942.

106. Zhang, J.H., Chung, T.D. and Oldenburg, K. R. (1999) A simple statistical parameter for use in evaluation and validation of high throughput screening assays. *J Biomol Screen* 4:67-73

Index*

---

*Page numbers with t and f represent table and figures, respectively